Low Energy Particle Accelerator-Based Technologies and Their Applications

Low Energy Particle Accelerator-Based Technologies and Their Applications

Vlado Valković

CRC Press
Taylor & Francis Group
Boca Raton London New York

CRC Press is an imprint of the
Taylor & Francis Group, an **informa** business

Cover photo credit: 3SDH 1 MV Pelletron system with RF source and analysis endstation designed with the intended purpose of aiding in fusion research. It is capable of Ion Beam Analysis (IBA) techniques such as RBS, ERD, PIXE and NRA. Further detectors could be added to the endstation to allow for other techniques. Installed in Japan in 2014. Courtesy of National Electrostatics Corp.

First edition published 2022
by CRC Press
6000 Broken Sound Parkway NW, Suite 300, Boca Raton, FL 33487-2742

and by CRC Press
4 Park Square, Milton Park, Abingdon, Oxon, OX14 4RN

CRC Press is an imprint of Taylor & Francis Group, LLC

© 2022 Taylor & Francis Group, LLC

Library of Congress Cataloguing-in-Publication Data
Names: Valković, Vlado, author.
Title: Low energy particle accelerator-based technologies and their applications / Vlado Valković.
Description: First edition. | Boca Raton : CRC Press, 2022. | Includes bibliographical references and index. | Summary: "Low Energy Particle Accelerator-Based Technologies and Their Applications describes types of low energy accelerators, presents some of the main manufacturers, illustrates some of the accelerator laboratories around the globe and shows the examples of successful transfer of accelerators to needed laboratories. This is an essential reference to advanced students as well as physicists, engineers and practitioners in accelerator science"-- Provided by publisher.
Identifiers: LCCN 2021061967 | ISBN 9780367456320 (hbk) | ISBN 9781032276946 (pbk) | ISBN 9781003033684 (ebk)
Subjects: LCSH: Particle accelerators.
Classification: LCC TK9340 .V35 2022 | DDC 539.7/3--dc23/eng/20220124
LC record available at https://lccn.loc.gov/2021061967

ISBN: 978-0-367-45632-0 (hbk)
ISBN: 978-1-032-27694-6 (pbk)
ISBN: 978-1-003-03368-4 (ebk)

DOI: 10.1201/9781003033684

Typeset in Palatino
by MPS Limited, Dehradun

In memory of Gerry Phillips and Jerry Duggan,

my friends, colleagues and teachers,

my heroes.

Contents

Preface

Particle accelerators are one of the most versatile instruments designed by physicists. From its inception, as the cathode ray tube by J.J. Thomson who used it to discover the electron, to the present giant colliders, it is intimately associated with the major milestones of nuclear and particle physics.

Particle accelerators, conceived in the 1930s and developed in the 1940s, have contributed significantly to the understanding of physics laws and the structure of matter in the 1950s and 1960s. Originally developed as tools for frontier physics, particle accelerators provide valuable spinoff benefits in applied research and technology. Today, they and their products are used in almost all branches of high technology and modern medicine.

Here we shall mention some numbers, as reported by prof. Edgecock, R. 2017. Industrial Applications of Particle Accelerators. Presentation at ARIES Meets Industry – accelerator application to the ship exhaust gases treatment. 1 December 2017, CERN, Switzerland. There are some 40,000 accelerators in the world, about half are <5 MeV, nearly all the rest are <20 MeV. About 2/3 are electron accelerators, 1/3 are ion accelerators. Accelerators are used for a variety of every day applications: energy-few; environment-few; health >15,000; industry >21,000; security ~1,000. Most of the accelerators are commercially manufactured, accounting to around $5 \times 1,011$ US$ annually.

The first working particle accelerator came to existence in 1931, it was built by Ernest O. Lawrence, winner of the 1939 Nobel Prize for Physics. The Lawrence cyclotron was first used in clinical trials at the Lawrence Berkeley National Laboratory in 1954.

The underlying principles of particle accelerators are as follows: In an electric field a charged particle feels a force and is accelerated in the direction of the electric field. The energy the particle gains depends directly on the electric field that in turn depends on the potential difference. A charged particle traveling through a magnetic field feels a force at right angles both to its direction of motion and to the direction of the field. This does not increase the particle's speed (and hence its energy) but it does change the particle's direction. The result is that, in a uniform magnetic field, the particle follows a circular path; with a radius that depends on the strength of the field and on the particle's momentum. The key point here is that the charged particle must be moving through the magnetic field to feel a force.

The two basic designs which were used to accelerate particles in the 1940s are still used today: the linac (linear accelerator) and the cyclotron.

The author of this book is an old (born in 1939.) professor of nuclear physics who received his Ph.D. degree back in 1964 with the thesis title, "Nuclear Reactions with 14.4 MeV neutrons". In my work, a home-made Cockcroft–Walton accelerator was used as a neutron generator by accelerating beam of deuterons (d) up to the energy of 150 keV to bombard tritium ^{3}H target to produce 14.4 MeV neutrons by nuclear reaction:

$$d + {}^3H \rightarrow {}^4He + n.$$

This was my first accelerator, a robust and trustworthy machine, running only nights and weekends so not to irradiate anybody else in the physics building. I learned a lot doing nuclear reaction experiments with this machine and was soon ready to move elsewhere.

My next station was T. W. Bonner Nuclear Laboratories, Physics Department at Rice University, Houston, Texas, USA. Laboratory was equipped with an EN Tandem Van de Graaff accelerator (6 MV terminal) and single-stage machine (5 MV terminal). With lots of patience and steady hands, I was able to talk it into the delivery of 13 MeV protons on my target. In addition to doing experiments in Houston Lab, my boss, at that time prof. G. C. Phillips (1922–1998), would take his team to experiments in Brookhaven National Lab. on synchrotron called Cosmotron, which was running in GeV region, just before it ceased operation in 1966. With this experience, the group was ready to move to LAMPF (Los Alamos Meson Physics Facility), Los Alamos, New Mexico producing pions by high-intensity proton beams of variable energy up to 800 MeV. At that time LAMPF director was Dr. Louis Rosen (1918–2009).

In addition to National Laboratories in the USA in this period, I have also participated at experiments in Institut de Physique Nucleaire, Orsay, France at 156 MeV synchro-cyclotron (studying p-induced three-body break-up of ^{3}He with Dr. T. Yuasa (1909–1980)). In Italy, I worked on many applications of PIXE and XRF with prof. Guiliano Moschini (1940–2016) at Van de Graaff accelerators in Istituto Nazionale di Fisica Nucleare (INFN) at Legnaro National Laboratories and at the Synchrotron Radiation Facility at the INFN Frascati National Laboratory. The work resulted in a number of papers especially in the field of applications to bio-sciences, in particular to medicine.

During the year 1984, me and my co-workers moved the tandem EN accelerator from Houston to Zagreb, Croatia. T. W. Bonner Nuclear Laboratories had shut down both of its accelerators, they were dismounted and moved: EN to Zagreb and single-stage CN to UNAM, Mexico City. Both machines were successfully reassembled and running still today (2021).

The seven years (1989–1996) I have spent in Vienna, Austria working at International Atomic Energy Agency (IAEA) as a head of Physics-Chemistry-Instrumentation (PCI) Laboratory mainly promoting the use of accelerator-based technologies and the use of nuclear analytical techniques in various applications. During my tenure at IAEA, I have established and edited Accelerator Newsletter (2004–2007) disseminating the information on accelerator-based technologies developments as well as information on new developments in accelerator manufacturing, new ideas of accelerator applications, etc. This has been followed by, today existing, Accelerator Knowledge Portal (AKP) IAEA's community-driven website for the benefit of accelerator scientists, accelerator users and service providers worldwide (https://www.iaea.org/resources/databases/accelerator-knowledge-portal). This knowledge portal offers not only a database of MV particle accelerators in the world, but it has several networking and community features attempting to bring together the accelerator community, as well as provide information to accelerator users and policy makers. During this period, I also assisted in the implementation of numerous IAEA accelerator-based projects, especially in the "third world" countries, and prepared the terrain for the establishment of an Agency beam-line at near-by accelerator (Rice Univ. EN tandem in Zagreb).

Upon return to Zagreb, I devoted most of the time to the development of inspection and security applications of small (portable) neutron generators, especially the one that detected also associated alpha particles producing so-called tagged neutrons. Attending numerous conferences in various roles (being an organizer, scientific committee member, session chairperson or simply speaker) I have broadened my knowledge of the field so to feel capable to put my experience and views on the paper and write this book.

One of the first industrial applications of accelerators was the sealing of potato chip bags and milk cartons. Another example: at the end of the last century, almost every household had a mini-accelerator – a Cathode Ray Tube (CRT) television set. In such a TV, the video signal was transmitted through a circuit that fires a beam of electrons down a long CRT.

It is understandable that the complexity and dimensions of particle accelerators lead to the erroneous idea that they must be reserved exclusively for scientific purposes. Instead, accelerators have been winding their way out of research labs and into the industry for decades now. The new applications of accelerators and accelerator-related technologies are continuing to arise in diverse fields.

In the medical sector, accelerators are used to fight tumors (including cancer) with both electron and proton radiation. In contrast to traditional electro irradiation, protons have stronger penetrating power and can go through tissues causing little damage to surrounding organs, but killing malignant cells. Around half of all cancer patients require radiation therapy. Currently, there are several tens of thousands of particle accelerators across the globe, with over 40 million medical patients having benefited, either by diagnosis or by treatment, from almost 60 years of medical research using linear accelerators.

Accelerators are also widely used by the electronics industry, especially the microchip industry. This activity relies on a special technique called doping, in which boron and phosphorus ions are inserted into layers of silicon by means of accelerators. Since ions are positively charged, accelerators can direct ion beams with electromagnetic fields. Thus, the ions penetrate the surface of the silicon wafer and are deposited at desired locations inside it. This changes the conductivity of the material that affects the performance of the chip. As a result, the industry has been able to produce smaller, more powerful computer chips that have brought personal computers to us. Moreover, energy, food, plastic, clean-tech industries, national security and cultural heritage organizations also use accelerator-based installation for their specific needs. The potential of the technologies is high and might find further still-unimaginable applications.

It comes as no surprise that the demand for accelerator-based installations among research centers and, in particular, in different industries is growing progressively. There are hundreds of examples of how particle accelerators or related technologies have improved consumer products, assisted in medical progress, and benefited national security and cultural heritage – not to mention the outstanding scientific leap they have helped achieve. Accelerator technology is truly an enabler of innovation in many areas. Despite its complexity and – to a certain extent – its exclusivity, the world of accelerator technologies is much closer than we imagine.

Many industrial applications of particle accelerators are particularly useful for the automotive industry. Indeed, components such as tyres, foam, ball bearings, gears, camshafts and tie-rod ends are produced using either electron beam thermal processing or irradiation. Modern ion implantation systems make the advanced electronic systems in cars possible.

Surface hardening can benefit any object that undergoes stress, extreme temperature or experiences wear. Most of the moving parts in a car could benefit from surface hardening and from new methods like ion implantation and material irradiation. Other potential advances are in the area of alternative fuels. Advances in hydrogen fuel and hydrogen fuel cell technology could see hydrogen-powered cars become a reality in the future. Research into this is being carried out around the world and synchrotron radiation generated by particle accelerators is one of the key tools. In addition, biofuel researchers have found that treating the fuel with electron beams can improve its quality.

In order to cover the past and the present of this wide and exiting subject, this book is organized in six chapters, each containing many related subjects and numerous references.

Chapter 1 describes the various types of low-energy particle accelerators and some of the laboratories housing them. Electrostatic accelerators Van de Graff type, are described in details including the look on the Pelletron and its manufacturer NEC, National electrostatic Corporation. The cyclotron, linear accelerators, electron beam machine: betatron, neutron generators and their manufacturers are discussed next. Examples of worldwide distribution of accelerator are demonstrated by numerous examples of laboratories. Some examples of accelerators transfers from one laboratory to another are described together with the description of the role of IAEA in promoting accelerator-based technology.

Chapter 2 describes the medical and biological applications of low-energy accelerators. Charged particle beams are used in medical institutions for both diagnostics and therapeutical applications. Diagnostic applications include the use of nuclear analytical techniques for element analysis, the use of different radioisotopes and especially the use of positron emitters. Therapeutical applications are not limited only to radiotherapy, but also include a broad spectrum of other activities. One of the most exciting developments in accelerator technology is the use of carefully shaped hadron beams (protons and heavier ions) to kill cancer cells. Proton therapy is already well established, while carbon-ion therapy is also now being taken forward in several centers. Today, particle therapy is the expanding radiotherapy treatment option of choice for cancer. Some of the world's best equipment manufacturers and their products, dedicated electron or particle accelerator-based irradiation facilities, are presented and described in some details. This chapter also contains parts describing the manufacture and use of biomaterials as well as trace element analysis of biological material by PIXE and AMS.

Chapter 3 deals with various ion beam analyses methods and describes their analytical applications. When a sample is bombarded with a beam of charged particles a number of processes takes place, most of them were developed into analytical methods. Some of them are Rutherford Backscattering (RBS), Proton-Induced X-ray Emission (PIXE), Charged Particle Activation Analysis (CPAA) or Nuclear Reaction Analysis (NRA), Secondary Ionization Mass Spectrometry (SIMS), Proton Induced Gamma-Ray Emission (PIGE), Accelerator Mass Spectrometry (AMS). A common name for all of the techniques is Nuclear Analytical Techniques (NAT). They are described in details in this chapter together with appropriate QA/QC procedures. Successful applications of NATs require the availability of adequate reference materials in addition to some specific instrumental components. For example, some ion sources used in particle accelerator systems dedicated to ion beam analysis techniques are described.

Chapter 4 describes some of the industrial applications. The use of accelerators in material modification and subsequent analysis has been a fast-growing area of accelerator-based technologies. Numerous applications of ion implantation technology have been transferred from the research laboratory to the industry. It is an accepted view that a new industrial revolution will be brought about by the establishment of advanced material processing and machining technologies that can create new materials down to an atomic and molecular level. This could be accomplished by accelerator-based technologies through the development of high-energy ion beams that are focused, clustered and wide-ranging. In this chapter, the radiation processing by electron beams is discussed. Electron beam modification of semiconductor devices, cleaning of flue gases of coal-burning power stations, food irradiation and radiation sterilization are considered. In addition, some other applications are considered in particular electron beam welding. Ion beams are very useful when it comes to modification of materials, micromechanical manufacture. Accelerator applications are numerous in many fields of manufacture; example of the automobile industry is discussed in some details. Of special interest are new cluster ion beams applications.

Chapter 5 is presenting some of the so called miscellaneous applications like cargo inspection, detection of explosive, radioactive waste containers survey and more. Accelerator-based technology, both methodology and products, are so widespread in todays society that it is difficult to find a pore in the network of everyday life where they are not present. In this chapter some additional applications, not discussed so far, of the accelerator-based techniques have been described. Terrorism is a major threat to the 21st century civilization and an enduring challenge to human ingenuity. Nuclear analytical has been successfully used in combatting different terrorist activities, illicit trafficking of treat materials (chemicals, explosives), drugs, cigarettes, counterfeit items, animals and humans. Nuclear monitoring and safeguards activities are described in some details, attention paid to the radioactive waste containers survey. The list of scientific and technical papers on the applications of nuclear techniques to cultural heritage studies is enormous, therefore discussion of only some works has been presented, and number of references is given to guide the reader to the more detailed texts. Accelerator-based analytical techniques are used in monitoring environmental pollution and for identifying the pollution sources. Because of their multi-elemental capabilities, and the possibility of measuring concentration profiles, they have been extensively used for air pollution studies. Neutral particle beams (NPBs) involve accelerating streams of atomic or subatomic particles to nearly light speed and shooting them downrange at a target. Neutral particle beams have technological and engineering challenges similar to those of other directed energy weapons. Apparently active in this field are the Department of Defenses in USA, Russia and China.

Chapter 6 contains the description of new trends in accelerator technology as well as new trends in the applications. New trends in accelerator developments, including tabletop accelerators, laser-plasma accelerators, autonomous particle accelerators, as well as new electron sources, are presented and described in some details. New trends in accelerator applications should be carefully monitored because they can develop into a major driving force for industrial and other development. In particular, the following subjects could develop in the near future: micro machining using ion beams; production of new materials; applications in the nuclear fuel cycle (accelerator-assisted fission plants, heavy iron inertial fusion, incineration of nuclear waste); and applications in the field of medicine.

Acknowledgments

In addition to many of my colleagues in all of the laboratories I have worked, whose names are indicated as coauthors of publications listed in the literature, cited in the text or mentioned as additional reading, I need to mention people who helped me to put all of this material in the manuscript that resulted in this book. First of all my thanks go to Dr. Jasmina Obhodas, Dr. Karlo Nađ and Mr. Andrija Vinković who made the drawings shown in this book. Their drawings were complemented with so many photos from the actual laboratories, prof. Eduardo Andrade, UNAM, Mexico City, Mexico, prof. Lowry Conradie, iThemba Labs, South Africa and others.

I am also grateful to accelerators manufacturers and persons with whom I contacted, namely Ms. Christelle Lucchetti from IBA Industries; Dr. Robert W. Hamm from Technical Enterprises, Pleasanton CA, USA; Aymeric Harmant, Global Marketing Director Proton Therapy, IBA, Louvain-la-Neuve, Belgium; Morgane Delépine, Marketing Associate, IBA RadioPharma Solutions; Manuel Gnida, Media Relations Manager, SLAC National Accelerator Laboratory; Ivan Pongrac, Area Sales Manager, Varian Medical Systems, Budapest, Hungary; Henri van Oosterhout, Managing Director, High Voltage Engineering Europa BV; Vasile Sabaiduc, Director of Cyclotron Operations, Accelerator Technology Engineer Manager, Best Cyclotron Systems, Vancouver, B.C., Canada; Koichi Oikawa, General Manager/Group Leader, Medical System Sales Group, Marketing & Sales Dept., Industrial Equipment Division, Sumitomo Heavy Industries, Ltd., Tokyo, Japan; Mr. Alexis Albin, Directeur Business Unit Linac et Produits RF, PMB Alcen.

Special thanks to Ms. Stephanie E. Stodola, Manager, Customer Relations, National Electrostatics Corp., Middleton, WI, USA, for providing figures (including one on the book cover) of NEC accelerators and other material used in this book.

Author

Dr. Vladivoj (Vlado) Valković, a retired professor of physics, is a fellow of the American Physical Society and Institute of Physics (London). He has authored 22 books (from *Trace Elements*, Taylor & Francis, 1975, to *Radioactivity in the Environment*, Elsevier, 1st Edition 2001, 2nd Edition 2019), and more than 400 scientific and technical papers in the research areas of nuclear physics, applications of nuclear techniques to trace element analysis in biology, medicine, environmental research. He has life-long experience in the study of nuclear reactions induced by 14 MeV neutrons. This research has been done through coordination and works on many national and international projects, including US—Croatia bilateral, NATO, IAEA, EU-FP5, FP6 and FP7 projects.

He has worked as a professor of physics at Rice University, Houston, Texas; at IAEA, Vienna, Austria, as a Head of Physics-Chemistry-Instrumentation Laboratory; and at the Institute Ruder Boskovic, Zagreb, Croatia, as Laboratory head and scientific advisor.

1

Low Energy Particle Accelerators and Laboratories

1.1 Introduction

Particle accelerators, conceived in the 1930s and developed in the 1940s, have contributed significantly to the understanding of physics laws and the structure of matter in the 1950s and 1960s. Originally developed as tools for frontier physics, particle accelerators provide valuable spinoff benefits in applied research and technology. Today, they and their products are used in almost all branches of high technology and modern medicine.

Particle accelerators can be split into two fundamental types, electrostatic accelerators and oscillating field accelerators. Electrostatic accelerators, such as the Cockcroft–Walton (CW) accelerator and the Van de Graaff accelerator, make use of what is known as an electrostatic field. Electrostatic fields are simply electric fields that do not change with time. The main disadvantage of using electrostatic fields is that very large electric fields need to be generated to accelerate particles to experimentally useful energies, which could be difficult and dangerous to maintain. This disadvantage led to the development of the second type of accelerator: the oscillating field accelerator. This type of accelerator requires electric fields that periodically change with time.

Chao and Chou (2016a, 2016b) in volume 9 of the Reviews of Accelerator Science and Technology introduce a number of advanced accelerator concepts (AAC) – their principles, technologies and potential applications. For the time being, none of them stands out as a definitive direction in which to go. But these novel ideas are in pursuit and look promising. Furthermore, some AAC requires a high power laser system. This has the implication of bringing two different communities – accelerator and laser – to join forces and work together. It will have profound impact on the future of this field. In this special volume, also included are two special articles, one on "Particle Accelerators in China" which gives a comprehensive overview of the rapidly growing accelerator community in China. The other features the person-of-the-issue who is well-known nuclear physicist Jerome Lewis Duggan (1933–2014), a pioneer and founder of a huge community of industrial and medical accelerators in the US.

Many details on research and developments of accelerators, their use in industry, teaching and research can be found in proceedings of many conference series. Among the best known are:

- Biannual US Particle Accelerator Conferences (PACs) publishing proceedings for many years as a part of IEEE Transactions on Nuclear Science, recently as IEEE Conference Proceedings.

DOI: 10.1201/9781003033684-1

1

- Biannual linac conferences, proceedings published by a host laboratory.
- Biannual conference series Application of Accelerators in Research and Industry, CAARI. The CAARI Conference series brings together scientists, engineers, professors, physicians and students from all over the world who use particle accelerators in their research and applications. The biennial series that began in 1968 as a Conference on the Use of Small Accelerators for Teaching and Research by Jerry Duggan, while he was a staff member at Oak Ridge Associated Universities. In 1974, later it is known as the International Conference on the Application of Accelerators in Research and Industry.
- A European Particle Accelerator Conference (EPAC) series started in 1988 ant it is now a triannual event. EPAC is now rotating with the Asian Particle Accelerator Conference (APAC).
- Russia has also a long-standing biannual particle accelerator conference series.

These conference proceedings constitute the bulk of the accelerator technology literature, along with many laboratory internal reports and workshop proceedings. In addition, there are two active schools: USPAC particle accelerator school and the CERN Accelerator school series. In addition, American Physical Society has established its Division of Physics of Beams and Physical Review has a special topic component on Accelerators and Beams (PRSTAB).

At this point, we should also give a credit to Rutherford's contribution to science among which we chose to make reference to article by Campbell (2019) celebrating proton centenary who described the events leading to discovery of the proton, published in 1919 (Rutherford 1919). Rutherford studied long-range hydrogen particle recoils in several media and was surprised to find that the number of these "recoil" particles increased when air or nitrogen was present. He deduced that the alpha particles had entered the nucleus of nitrogen atom and a hydrogen nucleus was emitted. This marked the discovery that the hydrogen nucleus – or the proton (the name coined by Rutherford) – is a constituent of larger atomic nuclei. The necessary measurements were done by a very simple apparatus that contained an alpha emitter, metal absorber and the ZnS scintillation screen.

The publication of results obtained in 1917 on this simple apparatus was delayed because of the World War I situation until 1919. In the year 1919, Rutherford produced four papers about the light-atom work done. In the fourth (see ref. Rutherford 1919), he wrote "we must conclude that the nitrogen atom disintegrated … and that the hydrogen atom which is liberated formed a constituent part of the nitrogen nucleus". In 1920, Rutherford proposed building up atoms from stable alphas and H ions. He also proposed that a particle of mass one but zero charge had to exist (neutron) to account for isotopes (Campbell 2019).

According to Bryant (1994), the early history of accelerators can be traced from three separate roots. Each root is based on an idea for a different acceleration mechanism and all three originated in the 1920s. All events responsible for this developments are shown in Table 1.1. The first root has its origin in natural progression from atomic physics to nuclear physics and the need for higher energies and higher intensities than those provided by natural radioactive sources. In the 1920s, the electrostatic machines available were far from reaching the necessary voltage. The situation changed when Gurney and Condon (1928) and Gamov (1928) independently predicted tunneling. According to their calculation, it appeared that energy of 500 keV might be sufficient to split the atom. It was Rutherford who immediately encouraged Cockcroft and Walton to start designing a 500 keV particle accelerator.

TABLE 1.1

Particle accelerator three historical roots

Roots	Year	Person and event	Remarks
First root: Direct-voltage accelerators	1895	Philipp Lenard (1862–1947). Electron scattering on gasses < 100 keV electrons	The Nobel Prize in Physics 1905, for his work on cathode rays
	1914	James Franck (1882–1964) and Gustav Ludwig Hertz (1887–1975). Excitation of electron shells by electron bombardment	The Nobel Prize in Physics 1925, for their discovery of the laws governing the impact of an electron upon an atom
	1906–1919	Ernest Rutherford (1871–1937), Develops theory of atomic structure, induces nuclear reaction with natural alphas.	The Nobel Prize for Chemistry 1908, Natural alpha particles of several MeV
	1928	Gurney and Condon (1928), and Gamov (1928) predicted tunneling	Motivation to build hundred of keV particle accelerators
	1928	Sir John Douglas Cockcroft FRS (1897–1967) and Ernest Thomas Sinton Walton (1903–1995)	Start designing an 800 kV generator encouraged by Ruthrford
	1932	Cockcroft and Walton performed the measurement of the reactionp + ^{7}Li → ^{4}He + ^{4}He with only 400 keV protons	The Nobel Prize for Physics 1951, "for their pioneering work on the transmutation of atomic nuclei by artificially accelerated atomic particles".
Second root: Resonant acceleration	1924	Gustav Ising (1883–1960), proposed time varying fields across drift tubes.	Invented so called resonant acceleration.
	1928	Wideröe has built a proof of principle linear accelerator	Made 50 keV K ions with 1 MHz, 25 kV oscillator
	1929	Ernest Orlando Lawrence (1901–1958)	Inspired by Ising and Wideröe conceives the cyclotron
	1931	Milton Stanley Livingston (1905–1986)	Demonstrate the cyclotron by accelerating H$^+$ to 80 keV
	1932	Lawrence's cyclotron makes 1.25 MeV protons for nuclear reactions	The Nobel Prize for Physics 1939, "for the invention and development of the cyclotron ..."
Third root: Betatron mechanism	1923	Rolf Widerøe (1902–1996)	At that time student, draws the design of the betatron
	1927	Rolf Widerøe makes a model betatron which does not work	
	1940	Donald William Kerst (1911–1993)	Re-invents the betatron; builds first working machine: 2.2 MeV electrons
	1950	Kerst builds the world's largest betatron: 300 MeV	

Four years later, in 1932, Cockcroft and Walton succeeded in splitting the lithium atom with 400 keV protons, this was worth their Nobel prize for physics in 1951. At the same time, Van de Graaff invented an electrostatic generator for nuclear physics research band later in Princeton he built his first machine which reached the potential of 1.5 MV (Van de Graaff 1931).

Alternative to DC accelerators has been proposed by Ising (1924). He planned to repeatedly apply the same voltage to the particle using alternating fields. His invention became the underlying principle of all of today's high-energy accelerators, known as resonant acceleration. Ising (1924) suggested accelerating particles with a linear series of conducting drift tubes and Widerøe built one in 1928 (see Widerøe 1928).

Betatron is insensitive to relativistic effects and is therefore ideal for accelerating electrons. The betatron has also the great advantages of being robust and simple, the only active element being power converter that drives the inductive load of the main magnet. The focusing and synchronization of beam energy with the field level are both determined by the geometry of the main magnet. Widerøe has never published his idea, it was left in his laboratory notebook and re-appeared only when Kerst (1941) built the first machine of this type. In the same issue of Phys. Rev., Kerst and Serber (1941) published a paper on the particle oscillation calling them "betatron oscillation" a term that got to be universally accepted for such oscillations in all devices.

There were no new ideas for acceleration mechanisms until the 1960s when a collective acceleration was proposed for the acceleration of heavy ions (James et al. 1966). The race for ever higher energies of the accelerated particles has started and it is not yet finished.

Ion beams are utilized in many complementary processes to determine the elemental composition of samples. The main techniques are: Rutherford Backscattering (RBS), Proton Induced X-ray Emission (PIXE), Charged Particle Activation Analysis (CPAA) or Nuclear Reaction Analysis (NRA), Secondary Ionization Mass Spectrometry (SIMS) and Particle Desorption Mass Spectrometry (PDMS). While RBS is well adapted to the study of heavy elements in a light substrate which is the case of semiconductor research (Si substrate), NRA is better adapted to studies of the behaviour of light elements in heavy substrates (metals) and finds, therefore, a natural field of application in metallurgy. It is being used in particular for understanding the structure and features of high-Tc superconductors, materials that lose their resistance at much higher critical temperatures than their metal alloy counterparts. It makes it possible to characterize unambiguously what a sample really looks like and not what it was intended to be before the constituents were made to react. CPAA finds its field of application in two areas: ultra-low concentrations and wear studies. It is applicable to most elements and allows trace elements to be identified at the ppb (parts per billion i.e., 10^{-9}) level. One can determine the effect of impurities such as C, N or O in metals, monitor the elaboration process and detect low-level contaminants.

Ion beams are used in a wide energy range (1–45 MeV) allowing depth analysis ranging from microns to millimeters. CPAA is also a sensitive and fast technique for wear studies (corrosion and erosion). One activates a thin surface layer and for suitable isotopes, the loss of activity will correspond to the loss of matter. The method was reported to be used to monitor online industrial processes. It has also been applied to study the effect of pH on the corrosion rate in nuclear reactors.

The utilization of small-spot-size ion beams, also called nuclear microprobes, in the scanning mode has transformed the PIXE technique from an analytical tool into an imaging device. It permits a map of the elements and their distribution in the studied sample to be obtained so that the device could be described as being a nuclear microscope. The elemental map can be compared with the structure given by optical or electron microscopes. A compromise must be found between resolution requiring a small spot size and sensitivity which is directly related to beam intensity; hence the requirement of high brightness.

Recent progress has allowed the spot size to be reduced to the micron level. Reported applications of this technique include the mapping of structures in multilayer semiconductor devices to monitor the manufacturing process, the study of high-Tc superconductor compound structures, the analysis of weld failures, etc. The combination of RBS, which allows the depth profile to be determined, with PIXE can give a three-dimensional picture of the element distribution in the sample. While with ion beam analysis, the accelerator is used to bombard the sample with ions and detect the induced atomic or nuclear processes, in accelerator mass spectrometry (AMS) the constituents of the sample are ionized, accelerated and identified by mass spectrometry. The high sensitivity of AMS finds applications in the semiconductor industry. Semiconductor devices are rapidly degraded by even a small concentration of some impurities which can be readily detected by AMS. Up to now, this was essentially studied by Secondary Ion Mass Spectrometry (SIMS). AMS gives a dramatic improvement of two orders of magnitude in sensitivity. The sample is ionized by a cesium beam.

Another application of accelerators in material science is the study of radiation damage. It is of particular interest in studies of structural material for a future fusion power generator or for satellites and space systems.

1.2 Electrostatic Accelerators

Electrostatic accelerators are an important and widespread subgroup within the broad spectrum of modern, large particle acceleration devices. They are specifically designed for applications that require high-quality ion beams in terms of energy stability and emittance at comparatively low energies (a few MeV). Their ability to accelerate virtually any kind of ion over a continuously tunable range of energies make them a highly versatile tool for investigations in many research fields including, but not limited to atomic and nuclear spectroscopy, heavy ion reactions, accelerator mass spectroscopy as well as ion-beam analysis and modification. To obtain details of construction and operational principles as well as their maintenance, see the book edited by Hellborg (2005).

The machines in this group of accelerators are all using direct voltage, i.e., voltage gradient is used to accelerate charged particles (electrons or ions). The group includes:

- Van de Graaff accelerators: they use a charge carrying belt or chain. Energy range covered is from 1 to 15 MeV at currents from a few nA to a few mA.

- Dynamitron and Cockcroft Walton generator: those are basically voltage multiplier circuits at energies up to 5 MeV and currents up to 100 mA.

- Inductive Core Transformer (ICT): this is a transformer charging circuit with energies to 3 MeV at currents to 50 mA.

1.2.1 Cockcroft–Walton and Other DC Accelerators

Sir John Douglas Cockcroft FRS (1897–1967) was awarded the 1951 Nobel Prize for Physics together with Ernest Thomas Sinton Walton (1903–1995) "for their pioneering work on the transmutation of atomic nuclei by artificially accelerated atomic particles". Douglas Cockcroft was born on 27 May 1897 in Todmorden, straddling the Lancashire–Yorkshire border in northern England. In his early years, he experienced a varied educational background. He studied mathematics at Manchester University in 1914–1915, but the First World War interrupted his studies with service in the Royal Field Artillery. After the war, he returned instead to the College of Technology in Manchester to study electrical engineering. Later he joined the Metropolitan Vickers ("Metrovick") Electrical Company as an apprentice for two years, but subsequently went to St John's

College, Cambridge and took the Mathematical Tripos in 1924. This wide-ranging education served him well in later years. Nowadays, modern accelerator science and engineering rely on such a broad application of skill and innovation.

Today's Cockcroft Institute of Accelerator Science & Technology, located at Daresbury Laboratory adjacent to the Daresbury Innovation Centre is named after Sir John Cockcroft FRS as he is regarded as the pioneer of modern accelerator research. The CW generator, or multiplier, is an electric circuit that generates a high direct current (DC) voltage from a low-voltage alternating current (AC) or pulsing DC input. The circuit was discovered in 1913, by Heinrich Greinacher, a Swiss physicist (see Greinacher 1914). For this reason, this double cascade is sometimes also referred to as the Greinacher multiplier.

Cockcroft and Walton used this circuit in 1932 to power their particle accelerator, performing the first artificial nuclear disintegration in history (Cockcroft and Walton 1932a, 1932b). They used this voltage multiplier cascade for most of their research (Fig. 1.1).

Cockcroft and Walton developed a voltage-multiplying circuit that used capacitors (components used to store energy) to produce high voltage direct current (DC) from a much lower AC. They erected a column of rectifier diodes (which allow the electrical current to move forwards but prevent it from moving back again) and capacitors to produce a DC voltage four times greater than the AC voltage.

On 14 April 1932, Walton set up the voltage multiplier apparatus and bombarded lithium with high-energy protons. By early 1932, they had achieved an output voltage of 600 kV, which they used to accelerate protons. The target was a sheet of mica coated with lithium. When the protons struck the lithium target, the reaction

$$p + {}^7\text{Li} \rightarrow {}^4\text{He} + {}^4\text{He}$$

created two alpha particles flying off in roughly opposite directions. The "swift protons", as Cockcroft called them, had made lithium nuclei disintegrate. Cockcroft and Walton's

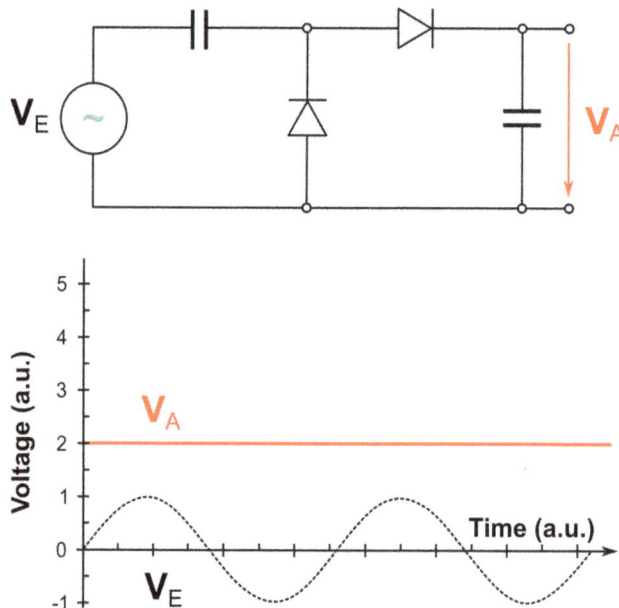

FIGURE 1.1
A circuit diagram for a Cockcroft–Walton generator.

research was published later that month in *Nature*, detailing how they had achieved this momentous scientific breakthrough (Cockcroft and Walton 1932c, 1932d). They sent a letter to *Nature* that same night announcing the first artificial disintegration of an atomic nucleus – the splitting of an atom – and the first nuclear transmutation of one element (lithium) into another (helium).

They subsequently accomplished the same feat with carbon, nitrogen and oxygen atoms, using protons, deuterons and alpha particles to produce radioactive isotopes. With their work, they open "a new and fruitful field of research" that had "profoundly influenced the whole subsequent course of nuclear physics [and] stands out as a landmark in the history of science". However, the release of energy was too gradual to be of much practical use.

Many countries established nuclear institutes in postwar years (1945–1950). I shall mention the one I am most familiar with: Institure Ruđer Bošković in Zagreb, Croatia at that time part of Yugoslavia, established in 1950. A group of physicists and electronics enthusiasts led by professor Mladen Paić (1905–1997) built-up in 1956 a homemade accelerator, 200 kV CW. The accelerator was dedicated to the neutron production via d+t→α+n reaction (Fig. 1.2).

FIGURE 1.2
Homemade Cockcroft–Walton 200 kV accelerator at the Rudjer Bošković Institute which was used mainly as neutron generator.

The accelerator was called simply "neutron generator" because its prime purpose was production of 2.5 MeV neutrons (by d+d→n+^{3}He reaction) and 14 MeV neutrons (by d + t → α + n reaction). The energy of deuterons was up to 200 keV with current of I = 2 mA. The maximum neutron beams intensities were 10^8 n/s for 2.5 MeV neutrons and 10^{10} n/s for 14 MeV neutrons.

In 1970, a 300 kV DC accelerator (deuterons up to 300 keV of energy, with beam currents up to 2.5 mA) was acquired from Texas Nuclear, USA to be used as a "new neutron generator" so the old one was abounded by nuclear physicists. Before the final shut down it was used for several years by solid-state physicists in their work on the study of properties of semiconductors. For that type of work, the accelerator had to be equipped by new ion source and a high-resolution magnetic analyzer. Today, this space is mainly office space while the 300 kV machine is in a new location.

The "new neutron generator", 300 kV Cockroft-Walton accelerator, uses a deuteron beam up to 300 keV of energy, with beam currents up to 2.5 mA, to produce 10^9 n/s in 4π of 2.7 MeV energy and up to 10^{11} n/s in 4π of 14.4 MeV energy.

The major components of the Neutron Generator are shown schematically in Fig. 1.3. Positive ions produced in a radio frequency type ion source are extracted by applying a potential across the ion source bottle. After extraction, the ions are focused by directly below ion source bottle. After leaving the gap lens, the ions enter the accelerating tube where they are accelerated through a gap lens situated at the entrance to the canal in the field of a potential of 150 kV. The 150 kV is distributed between 10 electrodes, so that an ion passing through the accelerating tube experiences a "kick" of 15 kV each time it passes between two electrodes. After leaving the accelerating tube, the ions drift through a potential free region (drift tube) until they fall on the target. A vacuum is maintained in the entire system to ensure proper functioning of the accelerator (Figs. 1.4 and 1.5).

1.2.2 Accelerator-Based X-Ray Sources

Accelerator-based X-ray sources have contributed uniquely to the physical, engineering and life sciences. There has been a constant development of the sources themselves as

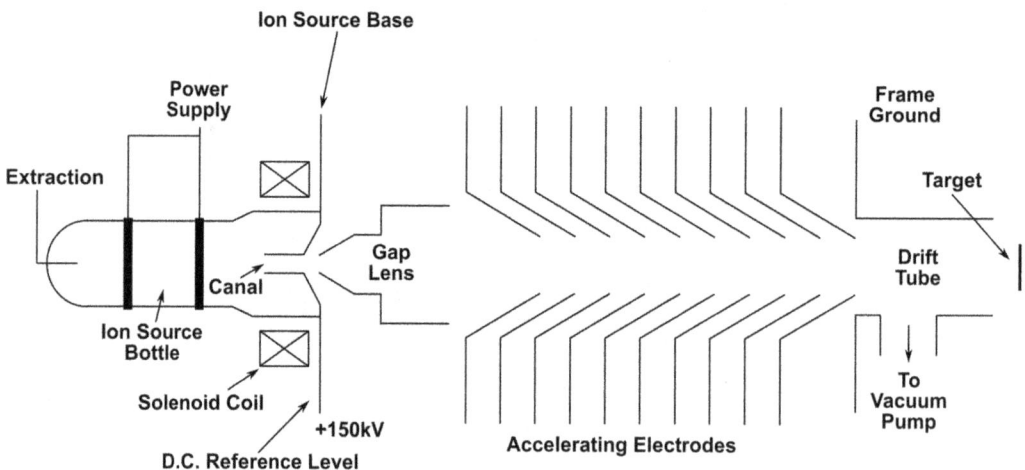

FIGURE 1.3
Schematic of Neutron Generator. Voltages shown are relative to ion source base.

FIGURE 1.4
Texas Nuclear Corporation 300 keV accelerator at Institute Ruđer Bošković, Zagreb, Croatia.

well as of the necessary X-ray optics and detectors. These advances have combined to push X-ray science to the forefront in structural studies, achieving atomic resolution for complex protein molecules, to meV scale dynamics addressing problems ranging from geoscience to high-temperature superconductors, and to spatial resolutions approaching 10 nm for elemental mapping as well as three-dimensional structures. Accelerator-based photon science is of great importance to optics, detectors and computation/data science as well as the source technology. There is a bright future for X-ray systems, integrating all components from accelerator sources to digital image production algorithms, and highlight aspects that make them unique scientific tools.

The Texas Nuclear Corporation neutron generator can be used as an X-ray machine, although the neutron generator is designed primarily to accelerate ionized beams of hydrogen and deuterium. We have seen, however, that it can be adapted to accelerate heavier charged particles such as ^{3}He ions. It is also an electron accelerator, and hence, an x-ray machine. The machine can be converted from a positive ion accelerator to an electron accelerator by simply reversing the polarity of the 150 kV (or 100 kV) power supply and the extraction and focusing power supplies in the high voltage terminal. The meter connections, of course, also have to be reversed. The HV power supply is provided with a polarity reversing switch. Conversion from a positive to negative ion accelerator can be accomplished in about a half-hour.[1] Operating at 150 kV, an electron beam current of 0.5 ma or greater can be obtained.

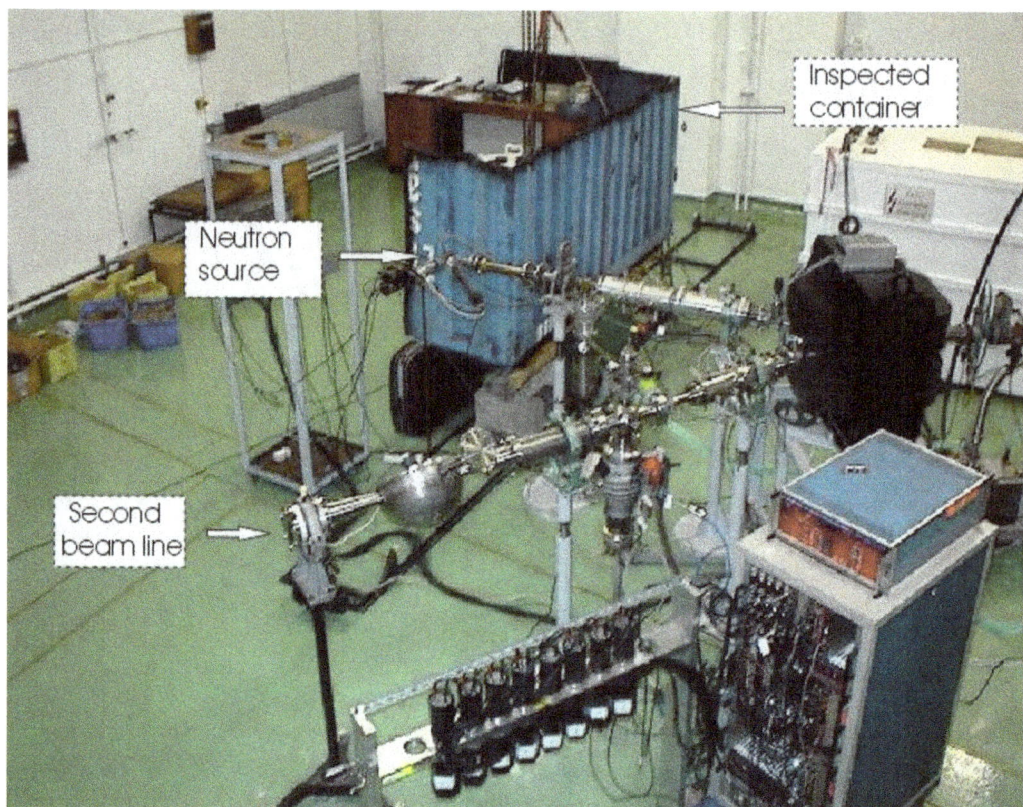

FIGURE 1.5
Target room. Neutron Laboratory at Institute Ruđer Bošković, Zagreb, Croatia.

X-rays are produced when an electron is stopped in the matter. The distribution of X-rays in energy varies from the energy of the electron to zero. The efficiency of X-ray production depends on the target and the energy of the incident electrons. A 500 μA beam of 150 Kev electrons falling on a gold target will produce a total yield of the order of 10^{13} X-rays/sec.

1.2.3 Van de Graaff Accelerator – Single Stage Machine

The most prominent of the accelerators in which charge is carried mechanically to the terminal is the electrostatic generator or Van de Graaff accelerator. A Van de Graaff accelerator consists of a large metallic sphere at the top of an insulating column. Within the column is a belt made from a conducting material pulled over two pulleys. One of the pulleys is attached to an electric motor driving the belt, at either end of the belt is a brush of metallic wires, the lower brush is attached to a voltage source which transfers a charge to the belt via the brush, the belt then carries the charge up to the second brush which will transfer the charge to the large metal sphere known as the electrode. The charge build-up generated in the electrode results in a potential difference between the electrode and the ground (see Fig. 1.6). A particle can be accelerated using this potential difference, from the electrode, to the ground. A Van de Graaff accelerator is a very big Van de Graaff

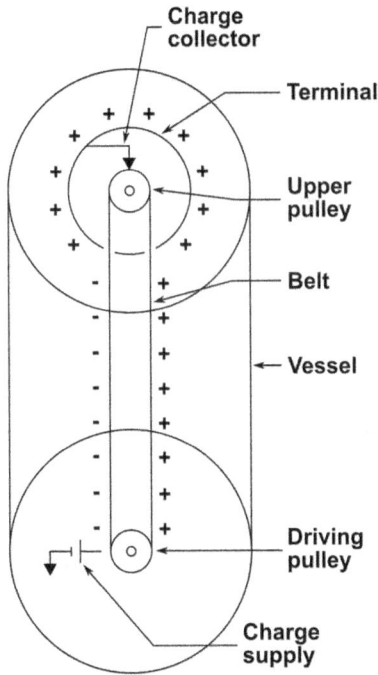

FIGURE 1.6
Schematic presentation of the single-stage electrostatic generator.

generator with an accelerator tube contained within it. By 1933, a Van de Graaff accelerator was in operation that could accelerate hydrogen ions to an energy of 0.6 MeV.

In this device, charge is carried by a moving belt (made of rubber, vulcanized fabric or another flexible material) to the high-voltage terminal. The belt is electrically charged by a brush or comb of metallic wires which is connected to a DC voltage source. The amount of electric charge sprayed onto the belt is controlled by the voltage V_C. The charge is then carried by the belt to the terminal electrode where the charge is transferred from the belt to the terminal electrode by a second brush or comb of metallic wires and accumulated on its external surface.

The resulting terminal voltage is a function of the diameter of the terminal electrode. The terminal behaves as a spherical capacitor with capacitance:

$$C = 4\pi\varepsilon\varepsilon_0 r = (1.11 \times 10^{-10}\,\text{F/m})r \tag{1.1}$$

where r is the radius of the terminal electrode. An air-insulated electrode with r = 1 m yields C = 111 pF. The terminal voltage, V, is proportional to the charge, Q, accumulated on the terminal electrode:

$$V = Q/C \tag{1.2}$$

The equilibrium voltage V_0 depends on the current I_{belt} transported by the belt, the beam current I_{beam} of the accelerating tube, the current I_R through all resistors and the current I_{cor} resulting from corona and tube discharge.

With a constant charging current the equilibrium voltage is reached asymptotically as given by the equation:

$$dV/dt = 1/C \ dQ/dt = 1/C(I_{belt} - I_{beam} - I_{cor} - I_R) \tag{1.3}$$

For low terminal voltages, I_{cor} is negligibly small so the above equation can be written as

$$dV/dt = 1/C(I_{belt} - I_{beam} - V/R) \tag{1.4}$$

R is the resistance of the resistors and insulating column. At this time we define equilibrium voltage V_0 to be

$$V_0 = R(I_{belt} - I_{beam}) \tag{1.5}$$

Resulting in the solution of the differential equation:

$$V(t) = V_0 + [V(0) - V_0]e^{-t/\tau} \tag{1.6}$$

Where time constant is defined as $\tau = RC$. This equation holds as long as the currents I_{belt} and I_{beam} and effective resistance R are constant and corona currents are negligible. The terminal will reach an equilibrium potential V_0 where the charge delivered by the belt is equal to the charge leaving the terminal electrode. Modern machines use fast closed-loop controls to achieve a fast adjustment and a high stability of the terminal voltage V. In such a way, the time dependence of V is determined by the characteristics of the closed-loop control.

Van de Graaff accelerators played an important role in nuclear physics development during the 1930s and are still in use in that field. They are now frequently used in industrial applications. The Van de Graaff accelerator was developed as a charged particle accelerator for physics research; its high potential is used to accelerate atomic nuclei to great speeds in an evacuated tube.

This accelerator was invented by Robert Jemison Van de Graaff, an American physicist, born on 20 December 1901 in Tuscaloosa, Alabama, USA. He attended the University of Alabama where he received a BS degree (1922) and an MS degree (1923) in mechanical engineering. After spending two years at the Sorbonne in Paris (1924–1925) and while there, attended lectures by Marie Curie on radiation. In 1925 he went to Oxford University in England as a Rhodes Scholar. At Oxford, he received a BS in physics (1926) and a Ph.D. in physics (1928). While at Oxford, he became aware of the need that particles could be accelerated to speeds sufficient to disintegrate nuclei. By disintegrating atomic nuclei much could be learned about the nature of individual atoms. From these ideas, Robert Van de Graaff saw the need for a particle accelerator He invented the Van de Graaff generator back in October 1929 when at Princeton New Jersey, first published his idea in Van de Graaff (1931) (Fig. 1.7).

Many small Van de Graaff accelerators of conventional design, some pressurized, some air-insulated, were built in university laboratories in the 1950s and 1960s in support of local research and to provide experience in nuclear techniques for students. Records of these machines are sparse, often confined to internal reports, and most are no longer operating.

In 1946, Robert J. Van de Graaff. John Trump (Donald Trump's uncle) and Denis M. Robinson founded the High Voltage Engineering Corporation (HVEC) to produce Van de Graaff Generators; in 1960 Robert J. Van de Graaff left MIT to work full time for HVEC.

FIGURE 1.7
Robert J. Van de Graaff (1901–1967). (Courtesy of HVEE.)

1.2.3.1 Tandem Accelerator

The tandem Van de Graaff accelerator is a particle accelerator used in nuclear physics, which consists of two electrostatic acceleration sections of opposite polarity, the center point (the "terminal") of the Van de Graaff generator is kept on positive high voltage. Negatively charged ions are injected on one side, which strip off some of their electrons in a so-called "stripper" (a gas target or a thin film) in the terminal and are then accelerated further as positive ions on the other side. The ion source and reaction target can thus be kept at ground potential, and the accelerating voltage is used multiple times, depending on the charge state of accelerated ion, see Fig 1.8. In the past, a fast-rotating belt made of a rubber-fabric mixture was used as a means of transport for the cargo, today it is chains of short, mutually isolated metal cylinders ("Pelletron" method).

Single-ended and tandem systems exhibit several important differences. First, of course the difference in ion sources, as for a tandem system a negative ion source is needed. The development of these sources has led to the availability of the multi-cusp ion source that combines low beam emittance with high current (see e.g. Lee et al. 1999). Furthermore,

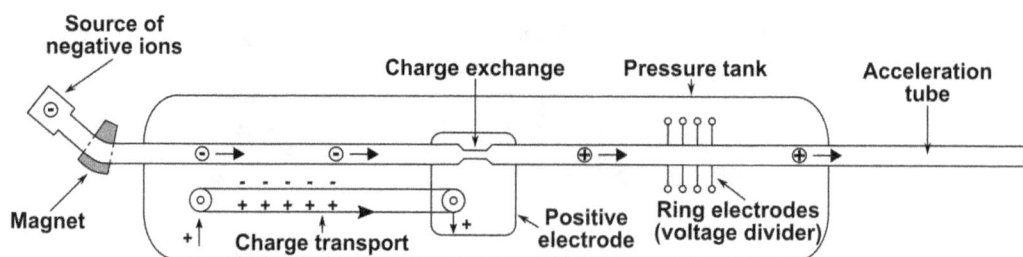

FIGURE 1.8
Schematic presentation of tandem configuration.

related to the stripping process needed with a tandem accelerator, energy straggling of the ions increases the energy spread and small-angle scattering of the ions tends to lower the beam brightness (Visser et al. 2005).

The acceleration tube between conductive levels and glass blocks is held in a self-supporting manner by a mechanical pre-tension of around 40 t. The chain of resistors divides the voltage at the terminal evenly (constant electric field in axial direction). In the radial direction (toward the tank), the field is that of a charged cylinder capacitor and decreases towards the outside. To avoid sparks each level is polished with potential rings closed toward the tank. The ion sources simply have to provide negatively charged ions pre-accelerated, which gain the energy eV in the first half of the tandem. In the terminal, the single negative ion beam is recharged into Z-fold charged positive ions by "stripping" on a thin carbon film. The achievable energy is $(Z + 1)$ eV, if V is the terminal voltage.

Review of the most commonly employed methods of making negative ions and the ion sources that are most suitable for use with tandem accelerators is given by Middleton (1974). Particular emphasis is placed on the production of negative ions by cesium beam sputtering and by direct extraction from a Penning source using a cesium seeded arc. Some more recent texts on negative ions sources can be found in Nishihashi et al. (1989), Diamond et al. (1996) and Faircloth and Lawrie (2018).

EN tandem Van de Graaff accelerator can be upgraded for AMS applications provided some additional equipment is acquired, including:

- Sputtering Ion Source with revolving cathode holder for multiple cathodes (for example ANIS sputtering source from DANFYSIK which has a holder for eight cathodes).

- Injection Analyzing Magnet with an isolated vacuum chamber and acceleration/deceleration at entrance/exit to allow fast switching between ^{14}C and ^{12}C injection (or other normalizing isotopes.

- Electrostatic Quadrupole Triplet to optimize injection optics.

- Analyzing Magnet with a wide exit pole to monitor ^{12}C, ^{13}C and ^{14}C simultaneously (or isotopes of other elements). Faraday cups with adjustable positioning for collecting neighbor isotopes.

- Wien Filter or Electrostatic Energy Analyzer.

- Extra Quadrupole Triplet Magnet to optimize optics on the high-energy side.

- Additional Beam Diagnostic Equipment to ensure optimum beam optics.

- $\Delta E/E$ Detector and/or Time of Flight System, Power Supplies for the ion source, magnets and electronics.

A large number of experiments has been done on low energy accelerators, in particular Van de Graaff accelerators and tandems. There are many textbooks, books, reports and scientific papers written on the subject of low energy nuclear physics and its applications using electrostatic accelerators as a tool. The study of atomic nuclei is central to our understanding of the world around us. Comprising 99.9% of the visible matter in the universe nuclei are, in multiple aspects, central to fundamental questions in physics, such as our understanding of the origin of the elements and how complex many-body quantum systems organize. Their properties depend sensitively on the number of protons (Z) and neutrons (N), and much of what we know about them comes from the measurement and characterization of their excitation modes and energy levels. Understanding nuclear properties, their role within the cosmos, and more broadly their application for society, requires measurements on elements and isotopes at the limits of their mass (N + Z), charge (Z), and β-decay stability (N–Z). In short, low energy nuclear physics research concerns itself with understanding the structure and stability of the nuclei in the atoms that make up the world around us, as well as the reactions which formed them in the cosmos.

There are many published titles on nuclear research with low energy accelerators (see e.g., Marion 1967, National Research Council 1999, Marwan and Krivit 2008 and others).

1.2.4 Pelletron

Van de Graaff accelerator depends for its operation on the deposition of a charge on a moving belt of insulating fabric. This charge is conveyed on the belt into a smooth, spherical, well-insulated metal shell, where it is removed, passing to the metal shell. The shell increases in potential until an electric breakdown occurs or until the load current balances the charging rate. Machines of this kind, properly enclosed, have produced potentials of about 13,000,000 volts (13 MV). In the Pelletron accelerator, the moving belt is replaced by a moving chain of metallic beads separated by insulating material. The belt is made of a series of insulated metal links, looking a little like a bicycle chain. The Pelletron accelerator at the Oak Ridge National Laboratory, Tenn., produces 25 MV and accelerates protons or heavy ions, which are then injected into an isochronous cyclotron for further acceleration.

A proposed model accelerator facility built around, for example, model 9SDH-2 pelletron with terminal voltage range 0.15–3.0 MV capable of accelerating singly charged ions to 0.2–6.0 MeV, and doubly charged ions to 0.3–9.0 MeV is shown in Fig. 1.9. This type of accelerator is manufactured by National Electrostatic Corporation (NEC) in the USA.

It is proposed that the accelerator will be equipped with two ion sources:

i. Negative ion duoplasmotron. The direct extraction, displaced intermediate electrode duoplasmotron ion source, produces high brightness negative ion beams from virtually all gaseous compounds. The NEC version of this popular and reliable ion source features all metal and ceramic, bakeable construction with no organic materials in the vacuum volume.

ii. SNICS. The Source of Negative Ions by Cesium Sputtering (SNICS) is compact, simple, versatile and easy to operate. It produces high current, high brightness negative ion beams from virtually all stable solid elements. It does not use toxic gases for beams such as B, P and As. Cathodes, inserted through an airlock, each contain a small amount of the solid material from which the beam is produced. Ionized cesium sputters this solid cathode material to produce tens of μA of negative ions.

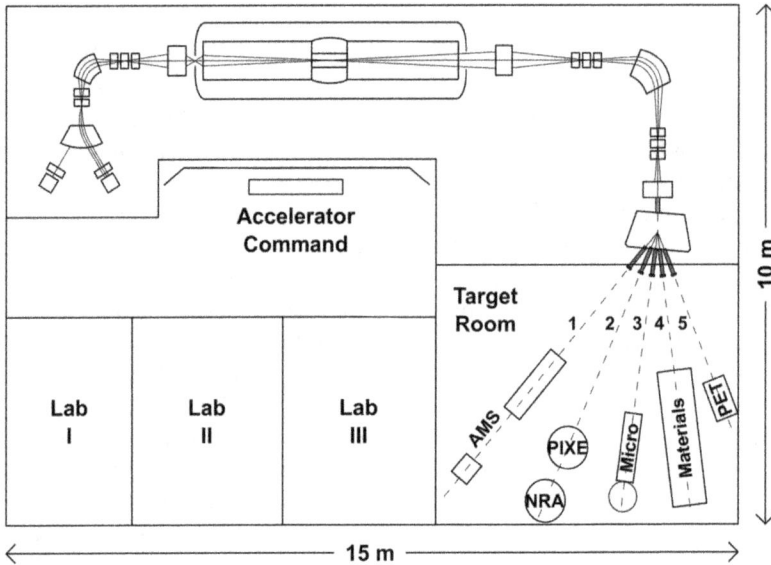

FIGURE 1.9
Accelerator facility – layout of the ground floor. A laboratory/office building of overall dimensions 10 × 15 m is required.

Beam handling system is made of NEC standard components: electrostatic quadrupole lenses, Faraday caps, slits, beam scanners, beam profile monitors, foil and target chambers. Five beam lines are proposed to be in the target room:

- Beam line #1: AMS measurements: This line requires an electrostatic lens for mass separation and detector systems (t-o-f, Δ, etc.).
- Beam line #2: Nuclear analytical methods, standard NEC equipment; PIXE chamber, chamber for RBS and channeling research, Si(Li) and other types of detectors.
- Beam line #3: Microbeam line. Available from several manufacturers, including slits, quadrupoles, scanning system and chamber (see Fig. 1.9).
- Beam line #4: Materials modification. Large chamber plus external beam facility.
- Beam line #5: Positron emission tomography arrangements.

1.3 The Cyclotron

In 1928, Ernest Lawrence of the University of California, inspired by the work of Wideröe, had the idea of utilizing a curved path for a particle accelerator. A magnetic field perpendicular to the plane of motion of an accelerated particle will result in the particle taking a curved path. By studying the simple relationship between the forces acting on the particle, Lawrence realized that the increase in the radius of the path taken by the particle is compensated for by the increased velocity of the particle if the magnetic field, the charge of the particle and the particles mass remain constant. With this in mind, he built a machine today known as a Cyclotron. It consisted of two hollow D-shaped electrodes alternatively charged to a voltage by an oscillator. The electrodes were separated by a small gap. When one of the electrodes is charged, a particle is accelerated across the gap into the other, where, under the influence of a magnetic field, it moves in a semi-circular path back to the surface of the electrode. Just as the voltage has charged the other electrode, the particle is again accelerated across the gap. As the speed of the particle increases, the radius of the semi-circular motion of the particle increases until the particles are eventually focused out of the Cyclotron as a high-energy beam.

The group of circular accelerators, one using magnetic field to maintain circular orbit includes:

- Cyclotrons: ion energies from 10 to 70 MeV at beam currents to several mA.
- Betatrons: electron energies to 15 MeV at few kW beam power.
- Rhodotron: electron energies from 5 to 10 MeV at beam power up to 700 kW.
- Synchrotron: electron energies up to 3 GeV and ion energies up to 300 MeV/amu.

In their article, Craddock and Symon (2008) described particle accelerators using magnets whose field strengths are fixed in time to steer and focus ion beams in a spiral orbit so that they pass between and can be accelerated by, the same electrodes many times. The first example of such a device, Lawrence's cyclotron, revolutionized nuclear physics in the 1930s, but was limited in energy by relativistic effects. To overcome these limits two approaches were taken, enabling energies of many hundreds of MeV/u to be reached: either frequency-modulating the rf accelerating field (the synchrocyclotron) or introducing an azimuthal variation in the magnetic field (the isochronous or sector-focused cyclotron). Both techniques are applied in fixed-field alternating-gradient accelerators (FFAGs), which were intensively studied in the 1950s and 1960s with electron models. Technological advances have made possible the construction of several proton FFAGs, and a wide variety of designs is being studied for diverse applications with electrons, muons, protons and heavier ions. All fixed-field accelerators offer high beam intensity, in some cases deliver beams of 2 mA. Synchrocyclotrons and most FFAGs operate in pulsed mode, but are capable of much higher pulse repetition rates (≤kHz) than synchrotrons (Craddock and Symon 2008).

1.4 Linear Accelerators

The first development in linear accelerators came from Rolf Wideröe in 1927 when he built a linear accelerator using an AC voltage and a series of drift tubes. In an AC, the flow of electric charge is periodically reversed, the flow of electric charge can be thought of as a series of peaks and anti-peaks of voltage. A charged particle acted on by an AC voltage would be accelerated from point X to point Y, during a peak, then when the current is reversed would be accelerated back from point Y to point X, during an anti-peak.

In his accelerator, Wideröe (1928) used a series of drift tubes to shield the particle being accelerated from the reversed electric field during an anti-peak preventing it from being decelerated (B). The particle then emerges from the drift tube just as the field returns to a peak, where it is further accelerated (A). As the particle gets faster and faster, the drift

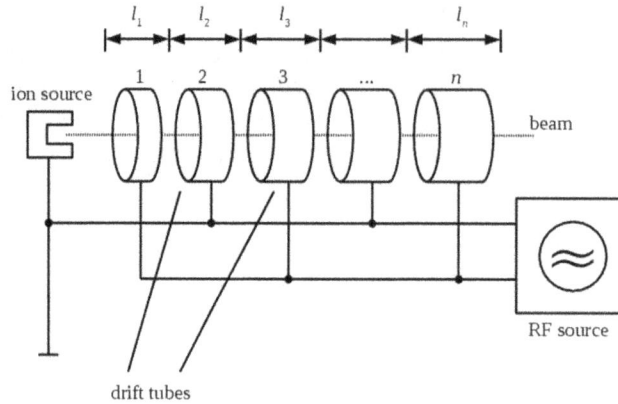

FIGURE 1.10
Widerøe linac with a series of longer and longer drift tubes.

tubes need to be longer and longer. Indeed the faster the particle moves, the greater distance it covers in the same amount of time. This is one of the limiting factors of linear accelerators, they need to be very long for particles to be accelerated to high energies (Fig. 1.10).

RF linacs use RF-generated voltage to accelerate "bunches" of charged particles.

- Electron linacs: standing wave cavities from 0.8 to 9 GHz, energies from 1 to 16 MeV at beam power to 50 kW.
- Ion linacs: all use RFQs at 100–600 MHz, energies from 1 to 70 MeV at beam currents up to mA.

Radio-frequency quadrupole (RFQ) linear accelerators appeared on the accelerator scene in the late 1970s and have since revolutionized the domain of low-energy proton and ion acceleration. The RFQ makes the reliable production of unprecedented ion beam intensities possible within a compact RF resonator which concentrates the three main functions of the low-energy linac section: focusing, bunching and accelerating. Its sophisticated electrode structure and strict beam dynamics and RF requirements, however, impose severe constraints on the mechanical and RF layout, making the construction of RFQs particularly challenging. The main beam optics, RF and mechanical features of a RFQ emphasizing how these three aspects are interrelated and how they contribute to the final performance of the RFQ is described in details in the paper by Vretenar (2013).

1.5 Electron Machine: Betatron

The physical principles governing the betatron were first described by Widerøe in a 1928 paper and put in to practice in 1940 by Donald Kerst. The development of the betatron was driven by the demand for high energy X-rays and gamma rays for medical and research use.

The betatron consists of the main ring, a doughnut-shaped vacuum chamber, known as the doughnut chamber, in which electrons, produced by an electron gun within the chamber, are accelerated. The chamber is set up between the two poles of an electromagnet driven by an AC current which results in a constantly changing magnetic field. The changing magnetic field means a changing magnetic flux (a component of the magnetic field passing through an area) across the doughnut chamber which produces an electromotive force that will accelerate the electrons.

The electrons in the chamber maintain a constant radius of orbit whilst being accelerated, due to the centripetal force generated by the particle motion, so long as the magnetic fields satisfy the betatron principle. Once accelerated, the electrons are directed out of the doughnut chamber, or inwards, towards a metal target to produce X-rays. The first betatron built by Kerst in 1940 was capable of producing 2.3 MeV electrons, but by 1950 he had built a betatron capable of producing 300 MeV electrons.

1.6 Some Examples of Laboratories

1.6.1 T. W. Bonner Laboratories, Rice University, Houston, Texas (1953–1994)

Rice's Nuclear Laboratory (later renamed the Bonner Laboratory in honor of Professor Tom W. Bonner (1910–1961), built in 1952–1953 to house a six-million volt Van de Graaff particle accelerator. In 1963, a 6.0 EN Tandem Van de Graaff accelerator was installed in the laboratory (Fig. 1.11).

Nuclear physics research started at Rice University with work done by Tom W. Bonner (1910–1961). He became an instructor in 1936, a professor in 1945 and chair of the Rice University Physics Department in 1947. He did important work in the development of high-pressure cloud chambers for the study of neutrons produced by accelerators. He invented a neutron-counter-ratio technique for the determination of neutron emission thresholds. He also invented a sphere-moderated neutron spectrometer (Bramblett et al. 1960, Bonner and Mills 1967, Houston 1965)).

The picture of Rice's Nuclear Lab as it looked just after it was completed in 1953, shown in (Kean 2011), looks the same. Buildings do not change much in ten years period. It is a little ironic, but there are far more pictures of its destruction in 1994 than of any other phase of its existence. After its destruction a new building (music hall) was built at the same location, today named a Duncan Hall.

Under the leadership of Gerry Phillips (Fig. 1.12) the T. W. Bonner Nuclear Laboratory, with a Van de Graaff accelerator and a Tandem Van de Graaff, has been a leader in the investigations of nuclear structure, the few nucleon problems and the cluster model. Advanced experimental techniques, including computer online data collection and processing with multi-dimensional inputs, allowed the complete study of three or more particles interacting in the final state of a scattering experiment. Modern time-of-flight technology is employed for neutron velocity measurements as well as charged particle mass identification.

By using the optical-pumping method (Colegrove et al. 1963), Baker and colleagues (Baker et al. 1969, Ohlsen et al. 1974) have produced beams and targets of polarized ^{3}He nuclei in order to study a large class of nuclear phenomena that had been previously unobservable. Experiments with polarized ^{3}He targets have helped to understand the structure of several

FIGURE 1.11
Two views of Rice's Nuclear Lab as it looked in 1966. The author of this book took both photos. The boy (right) is the author's son, then 3 years old.

few-nucleon systems. Biegert (1976) studied analysing powers and phase shifts in ^{3}He–^{3}He scattering. Rice team helped polarized ^{3}He^{+} source installation and operation on the Texas A&M 224-cm cyclotron during the mid-1970s. Polarization was produced by optical pumping. The source produced about 8 µA of ^{3}He^{+} at 16 keV with an emittance of 10 mm·mr·MeV$^{1/2}$. This beam was guided to the centre region of the cyclotron, accelerated without loss of polarization, stripped and focussed on target (100 nA on a 0.2 cm^{2} spot at 38 MeV). Beams were developed between 18.4 and 49.0 MeV (May and Baker 1985).

Numerous studies of nuclear reactions have been done. For example, the angular distributions of resolved n_{0} and n_{1} neutron groups from ^{11}B+d were measured at bombarding energies of 1.51 ± 0.05, 1.96 ± 0.05, 2.36 ± 0.04, 2.77 ± 0.04, 3.47 ± 0.04, 3.77 ± 0.04, 4.11 ± 0.04 4.68±0.03 MeV. They were obtained using a neutron spectrometer of low resolving power consisting of a polyethylene radiator with gas proportional dE/dx and scintillation E counters operating in coincidence to detect the recoil protons from the radiator. The shapes of the angular distributions do not allow an unambiguous specification of the reaction mechanism, which could involve both compound nucleus formation and direct interaction (Class et al. 1965).

Large number of both theoretical and experimental studies of nuclear reactions induced by charged particles accelerated at one of the two accelerators: EN Tandem Van de Graaff and 5.5 MV vertical Van de Graaff accelerator, including measurements of energy spectra, angular distributions and deductions of reaction mechanism, has been done during this period.

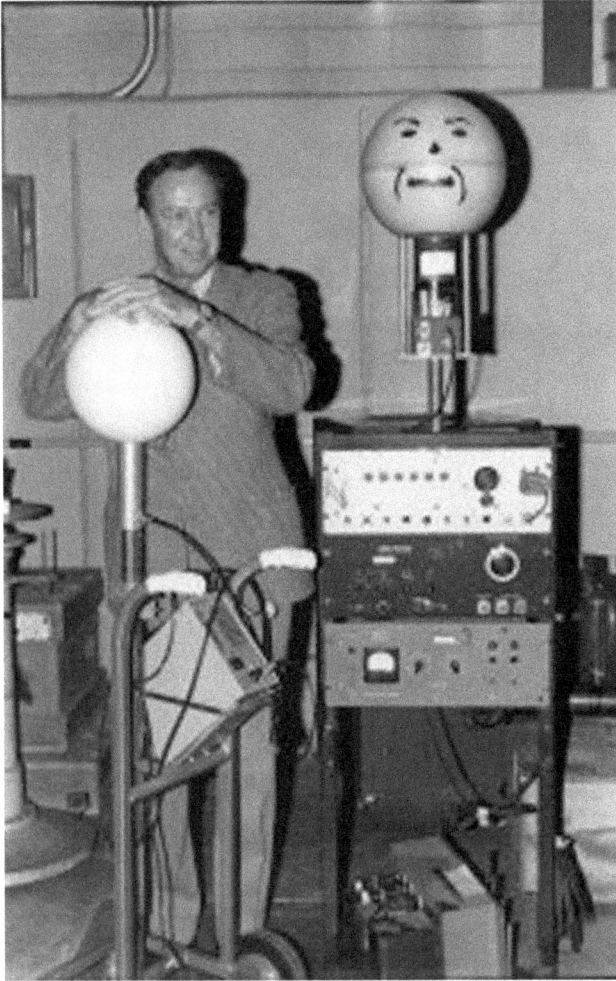

FIGURE 1.12
Prof. Gerry Phillips, Director of T. W. Bonner Nuclear Laboratories.

The Lab studies, both theoretical and experimental, of nuclear reactions with three particles in the final state induced either by neutrons or protons, have attracted much of the visibility. In particular, the identification of effects of final state interaction, quasi-free scattering, rescattering and charge-dependence of nuclear forces, on both low and medium bombarding energies (Baker et al. 1969, Beam and Valković 1972, Andrade et al. 1972, Bhasin et al. 1973, Duck et al. 1972, 1973, Duck and Valković 1973, Emerson et al. 1971, Niiler et al. 1969, 1970, Sweeney et al. 1971, Valković et al. 1967, 1968, 1969, 1970, 1971a, 1972, 1974, 1975, von Witsch et al. 1972).

Let us illustrate one such study as described by Valković et al. (1971a, 1971b). The simultaneous detection of the p–p and p–n coincidences from the $p + d + p + p + n$ reaction allows a trustworthy comparison of the p–p and p–n quasi-free scattering (QFS) cross-sections.

The spectra obtained are shown in Fig. 1.13 as a projection on the common proton energy axis.

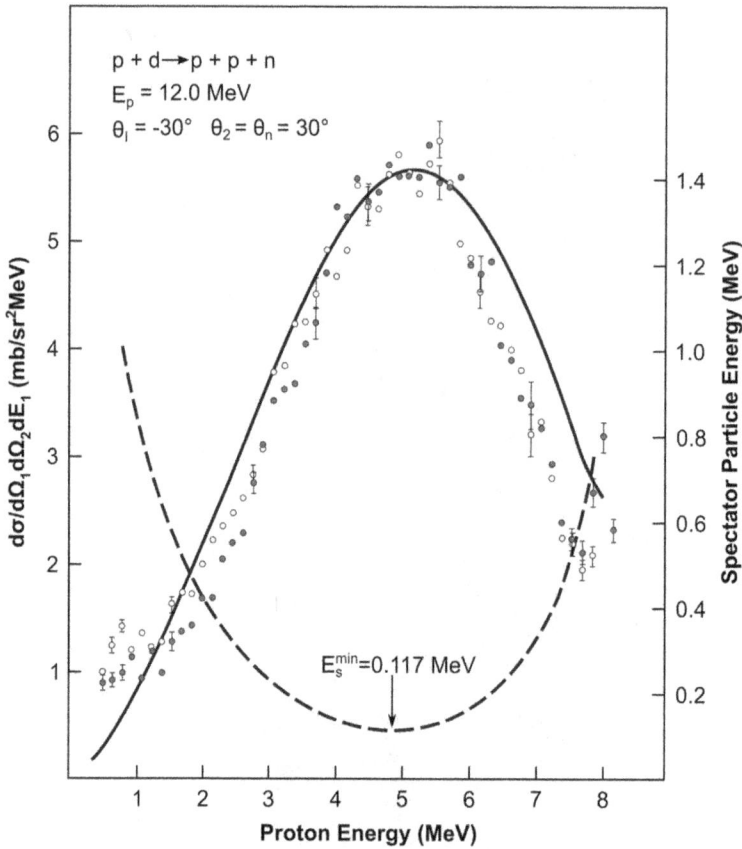

FIGURE 1.13

Proton–proton (solid dots) and proton–neutron (open circles) coincidence spectra projected on the common proton energy axis. The p–n spectrum is normalized to the p–p spectrum by N = 0.5. The dashed line shows the spectator particle energy, while the solid line represents the normalized (N = 0.2) impulse approximation prediction.

The experimental peak cross-section for p–p QFS is $(5.8 + 0.5)$ mb/sr^2 MeV while for n–p QFS it is (11.7 ± 0.5) mb/sr^2 MeV. The ratio of the observed n-p and p-p QFS contributions is much smaller than that reported by Peterson et al. (1970). Both experimental spectra, when normalized, have the same shape and are symmetrical around the minimum spectator particle energy. This fact indicates that the effect of Coulomb forces is negligible. Otherwise, p–p and p–n coincidence spectra would be somewhat shifted with respect to each other and the difference in absolute cross-section would be smaller if it were due to Coulomb penetrability only.

The development of charged particle and neutron detection methods, in particular position-sensitive detectors, was a large effort that resulted in visible results. Bonner Lab. was among first laboratories to start the development and application of analytical methods based on detection of proton-induced X-ray and gamma-ray emission, use of focused beams for measurement of concentration profiles and maps (Jackson et al. 1967, Andrade et al. 1979, 1990, Liebert et al. 1973, Plasek et al. 1973, Phillips et al. 1972, Plasek et al. 1976).

By using accelerator-based analytical techniques following has been initiated:

- Study of the role of chemical elements in biological systems (Valković et al. 1973, 1974, 1976, 1979, Alexander et al. 1974; Biegert et al. 1976, Biegert and Valković 1980, Rendić et al. 1976, Valković and Phillips 1976).
- Study of the environmental pollution by radionuclides, toxic elements, identification of pathways and evaluation of impact.
- Industrial applications of nuclear analytical methods, especially in metallurgy, coal industry and environmental safety.
- Quality assurance and quality control measures for nuclear analytical methods.

Although nuclear reactions with three particles in the final state occur very frequently in both high energy and nuclear physics, the understanding of these reactions has been limited, mainly because of the difficulties in the theoretical formulation of the problem.

1.6.2 USA

An outgrowth of immense investment in scientific research initiated by the US Government during World War II, the National Laboratories have served as the leading institutions for scientific innovation in the United States for more than seventy years. The Energy Department's National Labs tackle the critical scientific challenges of our time – from combating climate change to discovering the origins of our universe – and possess unique instruments and facilities, many of which are found nowhere else in the world. They address large scale, complex research and development challenges with a multidisciplinary approach that places an emphasis on translating basic science to innovation.

The United States Department of Energy currently operated seventeen national laboratories (see https://www.usa.gov/federal-agencies/national-laboratories) are presented in Table 1.2.

Let us mention only some individual laboratories:

Berkeley Lab was founded by Ernest Orlando Lawrence, a UC Berkeley physicist who won the 1939 Nobel Prize in physics for his invention of the cyclotron, a particle accelerator that opened the door to high-energy physics. It was Lawrence's belief that scientific research is best done through teams of individuals with different fields of expertise, working together. His "team science" concept is a Berkeley Lab legacy; today, a deep commitment to inclusion and diversity brings perspectives that inspire innovative solutions. For more information see https://www.lbl.gov/about/.

Jefferson Lab is a world leader in accelerator science. This expertise comes from the planning, building, maintaining and operating of the Continuous Electron Beam Accelerator Facility (CEBAF), the lab's primary particle accelerator, and the Low Energy Recirculator Facility, a test bed for a variety of technologies. CEBAF is based on superconducting radiofrequency (SRF) technology. It produces a stream of charged electrons that can be used to probe the nucleus of the atom. CEBAF was the first large-scale application of SRF technology in the world, and it is the world's most advanced particle accelerator for investigating the quark structure of the atom's nucleus. The CEBAF energy was recently upgraded to 12 GeV, and an additional experimental area was added to support the highest-energy experiments.

The Jefferson Lab Low Energy Recirculator Facility is powered by a smaller SRF accelerator. Formerly known as the Free-Electron Laser (FEL), the facility holds power

TABLE 1.2

USA national laboratories

Name	Location	Since	Presently operated by
Ames Laboratory	Ames, Iowa	1947	Iowa State University (since 1947)
Argonne National Laboratory	DuPage County, Illinois	1946	UChicago Argonne, LLC
Brookhaven National Laboratory	Upton, New York	1947	Stony Brook University (since 1998)
Fermi National Accelerator Laboratory	Batavia, Illinois	1967	Fermi Research Alliance (since 2007)
Frederick National Laboratory for Cancer Research	Frederick, Maryland	1972	Leidos Biomedical Research, Inc.
Idaho National Laboratory	Idaho Falls, Idaho	1949	Battelle Memorial Institute (since 2005)
Lawrence Berkeley National Laboratory	Berkeley, California	1931	University of California (since 1931)
Lawrence Livermore National Laboratory	Livermore, California	1952	Lawrence Livermore National Security, LLC (since 2007)
Los Alamos National Laboratory	Los Alamos, New Mexico	1943	Triad National Security, LLC (Since 2018)
National Energy Technology Laboratory	Pittsburgh, Pennsylvania	1910	Department of Energy
	Morgantown, West Virginia	1946	
	Albany, Oregon	2005	
National Renewable Energy Laboratory	Golden, Colorado	1977	Alliance for Sustainable Energy, LLC (since 2008)
Oak Ridge National Laboratory	Oak Ridge, Tennessee	1943	UT–Battelle (since April 2000)
Pacific Northwest National Laboratory	Richland, Washington	1965	Battelle Memorial Institute (since 1965)
Princeton Plasma Physics Laboratory	Princeton, New Jersey	1951	Princeton University
Sandia National Laboratories	Albuquerque, New Mexico	1948	Honeywell International (since 2017)
	Livermore, California		
Savannah River National Laboratory	Aiken, South Carolina	1952	Savannah River Nuclear Solutions, LLC (since 2008)
SLAC National Accelerator Laboratory	Menlo ParkCalifornia	1962	Stanford University
Thomas Jefferson National Accelerator Facility	Newport News, Virginia	1984	Jefferson Science Associates, LLC (since 2006)

records in the production of infrared, ultraviolet and terahertz laser light. The FEL was used in a variety of scientific studies, such as developing processes for producing high-quality carbon and boron-nitride nanotubes, identifying laser light wavelengths for use in medical treatments and in micromachining studies.

The Jefferson Lab Accelerator Division is responsible for delivering high-quality electron beams for experiments, using a sophisticated computer system to control hundreds of thousands of hardware components, including complex cryogenic, microwave, vacuum and magnet systems that comprise the accelerator. The division also pursues a broad program of theoretical and experimental research in accelerator and beam physics.

The Cornell Laboratory for Accelerator-based Sciences and Education is housed in three buildings on Cornell University's Ithaca campus. The Cornell Laboratory for Accelerator-based

ScienceS and Education (CLASSE) studies the world and universe using X-rays and high-energy electrons and proton beams. Among others, they study the folding of long DNA strands at CLASSE's CHESS national x-ray user facility; also, they study the Higgs boson with the Large Hadron Collider, and study the early universe with the Atacama Cosmology Telescope.

CLASSE also advances the frontiers of beam science and develops the technology needed to produce and accelerate ultra-bright, high power beams for research, medicine and industry. They study the behaviour of very compact beams using the Cornell Electron Storage Ring (CESR), develop high power electron sources and advance the superconducting technology that accelerates particles to near light speed. Most of the work done is directed at an Energy Recovery Linac, a special kind of accelerator invented at Cornell that would produce X-ray beams 1000 times brighter than those at today's synchrotron sources.

The workhorse of the *Cambridge Laboratory of Accelerator Studies of Surfaces (CLASS)* is the General Ionex 1.7 MV tandem ion accelerator, or Tandetron for short. The Tandetron is a completely solid-state ion accelerator (i.e., no moving parts) based on a parallel-fed cascade generator with a CW voltage multiplier. The terminal control electronics use a generating voltmeter for feedback and keep the terminal ripple down to a few kV, ensuring well-defined ion beam energies for confident interrogation of surfaces.

The Tandetron uses N_2 gas for electron stripping at the terminal. Depending on the electron stripping cross-sections, various charge states emerge at the high energy end of the accelerator. A selector magnet at the high-energy end, one can choose which charge state beam to use, so to have the ability of using ion beams that are +4 or +5 ions. This allows the utilization of beams with energies up to 10.5 MeV. The accelerator can operate with relatively high beam currents (<100 µA) for greater flexibility with IBA at the target.

Of interest is to mention plasma-surface interaction research at the CLASS. The material requirements for plasma-facing components in a nuclear fusion reactor are some of the strictest and most challenging. These materials are simultaneously exposed to extreme heat loads) and particle fluxes ($>10^{24}$ m^{-2} s^{-1}) while also undergoing high neutron irradiation (10^{18} neutrons/m^2 s). At CLASS, many of the most important issues in plasma-surface interaction research, such as plasma-driven material erosion and deposition, material transport and irradiation and hydrogenic retention were investigated with the use of a 1.7 MV tandem ion accelerator. IBA was used to investigate and quantify changes in materials due to plasma exposure and ion irradiation was used as a proxy for neutron irradiation to investigate plasma-surface interactions for irradiated materials (Wright et al. 2011).

1.6.3 iThemba Laboratory for Accelerator-Based Sciences, Johannesburg, South Africa

The iThemba Laboratory for Accelerator-Based Sciences (LABS) is a non-profit national research institution of the National Research Foundation of South Africa. It is highly multidisciplinary, as it is dedicated to physical, medical and biological sciences. The laboratory is based at two sites 1600 km apart, one in Cape Town and the other in Johannesburg. The main lines of research pursued are: Nuclear physics and materials research; Radiation therapy and cancer research; Production of unique radioisotopes.

At the moment iThemba LABS have only 2 electrostatic accelerators, namely the 3 MV tandetron and the 6 MV EN tandem accelerator now being used for AMS. The 6 MV CN Van de Graaff accelerator was decommissioned just before installing the 3 MV tandetron accelerator. The tandetron was installed in the same building which housed the 6 MV Van de Graaff accelerator (Figs. 1.14–1.16).

FIGURE 1.14
The 6 MV iThemba LABS tandem accelerator used for AMS, viewed from the injection side. (Picture reproduced courtesy of NRF: iThemba LABS.)

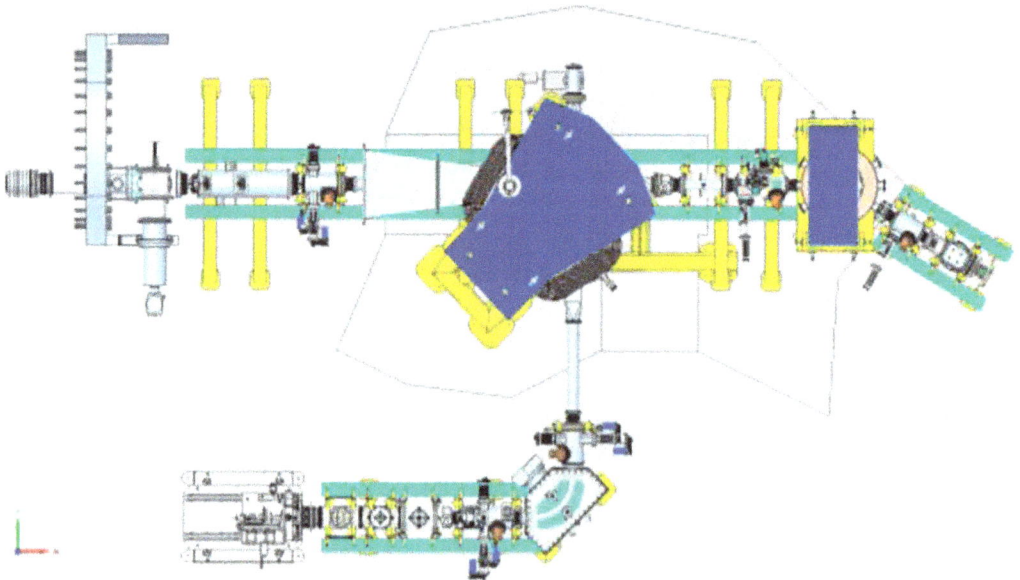

FIGURE 1.15
A view of the injection system for 6 MV EN tandem accelerator of iThemba LABS. (Picture reproduced courtesy of NRF: iThemba LABS.)

FIGURE 1.16
Control room 6 MV Van de Graaff 2003 iThemba LABS (top), Control room & MV CN Van de Graaff 2014 LABS (bottom). (Pictures reproduced courtesy of NRF: iThemba LABS.)

Research at the Johannesburg site is based around a refurbished 6 MV tandem accelerator that can produce a wide range of ion beams. A user-driven environmental isotope laboratory is also installed at Johannesburg, providing analytical services to research activities in: isotope hydrology – water resources assessment; pollution studies; archeometry; medical and biological research.

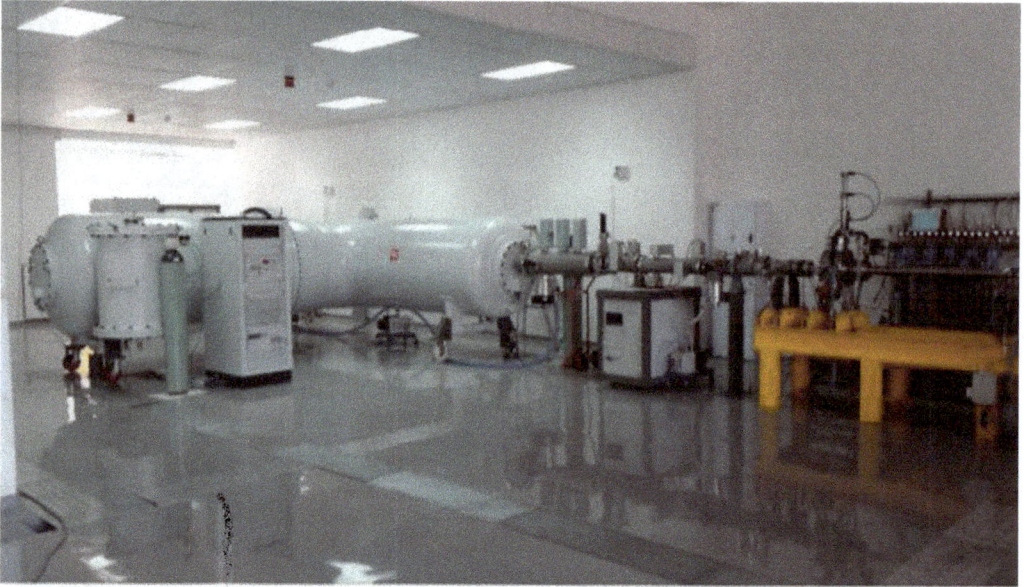

FIGURE 1.17
3 MV Tandetron at iThemba LABS. (Picture reproduced courtesy of NRF: iThemba LABS.)

The main analytical techniques used in materials research with IBA at iThemba LABS are micro-PIXE, (proton) RBS and heavy ion ERDA. The accelerator uses a sputter ion source as well as an additional source for He^+ RBS analyses. The research activities developed include: depth profiling of transition metal-based ceramics for applications in thin hard coatings; analysis of "smart" optical coatings; impurity diffusion studies of nuclear materials; elemental distribution in atmospheric aerosols for power plant monitoring; imaging mineral phases in geological samples using micro-PIXE; proton irradiation effects on diamond and graphite materials; energy loss and energy loss straggling in ceramic materials (Figs. 1.17 and 1.18).

1.6.4 Accelerators in Europe (EU)

There are several large accelerators in European countries, as well as accelerator centers, all listed in ELSA (2021). Actually, the page lists particle accelerators around the world, the list does not include accelerators that are used for medical or industrial purposes only. Here we shall present a list of small accelerators in Europe only, see Table 1.3.

Next, we shall present some of the European laboratories whose activities are concentrated around small particle accelerators.

The Institute for Applied Physics and Metrology of the Universität der Bundeswehr München focuses on applications based on three main experimental facilities:

- Scanning positron microscopy with positron annihilation.
- Scanning nuclear microbeam SNAKE, including hydrogen microscopy and directed irradiation of living cells with single ions.
- ERDA with a Q3D magnetic spectrograph.

The ion beam-based activities of the institute take place at the 15 MV tandem accelerator of the Maier-Leibnitz Laboratory in Garching, close to Munich. The accelerator laboratory is

FIGURE 1.18
High energy beam line of the tandetron accelerator. (Picture reproduced courtesy of NRF: iThemba LABS.)

operated by the Ludwig Maximilian University of Munich and the Technical University of Munich. It performs basic and interdisciplinary research concerning the interactions of ion beams with matter, nuclear physics, ion irradiation and IBA. These study areas involve:

- Interaction of energetic ions in matter (energy loss, straggling, sputtering).
- IBA methods (ERDA, RBS, coincident proton–proton scattering, PIXE, PIGE and STIM).
- Modification of materials using MeV ions.
- Accelerator mass spectrometry.
- Irradiation of human cells (medical applications).
- Irradiation of new nuclear fuel material.

The institute was part of the European Union FP7 project SPIRIT and provides transnational access. Within the transnational access scheme, scientists from the European Union and associated countries can apply for beam time at any of the beamlines dedicated to IBA and cell irradiation. The experimental facilities of the laboratory include:

- A Q3D magnetic spectrograph (for nuclear physics and high resolution ERDA and RBS);
- A micro beam SNAKE (cell irradiation, proton–proton scattering, PIXE, PIGE and RBS);
- A gas filled magnet for accelerator mass spectrometry;
- Various irradiation stages;
- Different beamlines for nuclear physics experiments.

Munich tandem accelerator is equipped with the Q3D magnetic spectrograph. The Q3D magnetic spectrograph has a large dispersion (dE/Edx) $\approx 2 \times 10^{-4}$/mm), an excellent

TABLE 1.3

Small accelerators in Europe

Country	Town	Type	Terminal voltage (MV)	Accelerated ions
Austria	Linz	Pelletron	1.7	L
	Vienna	Pelletron-tandem	3	L+H
Belgium	Geel	VdG	7	L
		VdG	3.7	L
	Liège	VdG	3	L+H
		VdG	2.3	L+H
	Louven	Pelletron	1.7	L
		VdG	4	L
Croatia	Zagreb	Tandem	6.5	L+H
Denmark	Aarhus	Tandem	6	L+H
	Copenhagewn	Tandem	9.5	L
Finland	Helsinski	VdG	2.5	L+H
		Tandem	10	L
France	Gradignan	VdG	4	L
	Louvre-Paris	Pelletron	2	L
	Orsay	Tandem	13	L+H
	Saclay	Tandem	9.5	L+H
	Strasbourg	VdG	6	L+H
		VdG	7	L+H
		Tandem	16	L+H
	Villeurbanne	VdG	2.5	L+H
		VdG	4	L+H
		Electrostate	1	L+H
Germany	Berlin	Pelletron	8	L
	Bochum	Tandem-Dynamitron	4	L+H
	Cologne	Tandem	9	L+H
	Darmstadt	Linac	10	H
	Frankfurt	Spiral-resonator	2	L+H
		VdG	2.5	L+H
		VdG	7.2	L+H
	Freiburg	VdG	7.5	L+H
	Giessen	Tandem	1	L
	Heildeberg	Tandem	6	L+H
		Tandem	13	L+H
		Linac	22	L+H
		Pelletron	4	L
	Marburg	VdG	5	L+H
	Munich	Tandem	13	L+H
		Linac	5	H
	Stuttgart	Dynamitron	4	L+H
	Tübingen	VdG	3	L
Greece	Athens	Tandem	5.5	L+H
Hungary	Budapest			
	Debrecen	Tandetron	2	L
		MGC-20E, cycl.	K = 20	L

TABLE 1.3 (Continued)

Small accelerators in Europe

Country	Town	Type	Terminal voltage (MV)	Accelerated ions
		Homemade VdG	5	L
Italy	Catania	VdG	2.8	L
		Tandem	15	L+H
	Florence	VdG	2.8	L
	Naples	Tandem	3	L+HL
	Padova	VdG	2	L
		VdG	7.5	L
		Tandem	16	H
Netherlands	Groningen	VdG	6	L
	Utrecht	VdG	3	L
		VdG	1	L
		Tandem	7	L+H
Portugal	Lisbon	VdG	2	L
		VdG	2	L
Romania	Bucuresti- Mägurele	Tandetron	3	L+H
Slovenia	Ljubljana	VdG	2	L
Spain	Madrid	Tandetron	5	L
Sweden	Lund	Tandem	3	L+H
	Uppsala	Tandem	6.5	L+H
Switzerland	Basel	Cockcroft–Walton	3	L
	Neuchatel	VdG	3	L
	Zurich	VdG	5.5	L
		Tandem	6.5	L+H

Note: L – light ions (m ≤ 4); H – heavy ions.

intrinsic resolution ($\Delta E/E = 2 \times 10^{-4}$), together with a large solid angle of detection (with a maximum value of up to 14.3 msr). Furthermore, the kinematical shift can be corrected up to the fourth order with a magnetic multipole element. In routine operation, the kinematic shift is corrected up to the third order only, leading to an overall energy resolution of 7×10^{-4}, even when a 5 msr solid angle is used. An uncorrected kinematical shift would lead to a 6% energy spread at a 15° mean scattering angle. The multipole element is adjusted to focus the recoil ions scattered from a certain depth as well as possible in a given position on the focal plane of the Q3D. The ions are thus identified, and their position is measured. Thin foil targets are mounted for the stopping measurements perpendicular to the incident beam. After passing the thin foils, the energy loss of the ion beam is analyzed with the Q3D spectrograph at a 0° scattering angle.

The ion beam-based activities of the Instituto Superior Técnico/University of Lisbon take place in the Accelerator and Radiation Technologies Laboratory. The Ion Beam Laboratory is devoted to research and applications of ion beams in materials characterization and materials synthesis. The physical processes involved in IBA are also studied, with a strong effort in the experimental determination of stopping powers and scattering cross-sections. Several groups use the facilities available in different main areas of study, including: Materials science; Earth and environmental sciences; Cultural heritage; Life and health sciences; Nuclear physics; Materials for fusion.

The experimental facilities of the laboratory include:

- A 2.5 MV Van de Graaff accelerator, which is mainly used for: Rutherford backscattering; ERDA; Particle induced X-ray emission (PIXE); Particle induced gamma-ray emission (PIGE); Nuclear reaction analysis.
- An ion microprobe connected to the Van de Graaff accelerator, which includes: A standard micro beam analysis chamber; and an external beam set-up.
- A 3 MV tandem accelerator, which is mainly used for: High resolution PIXE; Accelerator mass spectrometry.
- An X-ray diffraction laboratory with several set-ups, including: A high-resolution line; A high-temperature set-up; A commercial Brucker D-8 spectrometer.

Most of the experiments undertaken at the Instituto Superior Técnico were performed with a "universal chamber" installed in one beamline of the Van de Graaff accelerator. This beamline had two experimental chambers. The beam is defined by two collimators, located 2.15 m apart, leading to a low angular dispersion. The final collimator, which defines the beam shape, is rectangular with a height of 0.6 mm and a variable width, which in this work was fixed at 0.6 mm. The beam fluence is measured, to an accuracy of 2%, with a transmission Faraday cup that periodically intercepts the primary beam. The chamber has three detectors, which can be used simultaneously. One is an annular detector around the beam, which has poor energy resolution. A movable detector is normally located at forward scattering angles and is used for ERDA experiments. The samples are inserted into the chamber without breaking the vacuum, leading to pressures below 10^{-7} mbar during experiments. The sample holder is connected to a two-axis goniometer, defining the angle of incidence with a precision of 0.02°.

"Horia Hulubei" National Institute for Physics and Nuclear Engineering – IFIN-HH, Măgurele, Romania has a new ion beam facility based on a 3 MV Tandetron (Burducea et al. 2015). The Institute has a long history of accelerator-based physics. A Soviet made U120 cyclotron was installed in 1956. In 1974, an FN tandem accelerator (upgraded later to 9MV) was brought in from HVEC. The Institute also operates 1 MV tandem dedicated to multi-element mass spectrometry The research infrastructure of IFIN-HH has continuously developed in order to become a representative Romanian institute for the 21st century physics at the European and international level, in fundamental and applied nuclear physics.

In 2015, a 3 MV Tandetron™ accelerator system has been installed and commissioned (Burducea et al. 2015). The accelerator is equipped with two ion sources, the HVE Model 860A Cs sputter type ion source and the HVE Model 358 duoplasmatron type ion source. The 860A ion source can produce ions of almost every element of the periodic table except the noble gases. The HVE Model 358 is characterized by a two-stage discharge. Applications are mainly for light element ion beams, namely hydrogen and helium. It has three endstations: IBA endstation, implantation endstation and cross-section measurement endstation.

MTA Atomki Institute, Debrecen, Hungary was established in 1954 as an accelerator center. The study of nuclear physics using particle accelerators has been one of the main research fields. The application of particle accelerators were in the fundamental investigation of the structure of the atomic nucleus, the study of nuclear reactions with astrophysical importance, elemental analysis of airborne aerosols and in other research fields like geology, materials science, biomedicine and in the study of museum artefacts.

The existing high-energy particle accelerators include a MGC-20E cyclotron (K = 20) and two homemade Van de Graaff accelerators, 1 MV and 5 MV, 30 and 45 years old, respectively. Recently, 2 MV Tandetron accelerator manufactured by HVEE was installed (Rajta et al. 2018).

1.6.5 Accelerators in the UK

Table 1.4 shows locations of accelerators discussed here within the UK. Some of them have worldwide reputation in different fields of applications of nuclear technology.

1.6.6 Accelerators in Asia Pacific Region

The Asian Nuclear Physics Association (ANPhA) (http://ribf.riken.jp/ANPhA/) was established in 2009 in Beijing, where representatives of the first four member countries of ANPhA gathered together. ANPhA is the central organization representing nuclear physics in Asia Pacific and currently consists of eleven member countries and regions, i.e., Australia, China, Hong Kong, India, Japan, Kazakhstan, Korea, Mongolia, Myanmar, Taiwan and Vietnam. In Asia Pacific region, many advanced accelerator facilities have been constructed. Some of them are world top class. ANPhA is now preparing a list of Asia-Pacific accelerator facilities available for nuclear physics experiments.

The list of major accelerator facilities for nuclear physics in Asia Pacific was prepared and published in 2019 (Tanaka 2020), see Table 1.5.

Let us shortly describe the situations in some of the countries in the region.

- India: Bhabha Atomic Research Center, BARC, is well known center in the area. Here are some of the new developments at the institute:
  - Accelerator-based Large Scanning Systems: Accelerators-based technology for Radio-graphing and scanning the big and large containers is being developed in the division. This is based on the 10 MeV RF electron linac. The electrons thus produced will be used for producing X-rays which in turn in conjunction with tomography, will be used for imaging the containers.
  - SC/NC Linac Cavity and Structures: The division has an elaborate R&D programme of designing and developing RF & Microwave structures. Since efficiency of the accelerators is one of the main concern, both types of structures, Normal Conducting (NC) & Super Conducting (SC), are being studied.

TABLE 1.4

Small accelerators in the UK

Town	Type	Terminal voltage (MV)	Accelerated ions
Harwell	VdG	6	L+H
	VdG	3.4	L
	Tandem	7	L+H
Manchester	VdG	7	L
Surrey	VdG	2	L
Doresbery	Tandem	20	L+H
Oxford	Tandem	10	L+H
	Tandem	6.5	L+H

Note: L – light ions (m ≤ 4); H – heavy ions.

TABLE 1.5

List of some accelerators in Asia Pacific Region

Location	Institution	Facility acronym	Type of accelerator
Canberra Australia	Australian National University (ANU). Heavy ion Accelerator Facility		15 MV Tandem accelerator + superconducting Linear Accelerator
Beijing, China	Beijing Tandem Accelerator Nuclear Physics National Laboratory	BTANL	15 MV tandem accelerator, 100 MeV 20 μA proton cyclotron, ISOL
Shanghai, China	Shanghai Laser Electron Gamma Source	SLEGS	0.4–20 MeV BCS γ ray source based on Synchrotron Radiation Facility
Jinping, China	China Jinping Underground Laboratory, Jinping Underground Nuclear Astrophysics Experiment	CJPL/JUNA	400 kV accelerator (H, He). Max. energy 400 kV q. Beam intensity <2.5 emA
Lanzhou, China	Heavy Ion Research Facility in Lanzhou	HIRFL	SSC cyclotron: K=450 and full ion acceleration; CSRm booster synchrotron 12.2 Tm
Huizhou, China	China initiative ADS	CIADS	The 250 MeV and 10 mA CW mode superconducting proton linac
Munbai, India	Bhabha Atomic Research Centre – Tata Institute of Fundamental Research	BARC-TIFR	14 MV heavy ion tandem + superconducting linac
New Delhi, India	Inter-University Accelerator Centre	IUAC	15MV heavy ion tandem + superconducting linac
Kolkata, India	Variable Energy Cyclotron Centre	VECC	VECC K130 cyclotron (p,α); K500 superconducting cyclotron
Spring-8 site, Hyogo, Japan	Laboratory of Advanced Science and Technology for Industry	NewSUBARU	Laser Compton Scattering Gamma-ray Beam Source (1–76 MeV)
Fukuoka, Japan	Kyushu University, Center for Accelerator and Beam Applied Science		FFAG synchrotron and tandem accelerator
Tokai, Ibaraki, Japan	Japan Atomic Energy Agency, Tandem Accelerator Facility		20 MV tandem accelerator and superconducting linac booster
Tsukuba, Ibaraki, Japan	University of Tsukuba, Tandem Accelerator Complex	UTTAC	6 MV tandem accelerator; 1 MV Tandetron accelerator
Sendai, Japan	Tohoku University, Cyclotron and Radioisotope Center	CYRIC	K110 and K12 cyclotrons
Sendai, Japan	Research Center for Electron-Photon Science, Tohoku University	ELPH	60 MeV high-intensity electron linac
Gyeongsangbuk-do, Korea	Korea Multi-purpose Accelerator Complex	KOMAC	100 MeV and 20 MeV proton linac
Seoul, Korea	Korea Institute of Science and Technology, The Accelerator Lab.		2 MeV and 6MV tandetron accelerators
Seoul, Korea	Heavy Ion Medical Accelerator at Korea Institute of Radiological and Medical Sciences	KIRAMS	AVF cyclotron for 50 MeV protons

TABLE 1.5 (Continued)

List of some accelerators in Asia Pacific Region

Location	Institution	Facility acronym	Type of accelerator
Jeollabuk-do, Korea	Advanced Radiation Technology Institute		15–30 MeV 500 mA proton cyclotron
Hsinchu, Taiwan	Graduate Institute of Nuclear Science, National Tsing Hua Univ	INS/NTHU	3MV Van de Graaff (KN); 3 MV tandem (NEC 9SOH-2); 500 kV open air accelerator
Hanoi, Vietnam	Tandem machine at Hanoi University of Natural Science		1.7 MV tandem Pelletron
Hanoi, Vietnam	Military Central Hospital		30 MeV 300 μA proton cyclotron

This is being done to cater to the needs of industrial electron accelerators as well as high current proton accelerators. The microwave structure frequencies lie in the S band whereas the RF ones are around 700 MHz.

- 14 MeV Neutron Source: A 250 keV, 10 mA, d+t based source is being designed and developed to provide a neutron yield of about 10^{12} n/sec. This will in principle be used for conducting experiments with a reactor in the sub critical state. It will primarily help in simulating the ADS system and validate the simulation models.

   For more information on accelerator programs in India see paper by Bhandri and Dey (2011). They showed that accelerators play an essential role in medical diagnostics and cancer therapy, industrial processing and will, possibly, play a very important role in solving the future energy problem in the form of Accelerator Driven Subcritical Systems.

- Japan: Accelerators have a long history in Japan starting from the 1930s. Two Cockcroft–Walton accelerators of Japan were completed by Prof. Arakatsu of Taihoku University and a few month later Prof. Kikuchi of Osaka University in 1934. In 1937, Dr. Nishina of RIKEN, a father of Japanese quantum physics completed a small cyclotron and then began to construct a large cyclotron with help of Prof. Lawrence. Prof. Kikuchi and later Prof. Arakatsu who moved from Taihoku University to Kyoto University also began to construct cyclotrons at Osaka and Kyoto, respectively. The Nishina laboratory of RIKEN was a leading one in this "Heroic" age of Japanese accelerator history (Inoue 1998).

   By the end of century, Japan has become a superpower in the accelerator world, it had a large number of accelerators spread around country and application fields (Table 1.6).

   Japan Radioisotope Association (JRIA) is publishing yearly statistics on the use of radiation in Japan We present here the data for 2016 (JRIA 2016), see Table 1.7 from https://www.jrias.or.jp/e/pdf/the_use_of_radiation_2016.pdf).

   Here is a description of country distributions for some specific types of accelerators:

   - Tandem accelerators: At JAERI-Tokai a super-conducting linac is working as an energy booster for a 20 MV pellet on tandem accelerator. The maximum accelerating energy of the linac is 30 MeV/q. High energy nuclear spectroscopy experiments with high resolution are performed by using this system.

TABLE 1.6

Number of accelerators in Japan (as of 31 March, 1997). Electron accelerators with energy less than 1 MeV are omitted

Accelerator type	Hospitals & Clinics	Educational organization	Research Institutions	Industry	Other	Total
Cyclotrons	17	–	16	18	1	52
Synchotrons	–	2	18	5	–	25
Linear	624	7	44	81	3	759
Betatrons	10	–	1	3	–	14
Electrostatic	–	15	26	11	–	52
C.-W.	–	27	33	35	2	97
Transformer	–	1	17	5	–	23
Microtons	22	3	–	4	–	29
Total	673	55	155	162	6	1051

TABLE 1.7

Number of Radiation Generators in use in Japan (as of 31 March 2016, from JRIA 2016)

Radiation generators Category	Total number	% of total	Hospitals Clinics	Education Institute.	Research Institute	Private Co.	Other
Cyclotrons	233	13.6	155	4	22	20	2
Synhotrons	44	2.6	11	3	26	4	–
Linear acc.	1 292	75.6	1 107	26	65	63	31
Betatrons	2	0.1	–	1	1	–	–
Van de Graaff acc.	35	2.0	–	13	21	1	–
Cockcroft–Walton acc	80	4.7	–	17	29	34	–
Transformer type acc.	14	0.8	–	–	6	8	–
Microtons	6	0.4	1	3	2	–	–
Plasma generators	2	0.1	–	–	2	–	–
Total	1 7068	100	1 274	67	174	160	33

Tsukuba University 12 MV tandem accelerator has also a small booster linac with energy of 2 MeV/q.

Two tandem Van de Graaffs of Kyoto University and University of Tokyo have been replaced by pelletron tandems though the original high pressure vessels are again used in both cases. The terminal voltages have been improved from 5 MV to 8 MV for Kyoto and from 5 MV to 6 MV for Tokyo.

The laboratory of Kyushu University 10 MV pelletron tandem which is a unique high-gradient field home-made machine works as an accelerator center of Kyushu area. These modern tandem accelerators have been constructed originally for nuclear physics but recently they are applied for not only nuclear physics but also many other fields such as AMS ^{14}C dating. And many small tandems have been constructed for material science and in particular element analysis, for example, Rutherford backward scattering or PIXE with micro-beam.

- Cyclotrons: Besides the RIKEN and RCNP ring cyclotrons there are many cyclotrons in Japan. At JAERI-Takasaki a K = 110 AVF cyclotron of TIARA (Takasaki Ion Accelerators for Advanced Radiation Application) works for material science, bio-technology and other fields in combination with other small accelerators. The INS SF-cyclotron is now operated by CNS (Center for Nuclear Study), University of Tokyo, because a few divisions have remained with the cyclotron at the University when INS and KEK (national laboratory for high energy physics) have merged to become the new KEK (high energy accelerator research organization) in 1997. The CNS cyclotron still injects the beam into the electron cooler ring TARN II which belongs now to the new KEK-Tanashi.

  The TARN II was originally constructed to study of the design of a heavy ion synchrotron for nuclear physics project NUMATRON. Unfortunately, the project was not funded. But the TARN II has been effectively used for design study of HIMAC and is now operated for accelerator and beam physics. One of the excellent experiments of the beam physics with the TARN II is precise observation of dissociative recombination (DR) of the molecular ion $^4HeH^+$ with the ultra-cold electron beam whose initial transverse temperature of 100 meV is reduced to the order of 1 meV by changing the electron guiding field from 3.5 T to 35 mT with a superconducting magnet as an adiabatic expansion device.

  Cyclotrons for isotope production are operated at universities, companies and hospitals. Tohoku University Cyclotron and Radioisotope Center (CYRIC) has operated an AVF cyclotron, which will be replaced by a new larger one. However, a small cyclotron is enough to produce RI for PET in hospital. Therefore Tohoku University will continue RI production for PET by another compact cyclotron during the construction of the new larger cyclotron because needs for PET RI is very high. In fact there were 31 PET facilities in Japan already in November 1997. On the other hand some cyclotrons produce positron emitters such as ^{27}Si which is used as a positron beam source for material science though some electron linacs are also used for production of high intensity positron beam.

- Linacs: The first modern linear accelerator for nuclear physics in Japan is the 300 MeV Tohoku University electron linac. It has been also used for accelerator physics; for example, coherent radiation from the electron beam was discovered by this linac. On the other hand, the pulsed beam is not convenient for recent nuclear physics. Therefore a stretcher and booster ring, STB with highest energy of 1.2 GeV has been attached after a long term experience of a small stretcher ring.

  There are additional electron linacs for radiation physics, for examples, a 150 MeV S-band and a 38 MeV L-band linacs at ISIR, Osaka University, a 45 MeV linac at Hokkaido University, a 10 MeV PNC-OEC quasi-cw linac for transmutation study in collaboration with Nihon University, a 46 MeV L-band linac as a neutron source at KURRI, Kyoto University, the 35 MeV twin-linac at NERL, University of Tokyo, a 500 MeV ETL linac and a 100 MeV linac at ICR, Kyoto University.

  Most of accelerators in Japan are electron linacs at hospitals. There are also many industrial electron linacs for non-destructive inspection and radiation processing. On the other hand, there are no so many ion linacs. At KEK, NIRS and RIKEN there are injector linacs of 40 Mev protons, 6 MeV/u ions and 3 MeV/u ions, respectively. A 30 MeV/q super-conducting linac at JAERI is a booster of the tandem as mentioned above. A 7 MeV RFQ-DTL of ICR, Kyoto University, 5 MeV RFQ-DTL of KEK and 2 MeV RFQ of JAERI are all proton linacs for accelerator physics or testing future projects. At RLNR, Tokyo Institute of Technology a few small ion linacs have been constructed for material science. The maximum energy of the RLNR linacs is 3.4 MeV/u. Their accelerating structures are IH (Inter-digital-H) and RFQ. Large number of linacs for MeV-ion implantation is present in industry. A linac system composed of a 25.5-MHz split coaxial RFQ (SCRFQ) linac and a 51-MHz IH linac has been completed at INS and is in operation at the KEK-Tanashi. This linac has been developed as a prototype of the ISOL post accelerator for E-arena of Japan Hadron Facility (JHF) project. The ^{19}Ne^{2+} ion beam whose half-life is 17.3 sec has been successfully accelerated. The maximum energy of the system is 1.05 MeV/u.

The Research Center for Nuclear Physics (RCNP) was founded in 1971 at Osaka University to promote nuclear physics research using a variable-energy, multiparticle azimuthally varying field (AVF) cyclotron with a K-value of 140 MeV. The maximum kinetic energy of a charged particle accelerated by the AVF cyclotron is given by $140 \times (Q/A)^2$ MeV/u, where Q and A are charge state and mass number of the charged particle, respectively. Construction of the AVF cyclotron facility was finished in 1973 and nuclear physics experiments started from 1976. A ring cyclotron with a K-value of 400 MeV (Miura et al. 1992) was completed in 1991 to increase the maximum energy of proton and heavy ion beams up to 420 and 100 A MeV, respectively.

The AVF cyclotron in a standalone operation mode is mainly used for radio isotope (RI) production, educational experiments and detector developments. The AVF cyclotron also provides an ion beam to the ring cyclotron as an injector. Four kinds of ion sources, a 10 GHz permanent magnet type electron cyclotron resonance (ECR) ion source named NEOMAFIOS, an 18 GHz superconducting ECR ion source, a 2.45 GHz ECR proton source, a polarized proton and deuteron source, were developed to fulfill the requirements for the research in nuclear physics, radiochemistry, nuclear-medicine and interdisciplinary field (Nakano et al. 2019).

A high-quality proton beam with the energy spread of $\Delta/E \sim 10^{-4}$ is available for the precise measurement of the nuclear level structures. A 392 MeV proton beam is provided for production of secondary beams of muons and neutrons. The neutron flux with a broad energy spectrum, a white neutron beam, approximates the energy spectrum of cosmic ray neutrons observed on the sea level. The white neutron flux is particularly significant for accelerating a test to evaluate the radiation- induced soft errors and hard errors of semiconductor integrated circuits. Proton and helium ion beams with energies from 10 to 30 MeV are mainly used for production of radio isotopes.

Here is some of nuclear physics studies done at Cyclotron Facility. Light-ion beams, such as proton, deuteron, ^{3}He and alpha beams, are of the highest quality in the world with low emittance and without beam halo. These high-quality beams, together with high-resolution double arm spectrometers called "Grand Raiden" and "Large Acceptance Spectrometer" (LAS), enable performing high-resolution nuclear-reaction experiments, including zero-degree inelastic proton scatterings and charge exchange (^{3}He,t) reactions. A beam-line configuration has been constructed allowing the unreacted primary beam to be transported to a beam dump located about 20 m downstream after the target, thus realizing coincidence measurements at very forward scattering angles down to 4.5 degrees with auxiliary detectors placed around the target position, (Nakano et al. 2019).

- China: China started the research and education on nuclear science in the 1950s with the aim of developing nation's atomic energy systems. The first higher education unit on nuclear science was established at Peking University (PKU) in 1955. The first class was composed of 99 students selected from a large number of undergraduates who had already studied in many other science universities for 3 years. Therefore they got graduation just 1 year later and went to various places to initiate the nuclear science research and education in China. Since then the nuclear science unit at PKU, later on named as the Department of Technical Physics, has fostered more than 5000 graduates, among whom are 11 Academicians of the Chinese Academy of Science (CAS). The basic research on nuclear physics and technology has been strongly boosted since the establishment of the Ministry of Education (MOE) Key Laboratory of Heavy Ion Physics at Peking University in 1990.

  Based on intensive works for nearly twenty years, this MOE key laboratory was promoted to the State Key Laboratory of Nuclear Physics and Technology (SKLNPT), with an initial approval in 2007 and a formal establishment in 2010. Before 2007, the state key lab system in China was composed of hundreds of top-level labs in the nation, which were distributed in all fields of science and technology except the nuclear fields. Hence the formation of a state key lab in nuclear science at PKU was a real breakthrough. The state key lab and the education base has the advantages to receive regular operation budget from the national funding agencies. Actually there are about 80 staff members in SKLNPT, coming basically from the Department of Technical Physics, Institute of Heavy Ion Physics, Institute of Theoretical Physics, and Institute of Plasma Physics, of Peking University. The manpower of the laboratory is also strengthened by a large number of post-doctors and visiting scholars.

  The laboratory comprises an accelerator building with more than 7,000 m^2 floorage located on the eastern side of the University campus, and some offices and laboratory spaces in the technical physics building area. Based on the four well-running accelerators, namely the 2 × 6 MV tandem accelerator, the 4.5 MV Van De Graaff accelerator, the 2 × 1.7 MV tandem accelerator and the compact accelerator for ^{14}C AMS, the laboratory has been opened to other research teams nationwide and worldwide that use low energy ion beams to various research fields. Each year more than 3,000 h beam-time, together with some basic financial supports, are provided to the users. In addition, several laboratories are formed to support the advanced basic research, including the Subatomic and Particle Detection Laboratory, the Superconducting Accelerator Laboratory and the Laser Acceleration Laboratory. A teaching lab for nuclear physics experiment and nuclear electronics is also operated under the university teaching program for both undergraduate and graduate students (information from Wu 2013).

  The laboratory has been implementing many large national projects at the scientific and technological frontiers. The laboratory has built a broad research program covering four major research directions (divisions), namely the Radioactive Ion Beam (RIB) Physics, the Hadron Physics, the Advanced Particle Accelerator Techniques and the Applications of Nuclear Technology.

- Radioactive Ion Beam Physics: Groups in this division focus on experimental and theoretical studies of nuclear structure, reaction and decays of unstable nuclei. Nuclear detection techniques and nuclear electronics are implemented to support the experiments. For instance, the decay of the neutron-rich nuclei, such as ^{17}N and ^{21}N, have been measured and the complete decay diagrams were extracted (Li et al. 2009). The experiments were carried out at the Heavy Ion Research Facility in Lanzhou (HIRFL) with the radioactive ion beam-line RIBLL. A specially designed decay neutron spectrometer, called PKU neutron sphere, was constructed and applied in these experiments (Lou et al. 2009). The group has also been working on some direct reaction experiments. One typical example is the knockout reaction for ^{8}He, proposed and performed at RIKEN in Japan (Cao et al. 2012), which has resulted in a better understanding the exotic structure and reaction mechanism of this neutron drip-line nucleus. Afterwards, a new Multi-Neutron Correlation Spectrometer (MuNCoS) has been constructed at PKU and calibrated using the RIBLL beam line (You et al. 2013). This spectrometer is dedicated to the studies of the di-neutron and multi-neutron correlations, which are among the hottest topics nowadays in RIB physics.

- Advanced Particle Accelerator Techniques: The research in this division includes superconducting acceleration technology combined with free electron laser generation, the radio frequency quadrupole (RFQ) acceleration technology, and the newly developed laser-driven particle acceleration technology. Scientists and engineers in this division have proposed a new separate function RFQ (SFRFQ) concept and implemented the SFRFQ structure utilizing asymmetric diaphragms. An intense deuteron beam accelerator (2 MV) was constructed using the RFQ technique, which provides neutron beams via the (d, n) reaction. Neutron imaging method has been studied and applied using this facility which is named as Peking University Neutron Imaging Facility (PKUNIFTY). The fast neutron yield is 2.4×10^{11} n/s, and the spatial resolution for imaging is better than 0.4 mm.

  Superconducting acceleration has quite a long history at PKU. Scientists in this direction have developed special techniques to produce the superconducting cavities being used in the accelerator. The accelerating gradient of the third TESLA-type 9-cell superconducting RF cavity, developed using the domestic niobium materials, achieved a high record. Another 3.5-cell cavity for DC-SRF injector, made of large grain niobium, achieved the highest accelerating gradient of 23.5 MV/m. Now the self-produced cavities are assembled together to build a superconducting electron accelerator aiming at the generation of the FEL at THz range. The latest major effort in this division has been devoted to the laser-driven ion acceleration. A large project was approved to build a prototype proton accelerator CLAPA (15–100 MeV) based on the RPAPSA mechanism and plasma lens proposed from Peking University (Huang et al. 2011). This compact facility may be very useful for applications such as cancer therapy, plasma imaging and fast fusion ignition.

- Applications of Nuclear Techniques: Nuclear techniques have broad applications in various fields. The main focus of SKLNPT in this direction is the research on materials dedicated to the advanced nuclear power systems, the studies on biological systems with ion beams, neutron data measurements and evaluations, AMS and fusion physics related to International Thermonuclear

Experimental Reactor (ITER) project. Test of the newly fabricated metal-ceramic materials, such as Ti_3SiC_2 and Ti_2AlC, show that they may sustain a radiation damage up to hundreds of dpa while keeping basic structure and properties almost unchanged.

Si-based nano-porous membrane was fabricated with the combination of microfabrication technology and ion track techniques. A synthetic nanopore-DNA system based on single track-etched conical nanopores was produced and the synthetic ion channel can be gated by the conformation of DNA molecules. This work provides an artificial counterpart of switchable protein-made nano-pore channels (Xia et al. 2008).

There are four low-energy accelerators operated by SKLNPT and opened to the outside and inside users for ion beam applications. The 4.5 MV Van De Graaff accelerator is mostly used to accelerate protons and deuterons. Neutrons produced from the (p, n) and (d, n) reactions are used to study the neutron data (Zhang et al. 2011). The 2×1.7 MV tandem accelerator is usually applied to the ion beam analysis and implantations. There are two AMS systems installed in the accelerator building. One is a commercially produced compact system based on a Model 1.5SDH-1 Pelletron accelerator from NEC. For ^{14}C spectroscopy, an accuracy of about 0.3% has been achieved and the machine background can be reduced to a level lower than 0.03 pMC, corresponding to a measurable limit of 65 ka. Another one is an EN tandem-accelerator-based AMS system mainly used for ^{10}Be measurements. The PKU-AMS has made contributions to a Chronology Project for dating the old epoch around 1000–3000 B.C. (Xia–Shang–Zhou dynasties) and also the Project for Exploring the Chinese Early Civilization.

1.6.7 Examples of Transfers

There has been a number of successful transfers of accelerator operational for a period of time in one laboratory to another laboratory and continuing the operation. Here, we shall describe three of the cases: Rice University, Houston, Texas, the USA to the Institute Ruder Boskovic (IRB), Zagreb, Croatia; Rice University, Houston, Texas, USA to UNAM, Mexico City, Mexico; and Canberra, Australia to Gracefield, New Zealand.

1.6.7.1 *Rice University, Houston, Texas, USA → Institute Ruder Boskovic, Zagreb, Croatia (personal point of view)*

The whole operation of EN Tandem Van de Graaff accelerator transfer from T. W. Bonner Nuclear Laboratory, Rice University in Houston, Texas, USA to the IRB in Zagreb, Croatia is well documented in the two-volume publication "How the Institute Ruder Boskovic got Tandem Van de Graaff Accelerator" (some parts in Croatian, with original documentation in English). The two volumes (Volume I. 1983–1985 and Volume II. 1986–1988) are kept in several major libraries and are available from the author of this Handbook in e-form.

These two volumes contain almost all the correspondence in connection with the transfer of accelerator from Houston to Zagreb and the formation of Laboratory for Nuclear Analytical Methods at the IRB around this facility. All the documentation is presented chronologically, with some documents appearing more than once when attached to some correspondence. From the inspection of the letters, notes, minutes of meetings, newspaper articles and other documents one can easily notice that some personal names appear often, some only occasionally and some of the individuals who later build their carriers (on and around the accelerator) almost not at all.

In ex-Yugoslavia the IRB in Zagreb was assumed to be a national "accelerator center", mainly because of very successful homemade neutron generator, CW type, proton and deuteron accelerator, main personnel prof Mladen Paic (1905–1997) and dr Krsto Prelec (1927), and not so successful homemade cyclotron (never got the external beam!) although it was opened by that time president of Yugoslavia, marshal Tito! The ambitions of some local physicists were even bigger, they were thinking about Zagreb being "regional accelerator center". In spite of much time spent on meetings, talks and planning the idea never materialized.

In the 15 years period (1956–1971), the neutron generator at IRB was a machine that produced numerous scientific papers dealing with various aspects of neutron physics (for some details see Valković 2015) and many young nuclear physics scientists were educated in the laboratory. Some of the papers attracted the attention of scientists around the globe including USA. Among them was Professor G. C. Phillips from Rice University who visited Zagreb in the year 1964. He was surprised to see that some of the experimental spectra from reactions induced by 14 MeV neutrons resulting in three particles in the final state were fitted quite well by the "generalized density of states" theory developed at Rice University (PhD thesis of the author of this book). He realized that in spite of being on different continents, on opposite sides of Atlantic, Zagreb and Houston were very close in the nuclear physics field, in particular in the study of nuclear reactions with three particles in the final state!

G. C. Phillips invited me to come to the Rice University and work in his lab as a research associate. I accepted gladly the invitation and moved with my family (wife and son) to Houston, Texas toward the end of 1965. T. W. Bonner Laboratory has been already known as an established center for the study of low energy nuclear physics. I stayed there until 1977. with two one-year brakes (returning back home to Zagreb) working mainly on EN tandem. Those were very productive years and we published a lot of papers (see list of references and additional readings). In this period, I made a carrier at Rice, in this period I was promoted from research associate to assistant professor to associate professor to full professor and finally to full professor of physics with tenure. In this period, I brought some of my colleagues and ex-students (Dubravko Rendić, Đuro Miljanić, Jožica Hudomalj and Miroslav Furić) to work in the Laboratory, some stayed shorter, some longer periods of time (2–3 years).

In the beginning of the 1970s, the interest for low-energy nuclear physics slowly started diminishing together with the financing of nuclear laboratories. It was time for rather fast switch of interest towards medium and high energy physics or towards the applications. Gerry Phillips switched mainly to medium energy physics (LAMPF, Oak Ridge) and high energy physics (Brookhaven and Argonne National Laboratories). This was mainly so called "suitcase physics" which did not attract me, mainly because of frequent airplane trips and long separations from the family. So, my choice was applied physics on the accelerators we had in the lab. We were among the first group which has developed PIXE to be an analytical technique. Papers we produced in this period were published in the most prestigious journals.

The work on both Van de Graaff machines (tandem-horizontal and single-ended vertical) was funded by NSF, DOE and Welch Foundation. The experimental work on EN tandem was finished with my return to Zagreb (because of family reasons) and simultaneously vertical machine has stopped working. Slowly, the idea that the space occupied by accelerators and associated experimental rooms should be put into service of other Rice University needs. Finally, in the year 1983. G. C. Phillips, after some telephone discussion, wrote a formal letter inquiring about the interest of IRB, Zagreb to accept the accelerator (see Fig. 1.19). The date of this letter (7 December 1983) is the starting point of operation called the transfer of EN tandem Van de Graaff accelerator to the IRB in Zagreb.

RICE UNIVERSITY
HOUSTON, TEXAS
77001

T. W. BONNER NUCLEAR LABORATORIES
OFFICE OF THE DIRECTOR

December 7, 1983

Professor V. Valkovič
Institut "Ruder Boskovič"
P. O. Box 1016
41001 ZAGREB
Croatia, Yugoslavia

Dear Professor Valkovič:

I wish to inform you that Rice University, T.W. Bonner Nuclear Laboratories, will consider disposing of the EN Tandem Accelerator that it owns. We are pleased that your institution and one other have expressed interest in possibly acquiring this surplus equipment. In the case that the Institut "Ruder Boskovič" may wish to acquire this equipment, I suggest the conditions for its acquisition. We propose the following steps as a way to proceed:

1. Intent to Use for Research. We would require that the new owner declare its intention to use the equipment for education and research and not for immediate resale or salvage. A letter outlining your plans is sufficient.

2. Visit for Inspection and Shipping Arrangements. We suggest it is essential for a delegation to visit the Rice University T.W. Bonner Nuclear Laboratories to view the equipment and arrange for dismantling, packing, and shipment. We would aid in a consulting capacity, but could not supply any other support. Since several of your staff members are very familiar with the equipment, you are in a position to specify all the equipment that you may want. For example, you may wish to consider the gas-handling equipment and other auxilliary equipment such as a magnetic spectrometer, etc.

3. Decision. We realize that only after your study of the requirements of dismantling, packing, and shipping have been made can you decide if you wish to proceed. We are asking that you and the other institution make these studies as soon as possible.

Sincerely,

G. C. Phillips
Director

FIGURE 1.19
Letter from G. C. Phillips, dated 7 December 1983.

In addition to support of Institute (director general dr. Sergije Kveder) and Experimental Physics Department (chairman dr. Krunoslav Pisk) it was needed to obtain the support of the Republic of Croatia (at that time part of Yugoslavia) institutions in charge of scientific institutions and funding in Zagreb. The administration bodies in charge were the Committee for Science, Technology and Informatics (chairman prof. Uroš Pruško) and funding body for science, called SIZ (Secretary of SIZ-I dr Petar Colić) which fully supported the project. The admirable degree of synergy was achieved which made working on the project a real joy.

All done in Zagreb the next goal was to travel to Houston, dismount and pack the machine, transfer the package to port of Houston, embark everything on ship going to Rijeka, transfer everything from Rijeka to the Institute in Zagreb, build the housing around the accelerator tank, reassemble the accelerator and hope it will work again. Description of all of the actions together with accompanying documentation is saved on a CD, available upon request from the author of this book.

In my letter dated 9 January 1984, I defended the project by presenting to the possible funding agencies by presenting scientific, educational and geographic considerations. It is of interest to list the presented scientific considerations: The EN tandem Van de Graaff accelerator was planned to be used in a variety of interdisciplinary programs including,

- Ion implantation. It was proposed to study the effects of ions implanted into metals as well as the formation of new compounds by implantation of one or more ions.
- Surface analysis (corrosion).
- Testing of detectors for medium, high energy physics, particle and γ-ray astronomy, etc.
- Trace element analysis using PIXE spectroscopy. Development of micro-beam facility.
- Fundamental research in atomic, nuclear and other fields. Atomic: X-ray production, multiple processes, etc. Nuclear: Many particle breakups, nuclear reaction studies, applied neutron physics, instrumentation development.
- Short-lived tracer production for biological and medical applications.

Teaching nuclear physics and engineering can be done only in a nuclear physics laboratory. It is essential that some institutions preserve the ability to teach nuclear science. It is possible to foresee a manpower shortage in this field in years to come. The presence of the Van de Graaff accelerator in dedicated institutions will ensure the readiness for this service when the need arises.

With the letter dated 26 June 1984 we have proposed the "D-day" for the operation to be 24.08.1984 (accepted by G. C. Phillips) in which the team (IRB: Vladivoj Valković, Dubravko Rendić, Đuro Miljanić, Ivo Orlić, Leander Kukec and HVEC Ken Rogers) "invaded" the TWBNL. The detailed plan of activities was sent to all people supporting the project financially, by consulting, advising or any other way. In the same time, back in Zagreb, the team lead by Vinko Tomljenović made all the paper work needed for building construction licenses (including all required electrical installations as prepared by Stanko Orlić and Marijan Pavin).

The International Atomic Energy Agency (IAEA, Physics Section, Mr. Joe Dolničar) also supported this project of EN transfer to Zagreb. Their financial support through Reserve Fund project (YUG-1–009 – Tandem accelerator) in an amount of US $ 20,000 was extremely helpful to pay some travel and per diem expenses, packer's services, transport to the port of Houston and numerous materials needed in such a mission. Transport from port of Houston, Texas, USA to port of Rijeka, Croatia (at that time part of Yugoslavia) was done by the company Jugolinija, Rijeka (with the assistance of Capt. Vojinovic) covering the cost of shipping by m/v Velebit in October/November 1984. Transport of four containers and the tank from the port of Rijeka to the location at the Institute in Zagreb was done by trucks.

Still, a lot of paper work was needed to be done, including the Custom's declaration, donation letter (see Fig. 1.20). After this long journey accelerator came to the Institute ground and we had to start putting it together. The critical was the advice of Joseph F. Bromberger, President, HVEC Science Division (see Fig. 1.21) and again the help of Ken Rogers (HVEC) who consistently answered the question "How are you" with "Too early

RICE UNIVERSITY

P. O. BOX 1892

HOUSTON TEXAS

77251

T. W. BONNER NUCLEAR LABORATORIES
OFFICE OF THE DIRECTOR

September 12, 1984

-------- TO WHOM IT MAY CONCERN ----------

An EN Tandem Van de Graaff accelerator with associated spare parts and beam handling system is donated gratis to the Institut "Ruder Bosković", Zagreb, Yugoslavia, Department of Physics, Energy and Applications. This donation is based on the long scientific cooperation and scientific research between Institut "Ruder Bosković" and the Bonner Labs, Rice University over a period of 20 years. These collaborations have been partially supported under the auspices of the governments of the United States of America and Yugoslavia through the project whose title is: "Study in Few Particle Physics and Nuclear Reactions," (No. F6F005Y).

Value in United States dollars for Yugoslavia customs' purpose is $50,000.00.

Total weight of the accelerator system is approxiately 90 tons.

G. C. Phillips
Director & Professor of Physics
T. W. Bonner Nuclear Laboratories
RICE UNIVERSITY
Houston, TX 77251

FIGURE 1.20
Letter from G. C. Phillips, dated 12 September 1984.

HIGH VOLTAGE ENGINEERING CORPORATION

P.C. BOX 416, SOUTH BEDFORD STREET, BURLINGTON, MASSACHUSETTS 01803, U.S.A.

May 29, 1984

Dr. Vlado Valkovic
Professor of Physics
Ruder Boskovic Institute
Bijenicka 54
41001 Zagreb, Croatia, YUGOSLAVIA

Dear Dr. Valkovic:

I am replying to your letter of May 3, 1984 regarding the transfer of the Rice EN tandem accelerator to the Ruder Boskovic Institute. This accelerator should be in good condition and I think that you have made the right decision to acquire it.

You have also made a wise decision to involve my company in the disassembly and later assembly of the basic accelerator at your Institute. The one area which requires our expertise is the column construction and, in particular, the decompression and later compression of the column. This work can only be done by our field service engineers. It also requires special equipment.

Before proceeding, I should advise you that you cannot ship the tank with the column and acceleration tubes intact. I am sure that the glass structures would be broken by the time that they reached your Institute. In fact, the column and acceleration tubes cannot be subjected to cold temperatures either in shipment or in storage; otherwise they will become defective from thermal shock. Our general rule is to ship by sea between April and the end of September. For the remaining months we ship by air in heated compartments.

At Rice University, your crew of physicists and technicians can remove the column rings, spinning, resistors, belt spacers, acceleration tubes, drivemotor, alternator and stripper assembly. If the injector still exists, it should be disconnected and rolled back. There should be 20 feet of clear space at each end of the tank to roll out the column carriages.

About two years ago, the scientists at Auckland, New Zealand acquired the EN tandem at Canberra, Australia. A crew was sent to dismantle and later assemble the tandem. One of our field engineers was hired from our company to supervise the decompression and disassembly of the column. Based on that experience, I believe that you should plan to have one of our engineers in Houston for two weeks. During the assembly at Zagreb, you should plan on one engineer for three weeks. If it takes less or even more time, you only pay for the actual time the field engineer is in transit and on site.

I enclose a copy of our service labor rates for 1984. I estimate that these rates will increase by about six percent in 1985.

In the first part of this letter, I spoke about special equipment. For the disassembly at Rice and the later assembly at Zagreb, you will need column handling jigs and decompression equipment. It is possible that this equipment exists at Rice. If not, we will inquire at other EN installations and arrange to borrow it. We will prepay the shipping expense and invoice you at the actual cost.

Please keep us advised of the status of your program, particularly the period in which you will need our field engineer at Rice. We need as much notice as possible because we are installing a large tandem in Beijing, China which requires most of our field engineers.

Sincerely yours,

Joseph Bromberger

Joseph F. Bromberger
President, Science Division

FIGURE 1.21
Letter from Joseph F. Bromberger, 29 May 1984.

FIGURE 1.22
The column ready to be moved into the tank.

to know" since each day would bring something unpredictable (Figs. 1.22 and 1.23). During this period there were good and bad days. The bad days are mainly due to the excessive paper work which sometimes needed more steps than originally assumed. The good days were connected usually with the arrivals of auxiliary equipment as for example sputtering ion source obtained from Brookhaven (thanks to dr. Krsto Prelec).

It was Jerry Duggan, my friend and colleague of many years, see Fig. 1.24, participating at SPIE-2009 conference in Orlando, Florida when the picture was taken, who many years ago indicated to me the possibility of receiving tandem accelerator from the University of Texas in Austin, TX. Both of us were at that time attending the 4th International PIXE conference at Tallahassee, FL (9–13 June 1986) when Jerry Duggan suggested to me to contact Dr. Michael Rodgers, director of Center for Fast Kinetics Research at the University of Texas at Austin, what I did and after some more letters and telephone calls, I received two letters as shown in Figs. 1.25 and 1.26. So, we were ready to organize one more expedition to Texas! See Fig. 1.27 showing some goodies brought from University of Texas at Austin. This brought us closder to D-day, see Fig. 1.28.

Fortunately, we had more friends, one of them prof. Bogdan Povh (1932), director of Max Planck Institute for Nuclear Physics in Heidelberg, Germany, invited me to see their new PIXE microbeam facility and present my work in this area, especially applications in biology and medicine. Being impressed by our effort in transferring accelerators from Texas to Zagreb he made also some contribution! He donated to us several

FIGURE 1.23
Final test of the column. Up front Ken Rogers, HVEC, USA.

turbomolecular pumps and scattering chambers which they did not plan to use in the future. A very valuable donation to us since we did not have good pumps.

The ceremony of the official inauguration of EN Tandem Van de Graaff accelerator was held at the Institute on 26 November 1987 with the presence of the number of different dignitaries. However, it took another year to complete all the paper work and to obtain all required permits for normal continuous work.

1.6.7.2 Rice University, Houston, Texas, USA → UNAM, Mexico City, Mexico

The first Van de Graaff particle accelerator in Latin America was installed at the Universidad Nacional Autónoma de México (UNAM) in 1952. This event marked the beginning of experimental nuclear physics, exclusively for peaceful purposes, in Mexico. The acquisition of this accelerator was fundamental for placing other accelerators into operation, which were used for both research and the resolution of national problems.

In 1984, the Institute received a 5.5 MV Van de Graaff donated by Rice University (prof. Gerry Phillips). The transfer was done by the group of UNAM physicists and technicians

FIGURE 1.24
Prof. Jerome Lewis Duggan and the author of this book, Florida, 2009.

led by Prof. Eduardo Andrade. The process was documented by the movie camera, cut off frames are shown in Figs. 1.29–1.32.

1.6.7.3 EN Tandem from Canberra, Australia to Gracefield, New Zealand

A major development of accelerator applications in science and technology in New Zealand happen in the 1980s with the purchase of a 6-million volt tandem Van de Graaff, formerly owned by the Australian National University, to be used for accelerator mass spectrometry (AMS). The AMS facility became operational in 1987 and was the main tool of the Rafter Radiocarbon Laboratory. In May 2010, the laboratory acquired a compact AMS (CAMS) from NEC furthering its analytical capabilities.

At the end of the Second World War, the New Zealand Department of Scientific and Industrial Research (DSIR) established a small research team to explore the rapidly growing field of atomic science. This was the beginning of the Isotope Laboratory, led by Athol Rafter and Gordon Fergusson and attached to the Dominion Physical Laboratory. In 1957, the Isotope Laboratory was made into a separate division of the DSIR with Athol Rafter as Director, becoming the Institute of Nuclear Sciences (INS) in 1959.

In 1991 the old DSIR, founded in 1926, was dissolved, and the various divisions were organized into ten Crown Research Institutes (CRI). In this process, the INS was joined with the earth science components of the DSIR to become the Institute of GNS Science. In March 1993, the radiocarbon laboratory took the name Rafter Radiocarbon Laboratory to mark Athol Rafter's 80th birthday. Athol Rafter died in 1996, at 83 years of age.

THE UNIVERSITY OF TEXAS AT AUSTIN

Executive Vice President and Provost
Main Building 201 · Austin, Texas 78712-1111

February 27, 1987

Dr. Vlado Valkovic   AIR MAIL
Ruder Boskovic Institute
Post Office Box 1016
41001 Zagreb
Bijenicka 54
Croatia, Yugoslavia

Dear Dr. Valkovic:

    This letter is to inform you that the University of Texas at
Austin has agreed in principle to transfer the 12 MeV model EN-II
Tandem Van de Graaf Accelerator to your Institute. This will be
in the nature of a gift to the Ruder Boskovic Institute from the
University of Texas at Austin in the expectation that it will be
put to good use there and will contribute to the pursuit of
knowledge. It is understood that you will arrange and pay for
the removal, crating and shipping.

    Please note that this letter is a statement of the
University's intent only and is conditioned upon the satisfactory
conclusion of all necessary details related to the actual move.

    We anticipate no problem with the remaining items and have
every expectation that your suggested starting date of early May
could be met.

    It is our earnest hope that this proposed transfer will be
successfully completed and that the accelerator will serve as a
useful research tool in the Boris Rudevic Institute.

                              Sincerely yours,

                              Stephen A. Monti
                              Vice Provost

SAM:cs

cc: Dr. Michael A. J. Rodgers
     Director, Center for
     Fast Kinetics Research

FIGURE 1.25
Letter from Stephen A. Monti, 27 February 1987.

In New Zealand, there are two radiocarbon laboratories today: the GNS Science' Rafter Radiocarbon Laboratory in Lower Hutt and the University of Waikato's laboratory in Hamilton. The story of one of them follows. GNS Science is New Zealand's leading provider of Earth, geoscience and isotope research and consultancy services.

CENTER FOR FAST KINETICS RESEARCH

THE UNIVERSITY OF TEXAS AT AUSTIN

ENS Annex 16N · Austin, Texas 78712 · (512) 471-7583

August 11, 1987

Dr. Vlado Valković
Ruder Bosković Institute
41001 Zagreb
Bijenicka 54
Croatia
Yugoslavia

Dear Dr. Valković:

I can now report that this Center has received approval from the Central Administration at the University of Texas at Austin for the removal of the EN Tandem accelerator. This will be disposed as follows.

(i) The pressure vessel, accelerating column and rings will be transferred to the Lawrence Livermore Laboratory (Dr. J. C. Davis).

(ii) All other items within the pressure vessel and a selection of associated peripheral items will be transferred to the Rudex Bosković Institute in Yugoslavia.

We require that the costs of these transfers be wholly born by the recipients of the units.

As per our earlier communications we shall expect your team to commence removal operations on or around September 2, 1987. The CFKR staff will provide as much advice and assistance as possible, but as far as is feasible, our own research operations will need to continue during the removal period. Mr. Billy Naumann is our representative in this matter and you are free to contact him about matters of detail.

It is this University's wish that the accelerator parts donated to your Institute will further the progress of science in Yugoslavia.

I look forward to your coming to Austin.

Sincerely,

Michael A. J. Rodgers
Director

MAJR:ksb
c.c. S. A. Monti, Vice Provost
    R. E. Boyer, Dean
    Bob G. Sanders, Associate Dean
    M. A. Fox, Professor
    Billy K. Naumann
    Bobby Cook, Associate Vice President

FIGURE 1.26
Letter from Michael A. J. Rodgers, 11 August 1987.

FIGURE 1.27
Ion source, duoplazmatron, received together with many other parts from the University of Texas at Austin.

The 6 MV EN Tandem Accelerator (number EN 5) was purchased secondhand from ANU in 1980. The accelerator was built in 1959 and used very successfully at ANU for this period of time. For its home in New Zealand, new building was built at Lower Hutt in 1981–1982. Also, an in-house Cs sputter ion source was built at the same time. Finally, the Tandem was re-commissioned as an AMS Radiocarbon dating machine in 1985.

It was February 1980 when the first attempt to come to grips with the problems of transferring the EN tandem from Canberra to Gracefield was done. Rodger Sparks flew to Canberra on 4 February and found that the ANU technicians had already begun dismantling some beam lines, and that the entire tandem facility was available for removal. With the assistance of Neil Whitehead, who was in Canberra to attend a conference, an immediate start was made on dismantling the remaining beam lines and packing the myriad small (and heavy!) components into cardboard boxes in preparation for the actual shift to take place in April 1980. In this work, they were assisted by the ANU technical staff under John Harrison, and the cooperation they offered set the tone for the subsequent stages of the operation. At the same time, contact was established with the Canberra branch of the shipping company (Grace Bros. Ltd) and with the N.Z. High Commissioner's Office, so that planning could begin for the actual transport of the material. All these activities are described in an unpublished report "Transport of Accelerator Components" by R. J. Sparks, R. D. More and C. R. Pureell (Fig. 1.33).

In April of the same year, the "second wave", comprising Ray More, Chris Pureell and Rodger Sparks flew to Canberra to arrange the shipment to New Zealand of as many as possible of the available items making up the tandem facility. In the intervening period since

FIGURE 1.28
Ready to go!

February, the ANU staff had been busy providing clear access into the tandem building by opening up the side of the building and blasting away a two-foot thick concrete wall, thus making it possible to begin immediately dismantling and shifting the larger items of equipment (magnets, chiller, etc.) and packing them into containers and sea freighters for transport to Gracefield. The result of these labors was the dispatching to New Zealand of all the beamlines and vacuum systems, both analyzing magnet assemblies, two switching magnets and their supports, the accelerator control console, the power distribution board and transformer, the water chiller, the complete gas handling system, the gas storage cylinders and the negative ion source assembly. Remaining in Canberra was the accelerator itself comprising the pressure vessel, carriages and the column assembly, as well as a set of spare column sections and accelerator tubes. All the material shipped has since arrived in New Zealand in remarkably good condition. The packing program was as follows. During the first week, a container was packed with the equipment associated with the 6 beamlines. Both the X36 and the X52 magnets were rolled out and loaded onto sea freighters. Then the loaded container and the X52 magnet freighter were trucked to Sydney.

The second week involved packing the magnet platforms and beam stands onto freighters. The control console was prepared for removal and both the console and the power distribution board racks were moved out and into a container along with turbo

FIGURE 1.29
Dismounting the accelerator in Houston and transfer to Mexico City. (Frames are taken from movies courtesy of Prof. Eduardo Andrade Ibarra.)

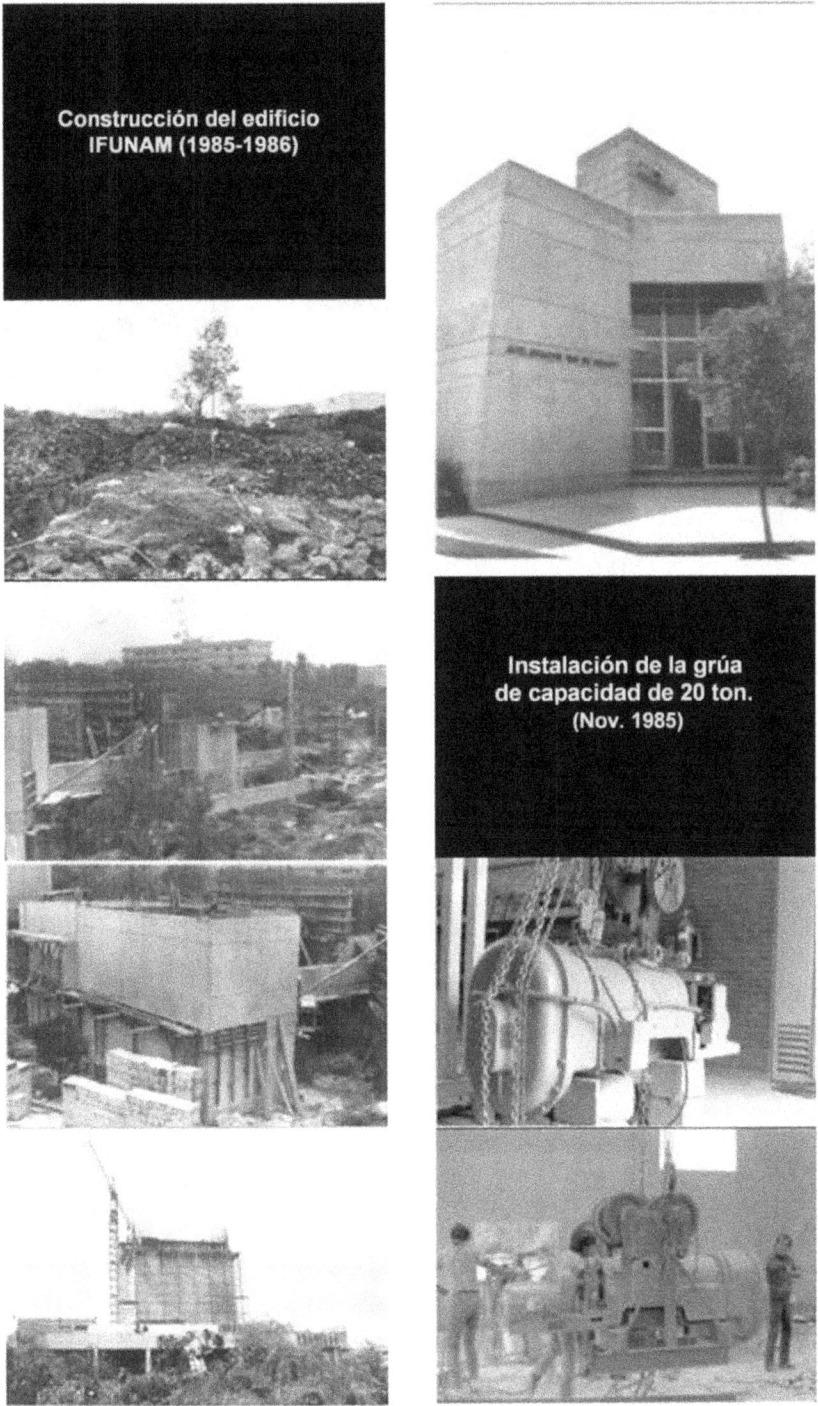

FIGURE 1.30

Construction of accelerator building and installation of 20 ton lift. (Frames are taken from movies courtesy of Prof. Eduardo Andrade Ibarra.)

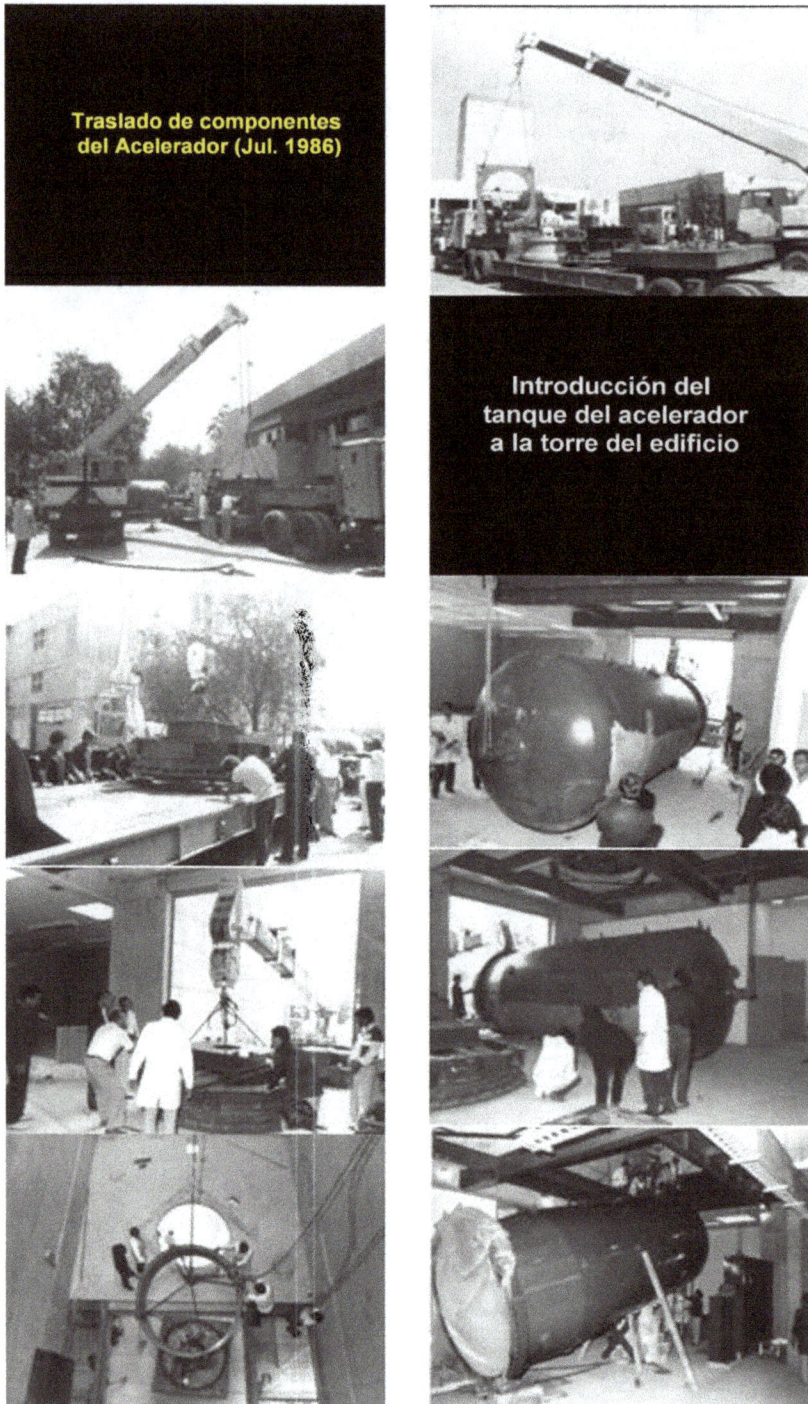

FIGURE 1.31
Transport of accelerator components (including the tank) into the building. (Frames are taken from movies courtesy of Prof. Eduardo Andrade Ibarra.)

Construcción de andamios similares a los usados por los albañiles para la instalación de la columna y la polea superior del acelerador

Inauguración edificio IFUNAM 26 de agosto de 1986

FIGURE 1.32
Completion of the assembly and inauguration. (Frames are taken from movies courtesy of Prof. Eduardo Andrade Ibarra.)

FIGURE 1.33
EN5 Tandem.

pumps, etc. The chiller was then brought out and loaded on a sea freighter and the gas storage cylinders were removed and loaded straight onto a truck. The third week involved disassembling all the equipment associated with the gas handling area and loading into a container. The ion source was rolled out and put onto a freighter and then both the chiller and ion source freighters were taken to Sydney. Finally, the last two containers were topped up with all remaining parts ready for removal the following week.

So in total 3 containers, 5 sea freighters and 2 lots of gas cylinders were sent off. For shipping, the gas cylinders were loaded onto 2 sea freighters. During the third week in May, one container and five sea freighters arrived in Wellington. The ion source and the X52 magnet were put in the experimental area while the rest of the equipment was unpacked and stored in the old Sunbeam factory nearby. The target room was then prepared and the blocks in the west wall removed, thence the X52 magnet platform and base were shifted in. However, the first attempt at installing a magnet half was unsuccessful and a frame was constructed for this operation.

Another two containers arrived in the second week of June. One container was unloaded on top site, while the gas handling container was unpacked at the Sunbeam factory store. During this operation, the flywheel on the compressor was damaged and will need replacing. The console was shifted into the control room while the transformer and power distribution racks were positioned in the stand-by generator room. Finally, the two magnet halves were successfully positioned in the target room using the special frame.

It was proposed to initially get the accelerator operating with the negative ion source supplied with the machine. However for use as an ultra-sensitive mass spectrometer a number of new items was required, the most important of these being a sputter ion source. It is expected that an ion source being developed by Dr. Harry Naylor at Auckland University would be able to be used. A high resolution 90 analyzing magnet has been ordered from ANAC and several smaller items are already in various stages of development.

McCallum (1980) described how the sputter ion source controls were acquired and put together. Information on items required for the floating (at high voltage) platform for the ion source was first obtained. Then a 3KVA 230V AC isolation transformer for 100–200 kV DC potential together with suitable standoff insulators was ordered. Controls and sensing related to the ion source were implemented using fiber optic couplings. A development kit has been obtained for evaluation so that designs could be based on hands-on experience as well as on published manufacturer's specifications.

1.6.8 Role of the IAEA

The IAEA through its technical co-operation programs, coordinated research projects, consultants and technical meetings, conferences promotes the peaceful use of radiation technologies. Because of the IAEA's support, some new technologies (such as hydrogel wound dressings, sulfur-free radiation vulcanized latex) were developed and transferred to Member States over the previous years. Other promising applications the IAEA has supported are flue gas and wastewater treatment. At an industrial fossil fuel power plant, electron accelerators exceeding 1 MW have been installed to remove SO_x and NO_x from flue gases – a further breakthrough in the large scale use of electron beam technology.

Since 1998, the IAEA has been organizing many meetings and coordinated research projects in which trends and new developments have been discussed and technical work conducted. Recent developments in the field were discussed by representatives from industry, universities and research institutes during the International Meeting on Radiation Processing, held in Avignon, France, in 2001; the last such meeting was held in September 2003 in Chicago, USA. New developments were also reported during the symposium on "Radiation Technologies in Emerging Industrial Applications" organized by the IAEA in Beijing, China, in 2000. More than 200 participants, representing over 30 countries attended that symposium. The main research and implementation of developments presented at these meetings concentrated on three fields: (1) medical and health-related applications, (2) environment pollution control and (3) polymers. Post-irradiation stability of radiation sterilized medical implants, the use of radiation processing for the sterilization or decontamination of pharmaceuticals and pharmaceutical raw materials, radiation synthesis and modification of polymers for biomedical applications have been studied. Since separation and enrichment technologies play a very important role in product recovery and pollution control, the possibilities of radiation synthesis of stimuli-responsive membranes, hydrogels and adsorbents are being investigated.

The IAEA report (2020a), in addition to the database of cyclotrons used for the production of medical radioisotopes, lists the number of other accelerator databases, including:

- The NuPECC Facilities Handbook: A pdf handbook listing the facilities located in the EU member countries that are part of the NuPECC network, mainly devoted to applications involving scientific research in fundamental physics. Achievable through the link http://www.nupecc.org/pub/hb12/hb2012.pdf.

- The Acceleratorer: A community-oriented point of contact for laboratories and scientists working in the field of ion accelerator. Achievable through the link http://www.acceleratorer.com. Created by Julien Demarche.

- Los Alamos Accelerator Database: A database of 210 facilities, including geographic area, country and city, facility URLs, accelerator classification, research applications, accelerated particles, beam energy and beam current. Achievable through the link http://laacg1.lanl.gov/laacg/accphys/db/accdbs/search_acc. phtml. Created by Julien Demarche.

- LINAC96 Conference Database: The LINAC96 Conference handbook contains a list of institutions around the world hosting a linear accelerator. Achievable through the link http://linac96.web.cern.ch/linac96/.

- ICNMTA2008 Microprobe Database: The ICNMTA 2008 conference web site provides a list of the Nuclear Microprobe installations (a sub-set of ion accelerators) around the world. Achievable through the link http://www.atomki.hu/ atomki/IonBeam/icnmta/microprobefac.html.

- The ECOS Chart of Stable Beams: This chart developed in the frame of ECOS collaboration contains data (isotope, energy, intensity, etc.) related to the currently available and/or future stable ion beams at some nuclear facilities in Europe. Achievable through the link http://u.ganil-spiral2.eu/chart-ecos/.

- AMDIS (IAEA, Nuclear Data Section): Atomic and Molecular Data Information System. AMDIS provides on-line computer access to atomic, molecular, plasma-material interaction and material properties data basis and to other information.

- AMBDAS (IAEA, Nuclear Data Section): A+M Bibliographic Data System. The system contains more than 33,000 bibliographic entries with information relevant to fusion research and development. The online system gives access to this bibliographic database with a menu-driven and user friendly environment.

As for all nuclear facilities, decommissioning is the inevitable end of an accelerator's life cycle. The evaluation of potential challenges (including the radiological exposure of workers and the public), characterization, dismantling techniques, generation and management of radioactive waste, costs and site reuse are all important aspects of the decommissioning process of any nuclear facility, including those housing accelerators. In many countries, accelerators are not regulated in the same way as nuclear installations such as nuclear reactors or nuclear fuel cycle facilities, yet staff at accelerators may have less knowledge of waste management and decommissioning than staff at other nuclear installations. In some cases, accelerators have been semi-abandoned owing to lack of interest or an incorrect perception that their decommissioning is a low priority activity. Under these circumstances, even the minimum requirements and strategies might have been disregarded in decommissioning, resulting in unnecessary costs, delays and, possibly, safety issues. IAEA (2020b) report provides practical information on the selection and implementation of decontamination and dismantling strategies and techniques for accelerators. It is written for those carrying out decommissioning with little or no experience in this discipline.

1.7 Accelerator Manufacturers

Here we shall present some of the world leading accelerator manufacturers. The accelerators they manufactured produced large amount of data which helped building our knowledge of low-energy nuclear physics and established the foundation of large number of applications of nuclear methods and technologies.

1.7.1 National Electrostatic Corporation

NEC provides industry-leading electrostatic ion beam accelerator systems and related components designed to expand the research goals of scientific and technical communities around the world. A private, employee-owned company, NEC was incorporated in Middleton, Wisconsin in 1965. To date, NEC has manufactured over 230 Pelletron systems, with their installations in over 50 countries around the world. These systems produce ion beams of essentially all stable nuclei with energies ranging from a few keV to hundreds of MeV. The Pelletron is the world's only commercially available accelerator that incorporates an all metal and ceramic acceleration tube, with no organic material in the vacuum volume (some of the models are shown in Figs. 1.34–1.38).

Most of the Pelletron accelerator systems in operation are listed in Table 1.8 with their model number, such as 20UR or 5SDH. Standard Pelletron accelerators have model numbers such as 5SDH-2 (1.7 MV tandem horizontal 2-chain Pelletron) or 4UD-4 (4 MV vertical 4-chain Pelletron). Other systems include high voltage decks without a Pelletron accelerator component, as well as former belt-driven Van de Graaff accelerators that have been retrofitted by NEC with a Pelletron chain-based charging system conversion. Systems specifically designed for accelerator mass spectrometry are noted in parentheses, including their standard line of low energy AMS systems such as SSAMS, CAMS and XCAMS. Additional system configurations noted in parentheses, such as 5S-MR10 and MAS1700, are designed specifically for ion beam analysis. An asterisk (*) next to the year of installation specifies that the system has since been moved from its original location to its current facility.

National Electrostatic Corp., 7540 Graber Road, P.O. Box 620310, Middletown, WI 52562, USA.

1.7.2 PMB – Alcen, France

PMB's main facility is located in the south of France, close to Aix-en-Provence, with more than 100 employees, 6,600 m^2 of factory. Since 1985, before its acquisition by Alcen, PMB's technical expertise was mainly focused on the brazing of ceramic/metal and metal/metal of complex assemblies for ultra-high-vacuum components. Today they are known for these products:

1. iMiGiNE: a robotized system capable of producing various radiopharmaceuticals used in PET imaging. It combines a cyclotron ISOTRACE with a radiochemistry room iMiLAB, complete with an automated QC system. It is characterized by superconducting magnet technology, external self-shielded beam. It is equipped with an automated target changer, and automated syringe dispenser, including high-precision activity measurement and traceability, for details see PMB-Alcen (2021).

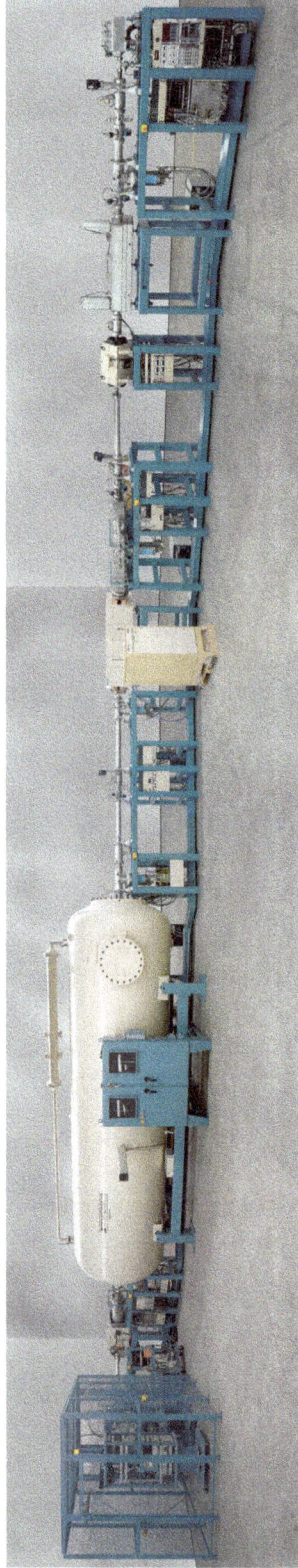

FIGURE 1.34
This is a picture of a 15SDH-2 5 MV Pelletron system with a Multi-Cathode SNICS (MC-SNICS) source (left) and AMS detector (right). Originally installed in the UK in 1998 and moved to Germany around 2015. (Courtesy of National Electrostatics Corp.)

FIGURE 1.35
A 5SDH 1.7 MV Pelletron system with RF source (left) and high-resolution analysis endstation (right). It is capable of Ion Beam Analysis (IBA) techniques such as RBS including High-Resolution RBS, ERD, PIXE and NRA. Further detectors could be added to the endstation to allow for other techniques. Installed in Israel in 2011. (Courtesy of National Electrostatics Corp.)

FIGURE 1.36
A 5SDH 1.7 MV Pelletron system with RF source (left) and high-resolution analysis endstation (right) Installed in Germany in 2018. Since this picture was taken, they have also added a heavy-ion source (MC-SNICS 20 sample), a high-energy switching magnet, and a single wafer implanter. (Courtesy of National Electrostatics Corp.)

FIGURE 1.37

This photo was taken on the second floor (US counting) looking up from the beam exit end of the accelerator. The blue tank houses the 5UD. The ceiling at the top of the photo is the floor of the room that contains the injection beamline, and a second room above that houses the ion sources and the 90° injection magnet. The bottom of the tank is about 2 meters above the floor of this room. The beam goes down from the tank through the floor to the ground floor target rooms, into the 90° analyzing magnet that bends the beam from the vertical direction to the horizontal, and sends it to the target rooms. The three massive white legs are designed with earthquakes in mind. Installed in Japan in 1992. (Courtesy of National Electrostatics Corp.)

FIGURE 1.38

A 3SDH 1 MV Pelletron system with RF source (left) and analysis endstation (right). It was designed with the intended purpose of aiding in fusion research. It is capable of Ion Beam Analysis (IBA) techniques such as RBS, ERD, PIXE and NRA. Further detectors could be added to the endstation to allow for other techniques. Installed in Japan in 2014. (Courtesy of National Electrostatics Corp.)

TABLE 1.8

NEC accelerators worldwide data from https://www.pelletron.com/resource-category/accelerators-around-the-world/

Institution	Laboratory	System	Voltage	Installed	City/Province	Country
Instituto de Física da Universidade de São Paulo (IFUSP)	Laboratório Aberto de Física Nuclear (LAFN)	8UD	8 MV	1971	São Paulo	Brazil
Australian National University (ANU)	Heavy Ion Accelerator Facility	14UD	14 MV	1973	Canberra	Australia
Sandia National Laboratories	Ion Beam Laboratory	1UEH	1 MV	1972	Albuquerque, NM	USA
University of Melbourne	Experimental Condensed Matter Physics (ECMP)	5U	5 MV	1974	Melbourne	Australia
Tokyo Institute of Technology (TIT)	Research Laboratory for Nuclear Reactors	3UH-HC	3 MV	1976	Tokyo	Japan
Oak Ridge National Laboratory	Accelerator Systems Group	25URC	25 MV	1979	Oak Ridge, TN	USA
Japan Atomic Energy Agency (JAEA)	Tokai Research and Development Center	20UR	20 MV	1978	Tokai	Japan
Comisión Nacional de Energía Atómica (CNEA)	Departamento Acelerador TANDAR	20UD	20 MV	1980	Buenos Aires	Argentina
Argonne National Laboratory	Argonne Tandem Linac Accelerator System	2UDHS	2 MV	1981	Argonne, IL	USA
National Institute of Advanced Industrial Science and Technology (AIST)	Neutron Standard Group	4UH-HC	4 MV	1981	Tsukuba	Japan
Tarleton State University	Accelerator and Materials Research Group	3SDH	1 MV	1983*	Stephenville, TX	USA
Tata Institute of Fundamental Research (TIFR)	BARC-TIFR Pelletron linac Facility	14UD	14 MV	1985	Mumbai	India
Chinese Academy of Sciences	Shanghai Institute of Applied Physics (SINAP)	4UH	4 MV	1984	Shanghai	China
University of California Santa Barbara	Materials Research Laboratory	6UE	6 MV	1983	Santa Barbara, CA	USA
Tokyo Institute of Technology (TIT)	Laboratory for Advanced Nuclear Energy (LANE)	5SDH-2	1.7 MV	1984	Tokyo	Japan

(Continued)

TABLE 1.8 (Continued)

NEC accelerators worldwide data from https://www.pelletron.com/resource-category/accelerators-around-the-world/

Institution	Laboratory	System	Voltage	Installed	City/Province	Country
United Kingdom Atomic Energy Authority (UKAEA)	Culham Centre for Fusion Energy	5SDH	1.6 MV	1984	Abingdon	UK
IBM	Thomas J. Watson Research Center	9SDH-2	3 MV	1986	Yorktown Heights, NY	USA
University of Jyväskylä	Accelerator Laboratory	5SDH-2	1.6 MV	1985*	Jyväskylä	Finland
Centre de Recherche et de Restauration des Musées de France (C2RMF)	Accélérateur Grand Louvre d'Analyses Elémentaires (AGLAE)	6SDH-2	2 MV	1987	Paris	France
Inter-University Accelerator Center	Pelletron Laboratory	15UD	15 MV	1985	New Delhi	India
Katholieke Universiteit (KU) Leuven	Ion and Molecular Beam Lab (IMBL)	5SDH-2	1.7 MV	1986	Leuven	Belgium
University of Florida	Materials Characterization/Analysis and Ion-Beam Reactions Group	5SDH	1.6 MV	1986*	Jacksonville, FL	USA
Fudan University	Key Laboratory of Nuclear Physics and Ion-beam Application (MOE)	9SDH-2	3 MV	1987	Shanghai	China
Los Alamos National Laboratory	Ion Beam Materials Laboratory	9SDH-2	3 MV	1986	Los Alamos, NM	USA
Navy Surface Warfare Center (NSWC)	Carderock Division	9SDH-2	3 MV	1989*	West Bethesda, MD	USA
Army Research Laboratory (ARL) Aberdeen Proving Ground	Rodman Materials Research Laboratory	5SDH-2	1.7 MV	1988*	Aberdeen, MD	USA
IBM	Almaden Research Center	3UH	3 MV	1987	San Jose, CA	USA
University of North Texas	Ion Beam Modification and Analysis Laboratory	9SDH-2	3 MV	1987	Denton, TX	USA
University of Houston	Ion Beam Laboratory	HV deck	0.2 MV	1987*	Houston, TX	USA
Oxford Microbeams Ltd.		5SDH-2	1.7 MV	1987	Oxford	UK
Institute of Physics Bhubaneswar	Ion Beam Laboratory	9SDH-2	3 MV	1988	Bhubaneswar	India
Korea Atomic Energy Research Institute (KAERI)	Korea Multi-purpose Accelerator Complex (KOMAC)	5SDH-2	1.7 MV	1987*	Gyeongju	Korea
University of Wisconsin Madison	Ion Beam Laboratory	5SDH-4	1.7 MV	1988*	Madison, WI	USA
University of California Santa Barbara	Institute for Terahertz Science and Technology	2UDHS	2 MV	1988	Santa Barbara, CA	USA

Institution	Department/Laboratory	Model	Year	Voltage	City	Country
Mechanical Engineering Laboratory		3SDH-4	1988	1 MV	Tsukuba	Japan
Kyushu University	Center for Accelerator and Beam Applied Science	8UDH	1989*	8 MV	Fukuoka	Japan
United States Naval Academy	Physics Department	5SDH	1989	1.7 MV	Annapolis, MD	USA
Academia Sinica	Institute of Physics	9SDH-2	1989	3 MV	Taipei	Taiwan
University of Houston	Ion Beam Laboratory	5SDH-2	1991	1.7 MV	Houston, TX	USA
Lawrence Livermore National Laboratory (LLNL)	Materials Science Division	4UH	1990	4 MV	Livermore, CA	USA
Alabama A&M University	Center for Irradiation of Materials (CIM)	5SDH-2	1990	1.7 MV	Normal, AL	USA
Union College	Ion Beam Analysis Laboratory	3SDH	1989	1 MV	Schenectady, NY	USA
Osaka University	Nano-Material & Nano-Characterization Lab	3SDH (3S-R10)	1989	1 MV	Osaka	Japan
Japan Atomic Energy Agency (JAEA)	Takasaki Advanced Radiation Research Institute (TARRI)	9SDH-2	1990	3 MV	Takasaki	Japan
Kanagawa University	Department of Mathematics and Physics	3SDH (3S-R10)	1990*	1 MV	Yokohama	Japan
University of Göttingen	Second Institute of Physics	9SDH-2	1989*	3 MV	Göttingen	Germany
University of Notre Dame	Nuclear Science Laboratory	9SDH-2	1991*	3 MV	Notre Dame, IN	USA
National Institute for Environmental Studies (NIES)	Tandem Accelerator for Environmental Research and Radiocarbon Analysis (TERRA)	3SDH (3S-R10)	1990	1 MV	Tsukuba	Japan
Universidade de São Paulo	Laboratório de Materiais e Feixes Iônicos	5SDH	1990	1.7 MV	São Paulo	Brazil
University of Louisiana at Lafayette	Louisiana Accelerator Center	5SDH-2	1991	1.7 MV	Lafayette, LA	USA
NASA Marshall Space Flight Center (MSFC)	Space Environmental Effects Team	7.5SH-2	1992	2.5 MV	Huntsville, AL	USA
NASA Marshall Space Flight Center (MSFC)	Space Environmental Effects Team	2SH-2	1992	0.6 MV	Huntsville, AL	USA
Lawrence Berkeley Lab	Accelerator Technology and Applied Physics Division (ATAP)	5SDH (5S-MR10)	1991*	1.7 MV	Berkeley, CA	USA
Ghana Atomic Energy Commission (GAEC)	Accelerator Research Centre (ARC)	5SDH-2	1991*	1.7 MV	Accra	Ghana

(Continued)

TABLE 1.8 (Continued)

NEC accelerators worldwide data from https://www.pelletron.com/resource-category/accelerators-around-the-world/

Institution	Laboratory	System	Voltage	Installed	City/Province	Country
University of Pennsylvania	Surface and Thin Film Analysis Facility	5SDH	1.7 MV	1991	Philadelphia, PA	USA
Nara Women's University	Radiation Physics Lab	5SDH-2	1.7 MV	1992	Nara	Japan
Synergy Health Sterilisation	Applied Sterilisation Technologies Facility	15SDH-2	5 MV	1992*	Harwell	UK
University of Tokyo	Micro Analysis Laboratory, Tandem Accelerator (MALT)	5UD	5 MV	1992	Tokyo	Japan
Universidad de Sevilla	Centro Nacional de Aceleradores	9SDH-2	3 MV	1997	Seville	Spain
Tokyo City University (TCU)	Atomic Energy Research Laboratory	5SDH (MAS1700)	1.7 MV	1993*	Tokyo	Japan
Institute for Physical and Chemical Research (RIKEN)	Nishina Center for Accelerator-Based Science	5SDH-2	1.7 MV	1993	Wakō	Japan
Inter-University Accelerator Center (IUAC)	Pelletron Accelerator RBS-AMS Systems (PARAS)	5SDH-2	1.7 MV	1993*	New Delhi	India
Tsing Hua University	Accelerator Laboratory	9SDH-2	3 MV	1995	Hsin-Chu	Taiwan
University of Minnesota	Characterization Facility	5SDH (MAS1700)	1.7 MV	1994	Minneapolis, MN	USA
Universidad National Autonoma de Mexico (UNAM)	Instituto de Física	9SDH-2	3 MV	1995	Mexico City	Mexico
University of Vienna	Isotope Research and Nuclear Physics Group	9SDH-2	3 MV	1995	Vienna	Austria
Texas A&M University	Microbeam Cell Irradiation Facility	9SDH-2	3 MV	1996*	College Station, TX	USA
Korea Institute of Science and Technology (KIST)	Chemical Analysis Center	6SDH-2	2 MV	1995	Cheongryang	Korea
University of Michigan	Michigan Ion Beam Laboratory	5SDH (5S-MR10)	1.7 MV	1995*	Ann Arbor, MI	USA
National Institute for Environmental Studies (NIES)	Tandem Accelerator for Environmental Research and Radiocarbon Analysis (TERRA)	15SDH-2 (AMS)	5 MV	1995	Tsukuba	Japan
Connecticut College	Accelerator Laboratory	3SH	1 MV	1995	New London, CT	USA

Institution	Department/Laboratory	Model	Voltage	Year	City	Country
Interuniversitair Micro-Elektronic Centrum (IMEC)	Materials and Components Analysis	6SDH	2 MV	1995	Leuven	Belgium
China Institute of Atomic Energy (CIAE)	Division of Radiometrology	5SDH-2	1.7 MV	1996	Beijing	China
Kobe University of Mercantile Marine	Laboratory of Particle Beam Engineering	5SDH-2	1.7 MV	1996	Kobe	Japan
Japan Atomic Energy Agency (JAEA)	Tono Geoscience Center	15SDH-2 (AMS)	5 MV	1997	Toki	Japan
Vanderbilt University	Institute for Space and Defense Electronics (ISDE)	6SDH-1	2 MV	1996	Nashville, TN	USA
Evans Analytical Group	Sunnyvale Lab	3SDH	1 MV	1996	Sunnyvale, CA	USA
University of Wisconsin Madison	Waisman Center	9SDH-2	3 MV	1997	Madison, WI	USA
Auburn University	Department of Physics	6SDH-2	2 MV	1997	Auburn, AL	USA
University of Oslo	Microsystems and Nanotechnology Laboratory (MiNaLab)	3SDH-2	1 MV	1997*	Oslo	Norway
University of North Texas	Ion Beam Modification and Analysis Laboratory	9SDH-2	3 MV	1997*	Denton, TX	USA
Helmholtz-Zentrum Dresden-Rossendorf (HZDR)	Felsenkeller Laboratory	15SDH-2 (AMS)	5 MV	1998*	Dresden	Germany
Universidade Federal do Rio de Janeiro	Laboratório de Colisões Atómicas e Moleculares (LaCAM)	5SDH	1.7 MV	1998	Rio de Janeiro	Brazil
National Council for Scientific Research (CNRS)	Lebanese Atomic Energy Comission (LAEC)	5SDH	1.7 MV	1998	Beirut	Lebanon
Lawrence Livermore National Labs (LLNL)	Center for Accelerator Mass Spectrometry	3SDH-1 (AMS)	1 MV	1999	Livermore, CA	USA
University of Arizona	Accelerator Mass Spectrometry Lab	9SDH-2 (AMS)	3 MV	2000	Tucson, AZ	USA
Japan Aerospace Exploration Agency (JAXA)	Tsukuba Space Center	6SH	2 MV	2000	Tsukuba	Japan
Australian National University (ANU)	Department of Electronic Materials Engineering (EME)	5SDH-4	1.7 MV	1999	Canberra	Australia
Japan Atomic Energy Agency (JAEA)	Tokai Research and Development Center	4UH-HC	4 MV	2000	Tokai	Japan
Uppsala University	Tandem Laboratory		5 MV	2001	Uppsala	Sweden

(Continued)

TABLE 1.8 (Continued)

NEC accelerators worldwide data from https://www.pelletron.com/resource-category/accelerators-around-the-world/

Institution	Laboratory	System	Voltage	Installed	City/Province	Country
University of Georgia	Center for Applied Isotope Studies (CAIS)	1.5SDH-2 (AMS)	0.5 MV	2000	Athens, GA	USA
Institute for Accelerator Analysis (IAA)	Shirakawa Analysis Center	9SDH-2 (AMS)	3 MV	2000	Shirakawa	Japan
University of Guelph	Guelph PIXE Group	9SH	3 MV	2001	Guelph, ON	Canada
University of Durham	Department of Chemistry	5SDH	1.7 MV	2001	Durham	UK
Adam Mickiewicz University	Poznan Radiocarbon Laboratory	1.5SDH-1 (AMS)	0.5 MV	2001	Poznan	Poland
Scottish Universities Environmental Research Centre (SUERC)	AMS Laboratory	15SDH-2 (AMS)	5 MV	2002	East Kilbride	UK
University of California Irvine	Keck-Carbon Cycle AMS Facility	1.5SDH (CAMS)	0.5 MV	2002	Irvine, CA	USA
Commissariat a l'Energie Atomique (CEA)	Laboratoire de Mesure du Carbone 14 (LMC14)	9SDH-2 (AMS)	3 MV	2002	Saclay	France
University of California Los Angeles (UCLA)	MegaSIMS Laboratory	3SDH-2 (AMS)	1.2 MV	2004	Los Angeles, CA	USA
Lund University	Radiocarbon Dating Laboratory	HV deck (SSAMS)	0.25 MV	2004	Lund	Sweden
Peking University	Institute of Heavy Ion Physics	1.5SDH-1 (CAMS)	0.5 MV	2004	Beijing	China
University of Campania Luigi Vanvitelli	Center for Isotopic Research on Cultural and Environmental Heritage (CIRCE)	9SDH-2 (AMS)	3 MV	2005	Caserta	Italy
Hope College	Department of Physics	5SDH	1.7 MV	2004	Holland, MI	USA
Paleo Labo	AMS Dating Facility	1.5SDH-1 (CAMS)	0.5 MV	2004	Kiryu	Japan
Woods Hole Oceanographic Institute	National Ocean Sciences Accelerator Mass Spectrometry (NOSAMS)	1.5SDH-1 (AMS)	0.5 MV	2005	Woods Hole, MA	USA

Institution	Department/Laboratory	Model	Voltage	Year	Location	Country
Accium BioSciences		1.5SDH-1 (CAMS)	0.5 MV	2006	Seattle, WA	USA
Commissariat a l'Energie Atomique (CEA)	Joint Accelerators for Nanoscience and Nuclear Simulation (JANNUS)	3UH-4	3 MV	2006	Saclay	France
Beta Analytic, Inc.	Radiocarbon Laboratory	HV deck (SSAMS)	0.25 MV	2005	Miami, FL	USA
Scottish Universities Environmental Research Centre (SUERC)	AMS Laboratory	HV deck (SSAMS)	0.25 MV	2006	East Kilbride	UK
Australian National University (ANU)	Department of Nuclear Physics	HV deck (SSAMS)	0.25 MV	2006	Canberra	Australia
Government College University	Centre for Advanced Studies in Physics (CASP)	6SDH-2	2 MV	2007	Lahore	Pakistan
Queen's University	^{14}CHRONO Centre	1.5SDH-1 (CAMS)	0.5 MV	2006	Belfast	UK
State University of New York (SUNY) Geneseo	Nuclear Structure Laboratory	5SDH	1.7 MV	2007	Geneseo, NY	USA
University of Michigan	Michigan Ion Beam Laboratory	HV deck	0.4 MV	2007	Ann Arbor, MI	USA
Obafemi Awolowo University	Center for Energy Research and Development (CERD)	5SDH	1.7 MV	2007	Ife	Nigeria
National Centre for Physics	Experimental Physics Directorate	5UDH-2	5 MV	2007	Islamabad	Pakistan
Pharmaron		HV deck (SSAMS)	0.25 MV	2008	Germantown, MD	USA
Toray Research Center (TRC)	Shiga Laboratory	3SDH	1 MV	2008	Ōtsu	Japan
Beta Analytic, Inc.	Radiocarbon Laboratory	HV deck (SSAMS)	0.25 MV	2008	Miami, FL	USA
Comisión Nacional de Energía Atómica (CNEA)	Centro Atómico Bariloche	5SDH	1.7 MV	2008	Bariloche	Argentina
Bar Ilan University	Ion Beam Analysis Laboratory	5SDH (5S-MR10)	1.7 MV	2011	Ramat Gan	Israel
Ecole Polytechnique	Laboratoire des Solides Irradiés (LSI)	7.5SHe-2	2.5 MV	2009	Palaiseau	France
Commissariat a l'Energie Atomique (CEA)	Joint Accelerators for Nanoscience and Nuclear Simulation (JANNUS)	6SDH-2	2 MV	2009	Saclay	France

(Continued)

TABLE 1.8 (Continued)

NEC accelerators worldwide data from https://www.pelletron.com/resource-category/accelerators-around-the-world/

Institution	Laboratory	System	Voltage	Installed	City/Province	Country
Institute of Geological and Nuclear Sciences (GNS)	Rafter Radiocarbon Laboratory	1.5SDH-1 (XCAMS)	1 MV	2010	Lower Hutt	New Zealand
Hanoi University of Science (HUS)	Department of Nuclear Physics	5SDH-2	1.7 MV	2010	Hanoi	Vietnam
Sandia National Laboratories	Ion Beam Laboratory	3UH-2	3 MV	2010	Albuquerque, NM	USA
Kyoto University	Quantum Science and Engineering Center	6SDH-2	2 MV	2010	Kyoto	Japan
Yamagata University	Center for Accelerator Mass Spectrometry	1.5SDH-1 (CAMS)	0.5 MV	2010	Yamagata	Japan
Beta Analytic, Inc.	Radiocarbon Laboratory	HV deck (SSAMS)	0.25 MV	2011	Miami, FL	USA
University of Notre Dame	Nuclear Science Laboratory	5U-4	5 MV	2012	Notre Dame, IN	USA
University of Colorado	Institute for Modeling Plasma, Atmospheres, and Cosmic Dust (IMPACT)	9SH	3 MV	2011	Boulder, CO	USA
Arnold Engineering Development Complex (AEDC)	Space Threat Assessment Testbed (STAT)	HV deck	0.15 MV	2011*	Arnold AFB, TN	USA
Instytut Fizyki Polskiej (IFP)	Laboratory of X-Ray and Electron Microscopy Research	3SDH-2	1 MV	2011	Warsaw	Poland
University of Manchester	Dalton Cumbrian Facility	15SDH-4	5 MV	2012	Manchester	UK
University of Georgia	Center for Applied Isotope Studies (CAIS)	HV deck (SSAMS)	0.25 MV	2011	Athens, GA	USA
East Carolina University	Accelerator Laboratory	6SDH-2	2 MV	2011	Greenville, NC	USA
Universidade Federal Fluminense (UFF)	AMS Radiocarbon Laboratory	HV deck (SSAMS)	0.25 MV	2011	Niterói	Brazil
University of Tennessee	Ion Beam Materials Laboratory	9SDH-2	3 MV	2012	Knoxville, TN	USA
Australian Nuclear Science and Technology Organisation (ANSTO)	Centre for Accelerator Science	18SDH-2	6 MV	2014	Lucas Heights, NSW	Australia
Australian Nuclear Science and Technology Organisation (ANSTO)	Centre for Accelerator Science	3SDH-1	1 MV	2013	Lucas Heights, NSW	Australia

Institution	Department/Laboratory	Model	Voltage	Year	City	Country
United States Military Academy (USMA)	Department of Physics and Nuclear Engineering	5SDH	1.7 MV	2012	West Point, NY	USA
National Renewable Energy Laboratory		3SDH (3S-MR10)	1 MV	2012	Golden, CO	USA
Weizmann Institute of Science	Dangoor REsearch Accelerator Mass Spectrometry (D-REAMS) Laboratory	1.5SDH-1 (CAMS)	0.5 MV	2013	Rehovot	Israel
Adam Mickiewicz University	Poznan Radiocarbon Laboratory	1.5SDH (CAMS)	0.5 MV	2012	Poznan	Poland
Guru Ghasidas University	Department of Pure and Applied Physics	9SDH-4	3 MV	2013	Bilaspur	India
University of Tokyo	Atmosphere and Ocean Research Institute (AORI)	HV deck (SSAMS)	0.25 MV	2013	Tokyo	Japan
Comenius University	Center for Nuclear and Accelerator Technologies (CENTA)	9SDH-2	3 MV	2013	Bratislava	Slovakia
Naval Research Laboratory (NRL)	Materials Science and Technology Division	HV deck (SSAMS)	0.4 MV	2013	Washington, DC	USA
Centre National de la Recherche Scientifique (CNRS)	Delegation Centre Poitou-Charentes	3U-2	3 MV	2014	Orleans	France
University of Tsukuba	Tandem Accelerator Center	18SDH-2 (AMS/ IBA)	6 MV	2014	Tsukuba	Japan
Guangzhou Institute of Geochemistry		1.5SDH-1 (CAMS)	0.5 MV	2014	Guangzhou	China
University of Michigan	Michigan Ion Beam Laboratory	9SDH-2	3 MV	2014	Ann Arbor, MI	USA
Idaho National Laboratory		1.5SDH-1 (ICAMS)	0.5 MV	2014	Idaho Falls, ID	USA
Lawrence Livermore National Labs (LLNL)	Center for Accelerator Mass Spectrometry	HV deck (SSAMS)	0.25 MV	2014	Livermore, CA	USA
Commissariat a l'Energie Atomique (CEA)	Joint Accelerators for Nanoscience and Nuclear Simulation (JANNUS)	7.5SH-2	2.5 MV	2014	Saclay	France
Tohoku University		3SDH-2	1 MV	2014	Sendai	Japan
National Institute for Environmental Studies (NIES)		1.5SDH-1 (CAMS)	0.5 MV	2014	Tsukuba	Japan

(Continued)

TABLE 1.8 (Continued)

NEC accelerators worldwide data from https://www.pelletron.com/resource-category/accelerators-around-the-world/

Institution	Laboratory	System	Voltage	Installed	City/Province	Country
Centre National de la Recherche Scientifique (CNRS)	Institut de Physique Nucléaire	4UH-4	4 MV	2014	Orsay	France
Xiamen University		HV deck	0.4 MV	2014	Xiamen	China
Inter-University Accelerator Center (IUAC)	Accelerator Mass Spectrometry Facility	1.5SDH-1 (CAMS)	0.5 MV	2015	New Delhi	India
National Institute for Fusion Science (NIFS)		3SDH (3S-MR10)	1 MV	2014	Toki	Japan
University of Tokyo Museum	Laboratory for Radiocarbon Dating	1.5SDH-1 (CAMS)	0.5 MV	2015	Tokyo	Japan
University of Manchester	Dalton Cumbrian Facility	7.5SH-2	2.5 MV	2015	Manchester	UK
Türkiye Bilimsel ve Teknolojik Araştırma Kurumu (TÜBİTAK)	Marmara Research Center	3SDH-1 (UAMS)	1 MV	2015	Gebze	Turkey
European Commission Joint Research Centre (EC-JRC)	Institute for Reference Materials and Measurements	4UD-4	4 MV	2016	Geel	Belgium
Center for Physical Sciences and Technology	Mass Spectrometry Laboratory	HV deck (SSAMS)	0.25 MV	2015	Vilnius	Lithuania
Penn State University	Energy and Environmental Sustainability Laboratories	1.5SDH-1 (CAMS)	0.5 MV	2016	University Park, PA	USA
University of Huddersfield	Ion Beam Centre (IBC)	HV deck	0.4 MV	2016	Huddersfield	UK
University of Illinois Urbana-Champaign	Materials Research Laboratory	3SDH	1 MV	2016	Urbana, IL	USA
Tianjin University	Institute of Surface-Earth System Science	1.5SDH-1 (XCAMS)	0.5 MV	2017	Tianjin	China
Korea Institute of Radiological and Medical Sciences (KIRAMS)		1.5SDH-1 (CAMS)	0.5 MV	2017	Seoul	Korea
Max-Planck-Institut für Mikrostrukturphysik	Nano-Systems from Ions, Spins and Electrons (NISE)	5SDH	1.7 MV	2017	Halle	Germany
Qingdao National Laboratory for Marine Science and Technology		1.5SDH-1 (XCAMS)	0.5 MV	2018	Qingdao	China

Institution	Facility	Model	Energy	Year	Location	Country
Swiss Federal Institute of Technology (ETH)	Laboratory of Ion Beam Physics	1.5SDH (AMS)	0.5 MV	1998	Zurich	Switzerland
Technische Universität München	Accelerator Mass Spectrometry Group (GAMS)	MP conversion	14 MV	1975	Garching	Germany
Brookhaven National Laboratory	Tandem Van de Graaff Facility	MP conversion	15 MV	1980	Upton, NY	USA
Comisión Nacional de Energía Atómica (CNEA)	Centro Atómico Ezeiza	FN conversion	9 MV	1982*	Buenos Aires	Argentina
Lawrence Livermore National Labs (LLNL)	Center for Accelerator Mass Spectrometry	FN conversion	10 MV	1987	Livermore, CA	USA
Duke University	Triangle Universities Nuclear Laboratory	FN conversion	10 MV	1990	Durham, NC	USA
Sandia National Laboratories	Ion Beam Laboratory	EN conversion	6 MV	1992	Albuquerque, NM	USA
Purdue University	Purdue Rare Isotope Measurement Laboratory	FN conversion	10 MV	1993	West Lafayette, IN	USA
University of Washington	Department of Nuclear Physics	FN conversion	10 MV	1994	Seattle, WA	USA
Florida State University	Superconducting Linear Accelerator Facility	FN conversion	9 MV	1995	Tallahassee, FL	USA
Australian Nuclear Science and Technology Organisation (ANSTO)	Centre for Accelerator Science	FN conversion	10 MV	1997	Lucas Heights, NSW	Australia
Kansas State University	James R. Macdonald Laboratory	EN conversion	7 MV	2000	Manhattan, KS	USA
University of Notre Dame	Nuclear Science Laboratory	FN conversion	10 MV	2000	Notre Dame, IN	USA
Institute of Geological and Nuclear Sciences (GNS)	Rafter Radiocarbon Laboratory	EN conversion	6 MV	2001	Lower Hutt	New Zealand
Université de Montréal	Canadian Charged Particle Accelerator Consortium	EN conversion	6 MV	2002	Montréal, QC	Canada
Western Michigan University	Department of Physics	EN conversion	6 MV	2003	Kalamazoo, MI	USA
Universität zu Köln	Institut für Kernphysik	FN conversion	10 MV	2004	Cologne	Germany

(Continued)

TABLE 1.8 (Continued)

NEC accelerators worldwide data from https://www.pelletron.com/resource-category/accelerators-around-the-world/

Institution	Laboratory	System	Voltage	Installed	City/Province	Country
iThemba Laboratory for Accelerator-Based Sciences (LABS)	Tandem Accelerator Mass Spectrometry (TAMS)	EN conversion	6 MV	2006	Johannesburg	South Africa
Horia Hulubei National Institute for R&D in Physics and Nuclear Engineering (IFIN-HH)	Tandem Accelerators Department	FN conversion	9 MV	2007	Măgurele	Romania
Swiss Federal Institute of Technology (ETH)	Laboratory of Ion Beam Physics	EN conversion	6 MV	2011	Zurich	Switzerland
Ohio University	Edwards Accelerator Lab	T2 conversion	4.5 MV	2011	Athens, OH	USA
Commissariat a l'Energie Atomique (CEA)	Direction des Applications Militaires (DAM)	EN conversion	6 MV	2012	Bruyères-le-Châtel	France
Istituto Nazionale di Fisica Nucleare (INFN)	Laboratori Nazionali del Sud	MP conversion	14 MV	2014	Catania	Italy

Source: Courtesy of NationalElectrostatic Corp.

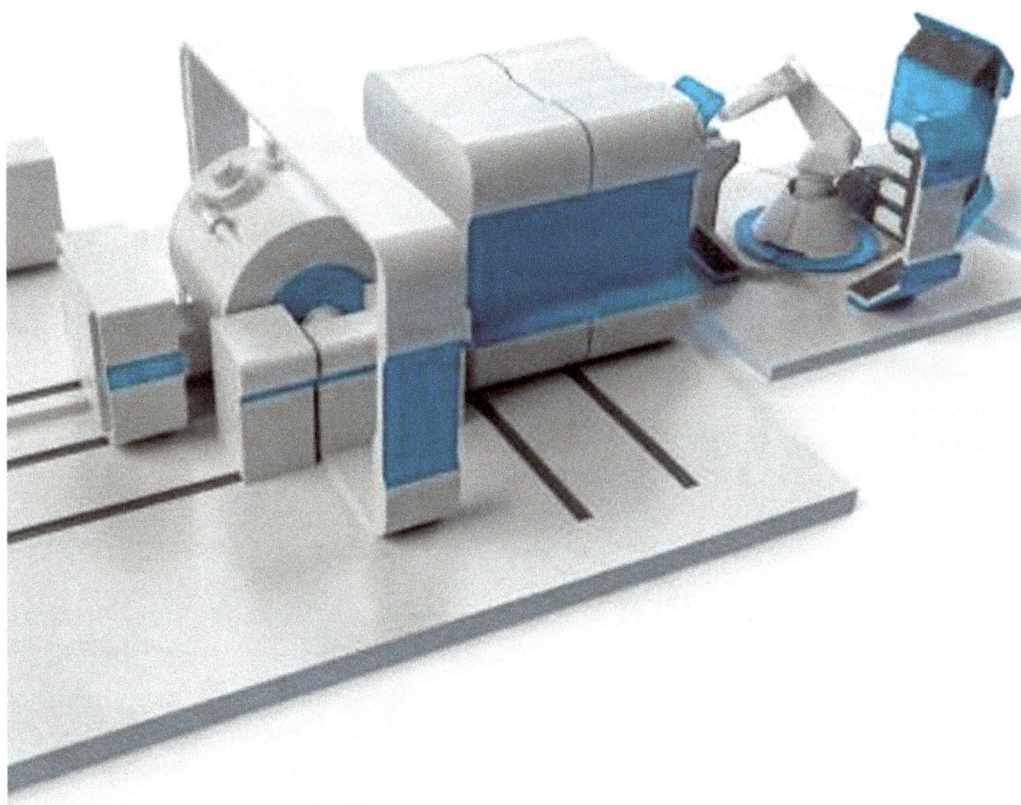

FIGURE 1.39
iMiGiNE – a robotized system combining a cyclotron ISOTRACE with a radiochemistry room iMiLAB. (With permission PMB-Alcen.)

2. iMiTRACE cyclotron: accelerates particles to 12 MeV: ideal energy for the production of ^{18}F, ^{11}C, ^{68}Ga and more; He-free Superconducting Magnet; Four External Targets; Self-shielded targetry; Lightweight and Compact; Reliable, specifically designed for imaging centers, the cyclotron grants them access to an on-site radiopharmaceutical production system. Reliable, stable and compact, iMiTRACE is entirely designed and manufactured at PMB. With a self-shielded, 4-port targetry and equipped with a helium-free superconducting magnet, iMiTRACE is the result of many technological improvements made to cyclotron OSCAR (Oxford Instruments) (Fig. 1.39).

   Designed for fully automated operation, from target selection and filling, to delivery, iMiTRACE is easy to control. It is equipped with an intuitive and user-friendly interface, designed to give all the necessary information, depending on the operator's level of expertise and training. Moreover, as the ion source and targets are external, the maintenance of the cyclotron is easier and quicker. This aspect of its architecture allows maintenance operations to be done with less equipment activation, due to the self-shielded targetry, and increased uptime. Easy to install in both new and existing facilities, iMiTRACE only requires 50-cm-thick walls, for details see PMB-Alcen (2021) (Fig. 1.40).

FIGURE 1.40
iMiTRACE cyclotron. (With permission PMB-Alcen.)

3. iMiLAB radiochemistry: Microfluidic-Based Synthesis; One Dose One Patient; Automated Syringe Filling; Self-shielded; Reduced Radiation Exposure; Multiple Radiopharmaceuticals Produced with ^{18}F, ^{11}C, ^{68}Ga and more. The radiochemistry room iMiLAB is compact and self-shielded. Entirely robotized and GMP compliant, the iMiLAB radiochemistry room utilizes micro-fluidic techniques capable of producing diversified radiopharmaceuticals in an automated synthesis box and in sterile cartridges with onboard precursors and solvents. It is capable of synthetizing multiple molecules on a same-day basis, with a decreased need in staff; for details see PMB-Alcen (2021).

4. ORIATRON Linacs: high-energy electron linear accelerators & high-energy X-ray NDT systems. PMB provides linacs with the following nominal energies: 3 MeV, 4 MeV, 6 MeV and 7 MeV. These standard ranges have several options catering for the specific requirements of each project. By means of simulations, calculations and study, PMB develops specific machines for particular applications. These cases have required the development of specific secondary collimators, beam flatteners and operation in the green zone (self-shielded machines) (Fig. 1.41).

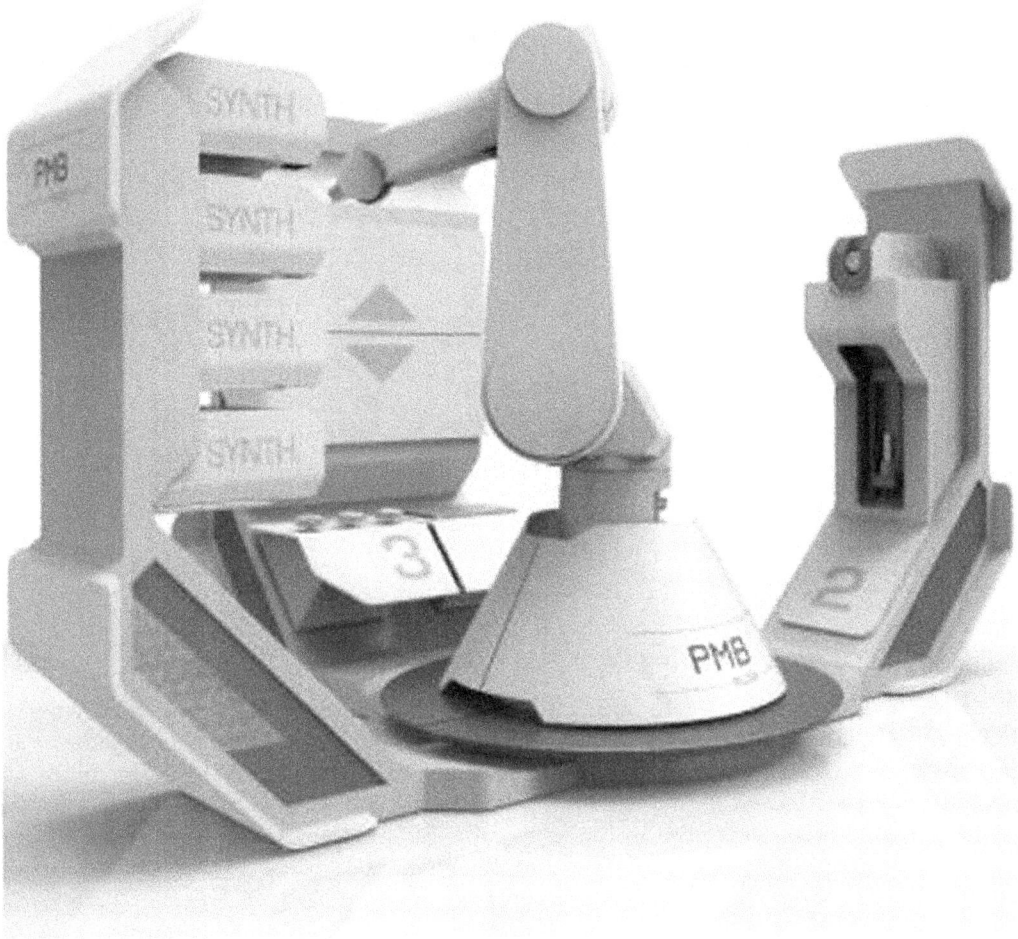

FIGURE 1.41
iMiLAB radiochemistry. (With permission PMB-Alcen.)

PMB uses 3D HFSS electromagnetic simulation software for the design and sizing of RF components. Egun code is used to simulate the movement of charged particles and is used for the sizing of electron guns.

The complete architecture is validated by simulation in beam dynamics. The machine high-voltage circuits are simulated using SPICE software. All PMB accelerators are fully tested at PMB before being installed at customer sites, demonstrating full compliance with specifications. For details see PMB-Alcen (2021) (Fig. 1.42).

5. RF components (RF windows, accelerating waveguides, etc.); High-power RF couplers; Brazed and welded mechanical assemblies with high technology coating for demanding applications (ultra-high vacuum, cryogenic temperature, etc.); Beamline components (electron guns, conversion targets, collimators, ion chambers, etc.); Components for X-ray tubes (anode and cathode insulators, beryllium center frames, filament carriers, etc.).

FIGURE 1.42
PMB-Alcen ORIATRON linac. (With permission from PMB-Alcen.)

1.7.3 High Voltage Engineering Corp., USA (1946–2005)

In 1946, Robert J. Van de Graaff, John Trump, and British engineering professor Denis M. Robinson established the HVEC with the aim to manufacture Van de Graaff accelerators. Van de Graaff served as the company's chief physicist and as later chief scientist. The following year, the company began its manufacturing operations and became a leading supplier of particle accelerator systems used in cancer therapy, radiography and nuclear structure studies. Van de Graaff resigned from MIT in 1960 but continued his research in nuclear physics at HVEC until his death in Boston, Massachusetts, on 16 January 1967. By this time, there were more than 500 Van de Graaff particle accelerators in more than 30 countries.

1.7.4 High Voltage Engineering Europa B.V.

High Voltage Engineering Europa B.V. (HVE, Amsterdamseweg 63, 3812 RR Amersfoort, the Netherlands) is one of the largest and most diverse manufacturers of particle accelerator systems for science and industry. In the period 1959–2005 it was a subsidiary of HVEC, USA; since 2005 until now is subsidiary of Aimland Technologies, NL. The company is specialized in the development and manufacture of ion beam and electron beam technology-based equipment. In addition to research-type accelerator systems HVE also manufactures industrial-type accelerator systems and sub-assemblies for semiconductor ion implantation systems and of electron beam processing systems.

Following products are supplied by HVEE:

- Ion accelerators: air insulated accelerators up to incl. 500 kV.
- Singletron single-ended accelerators up to incl. 6.0 MV (8.0 MV).
- Tandetron tandem accelerators up to incl. 6.0 MV (8.0 MV), DC as well as pulsed beams.
- Electron accelerators: beam energies up to incl. 6.0 MV (8.0 MV), beam power up to incl. 50 kW.

TABLE 1.9

Characteristics of single-stage Van de Graaff accelerators made by HVEC, electron accelerators

Type	AS-400	GS	AS-2000	KS-3000	KS-4000
Max. voltage (MV)	0.4	1.5	2.0	3.0	4.0
Max. current (μA)	100	1700	250	1000	1000
X-ray intensity at 1m (R/min)	1	210	85	1350	3500
Tank length (m)	1.25	2.75	2.36	3.56	4.73
Insulating gas pressure (atm)	7	25	20	20	20
Voltage stability (kV)	±10	±10	±10	±10	±10

TABLE 1.10

Characteristics of single-stage Van de Graaff accelerators made by HVEC, heavy-ions accelerators

Type	AN-400	AN-200	KN-3000	KN-4000	CN
Max. voltage (MV)	0.4	2.0	3.0	4.0	5.5
Max. current (μA)	150	150	400	400	70
Neutron flux 8neutrons/sec)	$>2\times10^{10}$	$>4\times10^{10}$	$>5\times10^{11}$	$>10^{12}$	$>5\times10^{11}$
Tank length (m)	1.25	2.75	2.36	3.56	4.73
Insulating gas pressure (atm)	7	25	20	20	16
Voltage stability (kV)	±5	±2	±2	±2	±2

In addition to bare accelerators the company can supply complete systems for the following activities:

- Systems for Ion implantation
  - Ion species H, He, B, P, As and others.
  - Beam energies 10 keV to 60 MeV and above.
  - Beam power up to incl. 25 kW.
  - Research as well as industrial endstations.
- Systems for ion beam analysis
  - RBS, PIXE, PIGE, NRA, ERD.
- Systems for accelerator mass spectrometry (AMS).
- Systems for micro-beam applications.
- Accelerator-based systems for neutron generation.

Şingletron single-ended accelerator systems (see Tables 1.9 and 1.10) are designed to produce a variety of highly stable ion beams for material modification and analysis (Fig. 1.43). They are designed to operate in a normal laboratory environment and can fit into single room. In addition, a wide range of analyzing/switching magnets, beamlines and endstations is available for various applications including:

FIGURE 1.43
HVEE Coaxial Singletron. (Courtesy of HVEE.)

- Rutherford backscattering Spectroscopy.
- Particle Induced X-ray Emission.
- Particle Induced Gamma-ray Emission.
- Nuclear Reaction Analysis.
- Micro- and Nano-beam applications.
- Ion beam modification.
- Ion beam mixing.

The central feature of the HVE Singletron concept is an SF_6 insulated, parallel fed, capacitively coupled CW type HV power supply characterized by low noise level, high terminal voltage stability and low terminal voltage ripple. Singletron HV power supply is a purely electronic power supply, it has no moving parts. As a result, there are no vibrations so ripple and stability values and dynamic behaviour are stable over long period of time.

The HVE Singletron accelerators are available in two versions: Coaxial and in-line Singletons. Coaxial Singletrons' HV power supply is built around the acceleration stage. In-line Singletrons are the preferred choice in applications where ripple specifications are of importance. In this type, the tank is split into two sections and the HV power supply is a separate self-supporting assembly suspended from the base of one of the pressure tank sections. The generator that supplies electrical power to the components in HV terminal is mechanically isolated from the accelerator stage and ion source to avoid that the generator vibrations are transmitted to the HV terminal and/or the ion source.

Both, coaxial and in-line Singletrons, are equipped with a spacious HV terminal that provides room for an ion source, all solid-state power supplies and up to four source feed gas systems. A dedicated Microsoft Windows-based software program provides user-friendly interfaces and allows automatic startup and shutdown and automated tuning, control and monitoring of the entire accelerator system (Fig. 1.44).

HVEE has developed a high-current, light-ion 3.5 MV single-ended accelerator system to meet the stringent requirements on beam intensity and stability of the LUNA-MV project at Laboratori Nazionali del Gran Sasso (LNGS), L'Aquila, Italy (Sen et al. 2019).

HVEE has designed, built and tested a 2 MV dual irradiation system that can be applied for radiation damage studies and ion beam material modification. The system consists of two independent accelerators which support simultaneous proton and electron irradiation (energy range 100 keV to 2 MeV) of target sizes of up to 300×300 mm^2, shown in Fig. 1.40 (Fig. 1.45).

Three-dimensional finite element methods were used in the design of various parts of this system. The electrostatic solver was used to quantify essential parameters of the solid-state power supply generating the DC high voltage. The magnetostatic solver and ray tracing were used to optimize the electron/ion beam transport (Podaru et al. 2013).

Tandetron accelerator systems are state-of-art tandem accelerator systems designed to produce MeV ion beams of virtually all periodic system elements for material modification and analysis. A built-in magnetic suppression system to reduce back-streaming electron energy gain ensures a virtually X-ray radiation-free operation. The HVE Tandetrons are available in three beam current versions: medium current (MC), medium current plus (MC$^+$), and high current (HC) see Fig. 1.47.

FIGURE 1.44
In-line Singletron.(Courtesy HVEE.)

FIGURE 1.45
Singletron, 2.0 MV ion & electron. (Courtesy HVEE.)

FIGURE 1.46
Tandetron accelerator: T-shape 1.0–3.0 MV, described in Accelerator Newsletter 1(3) October 1994. (Courtesy HVEE.)

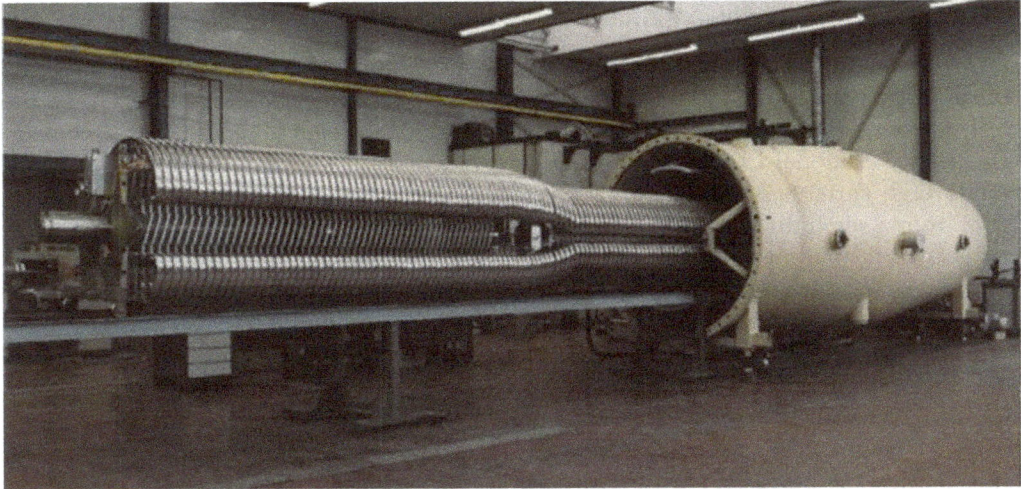

FIGURE 1.47
Tandetron coaxial accelerator: 4.0–6.0 MV. (Courtesy HVEE.)

A nanosecond pulsing system for H, D and He ions has been developed to satisfy the demands of a new neutron reference field (2 keV to 20 MeV) for neutron metrology and dosimetry. The system is capable of delivering ion energies of 0.2–4 MeV at target with currents of 50 and 8 µA in DC and pulsed mode, respectively Mous et al. (2004a) (Fig. 1.46).

All the steps in the Tandetron series development have been described in scientific literature. Gottdang et al. (2002) described the HVEE 5 MV Tandetron, Mous et al. (2004b) described a range of high-current Tandetron accelerator systems with terminal voltages of 1–6 MV. The first HVEE Tandetron with a nominal thermal voltage of 5 MV has been put into operation at the Universidad Autonoma de Madrid, Spain (Mous et al. 2003).

Besides MC, MC⁺, and HC Tandetrons HVEE offers a range of Coaxial Tandetrons capable of producing even higher beam currents, Fig. 1.47 shows such an accelerator. The three types of ion sources are available and any two can be mounted simultaneously on the dual-source injector system.

The central feature of the HVE Tandetron concept is SF_6 insulated, parallel fed, capacitively coupled CW type High Voltage power supply.

1.7.5 DANFYSIK

Since the company was founded in 1964, DANFYSIK has exported high technology equipment for research laboratories worldwide, and particularly for institutions studying physics in which particle accelerator technology is an important tool. All major particle accelerator laboratories in Europe and the USA are daily using beam analyzing and focusing magnet systems including power supplies delivered from DANFISIK.

More information can be obtained by writing to the company at

Danfysik A/S
Gregersensvej 8
DK-2630 Taastrup
Denmark; or by visiting their web page https://www.danfysik.com/en/.

1.7.6 AccSys Technology – A Hitachi Subsidiary

Here is how it all started. In the early 1970s, a revolutionary new linear accelerating ("linac") structure, the Radio Frequency Quadrupole (RFQ), was introduced to the western scientific community. Originally conceived by Russian inventors, researchers at Los Alamos National Laboratory (LANL) proved in 1979 that it was capable of accelerating high currents of protons to several MeV energy using only a 30 keV DC input beam from a small duoplasmatron ion source. The result was a compact accelerator that had great potential for commercial applications. Dr. Robert W. Hamm was a member of the Los Alamos development team, and designed the small duoplasmatron ion source, as well as performed many of the tests conducted on the prototype.

He and his wife, Dr. Marianne E. Hamm, along with two colleagues from LANL (Dr. James M. Potter and Kenneth R. Crandall) formed AccSys Technology, Inc. in 1985 following the successful National Cancer Institute Phase I SBIR grant they had completed through his wife's company, Technical Programming Services. The company develops and manufactures ion linear accelerators based on the RFQ structure they had helped developed at Los Alamos. During his 22 years tenure as President and CEO, AccSys was awarded eight patents on ion linac technology and designed and built a number of unique ion linacs for research, industrial and medical applications. These include the first permanent magnet quadrupoles focused drift tube linac (DTL), the first close-coupled RFQ/DTL system and the first dual energy RFQ system. The company also developed a unique variable energy RFQ, the first portable RFQ using a 600 MHz structure, the first ^{18}F isotope production targets for pulsed linac beams and a multi-tube rf amplifier based on triode tubes that is still in use today. AccSys was purchased by Hitachi Ltd. in 2007 and continues to provide these systems for customers worldwide.

Today, AccSys Technology, division of Hitachi Particle Engineering and Services, Inc. is a global leader in the production of ion linear (linac) accelerator systems. Address: AccSys Technology, division of Hitachi Particle Engineering and Services, Inc. 1177 Quarry Ln, Pleasanton, CA 94566, USA. Hitachi offers a spectrum of models for various applications. Their customized linac systems provide varying peaks and average currents at energies up to and above 100 MeV. Our ion linear accelerators can be used as synchrotron injectors for proton beam therapy, neutron generators and radioisotope production equipment for production PET (Positron Emission Tomography) isotopes.

Hitachi's history with North America dates back to 1926, when Hitachi first exported 30 electric fans to the United States. Hitachi America, Ltd. was established in 1959 as a regional subsidiary, and the first Hitachi manufacturing facility in the USA opened in 1977. As of 31 March 2020, Hitachi's presence and commitment to the US economy has grown to 17 major research and development (R&D) facilities, 52 main manufacturing sites, 75 Group Companies and over 21,200 employees.

AccSys is an integrated design and manufacturing company specializing in the development, production, installation and servicing of ion linear accelerator (linac) systems using the radiofrequency quadrupole (RFQ) linac and drift-tube linac (DTL). Their products include:

- PULSAR®PET Isotope Production systems: The PULSAR® PET Isotope Production systems are linac based proton accelerators designed to replace large and demanding cyclotron systems for the production of positron emitting isotopes. Sufficient amounts of fluorine-18 and carbon-11 can be produced for synthesis into compounds used in oncology, cardiology, neurology and molecular imaging. The radio-labeled glucose analog, FDG, can be synthesized and distributed for use in Positron Emission Tomography. PULSAR® linacs offer

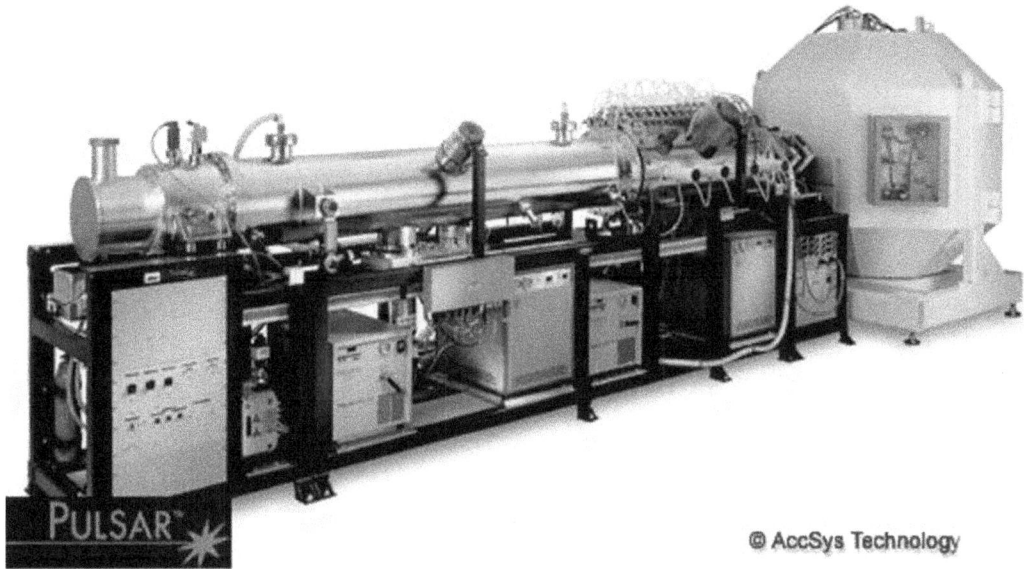

FIGURE 1.48
AccSys Pulsar®. (Courtesy of AccSys Technology.)

highly flexible productivity. High-production single targets or multiple target configurations are available (Fig. 1.48).

The PULSAR™ 7 system is cost-effective positron isotope and tracer production. The proton energy of 7.0 MeV chosen for the system is the optimum energy for the production of useful quantities (up to 1000 mCi) of ^{18}F using the modest beam current available and a bombardment time of 1 hour for ^{18}F. The system contains 3.5 MeV RFQ linac, 3.5 Drift Tube Linac, Ion injector system, 2 425 MHz Amplifiers, Target ladder with ^{18}F targets, Target Shield, Automatic control system, FDG synthesizer.

A complete PULSAR®-based positron radiotracer production facility can occupy less than 93 m^2 to make the most efficient use of valuable space. A complete mobile system, including the tracer production laboratory, can be provided in a standard 14.6 m trailer from Medical Coaches (see http://www.medcoach.com/news/mobilelinearaccelerator.html).

Compact size, light weight, less radiation and ease of installation mean that purchasing or leasing a PULSAR® system reduces installation time and minimizes the need for a costly special facility.

- LANSAR®: A versatile line of accelerators for use with customer-supplied targets, ideally suited for a wide range of research, industrial and medical applications. LANSAR® accelerators use AccSys' patented linac technology to provide compact systems designed for reliable long-term operation in neutron generation applications that utilize customer supplied targets. In a typical application a customer-supplied target is mated to an AccSys supplied linear accelerator.

  The LANSAR® family of linear-accelerator-based systems have been developed specifically for reliable long-term operation for non-destructive inspection applications in research and industrial environments. LANSAR® System Components include:

- Pulsed ion injector with a duo-plasmatron ion source provides a high current proton or deuteron beam.
- Linear accelerator (typically a single RFQ but a second stage DTL is used for beam energies greater than 4 MeV).
- Pulsed rf power system featuring compact and reliable multiple planar-triode technology.
- Vacuum system consisting of commercial cryo-pumps or turbo-molecular pumps with computer-automated controls.
- PC-based control system with built-in diagnostics.

Note

1 With models which use an ion type pump, the conversion is more difficult because ion pump magnets deflect the electron beam. Auxiliary magnets are needed to realign the beam and some magnetic shielding may be required.

References

Alexander, M. E., Biegert, E. K., Jones, J. K., Thurstorn, R. S., Valković, V., Wheeler, R. M., Wingate, C. A., and Zabel, T. 1974. Trace element analysis of seawater and fish samples by PIXE spectroscopy. *Int. J. Appl. Radiat. Isot.* 25 (1974): 229.

Andrade, E., Valković, V., Rendić, D., and Phillips, G. C. 1972. Angular distributions of quasi-free scattering contributions in deuteron break-up by protons and deuterons. *Nucl. Phys. A* 183 (1972): 145.

Andrade, E., Biegert, E. K., Valković, V., and Otte, V. A. 1979. Trace element concentration ratios in mice hair. *Int. J. Nucl. Med. Biol.* 6 (1979): 58–59.

Andrade, E., Feregrino, M., Zavala, E. P., Pineda, J. C., Jiménez, R., and Jaidar, A. 1990. Energy calibration of a 5.5 MV Van de Graaff accelerator using a time-of-flight technique. *Nucl. Instrum. Methods Phys. Res. Sect. A Accel. Spectrom. Detect. Assoc. Equip.* 287(1–2): 135–138.

Baker, S. D., McSherry, D. H., and Findley, D. O. 1969. Elastic scattering by Polarized ^{3}He. *Phys. Rev.* 178: 1616–1620.

Beam, J. and Valković, V. 1972. Triangle graphs and isospin violation in 3-particle final state reactions. *Phys. Rev. Lett. B* 41 (1972): 13.

Bhandri, R. K. and Dey, M. K. 2011. Applications of accelerator technology and its relevance to nuclear technology. *Energy Procedia* 7: 577–588.

Bhasin, V. S., Duck, I. M., and Valković. V. 1973. Isobar exchange effects in p-d backward inelastic scattering. *Phys. Lett. B* 44 (1973): 317.

Biegert, E. K. 1976. Analyzing powers and phase shifts in ^{3}He-^{3}He scattering. Master's Thesis, Rice University. https://hdl.handle.net/1911/104176

Biegert, E. K., Creig, M. E., Storck, R. L., and Valković, V. 1976. Effects of the Magnetic Field on Trace Element Concentration Factors. Proc. IV Conf. on Application of Small Accelerators, Denton, Texas, 1976. (Eds. J. L. Duggan and I. L. Morgan) pp. 99–101.

Biegert, E. K. and Valković, V. 1980. Acute toxicity and accumulation of heavy metals in aquatic animals. *Period. Biol.* 82 (1980): 25–31.

Bonner, T. W. and Mills, J. W. R. 1967. *Semiconductor Radiation Detector for use in Nuclear Well Logging.* U.S. Patent No. 3,312,823. Washington, DC: U.S. Patent and Trademark Office.

Bramblett, R. L., Ewing, R. I., and Bonner, T. W. 1960. A new type of neutron spectrometer. *Nucl. Instrum. Methods.* 9(1): 1–12.

Bryant, P. J. 1994. A brief history and review of accelerators. cds.cern.ch › record › files. CAS - CERN Accelerator School : 5th General Accelerator Physics Course, pp. 1–16. DOI: 10.5170/CERN-1 994-001.1

Burducea, I., Straticiuc, M., Ghiță, D. G., et al. 2015. A new ion beam facility based on a 3 MV Tandetron™ at IFIN-HH, Romania. *Nucl. Instrum. Methods Phys. Res.* B 359: 12–19.

Campbell, J. 2019. Rutherford, transmutation and the proton. *CERN Courier* 2019: 27–30. Cerncourier.com.

Cao, Z. X., et al. 2012. Recoil proton tagged knockout reaction for ^{8}He. *Phys. Lett.* B 707(1): 46–51.

Chao, A. W. and Chou, W., Eds. 2013. *Reviews of Accelerator Science and Technology – Reviews of Accelerator Science and Technology – Volume 5: Applications Of Superconducting Technology To Accelerators.* World Scientific, Singapore. 368 pages.

Chao, A. W. and Chou, W., Eds. 2008. *Reviews of Accelerator Science and Technology – Volume 1.* World Scientific, Singapore. 340 pages.

Chao, A. W. and Chou, W., Eds. 2010. *Reviews of Accelerator Science and Technology, Volume 3: Accelerators as Photon Sources.* World Scientific Publishing Co., Singapore.

Chao, A. W. and Chou, W., Eds. 2011. *Reviews of Accelerator Science and Technology, Volume 4:* World Scientific Publishing Co., Singapore.

Chao, A. W. and Chou, W., Eds. 2016a. *Reviews of Accelerator Science and Technology - Volume 8: Accelerator Applications in Energy and Security.* World Scientific, Singapore. 300 pages.

Chao, A. W. and Chou, W., Eds. 2016b. *Reviews of Accelerator Science and Technology, Volume 9: Technology and Applications of Advanced Accelerator Concepts.* World Scientific Publishing Co., Singapore.

Chao, A. W. and Chou, W., Eds. 2019. *Reviews of Accelerator Science and Technology, Volume 10:* World Scientific Publishing Co., Singapore.

Class, C. M., Price, J. E., and Risser, J. R. 1965. Angular distributions from the ^{11}B(d, n_0)^{12}C and ^{11}B (d, n_1)^{12*}C reactions for deuteron energies from 1.5 to 4.7 MeV. *Nucl. Phys.* 71(2): 433–440.

Cockcroft, J. D. and Walton, E. T. S. 1932a. Experiments with high velocity positive ions.(I) Further developments in the method of obtaining high velocity positive ions. *Proc. R. Soc.* A 136: 619–630.

Cockcroft, J. D. and Walton, E. T. S. 1932b. Experiments with high velocity positive ions. II. The disintegration of elements by high velocity protons. *Proc. R. Soc.* A 137: 229–242.

Cockcroft, J. and Walton, E. 1932c. Artificial production of fast protons. *Nature* 129: 242–24.

Cockcroft, J. D. and Walton, E. T. S. 1932d. Disintegration of Lithium by Swift Protons. *Nature* 129: 649.

Colegrove, F. D., Schearer, L. D. and Walters, G. K. 1963. Polarization of He^3 gas by optical pumping. *Phys. Rev.* 132(6): 2561–2572.

Craddock, M. K. and Symon, K. R. 2008. Cyclotrons and fixed-field alternating-gradient accelerators. *Rev. Accel. Sci. Technol.* 1(1): 65–97.

Diamond, W. T., Imahori, Y., McKay, J. W., Wills, J. S. C., and Schmeing, H. 1996. Efficient negative-ion sources for tandem injection. *Rev. Sci. Instrum* 67 (1404): 1996.10.1063/1.1146648

Duck, I. M., Valković, V., and Phillips, G. C. 1972. Off-energy shell effects in deuteron-induced deuteron break-up. *Phys. Rev. Lett.* 13 (1972): 875.

Duck, I. M., Phillips, G. C., and Valković, V. 1973. On the Possible Observation of Triangle Graph. *Proc. Int. Conf. Nucl. Phys.,* Munich (Ed.: J.de Boer and H.J.Mang), North Holland 1973 pp. 423.

Duck, I. M. and Valković, V. 1973. On the possible observation of the triangle graph through interference with pole graph. *Let. Nuovo Cim.* 8 (1973): 537.

ELSA. 2021. http://www-elsa.physik.uni-bonn.de/accelerator_list.html.

Emerson, S. T., Valković, V., Jackson, W. R., Joseph, C., Niiler, A., Simpson, W. D., and Phillips, G. C. 1971. Final state interactions in the ^{9}Be+p→d+2α reaction. *Nucl. Phys. A* 169 (1971): 317.

Faircloth, D. and Lawrie, S. 2018. An overview of negative hydrogen ion sources for accelerators. *N. J. Phys.* 20: 025007 (19 pp).

Franck, J. and Hertz, G. 1914. Über Zusammenstöße zwischen Elektronen und Molekülen des Quecksilberdampfes und die Ionisierungsspannung desselben. *Germ. Verh. Dtsch. Phys. Ges.* 16: 457–467.

Gamov, G. 1928. Quantum theory of the atomic nucleus. *Z. Phys.* 51(3–4): 204–212.

Greinacher, H. 1914. Das Ionometer und seine Verwendung zur Messung von Radium-und Röntgenstrahlen. *Phys. Z.* 15: 410–415.

Gottdang, A., Mous, D. J. W., and Haitsma, R. G. 2002. The novel HVEE 5 MV Tandetron^{TM}. *Nucl. Instrum. Methods Phys. Res. B* 190: 177–182.

Gurney, R. W. and Condon, E. U. 1928. Wave mechanics and radioactive disintegration. *Nature* 122: 439.

Hastings, J. B., Rivkin, L., and Aeppli, G. 2019. Present and future accelerator-based x-ray sources: a perspective. *Rev. Accel. Sci. Technol.* 10(1): 33–48.

Hellborg, R. (Ed.). 2005. *Electrostatic Accelerators, Fundamentals and Applications.* Springer, Berlin, Heidelberg.

Houston, W. V. 1965. Tom Wilkerson Bonner 1910–1961. *Biographical Memoirs of the NAS.* Washington, D.C: National Academy of Sciences.

Huang, H. Y., et al. 2011. Laser Shaping of a relativistic intense, short gaussian pulse by a plasma lens. *Phys. Rev. Lett.* 107 (2011): 265002.

Hyder, H. R. McK., and Helbory, R. 2005. Appendix: Electrostatic Accelerators – Production and Distribution. In Hellborg, R. Ed. 2005. *Electrostatic Accelerators, Fundamentals and Applications.* pp. 595–603. Springer-Verlag, Berlin, Heidelberg, Germany.

IAEA. 2019. Improvement of the reliability and accuracy of heavy ion beam analysis. Technical Reports Series No. 485. International Atomic Energy Agency. Vienna, Austria.

IAEA. 2020a. Accelerator Knowledge Portal. https://nucleus.iaea.org/sites/accelerators/knowledgerepository/Lists/OnlineResources/AllItems.aspx

IAEA. 2020b. Decommissioning of Particle Accelerators. IAEA Nuclear Energy Series No. NW-T-2.9. International Atomic Energy Agency, Vienna, Austria.

Inoue, M. 1998. Status of accelerators in Japan. Contribution to APAC98 Conference. https://accelconf.web.cern.ch/a98/APAC98/4B001.PDF.

Ising, G. 1924. Prinzip einer Methode zur Herstellung von Kanalstrahlen hoher Voltzahl. *Arkiv för Matematik, Astronomi och Fysik* 18(30): 1–4.

Jackson, W. R., Divatia, A. S., Bonner, B. E., Joseph, C., Emerson, S. T., Chen, Y. S., Taylor, M. C., Simpson, W. D., Valković, V., Paul, E. B., and Phillips, G. C. 1967. Method for neutron detection efficiency measurements and neutron-charged particle coincidence detection. *Nucl. Instrum. Methods* 55 (1967): 349–357.

Jackson, W. R., Valković, V., Emerson, S. T., Simpson, W. D., Joseph, C., Chen, Y. S., Taylor, M. C., and Phillips, G. C. 1971. The ^{2}H(p,np)p Reaction at 9.0 MeV. *Nucl. Phys. A* 166 (1971): 525.

James, G. S., Levy, R. H., Bethe, H. A., and Fields, B. T. 1966. On a new type of accelerator for heavy ions. *Phys. Rev.* 145(3): 925–952.

JRIA. 2016. https://www.jrias.or.jp/e/pdf/the_use_of_radiation_2016.pdf

Kean, M. 2011. Bonner Lab. Rice History Corner. Gleanings from the Rice University Archives. Posted August 2011. https://ricehistorycorner.com/2011/08/11/1694/

Kerst, D. W. 1941. The acceleration of electrons by magnetic induction. *Phys. Rev.* 60: 47–53.

Kerst, D. W. and Serber, R. 1941. Electronic orbits in the induction accelerators. *Phys. Rev.* 60: 53–58.

Lawrence, E. O. and Edlefsen, N. E. On the production of high speed protons 1930. *Science* 72: 376–377.

Lawrence, E. O. 1931. The production of high speed protons without the use of high voltages. *Phys. Rev.* 38: 834.

Lawrence, E. O. and Livingston, M. S. 1932. The production of high speed protons without the use of high voltages. *Phys. Rev.* 40: 19–35.

Lee, Y., Leung, K. N., Williams, M. D., Bruenger, W. H., Fallmann, W., Löschner, H., and Stengl, G. 1999. Multicusp ion source for ion projection lithography. Proceedings of the 1999 IEEE Particle Accelerator Conference, New York, 1999, (Eds: Luccio, A. and MacKay, W.). IEEE Catalog Number: 99CH36366. pp. 2575–2577.

Li, Z. H., et al. 2009. Experimental study of the β-delayed neutron decay of 21N. *Phys. Rev. C* 80(5): 054315.

Liebert, R. B., Zabel, T., Miljanić, Đ., Larson, H., Valković, V., and Phillips, G. C. 1973. X-Ray Production by Protons of 2.5-12 MeV. *Phys. Rev. A* 8: 2336.

Lou, J., et al. 2009. Performances of a β-delayed neutron detection array at Peking University. *Nucl. Instrum. Methods A* 606: 645–650.

Marion, J. (Ed.). 1967. *Nuclear Research With Low Energy Accelerators.* Academic Press.

Marwan, J. and Krivit, S. B., Eds. 2008. Low-Energy Nuclear Reactions Sourcebook. American Chemical Society, ACS Symposium Series, 998.

May, D. P. and Baker, S. D. 1985. The polarized ^{3}He beam on the Texas A&M cyclotron. *AIP Conf. Proc.* 131: 1–7. 10.1063/1.35327.

McCallum, G. J. (Ed.). 1980. Nuclear Physics Group Progress Report, January-June 1980. Institute of Nuclear Sciences Report INS-R-277. Department of Industrial and Scientific Research, Lower Hutt, New Zealand.

Middleton, R. 1974. A survey of negative ion sources for tandem accelerators. *Nucl. Instrum. Methods* 122: 35–43.

Miura, I., et al. 1992. Commissioning of the RCNP Ring Cyclotron. Proc. of the 13th Int. Conf. on Cyclotrons and their Applications. Vancouver, Canada, July 06–10. 1992. Paper I-01, pp. 3–10.

Mous, D. J. W., Gottdang, A., Haitsma, R. G., et al. 2003. Performance and Applications of the first HVE 5 MV Tandetron^{TM} at the university of Madrid. Applications of Accelerators in Research and Indusry: 17th Int'l Conference (EdsDuggan, J. L. and Morgan, I. L.), CP680: 999–1002.

Mous, D. J. W., Visser, J., and Haitsma, R. G. 2004a. A nanosecond pulsing system for MeV light ions using a 2 MV Tandetron^{TM}. *Nucl. Instrum. Methods Phys. Res. B* 219–220: 490–493.

Mous, D. J. W., Visser, J., Gottdang, A., and Haitsma, R. G. 2004b. A new range of high-current Tandetron^{TM} accelerator systems with terminal voltages of 1–6 MV. *Nucl. Instrum. Methods Phys. Res. B* 219–220: 480–484.

Nakano, T., et al. 2019. The research center for nuclear physics at Osaka University. *Nuclear Phys. News Int.* 29(4): 4–9.

National Research Council. 1999. *Nuclear Physics: The Core of Matter, The Fuel of Stars.* Washington, DC: The National Academies Press. 10.17226/6288.

Möller, S. 2020. *Accelerator Technology: Applications in Science, Medicine, and Industry (Particle Acceleration and Detection).* Springer Nature, Switzerland AG.

NEC. http://www.pelletron.com/resource-category/accelerators-around-the-world/

Niiler, A., Joseph, C., Valković, V., von Witsch, W., and Phillips, G. C. 1969. The p+d→p+p+n Reaction at 6.5≤Ep≤13 MeV. *Phys. Rev.* 182 (1969): 1083.

Niiler, A., von Witsch, W., Phillips, G. C., Joseph, C., and Valković, V. 1970. The d(p,d*)p cross section from the d(p,2p)n reaction. *Phys. Rev. C* 1 (1970) 1342.

Nishihashi, T., Tsuboi, H., Mihara, Y., et al. 1989. Negative ion source for the tandem accelerator. *Nucl. Instrum. Methods Phys. Res.* B37–B38: 205–207.

Ohlsen, G. G., Hardekopf, R. A., May, D. P., Baker, S. D., and Armstrong, W. T. 1974. Measurements of spin correlation and analyzing power in d-3He elastic scattering between 4 and 12 MeV. *Nucl. Phys.* A233: 1–8.

Peterson, E. L., Allas, R. G., Bondelid, R. O., Pieper, A. G., and Theus, R. B. 1970. Quasi-free scattering in the D(p,pn)p reaction from 15 to 50 MeV. *Phys. Lett. B* 31(4): 209–210.

Phillips, G. C., Duck, I. M., and Valković. V. 1972. Study of the ^{2}H(p,ppn) and ^{2}H(d,dpn) Reactions at low bombarding energies. Proc. Conf. Nucl. Structure Study with Neutrons, C.R.I.P., Budapest 1972, pp. 44–45.

Plasek, R., Miljanić, Đ., Valković, V., Liebert, R. B., and Phillips, G. C. 1973. Organic scintillator neutron detector efficiency: a comparison of experimental results with predictions. *Nucl. Instrum. Methods*. 111 (1973): 251.

Plasek, R., Valković, V., and Phillips, G. C. 1976. A study of neutron-proton final state interaction in proton induced deuteron break-up. *Nucl. Phys. A* 256 (1976): 189–204.

PMB-Alcen. 2021. https://www.pmb-alcen.com/en/systems/imigine, and https://www.pmb-alcen.com/en/accelerators/oriatron-linac.

Podaru, N. C., Gottdang, A., and Mous, D. J. W. 2013. Three dimensional finite element methods: their role in the design of DC accelerator systems. *AIP Cof. Proc.* 1525: 165–169.

Rajta, I., Vajda, I., Gyürky, Gy., Csedreki, I., Kiss, A. Z., Biri, S., Van Oosterhout, H. A. P., Podaru, N. C., and Mous, D. J. W. 2018. Accelerator characterization of the new ion beam facility at MTA Atomki in Debrecen, Hungary. *Nucl. Instrum. Methods Phys. Res. B* 880: 125–130.

Rendić, D., Holjević, S., Valković, V., Zabel, T. H., and Phillips, G. C. 1976. Trace Element concentrations in human hair by proton induced X-ray emission. *J. Investig. Dermatol.* 66 (1976): 371–375.

RnR Market Research. 2015. Global Linear Accelerator Market 2015–2021 Forecasts. http://www.rnrmarketresearch.com/contacts/purchase?rname=329342.

Rutherford, E. 1919. Collision of α particles with light atoms. IV. An anomalous effect in nitrogen. *Philos. Mag. Ser.* 6 37(222): 581–587.

Scharf, W. 1986. *Particle Accelerators and Their Uses* (Translated from the Polish by Lepa, E.). Harwood Academic Publishers.

Sen, A., Dominguez-Canizares, G., Podaru, N. C., Mous, D. J. W., Junker, M., Imbriani, G., and Rigato, V. 2019. A high intensity, high stability 3.5 MV Singletron™ accelerator. *Nucl. Instrum. Methods Phys. Res. B* 450: 390–395.

Sweeney, W. E., Valković, V., Rendić, D., and Phillips, G. C. 1971. On the Observation of re-scattering effects in the reaction ^{7}Li(d,nα)^{4}He. *Phys. Lett. B* 37 (1971): 183.

Tanaka, K. 2020. Major Accelerator Facilities for Nuclear Physics in Asia Pacific. Proc. 13th Int. Conf. on Nucleus-Nucleus Collisions, JPS Conf. Proc. 32: 010001 (8 pp).

Taylor, M. C., Valković, V., and Phillips, G. C. 1972. Study of the ^{9}Be(^{3}He,αα)^{4}He Reaction. *Nucl. Phys. A* 182 (1972): 558.

Thomas, A. W., Stuchbery, A. E., Liu, W., et al. 2020. Ten years of the Asian nuclear physics association (ANPhA) and major accelerator facilities for nuclear physics in the Asia Pacific Region. *Nuclear Physics News* 30(3): 3–45.

Valkovic, V. 2015. *14 MeV Neutrons – Physics and Applications*. CRC Press, Taylor & Francis Group, Boca Raton, London, New York.

Valković, V., Emerson, S. T., Jackson, W. R., and Phillips, G. C. 1967. Study of Some Effects in Reactions with Three Outgoing Particles:d+^{7}Li→n+α+ α and p+^{6}Li→p+d+ α. Proc. Internat. Conf. Nuclear Physics, Gotlinburg, Academic Press (1967) pp. 989–993.

Valković, V., Joseph, C., Niiler, A., and Phillips, G. C. 1968. Rescattering Effects in Reactions with Three Particles in the Final State. *Nucl. Phys.* A116 (1968): 497–515.

Valković, V., Joseph, C., Niiler, A., and Phillips, G. C. 1968. Ob Interferecionih effektah i javljenij pererasejanija v jadernih reakcijah s tremja casticama v konecnom sostojanii. *Izvestija Akad. Nauk SSSR* **32** No.12 (1968) 1976-1989; English translation: *Bull. Acad. Sci. USSR* **32** (1969)1820.

Valković, V., von Witsch, W., Rendić D., and Phillips, G. C. 1970. Proton-Proton Quasi-Free Scattering in the p+d→p+p+n Reaction for Ep = 4.5-13.0 MeV. *Phys. Lett. B* 33 (1970): 208.

Valković, V., Rendić, D., Otte, V. A., von Witsch, W. and Phillips, G. C. 1971a. Nucleon- nucleon quasi-free scattering in the p+d→p+p+n reaction at low bombarding energies. *Nucl. Phys. A* 166 (1971): 547.

Valković, V., Rendić, D., Otte, V. A., and Phillips G. C. 1971b. Comparison of p-p and n-p Quasi-Free Scattering in p+d→p+p+n Reaction. *Phys. Rev. Lett.* 26 (1971): 394.

Valković, V., Duck, I. M., and Phillips G. C. 1972. Coulomb effects in deuteron-induced deuteron break-up. *Phys. Lett. B* 42 (1972): 191.

Valković, V., Miljanić, Đ., Wheeler, R. M., Liebert, R. B., Zabel, T., and Phillips G. C. 1973. Variation in trace element concentrations along single hairs as measured by proton-induced X-ray spectroscopy. *Nature* 243 (1973): 543.

Valkovic, V. and Zyszkowski, W. 1994. Accelerators in science and industry: focus on the Middle East & Europe. *IAEA Bullet.* 1/1994: 24–29.

Valković, V., Liebert, R. B., Zabel, T., Larson, H. T., Miljanić, Đ., Wheeler R. M., and Phillips, G. C. 1974. Trace element analysis using proton induced X-ray emission spectroscopy. *Nucl. Instrum. Methods* 114 (1974): 573.

Valković, V., Liebert, R. B., Wheeler, R. M., Plasek, R., Zabel, T., and Phillips, G. C. 1974. Neutron-proton coincidences from $^{12}C(^{16}O,np)^{26}Al$ reaction. *Lett Nuovo Cimento* 10 (1974): 461.

Valković, V., Miljanić, Đ., Liebert, R. B., and Phillips, G. C. 1975. Energy dependence of the cross sections for the $d+^{10}B\rightarrow3\alpha$ reaction. *Nucl. Phys. A* 239 (1975): 260.

Valković, V., Zabel, T., and Phillips, G. C. 1976. *Trace Element Analysis in Biological Materials by Proton Induced X-Ray Emission Spectroscopy*. Proc. Int. Conf. Physics in Industry, Dublin (1976).

Valković, V., and Phillips, G. C. 1976. Trace Element Analysis by Proton Induced X-Ray Emission. Proc. IV Conf. on Application of Small Accelerators, Denton, Texas, 1976. (Eds.J. L. Duggan and I. L. Morgan) pp. 102–105.

Valković, V., Rendić, D., Biegert, E. K., and Andrade, E. 1979. Trace element concentrations in tree rings as indicators of environmental pollution. *Environ. Int.* 2 (1979): 27–32.

Van de Graaff, R. J. 1931. A 1,5000,000 Volt electrostatic generator. *Phys. Rev.* 38: 1919–1920.

Visser, J., Mous, D. J. W., Gottdang, A., and Haitsma, R. G. 2005. Considerations on accelerator systems requirements and limitations for µ-probe applications. *Nucl. Instrum. Methods Phys. Res. B* 231: 32–36

von Witsch, W., Ivanovich, M., Rendić, D., Valković, V., and Phillips, G. C. 1972. Decay of ^{12}C via the $^{11}B(p,2\alpha)$ Reaction. *Nucl. Phys. A* 180 (1972): 402.

Vretenar, M. 2013. The radio-frequency quadrupole. CERN Yellow Report CERN-2013-001, pp.207-223, see also arXiv:1303.6762v1 [physics.acc-ph].

Widerøe, R. 1928. Über ein neues Prinzip zur Herstellung hoher Spannungen. *Arch. Elektrot.* 21: 387–406.

Widerøe, R. 1984. Some memories and dreams from the childhood of particle accelerators. *Europhys. News* 15 (2): 9–11.

Wright, G., Barnard, H. S., Hartwig, Z.S., Stahle, P. W., Sullivan, R. M., Woller, K., and Whyte, D. G. 2011. Plasma-surface interaction research at the Cambridge Laboratory of Accelerator Studies of Surfaces. *AIP Conf. Proc.* 1336: 626–630. 10.1063/1.3586178.

Wu, J. 2013. State key laboratory of nuclear physics and technology at Peking University. *Nucl. Phys. News* 23(4): 5–9.

Xia, F., et al. 2008. Gating of single synthetic nanopores by proton-driven DNA molecular motors. *J. Am. Chem. Soc.* 130 (26): 8345–8350.

You, H. B., et al., 2013. Construction and calibration of the multi-neutron correlation spectrometer at Peking University. *Nucl. Instrum. Methods A* 728: 47–52.

Zhang, G. H., et al. 2011. Sm-149 (n, alpha) Nd-146 cross sections in the MeV region. *Phys. Rev. Lett.* 252502 (5 pages).

Additional reading

Livingston, M. S. 1969. *Particle accelerators: A brief history*. Harvard University Press, Cambridge, MA.

Miljanić, Đ., Zabel, T., Liebert, R. B., Phillips, G. C., and Valković, V. 1973. Quasi-Free scattering in the $^6Li(d,dd)^4He$ reaction at low bombarding energies. *Nucl. Phys. A* 215 (1973): 221.

Moore, C. P., Coker, W. R., Valković, V., Joseph C., and Sandler, J. 1967. Coincidence Spectra of the $^{92}Mo(d,np)^{92}Mo$ reaction. *Phys. Lett. B* 25 (1967): 468–469.

Niiler, A., Joseph, C., Valković, V., Spiger, R., Canada, T., Emerson, S. T., Sandler, J., and Phillips, G. C. 1969. Proton-proton bremsstrahlung at Ep=10 MeV. *Phys. Rev.* 178 (1969): 1621.

Rendić, D., Gabitzsch, N. D., Valković, V., and Phillips, G. C. 1971. d+^{11}B→3α+n Four-body break-up states in ^{9}Be. *Nucl. Phys. A* 178 (1971): 49.

Stock, R. 2013. *Encyclopedia of Nuclear Physics and its Applications.* Wiley & Sons.

Valković, V., Jackson, W. R., Chen, Y. S., Emerson, T., and Phillips, G. C. 1967. Three body break-up in ^{7}Li(d;n, α)^{4}He and ^{7}Li(d; α,α)n Reactions. *Nucl. Phys.* A96 (1967) 241–257.

Valković, V., Joseph, C., Emerson, S. T., and Phillips, G. C. 1967. Two particle coincidence spectra from the p+^{6}Li→p+d+ α reaction. *Nucl. Phys.* A106 (1967): 138–160.

Valković, V., Duck, I. M., Sweeney, W. E. Andrade, and Phillips, G. C. 1971. Nucleon-deuteron quasi-free scattering in the d+d→d+p+n reaction at low bombarding energies. *Phys. Rev. C* 4 (1971): 2289.

Valković, V., Duck, I., Sweeney, W. E., and Phillips, G. C. 1972. Three body break-up in the d+d→d +p+n reaction. *Nucl. Phys. A* 183 (1972): 126.

Valković, V., Gabitzsch, N., Rendić, D., Duck, I. M., and Phillips, G. C. 1972. Coincidence spectra from d+d→d+p+n and p+d→p+p+n reaction with neutron detection at zero degree. *Nucl. Phys. A* 182 (1972): 225.

Valković, V., Liebert, R. B., Plasek, R., and Phillips, G. C. 1973. (^{16}O,np) and (^{16}O,nα) Reactions on Some Light Nuclei. Proc. Int. Conf. Nucl. Phys., Munich (Ed.: J. de Boer and H. J. Mang), North Holland, pp. 412.

Valković. V. 1973. X-ray emission spectroscopy. *Contemp. Phys.* 14 (1973) 415–438: Part I; 439–462: Part II.

Valković, V., Liebert, R. B., and Phillips G. C. 1974. (^{16}O,np) and (^{16}O,nα) reactions on some light nuclei. *Nucl. Instrum. Methods* 122 (1974): 533.

Valković, V., Rendić, D., and Phillips, G. C. 1975. Elemental ratios along human hair as indicators of exposure to environmental pollutants. *Environ. Sci. Technol.* 9 (1975): 1150.

Wheeler, R. M., Liebert, R. B., Zabel, T., Chaturvedi, R. P., Valković, V., Phillips, G. C., Ong, P. S., Cheng, E. L., and Hrgovčić, M. 1974. Techniques for trace element analysis: X-ray fluorescence, X-ray excitation with protons and flame atomic absorption. *Med. Phys.* 1 (1974): 68.

2

Medical and Biological Applications of Low Energy Accelerators

2.1 Introduction

One of the most exciting developments in accelerator technology is the use of carefully shaped hadron beams (protons and heavier ions) to kill cancer cells. Proton therapy is already well-established, while carbon-ion therapy is also now being taken forward in several radiation therapy centers. The precision in dose delivery makes hadron therapy ideal for treating resistant tumors located in sensitive tissues. Other candidates for such a therapy, i.e., oxygen and helium ions, are also being studied. Diagnostic imaging using injected radioactive tracers is a well-established clinical procedure. Improved imaging methods are continually being explored, together with new isotopes giving better resolution, and advanced ion-beam technology that can combine imaging and therapy. A huge variety of radioisotopes is produced in nuclear facilities for injected therapeutic procedures. There is currently particular interest in highly targeted therapy combining isotopes with selected antibodies, and the development of pairs of isotopes of the same element that allow imaging and therapy to be carried out simultaneously – "theranostics".

Charged particle beams are used in medical institutions for both diagnostics and therapeutical applications. Diagnostic applications include the use of nuclear analytical techniques for element analysis, the use of different radioisotopes and especially the use of positron emitters. Therapeutical applications are not limited only to radiotherapy, but also include a broad spectrum of other activities (from the use of special materials to performing surgery). Medical applications of low energy accelerators include:

- The use of nuclear analytical methods and procedures for laboratory studies and routine measurements.
- Material productions and modifications to meet special requirements.
- Radioisotope productions and their applications in radiopharmaceuticals as well as positron emission tomography (PET).
- Radiotherapy with ions, based on improved understanding of the interaction of charged particles with living tissue.

Particle therapy is the expanding radiotherapy treatment option of choice for cancer. Its cost, however, is currently hindering its worldwide expansion. In addition, the ideal application of particle therapy is restricted by a series of unsolved technical challenges. Both the cost and technical limitations are directly traceable to dependence on the legacy of accelerators and their associated treatment possibilities. This chapter is written to

DOI: 10.1201/9781003033684-2

address these needs. First, a technical overview is presented of photon and particle therapy for cancer tumors. Second, the underlying limitations of the existing legacy systems are identified, especially those related to accelerators, and suggestions are made for current and future developments to address these shortcomings. The legacy systems referred to here are of the slow scanning variety using large circular accelerators.

The paper by Myers et al. (2019) makes a scientific comparison of the various types of accelerators currently used or being developed for particle therapy. The following procedure is pursued to perform a comparison between various types of accelerators:

- The parameters, which are pertinent to particle therapy accelerators ("specified parameters"), are identified from clinical efficiency and overall cost considerations.
- The range and values of "specified parameters" associated with each type of particle therapy accelerator are identified.
- A comparison is made on the best match between the various types of accelerators for each of the "specified parameters", i.e., the best in class accelerator, when compared to each criterion.
- Based on this match, an overall conclusion is made on the type of accelerator, which best fits, the needs for particle therapy.

The planning of cancer treatments is hampered by the fact that – depending on their cell biochemistry – patients respond differently to chemo- or immunotherapy. Nuclear physics can offer a more personalized approach known as theranostics. Using "matched pairs" of diagnostic and therapeutic isotopes (e.g., ^{64}Cu and ^{67}Cu, that combine with the same targeting vector), clinicians can tailor the radiation dose needed to maximize success. Nuclear physicists identified and produced a matched quartet of terbium isotopes (Müller et al. 2012), which provide a set of decay characteristics producing excellent tumor visualization and therapeutic efficacy. Theranostics is also possible in teletherapy, using high-energy proton beams for simultaneous proton radiography and treatment. Work to establish economic methods of production is underway.

Establishing a nuclear medicine facility is a major undertaking that requires careful planning, contributions from multiple stakeholders, the support and approval of the relevant authorities, secure funding and a detailed implementation strategy. Detailed strategic planning is particularly important in developing countries, where nuclear medicine may currently be unavailable, and the benefits and complexities of nuclear medicine imaging and therapy may not be clearly appreciated (IAEA 2020b). The accreditation of staff and their departments, with full documentation of procedures to international standards, will soon become a requirement, and this need is addressed in an International Atomic Energy Agency (IAEA) publication on quality management (IAEA 2015). This publication takes a systematic approach to the needs for nuclear medicine practice with regard to assessment, premises, human resources, equipment and quality assurance and quality control, medical physics and radio pharmacy support, radiation protection and safety and clinical applications.

2.2 Radioisotope Production

More than two-thirds of all radioactive nuclei known to man were discovered via accelerator-induced nuclear reactions. Reactors today manufacture many of them, but accelerator-produced isotopes are increasing in volume. For example, a cyclotron produces a range of radioisotopes for medical applications that is not available from a nuclear reactor. These radioisotopes can provide a better understanding of the chemical processes through which human diseases develop. Some radioisotopes have very short half-lives, measured in minutes, and must therefore be produced close to where they will be used (hospital). Radioisotopes, which have a longer half-life, can be produced on a routine basis by the cyclotron to provide nuclear medicine departments with the resources vital to the early diagnosis and treatment of a wide range of medical conditions, including cancer, coronary artery disease, stroke, severe trauma, epilepsy, asthma and infection.

IAEA is running an Accelerator Knowledge Portal, which contains the database of cyclotrons for radionuclide production. This database was created as a follow-up action to the older hard-copy "Directory of Cyclotrons" developed in 1983 and updated in 1998 and 2006 by the "Radioisotope Products and Radiation Technology Section, Division of Physical and Chemical Sciences, IAEA" and international experts. The database was established (IAEA 2020) and is currently under revision in response to the request of the Member States and worldwide interest in the installation and application of cyclotrons for medical radioisotope production. As of 3 June 2020 it contained 1278 entries. An interactive world map displaying the location of cyclotron facilities is also available. The number of facilities can be displayed on a country and city basis. Table 2.1 shows the list of countries having the highest number of cyclotrons dedicated to medical radioisotope production.

In addition, 5–7 cyclotrons for medical radioisotope production have the following countries: Egypt, Israel, Kazakhstan, Malaysia, Norway, Vietnam, Columbia, Czech Republic, Hungary, Switzerland and Pakistan. The list would not be completed without mentioning that additional 54 countries have 1–4 cyclotron facilities.

About 50 radioisotopes are used in medical diagnostic and treatment, in single-photon emission computed tomography (SPECT), PET and brachytherapy. Although cyclotrons are the machine of choice for radioisotope production, some linacs are also used. Both proton and deuteron beams are used. For PET, isotopes energies from 7 to 18 MeV and beam currents <200 µA are required, while for SPECT energies from 22 to 70 MeV with currents up to 2 mA are recommended (Table 2.2).

TABLE 2.1

Countries with the highest number of cyclotrons

Country	Number of cyclotrons	Country	Number of cyclotrons
USA	249	Australia	19
Japan	218	Brazil	14
China	175	Netherlands	13
Russia	60	Belgium	13
South Korea	51	Taiwan	11
Italy	46	Saudi Arabia	11
Germany	43	Iran	10
France	31	Finland	10
Canada	28	Denmark	10
UK	27	Sweden	9
India	25	Poland	9
Spain	21	Argentina	9
Turkey	20	Mexico	8

TABLE 2.2

Some common PET radionuclides

Isotope	Half-life	Positron energy (MeV)
^{11}C	20 minutes	0.385
^{13}N	10 minutes	0.492
^{15}O	2 minutes	0.735
^{18}F	110 minutes	0.250
^{38}K	8 minutes	1.216
^{62}Cu	10 minutes	1.315
^{64}Cu	12.7 hours	0.278
^{68}Ga	68.1 hours	0.836, 0.352
^{82}Rb	1.3 minutes	1.523, 1.157
^{124}I	4.2 days	1.691, 1.228
		1.509, 1.376

Here, we shall present some examples of research and development in this field. Nye et al. (2005) described radio halogen targetry as done at the University of Wisconsin. They have developed new target systems for the production of ^{18}F, ^{76}Br and ^{124}I. These new systems include the successful replacement of a double-foiled silver water target for the production of $[^{18}F]$-F⁻ with a niobium body, isolated from the cyclotron vacuum by a single grid supported entrance foil. The ^{18}F-fluoride target worked for over two years with maintenance-free operation. A new solid target system for the production of long-lived halogens, with specific reference to ^{124}I, was also introduced.

Johnson et al. (2005) also describe radioisotope production targets and modules. They are able to supply full radioisotope production systems that incorporate the accelerator, the beam lines, the targets and the radiochemistry in a unified package. The key component improvements include higher beam currents, more robust production targets and efficient radio synthesis modules.

The "classical" positron-emitting radionuclides include ^{15}O, ^{13}N and ^{11}C that possess unique properties for medical imaging. They are radionuclides of the fundamental elements of biological matter. They each possess short half-lives, which allow their use in designed radiotracers for clinical investigations with minimal risk, and they are readily able to be produced in sufficient activities by low-energy nuclear reactions. At present, several accelerator manufacturers offer production packages for these radionuclides emphasizing targetry with consideration of the cyclotron-extracted energies for nuclide production and online chemistry systems for the continuous production of specific precursors or radiotracers.

Following the installation and acceptance of the new cyclotron with the chemistry module for the preparation of ^{15}O labeled water, Finn et al. (2005) were forced to examine the design and the operation of the synthetic unit with a view toward the state of New York's regulations addressing the environmental pollution from radioactive materials. The chemistry module had to be refined with subtle modifications to ensure regulatory compliance.

Some time ago, Brookhaven National Laboratory (BNL) has purchased a new cyclotron for routine isotope production. Schueller et al. (2005) describe a system built to recover ^{18}O-enriched water from the ^{18}F target, and transport of the ^{18}F 50 m to a shielded dose splitter. The remoted system provides the operator with feedback on flow rates and radiation levels during processing. Recovery of the 2.6 ml of enriched water and transport of the ^{18}F to the radiochemistry labs takes under 10 minutes, with more than 80% of the activity arriving in the chemistry lab. Along the same line, Dehnel et al. (2005) describe beamline developments in commercial cyclotron facilities. This is of importance to radioisotope producers who have built their businesses around 30 MeV cyclotrons. An important aspect of these production facilities is the beamline system which must have a high transmission rate, low residual radiation and low maintenance requirements.

Therapeutic radiopharmaceuticals play a major role in today's nuclear medicine with a positive impact on the diagnosis and treatment of diseases. One area of application is radiation synovectomy (RSV). Previously, RSV agents were often simple colloids. More recently, matrixes labeled with short/medium range beta emitters have been developed. However, the lack of generic and peer-reviewed production, quality control as well as clinical application guidelines and recommendations, are a major concern for their application in patients. The publication (IAEA 2021) presents recommendations and suggestions for production, quality control and quality assurance procedures for its member state laboratories in charge of radiopharmaceutical production, with a focus on the latest RSV agents. It also proposes standard operating procedures for RSV application in patients.

Based on longstanding expertise, IBA RadioPharma Solutions supports hospitals and radiopharmaceutical distribution centers in two ways: with their in-house radioisotopes production; and by providing global solutions, from project design to the operation of their facility. In addition to high-quality technology production equipment (cyclotron solutions, targetry systems, synthesizers, control systems, etc.), IBA has developed in-depth experience in setting up current good manufacturing practice (cGMP) radio-pharmaceutical production centers.

IBA's Cyclone® KIUBE cyclotron (shown in Fig. 2.1) offers the highest production capacity enabling increased diagnostic capabilities. The Cyclone® KIUBE produces the widest range of radioisotopes, enabling it to produce FDG (the most commonly used radiopharmaceutical for cancer diagnosis), ^{68}Ga for the diagnosis of neuroendocrine tumors, and ^{64}Cu for a more accurate diagnosis of prostate cancer.

IBA's Cyclone® KIUBE cyclotron is a fixed-energy cyclotron that accelerates negative ions up to 18 MeV. It offers the highest production capacity with a PET cyclotron and thus enabling increased diagnostic capabilities. The Cyclone® KIUBE produces the widest range of radioisotopes; ^{18}F for oncology, neurology and cardiology imaging, ^{68}Ga for prostate cancer imaging (^{68}Ga-PSMA) and for neuroendocrine tumors imaging, ^{64}Cu for prostate cancer and many other radioisotopes. Detailed information is available at https://www.iba-radiopharmasolutions.com/resources that among other information contains a large list of relevant publications, see some of them: van der Meulen et al. (2020), Košťál et al. (2019), Allonia et al. (2019) and Sadeghi et al. (2014). For additional reviews on cyclotrons and their use for radioisotope production see also Schmor (2011), IAEA (2008) and Schaeffer et al. (2015).

2.3 Positron Emission Tomography

In nuclear medicine imaging, gamma cameras and positron emission scanners detect and form images from the radiation emitted by the radiopharmaceuticals. There are several techniques of diagnostic nuclear medicine: (a) Gamma camera performs both scintigraphy, as a 2-D images. (b) SPECT as a 3-D tomographic technique that uses data from

FIGURE 2.1
IBA's Cyclone® KIUBE. (Courtesy IBA.)

many projections and can be reconstructed in different planes. (c) PET uses coincidence detectors to image annihilation photons derived by positron-emitting radiopharmaceuticals. (d) Multimodality imaging exploits SPECT and PET images superimposed to computed tomography (CT) or MRI for a detailed anatomical localization. This practice is often referred to as hybrid imaging.

In recent years, cadmium zinc telluride (CZT) detectors have been introduced as an alternative to traditional NaI(Tl) crystals coupled with photomultiplier tubes, and now are available in large field of view SPECT systems as well. Due to the absence of the PMTs, CZT systems have thinner and lighter heads and allow for a very close positioning of the detector to the patient, resulting in an excellent spatial resolution. Moreover, the good energy resolution of CZT makes possible an increased sensitivity and scatter rejection compared to conventional NaI(Tl)-based cameras. Despite the different solutions adopted in terms of detectors, configuration and choice between dedicated or general-purpose systems, state-of-the-art SPECT cameras have in common the potential for substantial increase in count sensitivity with no loss of spatial resolution. This results in the potential for acquiring a SPECT scan with a standard activity in a fraction of the time, or, alternatively, in taking advantage of the higher sensitivity for reducing the injected activity, while maintaining a good image quality. Iterative reconstruction algorithms that may include accurate modeling of the detector–collimator system, making possible a recovery of spatial resolution and thus improved image quality (IAEA 2015), further enhance these features.

PET is a non-invasive procedure that is used for imaging tissues and organs of the body and to monitor their functioning. It is based on the in-vivo detection of positron-emitting radioisotopes, which are introduced as tracers into the organ, or tissue of interest. The PET technique offers the unique possibility of studying metabolic and physiologic processes *in vivo* in humans and animals, without disturbing the investigated system, due to its non-invasive properties. This non-invasive imaging technique is based on the use of compounds of exogenic or endogenic origin labeled with short-lived positron-emitting radionuclides such as ^{11}C ($t_{1/2}$ = 20.4 min), ^{13}N ($t_{1/2}$ = 10 min), ^{15}O ($t_{1/2}$ = 2 min) and ^{18}F ($t_{1/2}$ = 110 min). PET has evolved as a major non-invasive technique for *in vivo* studies of biochemical and physiological processes. Apart from being a valuable research tool, PET has also become important in clinical applications where it can be used to distinguish between normal and diseased states.

The facility, a PET center, can be located either within the clinical area of the hospital or at a nearby institution. A convenient accelerator for this goal is a cyclotron, which can produce a range of radioisotopes for medical applications, which are not available from a nuclear reactor. These radioisotopes can provide a better understanding of the chemical processes through which human diseases develop. Some radioisotopes have very short half-lives, measured in minutes, and must therefore be produced close to where they will be used.

PET and PET/CT are emerging as important imaging techniques and their popularity is growing within the medical community. However, PET has been a useful research tool for many decades its real growth into clinical applications has occurred in the last one decade or so. Currently, its major use is in oncologic imaging. However, it has a multitude of clinical applications in cardiology, neurology and psychiatry as well. In oncologic imaging, a major advantage of PET is that a single whole-body examination can provide accurate assessment of disease activity and spread. PET/CT amalgamates the functional information of PET with the structural details of the CT scan, thus greatly aiding in accurate staging, therapy response assessment and early detection of recurrent disease (Anand et al. 2009).

PET is a diagnostic imaging procedure used regularly to acquire essential clinical information. The PET-CT hybrid, which consists of two scanning machines: PET scanner and an X-ray CT. At present, these represent the technological hierarchy of Nuclear Medicine, occupying an important position in diagnostics. In fact, PET-CT has the capability to evaluate diseases through a simultaneous functional and morphostructural analysis. This allows for an earlier diagnosis of the disease state, which is crucial for obtaining the required information to provide a more reliable prognosis and therapy. Presently, the most frequently used PET radiotracer fluorodeoxyglucose (^{18}FDG) has a major role in oncology. Useful information is being regularly obtained by using both ^{18}FDG and a selection of radiotracer compounds to evaluate some of the most important biological processes (Kitson et al. 2009).

The Biomarker Generator produces PET Biomarkers and Radioisotopes "On Demand" and encompasses an entire PET production lab in a 30 m^2 room. It includes a self-shielded micro-cyclotron for producing the positron-emitting isotopes of ^{18}F and ^{11}C and a microchemistry system for labeling specific molecules with the positron-emitting isotopes. The uniqueness of this system is that it is much smaller than conventional PET cyclotrons, easier to install and simple to operate (see http://www.amv-europe.com).

In cardiology, a PET scan of the heart is a non-invasive nuclear imaging test using radioactive tracers. It is used to diagnose coronary artery disease and damage following a heart attack. PET scans are also used to define the best therapy treatment. Major technological breakthroughs were achieved in the diagnosis of coronary heart disease through PET. IBA's 70 MeV cyclotron enables the production of ^{82}Rb while the Cyclone® KIUBE produces ^{13}N-Ammonia – both are used for non-invasive myocardial perfusion tests.

The evaluation of brain functionality with PET molecular imaging is playing an increasingly important role in the positive diagnosis of neurodegenerative diseases, in particular dementias and Parkinsonian syndromes. Amyloid PET imaging offers a diagnostic accuracy of 90% in the diagnosis of Alzheimer's disease. Several tracers have received marketing approval for this indication, including ^{18}F-florbetaben.

For an extensive list of diagnoses, tracers and radiotherapeutics see Table 2.3 (after https://www.iba-radiopharmasolutions.com/radioisotopes).

A number of reviews of the clinical application of both PET and PET/CT can be found in the literature, see for example: Brady et al. 2008; Costouros and Hawkins 2009; Kim et al. 2013; Mahajan and Cook 2017; Jadvar 2021.

2.4 Radiotherapy

Stereotactic radiosurgery (SRS) uses many precisely focused radiation beams to treat tumors and other problems in the brain, neck, lungs, liver, spine and other parts of the body. It is not surgery in the traditional sense because there is no incision. Instead, SRS uses 3D imaging to target high doses of radiation to the affected area with minimal impact on the surrounding healthy tissue. Like other forms of radiation, SRS works by damaging the DNA of the targeted cells. The affected cells then lose the ability to reproduce, which causes tumors to shrink.

There are several places where one can find up-to-date information about radiotherapy. Here is an organization for those interested in proton, light ion and heavy charged particle radiotherapy: Particle Therapy Co-Operative Group.

Particle therapy facilities in clinical operation (last update: February 2021). Information on technical equipment is available. Also, one can log in to get the newest data on patients treated/statistics: https://www.ptcog.ch/index.php/facilities-in-operation.

The IAEA maintains a register of radiotherapy hospitals and clinical institutions having radionuclide and high-energy teletherapy machines: IAEA, Directory of Radiotherapy Centers (DIRAC), https://www.iaea.org › resources › databases › dirac; see also IAEA, 2020a.

It is likely that, unless dramatic progress is made in cancer prevention or cure, radiotherapy (i.e., the selective destruction of cancer tissues by the use of ionizing radiations) will remain one of the pillars in cancer therapy. The most common irradiation facilities in hospitals nowadays are still ^{60}Co sources and electron accelerators (e.g., Japan has more than 620 linear electron accelerators devoted to medical applications). However, among the

TABLE 2.3

Diagnosis tracers and radiotherapeutics https://www.iba-radiopharmasolutions.com/radioisotopes

Function/organ/disease	Diagnosis tracers	Radio-therapeutics
Whole body		
Hypoxia imaging	^{18}F, ^{64}Cu	–
Infectious diseases	^{99m}Tc, ^{67}Ga	–
Primary tumors metastases	^{18}F, ^{11}C	^{225}Ac
Brain	^{18}F, ^{99m}Tc, ^{123}I, ^{13}N, ^{15}O, ^{11}C, ^{68}Ga, ^{64}Cu	^{67}Cu
Stroke imaging	^{99m}Tc, ^{18}F, ^{15}O	–
Epilepsy	^{123}I, ^{18}F	–
Alzheimer's disease	^{18}F, ^{11}C	–
Parkinson diseases	^{18}F, ^{123}I	–
Tauopathies	^{18}F	–
Leukemia	–	^{225}Ac
Non-Hodgkin Lymphoma	^{111}In	–
Lung ventilation	^{68}Ga, ^{99m}Tc	–
Lung perfusion imaging	^{99m}Tc	–
Salivary gland imaging	^{99m}Tc	–
Lachrymal tract imaging	^{99m}Tc	–
Imaging of the thyroid	^{123}I, ^{99m}Tc, ^{201}Tl	–
Blood studies	^{99m}Tc	–
Cardiac diseases	^{99m}Tc, ^{201}Tl, ^{13}N, ^{15}O, ^{18}F, ^{82}Rb	–
Breast cancer	^{99m}Tc, ^{18}F, ^{89}Zr, ^{64}Cu	^{67}Cu
Spleen diseases	^{99m}Tc	–
Biliary function	^{99m}Tc	–
Bone scintigraphy	^{99m}Tc, ^{18}F	–
Cervix cancer	^{99m}Tc, ^{64}Cu	^{67}Cu
Deep vein thrombosis	^{99m}Tc	–
Liver imaging	^{99m}Tc, ^{13}N, ^{18}F	–
Gastro-intestinal absorption	^{99m}Tc	–
Neuroendocrine tumors	^{99m}Tc, ^{111}In, ^{18}F, ^{68}Ga, ^{64}Cu	^{67}Cu
Renal filtration studies	^{99m}Tc, ^{123}I. ^{89}Zr, ^{124}I	
Adrenal scintigraphy	^{99m}Tc, ^{18}F, ^{123}I	
Bladder imaging	^{99m}Tc	
Prostate cancer	^{18}F, ^{68}Ga, ^{11}C, ^{111}In, ^{64}Cu	^{225}Ac, ^{67}Cu

possible radiations usable in radiotherapy (X-rays, γ-rays, electrons, protons, heavy ions, π-mesons and neutrons) it is generally agreed that high-energy protons exhibit the best ballistic specificity, i.e., the best ratio of the dose delivered into the tumor, compared to the dose delivered to neighboring tissues. Particle therapy is being applied at more than twenty locations around the globe. Neutron therapy has also been performed for many years and the new hope is with boron neutron capture therapy (BNCT) in which a tumor is loaded with a boron-dope compound and then irradiated by epithermal neutrons.

The proton radiotherapy becomes a treatment method of growing importance. The unique properties of protons, which lose their energy forming a Bragg curve without nuclear fragmentations, enable a precise delivery of a high dose of radiation to the tumor region and the simultaneous spare of critical organs and healthy tissues. The steep distal fall-off of proton beams offers the best chance to avoid unnecessary radiation exposure in the treatment of uveal melanoma, the most common human intraocular tumor in adult patients.

The earliest treatment for uveal melanoma was the removal of the eye. Today, enucleation has been supplanted by radiotherapy as the standard of care for patients with uveal melanoma, offering patients preservation of an intact eye and, in many cases, preservation of visual function. Both, brachytherapy and external beam irradiation are used, choice of treatment depends on the availability of treatment facilities, and recommended proton facilities have been often lacking. According to Gragoudas (2006), the introduction of radiotherapy represented an important advance in the treatment of uveal melanoma, and today radiotherapeutic plaques and external beam irradiation are the most commonly used therapeutic modalities. The physical properties of proton beams make it possible to deliver high doses of radiation to the tumor with relative sparing of adjacent tissues.

There are more than 20 proton centers worldwide, including facilities in Canada, Europe, Japan, South Korea, South Africa and the United States. In the United States, the first facility to treat uveal melanomas (Gragoudas et al. 1977) was the Harvard Cyclotron (HCL) in Cambridge, Massachusetts. In April 2002, the Northeast Proton Therapy Center (NPTC) at Massachusetts General Hospital in Boston became operational, and treatments were transferred from the HCL to the NPTC. At this new facility, more than 1000 patients could be treated per year, and one of three treatment rooms is dedicated to the treatment of eye tumors. The Crocker Nuclear Laboratory at the University of California, Davis, and the Loma Linda University Medical Center also offer proton therapy. In 2004, the Midwest Proton Radiotherapy Institute began operations, and by the end of 2006, two additional proton centers became available (at Shands Medical Center, Jacksonville, Florida, and M. D. Anderson Cancer Center, Houston, Texas) to treat patients with eye tumors and other malignancies.

Proton beam radiotherapy (PBT) of uveal melanoma can be administered as primary treatment, as salvage therapy for recurrent tumor, and as neoadjuvant therapy prior to surgical resection (Damato et al. 2013). Kim et al. (2018) evaluated the clinical outcomes and complications of proton beam therapy (PBT) in a single institution in Korea and quantitatively analyzed the change in tumor volume after PBT using magnetic resonance imaging (MRI). The local control rate and complication profile after PBT in patients with choroidal melanoma in Korea were comparable with those reported in a previous PBT series. The change in tumor volume after PBT exhibited a gradual regression pattern on MRI.

According to Mishra and Daftari (2016), PBT of uveal melanoma and other malignant and benign ocular tumors has shown tremendous development and success over the past four decades. Proton beam is associated with the lowest overall risk of local tumor recurrence in uveal melanoma, compared with other eye-conserving forms of primary treatment. Proton beam is also utilized for other malignant and benign tumors as primary, salvage or adjuvant treatment with combined modality therapy. The physical characteristics of proton therapy allows for uniform dose distribution, minimal scatter and sharp dose fall off making it an ideal therapy for ocular tumors in which critical structures lay in close proximity to the tumor. High radiation doses can be delivered to tumors with relative sparing of adjacent tissues from collateral damage. Proton beam therapy for ocular tumors has resulted in overall excellent chances for tumor control, ocular conservation and visual preservation.

Lane et al. 2015 included in their analysis a group of 3088 patients with choroidal and ciliary body tumors who were subsequently treated with proton beam therapy in the period January 1975 to December 2005. In this large series of patients with ocular

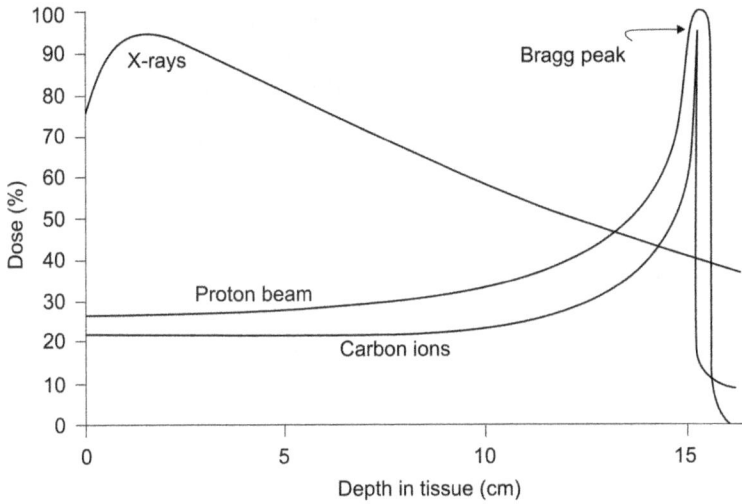

FIGURE 2.2
Depth-dose curves for X-rays and charged particles (protons and carbon).

melanoma treated conservatively with proton beam irradiation, the cumulative melanoma-related mortality rates continued to increase up to 23 years after treatment. Annual rates decreased considerably (to <1%) 14 years after treatment. Information regarding the long-term risk of dying of uveal melanoma may be useful to clinicians when counseling patients.

As it can be seen in Fig. 2.2, particle therapy exploits a favorable depth–dose curve. Charged particles deposit most of their initial energy when they slow down, close to the end of the range (Bragg peak) within the tumor target, while X-ray energy decreases exponentially with dose. Particles at high energy (entrance, normal tissue) have linear energy transfer (LET) similar to X-rays, while the LET becomes high when the ions slow down around the Bragg peak. This provides radiobiological advantages such as an increased relative biological effectiveness (RBE) and reduced oxygen enhancement ratio (OER) in the tumor.

2.5 Brachytherapy

Ionizing radiation damages the DNA of the malignant cells, most commonly by causing double-strand breaks, which may be misrepaired and disrupt the integrity of the chromosome. The effects of this damage become manifest during mitosis, at which point the cells cannot successfully replicate. Other consequences of irradiation include changes in growth factors and signal transduction pathways, apoptosis and the regulation of the cell cycle. Damage to cells, particularly those that rarely divide and are highly differentiated may also occur secondary to disruptions in their vascular supply.

Brachytherapy treats cancer by placing radioactive sources directly into or next to the area requiring treatment. One example is the implant of some ^{125}I inside the prostate gland, which deliver the required dose during the entire period they are active. More commonly, however, a radioactive source is inserted into the body and removed when the time calculated for the delivery of a specified radiation dose has elapsed.

Treatment can be delivered according to the dose rate as:

- Low dose rate brachytherapy, which utilizes ^{137}Cs and Ir sources, among others. In this technique, an applicator is placed in the cavity or inside the tumor and the source is fed into the applicator, once the patient is in a shielded room. They remain in isolation until the source is removed (usually 12–24 hours). This process often requires hospital admission.
- High dose rate brachytherapy, which can be delivered with miniaturized sources of ^{60}Co or ^{192}Ir, which allows a dose rate greater than 12 Grays per hour, accounting for short times of treatment. For this reason, this therapy can be administered as an outpatient treatment.

Brachytherapy is a key component of radiation treatment for gynecological cancers. Other indications for brachytherapy include: prostate, breast, soft tissue sarcomas, some head and neck tumors and skin cancers (see IAEA 2017, IAEA 2015b, IAEA 2015c). For cervical and skin cancers, it has become a standard therapy for more than 100 years as well as an important part of the treatment guidelines for other malignancies, including head and neck, skin, breast and prostate cancers.

2.5.1 Prostate Brachytherapy

In the case of brachytherapy, treatment for prostate cancer a radioactive source is placed inside the prostate. The radioactive source releases radiation to destroy the prostate cancer cells. The source might stay inside your prostate for a longer period of time, case of low dose rate brachytherapy. The radiation is gradually released over a number of months. Alternatively, the treatment could be a high dose rate of brachytherapy, where the radioactive source stays inside the prostate for about only 15–40 minutes. The source is afterward removed and there is no radiation left inside the body (see Koukourakis et al. 2009, and references therein).

There are various ways to treat prostate cancer with brachytherapy. The real-time prostate solutions workflow, as done by Elekta, the treatment is provided in a single session as shown in Fig. 2.3.

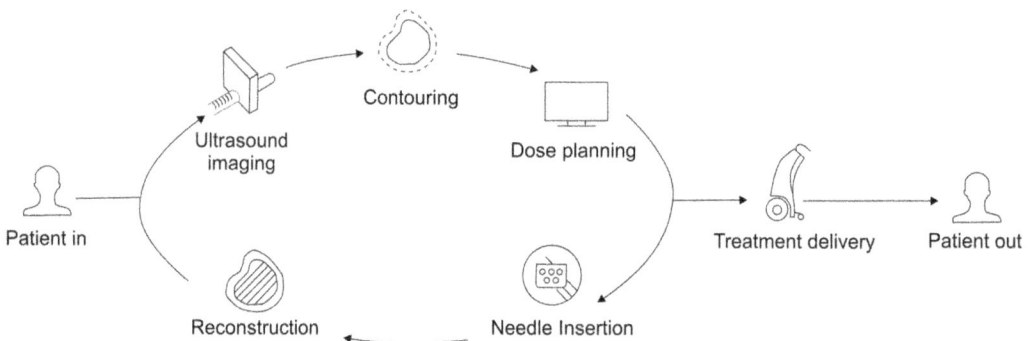

FIGURE 2.3
Elekta: genuine real-time workflow. The patient is not moved between procedure steps. (Image courtesy of Elekta. Source: https://www.elekta.com/company/gallery.)

2.5.2 Brachytherapy to the Cervix, Womb or Vagina

In these cases, the hollow tubes (applicators) are inserted into the vagina to the area where the cancer is under general anesthetic. Afterward, the tubes are attached to a brachytherapy machine. The radioactive source moves from the machine through the tubes to the cancer. The radioactive source may stay inside body for about 10–20 minutes. After the treatment, the source moves back into the machine. Patients must have several treatments as an inpatient or outpatient. For controversies in high-dose-rate versus low-dose-rate brachytherapy for the treatment of cervical cancer see the article by Stewart and Viswanathan (2006) (Fig. 2.4).

For brachytherapy for other cancer types (eye cancer, breast cancer, etc.) see for example https://www.cancerresearchuk.org/about-cancer/cancer-in-general/treatment/ radiotherapy/internal/radioactive-implant-treatment/what-is-brachytherapy, or https:// www.targetingcancer.com.au/radiation-therapy/brachytherapy/ or https://www. mayoclinic.org/tests-procedures/brachytherapy/about/pac-20385159.

An interesting article by Bensaleh et al. (2009) presents the review of Mammosite brachytherapy: advantages, disadvantages and clinical outcomes. The MammoSite Radiation Therapy System (RTS) has become the most widely used brachytherapy method used in the treatment of breast cancer, due to its ease of use, short learning curve and requirement of only

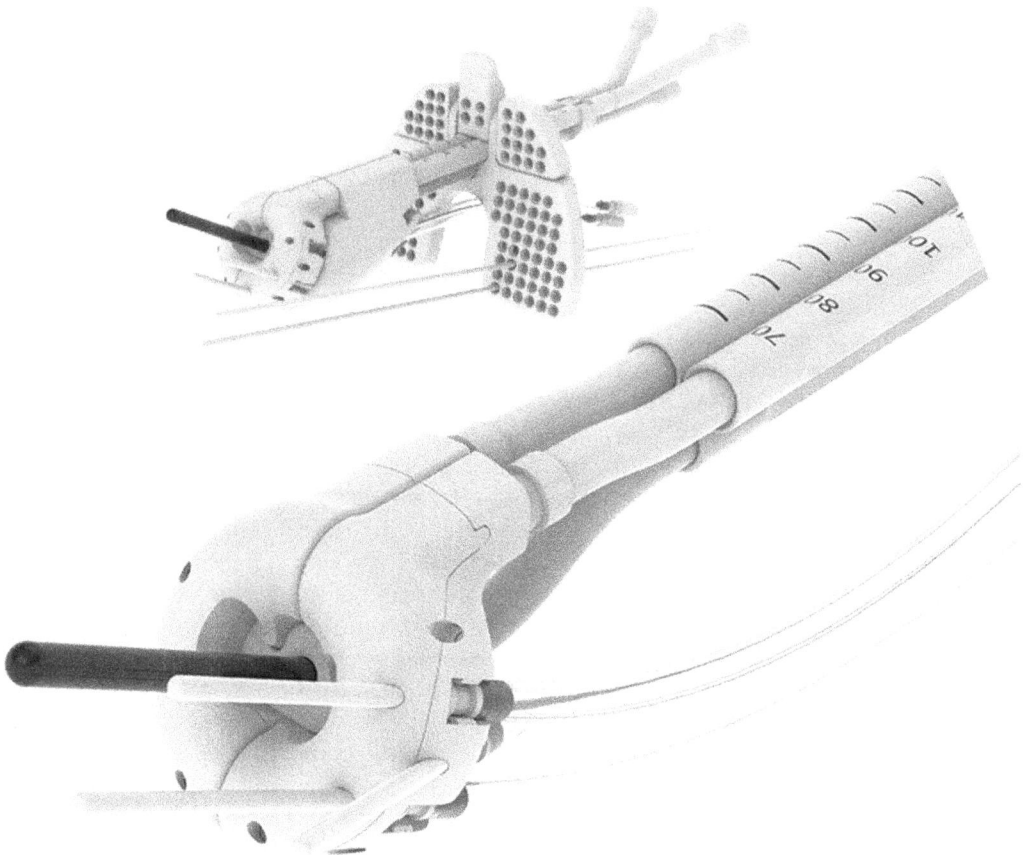

FIGURE 2.4
Advanced Gynecological Applicator Venezia™ by Elekta. (Image courtesy of Elekta. Source: https://www. elekta.com/company/gallery.)

one interstitial path through the breast skin (Streeter et al. 2003). The RTS has become the most widely used brachytherapy method due to its ease of use, short learning curve, and requirement of only one interstitial path through the breast skin. The dosimetry is simple, one source position in the middle of the MammoSite balloon catheter. The MammoSite® radiotherapy system is an alternative treatment option for patients with early stage breast cancer to overcome the longer schedules associated with external beam radiation therapy. The device is placed inside the breast surgical cavity and inflated with a combination of saline and radiographic contrast to completely fill the cavity. The treatment schedule for the MammoSite monotherapy is 34 Gy delivered in 10 fractions at 1.0 cm from the balloon surface with a minimum of 6 hours between fractions on the same day (Bensaleh et al. 2009).

According to Skowronek (2017), brachytherapy has increased its use as a radical or palliative treatment, and become more advanced with the spread of pulsed-dose-rate and high-dose-rate after loading machines. In addition, the use of new 3D/4D planning systems has additionally improved the quality of the treatment. Skowronek (2017) presented short summaries of current studies on brachytherapy for the most frequently diagnosed tumors. Data presented in his manuscript should help especially young physicians or physicists to explore and introduce brachytherapy in cancer treatments.

2.6 Proton Therapy

Proton therapy is delivered to selected cancer patients presenting with rare tumors, for which a dose-escalation paradigm and/or a reduced dose-bath to the organs at risk is pursued. It is a costly treatment with an additional cost factor of 2–3 when compared to photon radiotherapy.

The European Particle Therapy Network (EPTN) was created in 2015 to answer the critical European needs for cooperation among protons and carbon ions centers in the framework of clinical research networks (see Weber et al. 2020).

Proton Therapy Today is the online magazine for proton therapy accessible at http://www.proton-therapy-today.com/where-to-get-pt/. Proton Therapy Today is an editorial website that gathers serious, trustworthy stories and information about the "world of proton therapy".

As the most advanced form of radiation therapy, proton therapy helps limit side effects of cancer treatment and is preferably used to fight cancers seated where side effects are particularly unwanted (brain or prostate cancers, pediatric cancers, etc.) This type of treatment offers patients a more efficient and compassionate way of fighting against cancer.

Photon-based radiotherapy deposits most of its energy before reaching the tumor while proton therapy deposits most of its energy inside the tumor, as shown in Fig. 2.5.

However, today, proton therapy accessibility is still low, though growing, which puts it at the center of a lively debate. Many people, patients, physicians, nurses, physicists, researchers and manufacturers are working diligently to develop and make this treatment option accessible to more patients worldwide. "Proton Therapy Today" aims at telling the stories of these people worldwide, their experience with proton therapy and the journey of cancer patients and their family towards hope and recovery. This website is willing to put these modern-day heroes under the spotlight, so that their everyday efforts may be acknowledged and passed on to more people.

In addition to the radiobiological studies mentioned above, medical physics research is also needed to tackle the problem of range uncertainty, which currently makes proton

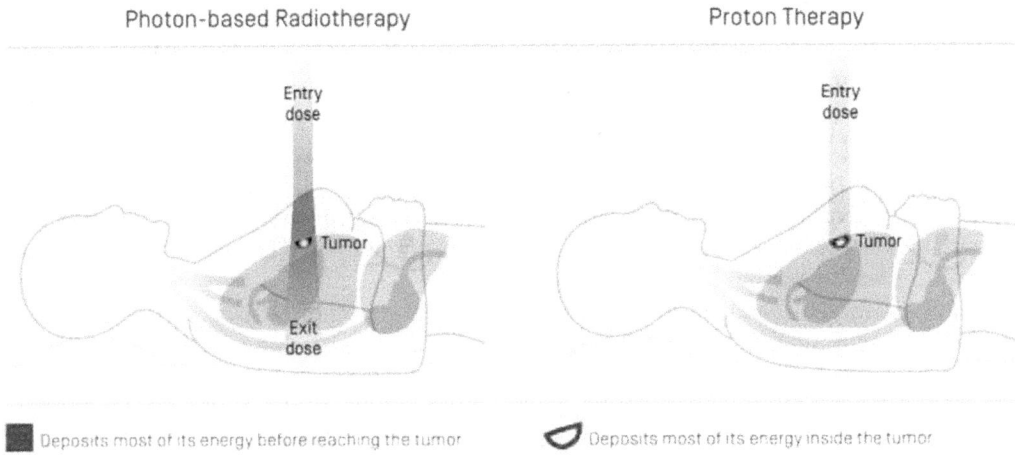

Photon-based Radiotherapy   Proton Therapy

■ Deposits most of its energy before reaching the tumor   ▽ Deposits most of its energy inside the tumor

FIGURE 2.5
Illustration of difference in photon-based radiotherapy and proton therapy. Photon-based radiotherapy deposits most of the energy before reaching the tumor. Proton radiotherapy deposits most of the energy inside the tumor.

therapy less robust than conventional radiotherapy, and makes the treatment of moving targets problematic. The full potential of the Bragg peak has still not been exploited. If the accuracy of proton therapy is increased, the Bragg peak could be viewed as a non-invasive scalpel and potentially used to treat diseases other than cancer, in cranial and extra-cranial targets such as the heart (Durante 2019).

Table 2.4 shows a list of clinical institutions offering proton therapy today. Those are the active centers of proton therapy, source PTCOG data.

TABLE 2.4

Carbon beam radiotherapy facilities

No.	Country	Town	Therapy center	Beam direction
1	Austria	Wiener Neustadt	MedAustron	2 horiz. and 1 verticalfixed beam[a]
2	China	Shanghai	SPHIC	3 fixed beams[a]
3	China	Wuwei, Gansu	Heavy Ion Cancer Treatment Center	4 fixed beams[a]
4	Germany	Heidelberg	Heildelberg Ion Therapy Center	2 fixed beams, 1 gantry[a]
5	Germany	Marburg	MIT	3 horiz., 1 45deg. fixed beams[a]
6	Italy	Pavia	CNAO	3 horiz., 1 vertical, fixed beams
7	Japan	Chiba	HIMAC	horiz.[b], vertical[b], fixed beams, 1 gantry
8	Japan	Hyogo	HIBMC	horiz.,vertical, fixed beams
9	Japan	Gunma	GHMC	3 horiz., 1 vertical, fixed beams
10	Japan	Tosu	SAGA-HIMAT	3 horiz., vertical, 45 deg., fixed beams
11	Japan	Yokohama	i-Rock Kanagawa Cancer Center	4 horiz., 2 vertical, fixed beams
12	Japan	Osaka	Osaka Heavy Ion Therapy Center	3 fixed beams, 6 ports[a]

[a] With pencil beam scanning.
[b] With spread beam and pencil beam scanning.

The National Association for Proton Therapy [NAPT, PO Box 7801Mclean, VA 22106] is an independent non-profit organization founded in 1990 to educate and raise public awareness about the clinical benefits of proton therapy. It serves as a resource center for patients, physicians, health care providers, insurers, academic medical centers, hospitals, cancer centers, Medicare, the US Congress, health care agencies and the news media. The NAPT's supporting members include 37 of the nation's leading cancer centers, a number of whom are National Cancer Institute-designated comprehensive cancer centers and National Comprehensive Care Network (NCCN) members (Table 2.5). The proton therapy centers in the USA are shown in Table 2.6.

Proton therapy, also called proton beam therapy, is a type of radiation therapy. It uses protons rather than X-rays to treat cancer. People usually receive proton therapy in an outpatient setting. This means they do not need to have treatment in the hospital. The number of treatment sessions depends on the type and stage of the cancer.

Sometimes doctors deliver proton therapy in 1–5 proton beam treatments. They generally use larger daily radiation doses for a fewer number of treatments. This is typically called stereotactic body radiotherapy. If a person receives a single, large dose of radiation, it is often called radiosurgery.

Proton therapy requires planning. Before treatment, you will have a specialized CT or MRI scan. During this scan, you will be in the exact same position as during treatment.

Movement should be limited while having the scan. Therefore, you may be fitted with a device that helps you stay still. The type of device depends on where the tumor is in the body. For example, a person may need to wear a custom-made mask for a tumor in the eye, brain or head. He or she would also need to wear this device later for the radiation-planning scan.

During a radiation-planning scan, one will lie on a table and the doctor will figure out the exact places where the radiation therapy will be given on the body or the device. This helps make sure patient position is accurate during each proton treatment.

Proton therapy also may be used to treat these cancers:

- Central nervous system cancers, including chordoma, chondrosarcoma and malignant meningioma.
- Eye cancer, including uveal melanoma or choroidal melanoma.
- Head and neck cancers, including nasal cavity and paranasal sinus cancer and some nasopharyngeal cancers.
- Lung cancer.
- Liver cancer.
- Prostate cancer.
- Spinal and pelvic sarcomas are cancers that occur in the soft-tissue and bone.
- Noncancerous brain tumors.

PBT is being used increasingly to treat children with cancer. The chance of curing the cancer is no higher than with conventional radiotherapy but is likely to reduce the severity of the long-term side effects, although it will not eliminate them altogether.

For adults, the main use of PBT has been to treat cancers close to parts of the body that are very sensitive to the damaging effects of radiation. For example, PBT is used to treat certain cancers at the base of the skull, deep inside the head and close to the brain and cancers of the spine that are close to the spinal cord.

TABLE 2.5

Proton therapy centers worldwide

No	Country	Town	Therapy center	Website
1	Austria	Wiener Neustadt	MedAustron	https://www.medaustron.at/
2	Belgium	Leuven	UZ Leuven Particle Proton Centre,	https://www.uzleuven.be/en/proton-centre
3	Canada	Vancouver	TRIUMF	https://www.triumf.ca/proton-irradiation-facility
4	China	Zibo City	WPTC-Wanjie Proton Therapy Center	wjyl.wanjie.net
5	China	Shanghai	SPHIC	
6	Czech Republic	Prague	PTC Czech r.s.o.	https://www.ptc.cz/
7	France	Nice	Centre Laccassagne, CAL/IMPT	https://www.france-hadron.fr/
8	France	Orsay	CPO	https://institut-curie.org/liste/proton-therapy-center
9	France	Caen	Centre Cyclhad	normandie-protontherapie.com
10	Germany	Berlin	HMI	
11	Germany	Heidelberg	Heildelberg Ion Therapy Center	
12	Germany	Munich	Rinecker	
13	Germany	Dresden	Universitätsklinikum Carl Gustav Carus	
14	Germany	Essen	Westdeutsches Protonentherapiezentrum	
15	Germany	Marburg	MIT	
16	India	Chennai	Apollo Hospitals PTC	https://www.apollohospitals.com/departments/cancer/treatment/proton-therapy
17	Italy	Catania	Laboratori Nazionale del Sud	
18	Italy	Pavia	CNAO Pavia	
19	Italy	Trento	Agenzia provincial per la Protonterapia (ATreP)	
20	Japan	Hyogo	HIBMC, Hyogo Ion Beam Medical Center	https://www.hibmc.shingu.hyogo.jp/
21	Japan	Kashiwa	Japanese National Cancer center	https://www.ncc.go.jp/
22	Japan	Shizuoka	Shizuoka cancer center	https://www.scchr.jp/
23	Japan	Tsukuba	PMRC, Proton Medical Research Center	http://www.pmrc.tsukuba.ac.jp/
24	Japan	Fukui	Fukui Proton Cancer Center (FPCTF)	https://fph.pref.fukui.lg.jp/
25	Japan	Matsumoto, Nagano P.	Aizawa hospital, Aizawa Proton Therapy Center	http://w3.ai-hosp.or.jp/
26	Japan	Nagoya	Nagoya Proton Therapy Center	https://www.nptc.med.nagoya-cu.ac.jp/
27	Japan	Okayama	Tsuyama Chuo Hospital	http://en.tch.or.jp/

(Continued)

TABLE 2.5 (Continued)

Proton therapy centers worldwide

No	Country	Town	Therapy center	Website
28	Japan	Sapporo Hokkaido P	Hokkaido University Hospital	https://www.huhp.hokudai.ac.jp/
29	Japan	Ibusuki	Medipolis Medical Research Institute	http://www.medipolis-ptrc.org/
30	Japan	Koriyama	S. Tohoku Proton Therapy Center	http://www.southerntohoku-proton.com/
31	Japan	Hokkaido	PTC Teishinkai Hospital, Okayama	https://www.teishinkai.jp/thp/
32	Japan	Osaka	Hakuhokai Group Osaka PT Clinic	https://www.osaka-himak.or.jp/en/about/support/
33	Japan	Kobe	Kobe Proton Center	https://www.kobe-pc.jp/en/
34	Japan	Toyohashi	Narita Memorial Proton Center	https://iba-worldwide.com/proton-therapy/proton-therapy-centers/
35	Japan	Tenri City	Takai Hospital	
36	Japan	Kyoto	Nagamori Memorial Center of Innovative Cancer Therapy and Res	
37	Poland	Krakow	Instytut Fizyki Jadrowej PAN	https://www.ifj.edu.pl/
38	Russia	St Petersburg	Center of Nuclear Medicine	https://protherapy.ru/
39	Russia	Moscow	Institute for Theoretical and Exp. Physics	
40	Russia	Dubna	Joint Institute for Nuclear Research 2	
41	Russia	Obninsk	MRRC	
42	Russia	Dimitrovgrad	Federal HighTech Center of FMBA	
43	South Africa	Johannesburg	NRF-iThemba Labs	
44	South Korea	Ilsan-ro	Korean National Cancer Center	https://ncc.re.kr/
45	South Korea	Seoul	Samsug Hospital, Proton Therapy Center	http://www.samsunghospital.com/
46	Spain	Madrid	Quironsalud PTC	https://www.quironsalud.es/en/protonterapia
47	Spain	Madrid	Clínica Universidad De Navarra, CUN	https://www.cun.es/en/
48	Sweden	Uppsala	The Scandion Clinic	
49	Switzerland	Villingen	CPT, PSI	
50	Taiwan	Taipei		
51	Taiwan	Kaohsiung		
52	The Netherlands	Groningen	University Medical Center (UMCG)	https://www.umcgradiotherapie.nl/
53	The Netherlands	Delf	Holland PTC	
54	The Netherlands	Maastricht	ZON PTC	

TABLE 2.5 (Continued)

Proton therapy centers worldwide

No	Country	Town	Therapy center	Website
55	United Kingdom	Clatterbridge	The Clatterbridge Cancer Center	
56	United Kingdom	Newport, Wales	Rutherford Cancer Center, South Wales	
57	United Kingdom	Manchester	The Cristie Proton Therapy Center	
58	United Kingdom	Berkshire	Rutherford Cancer Centre Thames Velley	
59	United Kingdom	Northumberland	Rutherford Cancer Center North East	

TABLE 2.6

Proton therapy centers in the USA

No.	Therapy center	Location	Website
1	Ackerman Cancer Center Proton Therapy Center	Jacksonville, FL (904) 880-5522	ackermancancercenter.com
2	Beaumont Proton Therapy Center	Royal Oak, MI (248) 551-8402	www.beaumont.org
3	California Protons Cancer Therapy Center	San Diego, CA (858) 549-7400	www.californiaprotons.com
4	Cincinnati Children's Pediatric Proton Therapy Center/Proton Therapy at University of Cincinnati Medical Center	Liberty Township, OH CC 1-844-790-2866 UC 1-844-532-2326	www.cincinnatichildrens.org uchealth.com/
5	Emory Proton Therapy Center	Atlanta, GA (404) 778-3473	winshipcancer.emory.edu/proton
6	Francis Burr Proton Therapy Center at Massachusetts General Hospital	Boston, MA (617) 724-1680	https://www.massgeneral.org/cancer-center/radiation-oncology/treatments-and-services/proton-therapy/
7	Hampton University Proton Therapy Institute	Hampton, VA 1-877-251-6838	www.hamptonproton.org
8	Inova Schar Cancer Institute Proton Therapy Center	Falls Church, VA	
9	James M. Slater, MD Proton Treatment & Research Center at Loma Linda University Cancer Center	Loma Linda, CA 1–800-776–8667	www.protons.com
10	Maryland Proton Treatment Center	Baltimore, MD (410) 593–1202	https://mdproton.com/
11	McLaren Proton Therapy Center	Flint, MI	https://www.mclaren.org/main/proton-therapy-center
12	MD Anderson Cancer Center Proton Therapy Center	Houston, TX 1-866-632-4PTC (4782)	www.mdanderson.org/proton
13	MedStar Georgetown University Hospital Proton Therapy Center	Washington, DC (202) 444-4639	www.medstargeorgetown.org

(Continued)

TABLE 2.6 (Continued)

Proton therapy centers in the USA

No.	Therapy center	Location	Website
14	Miami Cancer Institute Proton Therapy Center at Baptist Health South Florida	Miami, FL (786) 596-2000	baptisthealth.net
15	Northwestern Medicine Proton Center	Warrenville, IL 1-877-887-5807	www.protoncenter.nm.org
16	Oklahoma Proton Center	Oklahoma City, OK 1–888-847–2640	www.okcproton.com
17	ProCure Proton Treatment Center – New Jersey	Somerset, NJ 1–877-967–7628	www.procure.com
18	Proton International at University of Alabama- Birmingham	Birmingham, AL	https://www.uabmedicine.org/proton
19	Provision CARES Proton Therapy Center – Knoxville	Knoxville, TN 1–855-566–1600	www.knoxvilleproton.com
20	Provsion CARES Proton Therapy Center – Nashville	Franklin, TN 1–844-742–2737	www.nashvilleproton.com
21	Roberts Proton Therapy Center at the University of Pennsylvania Health	Philadelphia, PA 1–800-789–7366	www.pennmedicine.org
22	S. Lee Kling Center for Proton Therapy Center at the Siteman Cancer Center	St. Louis, MO (314) 286–1222	siteman.wustl.edu
23	Seattle Cancer Care Alliance Proton Therapy Center	Seattle, WA 1–844-768–1239	www.sccaprotontherapy.com
24	South Florida Proton Therapy Institute	Delray Beach, FL 561-323-6498	https://sfpti.com
25	Texas Center for Proton Therapy	Irving, TX 1–844-544-0446	www.texascenterforprotontherapy.com
26	The Huntsman Cancer Institute at the University of Utah	801-585-0303 Salt Lake, UT	https://huntsmancancer.org/radiation
27	The Johns Hopkins National Proton Center	Washington, DC	https://www.hopkinsmedicine.org/
28	The Laurie Proton Therapy Center at RWJBarnabas Health	New Brunswick, NJ (732) 253-3176	rwjbh.org
29	The Marjorie & Leonard Williams Center for Proton Therapy at Orlando Health Cancer Institute	Orlando, FL (321) 841-8650	www.orlandohealthcancer.com
30	The Mayo Clinic Proton Beam Therapy Center – Arizona	Phoenix, AZ (480) 301-8000	www.mayoclinic.org/
31	The Mayo Clinic Proton Beam Therapy Center – Minnesota	Rochester, MN (507) 284-1511	www.mayoclinic.org
32	The New York Proton Center	New York, NY 833-697-7686	www.nyproton.com
33	University Hospitals Proton Therapy Center	Cleveland, OH 1–877-828-2978	www.uhhospitals.org
34	University of Florida Health Proton Therapy Institute	Jacksonville, FL 1–877-428-6560	www.floridaproton.com
35	Willis-Knighton Cancer Center	Shreveport, LA (318) 212-8300	www.wkhs.com

Protons were first investigated as radiographic probes as high-energy proton accelerators became accessible to the scientific community in the 1960s. Like the initial use of X-rays in the 1800s, protons were shown to be a useful tool for studying the contents of opaque materials, but the electromagnetic charge of the protons opened up a new set of interaction processes that complicated their use. These complications in combination with the high expense of generating protons with energies high enough to penetrate typical objects resulted in proton radiography becoming a novelty, demonstrated at accelerator facilities, but not utilized to their full potential until the 1990s at Los Alamos.

During this time, Los Alamos National Laboratory was investigating a wide range of options, including X-rays and neutrons, as the next generation of probes to be used for thick object flash radiography. During this process, it was realized that the charged nature of the protons, which was the source of the initial difficulty with this idea, could be used to recover this technique. By introducing a magnetic imaging lens downstream of the object to be radiographed, the blur resulting from scattering within the object could be focused out of the measurements, dramatically improving the resolution of proton radiography of thick systems. Imaging systems were quickly developed and combined with the temporal structure of a proton beam generated by a linear accelerator, providing a unique flash radiography capability for measurements at Los Alamos National Laboratory. This technique has now been employed at LANSCE for two decades and has been adopted around the world as the premier flash radiography technique for the study of dynamic material properties, see Merrill (2015).

Proton therapy is an advanced form of radiation therapy that uses a high-energy proton beam for cancer treatment. In contrast to conventional photon-based radiation therapy, proton beam delivers the majority of their destructive energy within a small range inside the tumor, known as the Bragg peak, thereby reducing adverse effects to adjacent healthy tissues. In IBA-equipped proton therapy centers, cyclotrons accelerate protons to an extremely high speed, generating a controlled beam, which is delivered very precisely in the treatment rooms, through a nozzle, to the targeted tumor. With proton therapy, there is significant potential to reduce side effects, improve overall outcomes in cancer treatment and offer a better quality of life to patients.

Technological advances in conventional external beam radiation therapy have led to a new approach: Intensity Modulated Radiation Therapy (IMRT). To reach the right amount of dose in the target, IMRT multiplies the number of beam incidences needed and spreads unnecessary dose in the surrounding healthy tissue. For the patient, this means receiving a much higher integral dose during IMRT treatment compared to a similar treatment in proton therapy. With proton therapy, you avoid depositing unnecessary dose in surrounding healthy tissues because the physics of protons enables the physician to deliver highly conformal treatments. Proton therapy delivers high-energy radiation doses directly to the tumor.

With advances in technology and the advent of new software, the radiation therapy market is expected to increase during the forecasting period. New techniques have changed the way treatments are planned and doses are delivered. The use of radiation therapy is expected to increase as the incidence of cancer increases. The greatest growth in radiation therapy will be driven by demand in developing countries, particularly India and China.

For example, ProBeam® 360° Proton Therapy System (made by Varian, A Siemens Healthineers Company) offers uncompromised clinical capabilities with ultra-high dose rates, a 360° gantry and exceptional precision, all within a 30% smaller footprint. The evolution of radiotherapy has pushed dose rates higher, and proton therapy is now leading this trend. The ProBeam 360° System features the most powerful particle accelerator available to treat cancer. High dose rates are used today to reduce treatment time, manage motion, and can improve treatment plan quality and conformity. We expect next-generation proton therapy to employ even higher and ultra-high dose rates.

RapidScan™ Technology: Fast Dose Delivery in <5 seconds per field. RapidScan revolutionizes motion mitigation, delivering each field within a single breath-hold for most patients. RapidScan, when used in conjunction with the Eclipse™ treatment planning system, simultaneously optimizes treatment plan quality and delivery time. The ProBeam 360° System with RapidScan simplifies conventional procedures required to treat moving targets, increases the number of patients who can comply with breath-hold treatments, reduces treatment time without sacrificing quality.

We shall discuss in some details the particle therapy in Japan where the pioneer work of the particle therapy by cyclotron began at NIRS in 1975. Fast neutrons and protons have been used for cancer treatments. The results for fast neutron therapy were not so excellent and the energy of the proton beam was not enough for the treatment of thick tissue. Higher energy proton therapy begun at the Tsukuba University with proton beams from the 500 MeV booster synchrotron of the 12 GeV KEK proton synchrotron.

After evaluations for the next dedicated machine, Japanese radiologists chose a heavy-ion synchrotron as the first priority machine. Thus the HIMAC project started. However, the HIMAC is a very expensive big machine, which is not suitable for a usual hospital. According to the experience of HIMAC, the carbon beam is preferable for therapy. Therefore, Hyogo prefectural government has decided to construct a smaller synchrotron, which accelerates heavy ions up to carbon for therapy at Harima Science Garden City. The building and the accelerator are under construction. Meanwhile, Fukui prefectural government has established the Energy Research Center Wakasa Bay where a 200 MeV proton synchrotron with a multi-purpose tandem accelerator as an injector is applied for proton therapy.

Many prefectural governments are considering to construct particle therapy accelerators. On the other hand, some universities such as Tsukuba University, Kyoto University and Osaka University also planned the construction of dedicated particle therapy accelerators. Tsukuba University has received the financial support to construct a dedicated proton therapy synchrotron. At Kyoto University, the ICR accelerator group has studied a combined function compact synchrotron for the medical group in collaboration with a company. A unique multi-feed untuned RF cavity has been developed by this accelerator group. This principle is applied to the Wakasa Bay and Tsukuba University synchrotrons through the Kyoto University project. After a decision made by the Ministry of Health & Welfare (MHW) the east hospital of National Cancer Center (NCC) of MHW suddenly began construction of proton therapy facility. The 230 MeV compact cyclotron has been completed. MHW's beginning of the proton therapy gave a great impact on advanced hospitals (see Inoue 1998).

Sakurai et al. (2016) compared proton beam therapy in Japan in the year 2000 and 2016, showing an increase in the number of proton facilities in 2016 when 11 proton facilities were available. In general, proton therapy has shown promise in treating several kinds of cancer. Studies have suggested that proton therapy may cause fewer side effects than traditional radiation, since doctors can better control where the proton beams deposit their energy. However, few studies have directly compared proton therapy radiation and X-ray radiation, so it is not clear whether proton therapy is more effective in prolonging lives. However, proton beam therapy is very useful for pediatric cancer, since the pediatric radiation dose to normal tissues should be reduced as much as possible because of

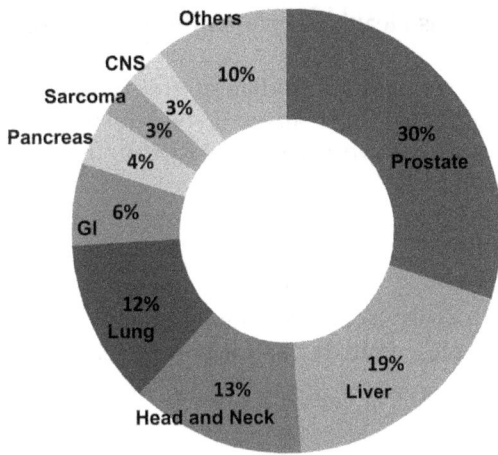

FIGURE 2.6
The numbers of patients treated by proton beam therapy, based on tumor site, in Japan (1979–2013). The data are provided by the Japan Clinical Study Group of Particle Therapy (JCPT), after Sakurai et al. (2016).

the effect of radiation on growth, intellectual development, endocrine organ function and secondary cancer development.

Proton beam therapy is a standard treatment for nasal and paranasal lesions and lesions at the base of the skull because the radiation dose to critical organs such as the eyes, optic nerves and central nervous system can be reduced with proton beam therapy. Regarding new proton beam therapy applications, experience with proton beam therapy combined with chemotherapy is limited, although favorable outcomes have been recently reported for locally advanced lung cancer, esophageal cancer and pancreatic cancer. Therefore, "chemoproton" therapy appears to be a very attractive field for further clinical investigations (Sakurai et al. 2016) (Fig. 2.6).

2.7 Equipment Manufacturers

Major market players include Varian (USA), Mitsubishi Heavy Industry (Japan), Elekta (Sweden), IBA-Worldwide Proton Therapy (Belgium), Accuray (USA), Philips, GE Healthcare, Toshiba, Varex, Shinva, Neusoft, Top Grade Healthcare, Huiheng Medical and Hamming.

MarketWatch is a website that provides financial information, business news, analysis and stock market data. Along with *The Wall Street Journal* and Barron's, it is a subsidiary of Dow Jones & Company, a property of News Corp. MarketWatch has released a global "Particle Accelerators Market" research report providing market size segment by type, segment by application, competition by company, company profiles. Also, the research report focuses on upstream, opportunities, market challenges, risks and influences factors analysis, market strategy analysis and trends, for details see https://www.marketwatch.com/press-release/particle-accelerators-market-share-2021-industry-overview-size-key-players-challenges-region-suppliers-and-forecast-to-2026. However, purchase of this Report has a price of 4000 US$ for a single-user license.

2.7.1 Varian, A Siemens Healthineers Company

Varian Medical Systems, Inc. has become recently a new business segment within Siemens. The acquisition of cancer care tech company by Siemens Healthineers was announced on

15 April 2021, which means that Varian and Siemens Healthineers are officially now one united company.

Varian hardware, software and professional services for radiation treatments are widely used on a global basis. There are more than 7,200 linear accelerators, 3,300 treatment-planning sites and 3,000 ARIA® sites in use worldwide. VARIAN offers the equipment and services in the fields of (Figs. 2.7–2.10):

- Radiosurgery involves the use of sophisticated technology to ablate tumors or other abnormalities with non-invasive, high-dose radiation.
- Radiotherapy uses high-energy radiation, usually X-rays, to damage cancer cells and treat tumors anywhere in the body where radiation treatment is indicated.
- Proton therapy uses beams of positively charged nuclear particles to deliver highly focused doses in the tumor.

Varian Commitment History of Proton Program: 1992 – Varian Develops Proton Treatment Planning; 1997 – Varian Provides Information Systems for Proton Centers; 2007 – Varian Purchases ACCEL Instruments; 2007 – Isochronous Superconducting Cyclotron becomes clinical at PSI; 2009 – Complete Treatment System Delivered to Rinecker Proton Center Munich, Germany (>1200 patients); 2011 – First US ProBeam System Delivered to Scripps Health in San Diego.

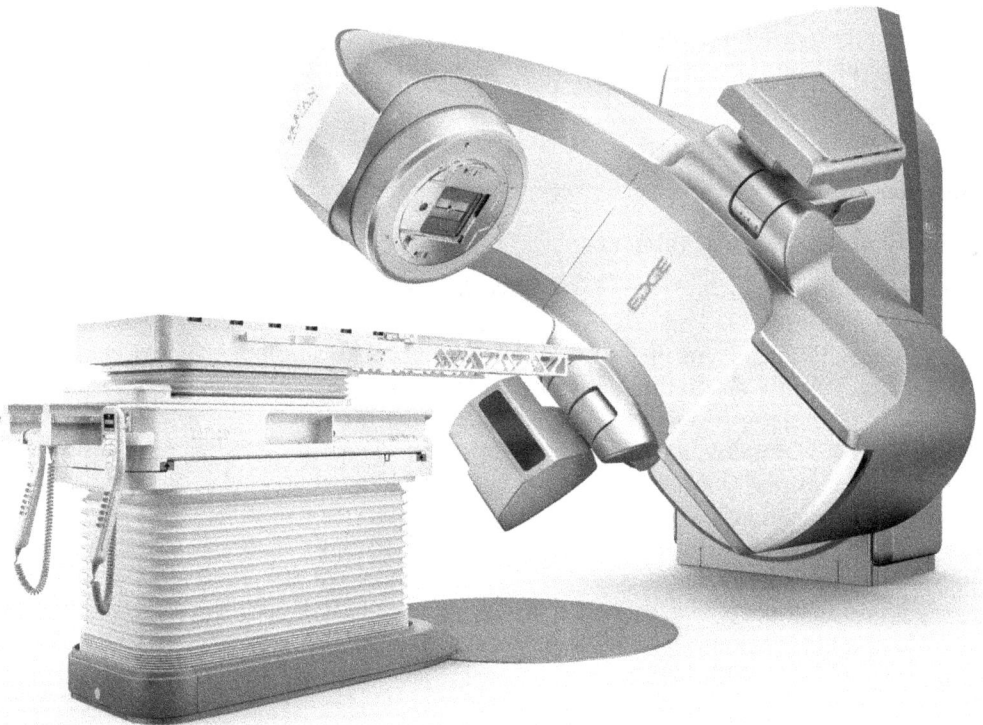

FIGURE 2.7
Edge system with integrated machine intelligence. (Courtesy of Varian, a Siemens Healthineers Company.)

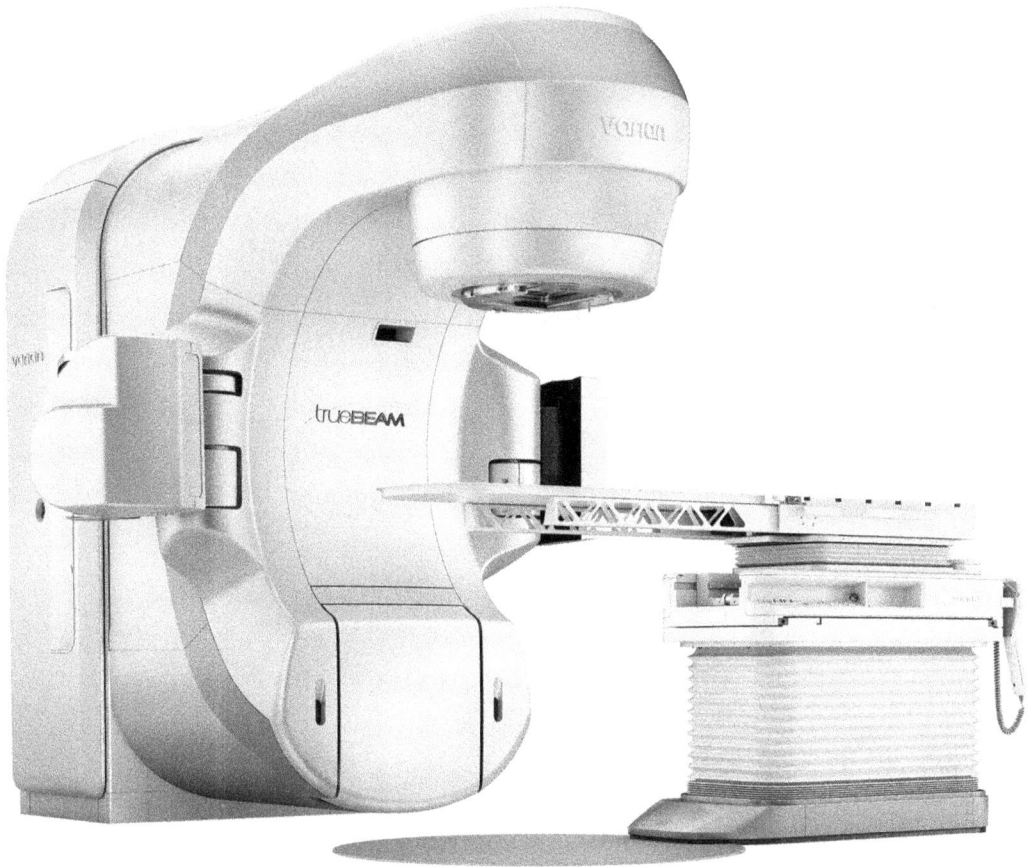

FIGURE 2.8
TrueBeam system is designed to treat cancer anywhere in the body. (Courtesy of Varian, a Siemens Healthineers Company.)

The proton therapy system is based on the use of AC 250 superconducting cyclotron (design: Schillo et al. 2001) capable of delivering 230–250 MeV protons, extracted current 800 nA, shown in Fig. 2.11. The ProBeam 360° system offers a comprehensive and customizable IMPT solution so you are able to tailor your single or multi-room center to meet your clinical, research and capacity needs. Treatment room options include 360° rotating gantries, fixed beam rooms or eye treatment rooms. Research rooms are available for non-clinical proton beam applications.

The ProBeam 360° system in addition to the superconducting cyclotron contains beam transport system that focuses, shapes and guides the beam to the treatment room, creating the small beam size necessary for IMPT and enabling clinicians to target very small spots across the full energy range. In addition, the 360° Rotating Gantry provides 360° of freedom to image and treat the patient from any angle without additional repositioning.

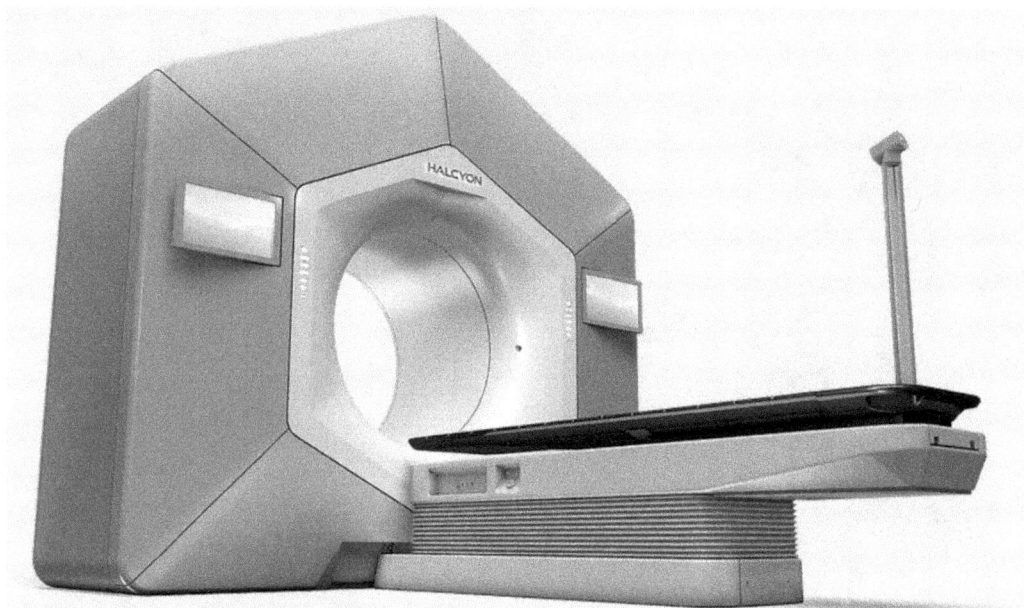

FIGURE 2.9
Halcyon system gives fast and high-quality treatments. (Courtesy of Varian, a Siemens Healthineers Company.)

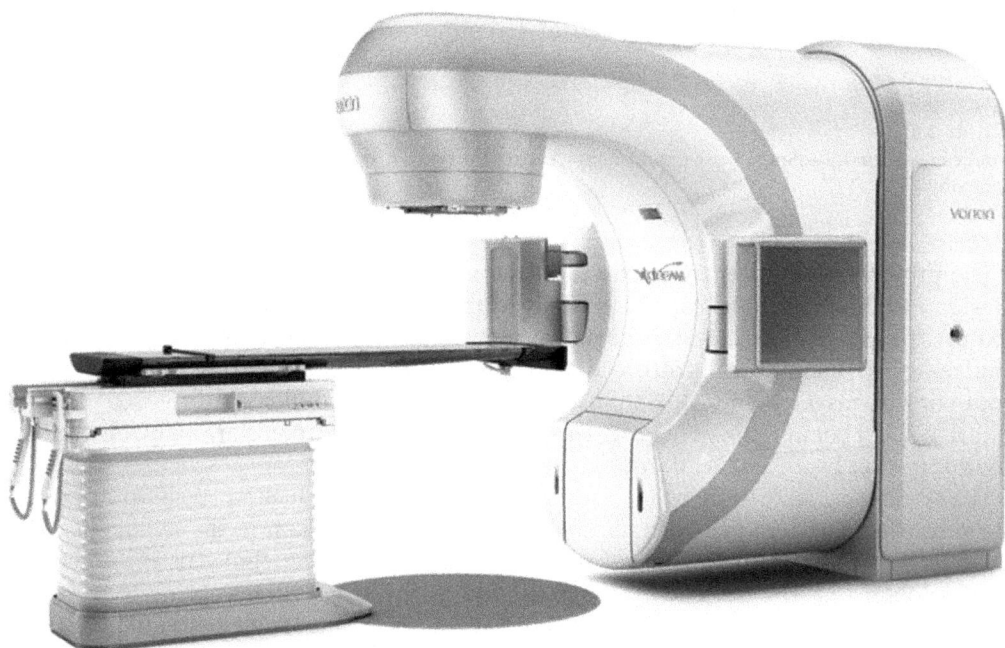

FIGURE 2.10
VitalBeam system enhances existing capabilities and throughput for patients and Easily customize the technology based on the specific needs of the clinic. (Courtesy of Varian, a Siemens Healthineers Company.)

FIGURE 2.11
AC 250 superconducting cyclotron used in proton therapy. (Courtesy of Varian, a Siemens Healthineers Company.)

2.7.2 Mitsubishi Heavy Industries

Mitsubishi Heavy Industries, MHI, Japanese giant, group of companies, is covering many areas including particle accelerators. Their Electron Beam Sterilization System is one such product well known in the medical field (see Urano et al. 2003). Nowadays, electron beam irradiation method is attracting attention as useful medical tools sterilization, which has the advantages such as safety, zero emission and high-speed processing. In this trend, in addition to present main object (the low-density medical tools like operation dress), other various objects are strongly desired that will be applied of electron beam sterilization method. Mitsubishi Heavy Industries, Ltd. has succeeded in the realization of high-efficiency continuous electron beam sterilization processing for high-density medical tools, vessel surfaces and so on. Further, compact electron beam systems are also developed, which can also be applicable for above-mentioned objects.

MHI has studied the electron beam sterilization characteristics, e.g., the correlation between survival fraction and DNA chain shortening.

1. High-density medical tools: Most dialyzer contains physiological saline with hollow fibers inside. Therefore, the density is higher than the present main objects for EB sterilization, and then uniform irradiation is difficult by previous EB irradiation techniques that irradiated from one or both directions. Noticing that

dialyzer is rotation symmetry, MHI developed a uniform irradiation technology inside it with the combination of EB irradiation from peripheral direction and the rotation them around the axis under conveyance. To confirm the validity of this method, an acrylic cylindrical container modeling dialyzer is irradiated by EB, which was filled with agar containing a sensitive pigment, which colour changes depending on the absorbed dose, and then the degree of discoloration on the cross-section was evaluated. The evaluated absorbed dose in the cylindrical container is uniform. The validity of this method has been also verified by the evaluation of absolute values of absorbed doses by film dosimeter, and of the fluidity of liquid in a cylindrical container. Henceforth, MHI intends to apply this for high-density medical tools sterilization systems with their various physical distribution technologies.

2. Inner surface of vessel: To sterilize the inner surface of vessels, such as storage containers of medicine or the like, the electron beam trajectory must be controlled suitably for each object. Seeing that the trajectory of an electron beam in the air can be controlled by a magnetic field, MHI developed a technology for applying a magnetic field simultaneously with irradiation. The Monte Carlo simulation results show that the spread of the electron beam trajectory in air caused by scattering with gas can be controlled by the magnetic field. The actual ratio of measured absorbed dose in inner surface of vessel also shows the uniform irradiation. The outer surface can also be irradiated uniformly by this technology (Tables 2.7 and 2.8).

TABLE 2.7

Mitsubishi heavy industries microwave acceleration type electron beam systems

Item	Small model	Medium model	Large model
Concept	Compact	Online	High power
Energy/MeV	1–10	1–10	10
Power/kW	2–4	3–6	14–31
Scanning width/cm	30–50	30–50	30–80
Throughout/kg/h	130–230	200–350	1000–1800
Acceleration system	C-band standing wave	S-band standing wave	S-band traveling wave
Required area/m^2	50–300	70–350	550

TABLE 2.8

Mitsubishi heavy industries electrostatic acceleration type electron beam systems

Item	Ultra-low energy	Low energy
Concept	Surface layer treatment	Surface treatment
Energy/MeV	80–120	120–300
Power/kW	15	36 or less
Irradiation width/cm	200 or less	200 or less
Processing speed at 25 kGy/m/min	40 or less (at 100 keV)	68 (at 300 keV)
Acceleration system	Electrostatic	Electrostatic
Required area/m^2	8	15

The Vero4DRT (marketed under the MHI-TM2000 Linear Accelerator System brand name) is high-precision image-guided radiotherapy equipment that was developed with a concept of sophisticatedly integrating beam delivery technologies and imaging technologies. The most distinguishingly unique features are the design of the entire structure, the configuration and the sub-devices mounted on the Vero4DRT, which are intended to realize dynamic tracking radiotherapy to deliver beams to tumors moving inside the body because of factors such as respiration. The Vero4DRT is equipped with their original ultra-compact C-band standing-wave accelerator, which delivers therapeutic beams (one-third the size of traditional accelerators and is being used in medical equipment for the first time anywhere in the world). The Vero4DRT also has an "irradiation head" containing accelerator mentioned above and a multi-leaf collimator (MLC) by which the X-ray beam is adjusted to fit the shape of the target tumor, and the irradiation head is mounted on a gimbal mechanism that can be swung. Thus, it has become possible to correct extremely small displacements of the irradiating direction and perform tracking radiation therapy onto a tumor moving by breathing.

The Vero4DRT has been installed at more than 18 hospitals worldwide, producing many clinically satisfied outcomes for cancer treatment.

In radiation therapy equipment, the accuracy of delivering radiation concentrated to the location of the cancer while avoiding irradiation to the healthy cells determines the accuracy of treatment. It is necessary to precisely align the diseased region with the precisely controlled irradiation area. To realize a mechanical accuracy of ±0.1 mm, which is the top level in the industry, the combination of the basic technologies accumulated through developments of various products including (1) micro-positioning technologies (as used in machine tools), (2) image-processing technologies (printing machines), (3) system control technologies (iron-making machines) and (4) ultra-compact accelerating structure technologies to generate X-rays should be combined. The equipment is comprised of a highly-rigid O-ring structure (a gantry) approximately 3.3 m in diameter, an irradiation head mounted on the swing mechanism (a gimbal), two perpendicular pairs of X-ray imaging systems (each consists of a kV X-ray generator and a flat-panel detector), and the patient's bed (a couch) movable in five axial directions.

In image-guided radiation therapy, the imaging systems installed on the rotatable gantry along the horizontal axis enable the three-dimensional capturing of a tumor from any direction, whereby exact alignment can be performed quickly. The rotation of the gantry along the vertical axis allows mechanical movements that had not been possible with conventional equipment, enabling non-planar irradiation without moving/rotating the couch. Not only does this reduce the patient's burden caused by couch rotation in existing equipment, but it also eliminates the patient's positional deviation and maintains the accuracy of patient positioning, which is vital for high-precision irradiation. Research for new treatment methods using this innovative gantry rotation technology is under way. Using its dynamic tracking irradiation function, the Vero4DRT can capture the movement of a tumor in real-time through the imaging systems and track the tumor using the swing mechanism of the interlocking irradiation head with the system. Irradiation can also be performed in combination with intensity-modulated radiation therapy. This functionality has been developed with the financial support of the Japan Society for the Promotion of Science, based on the Funding Program for World-Leading Innovative R&D on Science and Technology systemized by the Council for Science and Technology Policy. Clinical support has been provided by Kyoto University and the Institute of Biomedical Research and Innovation.

2.7.3 Elekta, Sweden

Prof. Lars Leksell (1907–1987) is the inventor of radiosurgery and developer of the advanced radiosurgical platform that bears his name, Leksell Gamma Knife®. Elekta was founded by Lars Leksell almost 50 years ago.

Elekta radiotherapy has three high-definition packages – now available across their family of digital accelerators – designed to ensure more patients could access high-quality treatment.

1. IntelliBeam is a high definition dynamic volumetric arc delivery technique only possible with the unique combination of Elekta digital accelerators and Monaco® treatment planning – allowing intelligent, high-definition and high-speed modulated treatment delivery.

2. SureStart is an advanced QA package that supports fast clinical start-up and allows you to optimize and maintain long-term performance of your digital accelerator in an efficient manner – enabling you to deliver treatments with confidence.

3. MOSAIQ* Plaza*, Elekta's patient centric, integrated software suite, is designed to work seamlessly with Elekta's radiotherapy systems to deliver comprehensive, cost-effective treatments. This technology helps drive efficient daily practice and connectivity to bring people and information together.

They produce three types of high definition (HD) digital accelerators:

1. Elekta Synergy HD is a digital accelerator for advanced IGRT, shown in Fig. 2.12. A trusted solution around the world. Close to 2,000 systems in clinical use in over 70 countries both mature and emerging markets, and about 1 million cancer patients helped annually through Synergy treatments. The unparalleled 50 cm × 26 cm largest IGRT field-of-view enables complete visualization of critical anatomical information to support clinical decisions at the time of treatment. Synergy's advanced MLC provides excellent beam shaping capabilities for dynamic delivery techniques and low leaf transmission protects organ-at-risk (OAR).

   Remote 24/7 support: Elekta IntelliMax proactive support and predictive maintenance help you avoid unplanned downtime to ensure departmental efficiency to boost your bottom line.

2. Elekta Infinity HD based on Elekta Infinity™ linear accelerator, as shown in Fig. 2.13.

3. Elekta Versa HD Advanced 4D image guidance for Lung SBRT with Symmetry, Optimized DCAT for Lung and Liver SBRT. Single isocenter high definition dynamic radiosurgery (HDRS) for multiple brain metastases. Lung SBRT treatment in less than 2 minutes with Versa HD (see Fig. 2.14).

2.7.4 IBA, Belgium

IBA is the world leader in the supply of PET & SPECT cyclotrons for radiopharmaceuticals production. Headquartered in Belgium, IBA installed and supports over 300 cyclotrons across the world. Based on 35 years of experience, IBA RadioPharma Solutions helps nuclear medicine department to design, build and operate PET center for the production of radiopharmaceuticals used for the detection and treatment of cancer and other critical diseases.

IBA is involved in the design, production and marketing of innovative solutions for the diagnosis and treatment of cancer and other serious illnesses, and for industrial applications

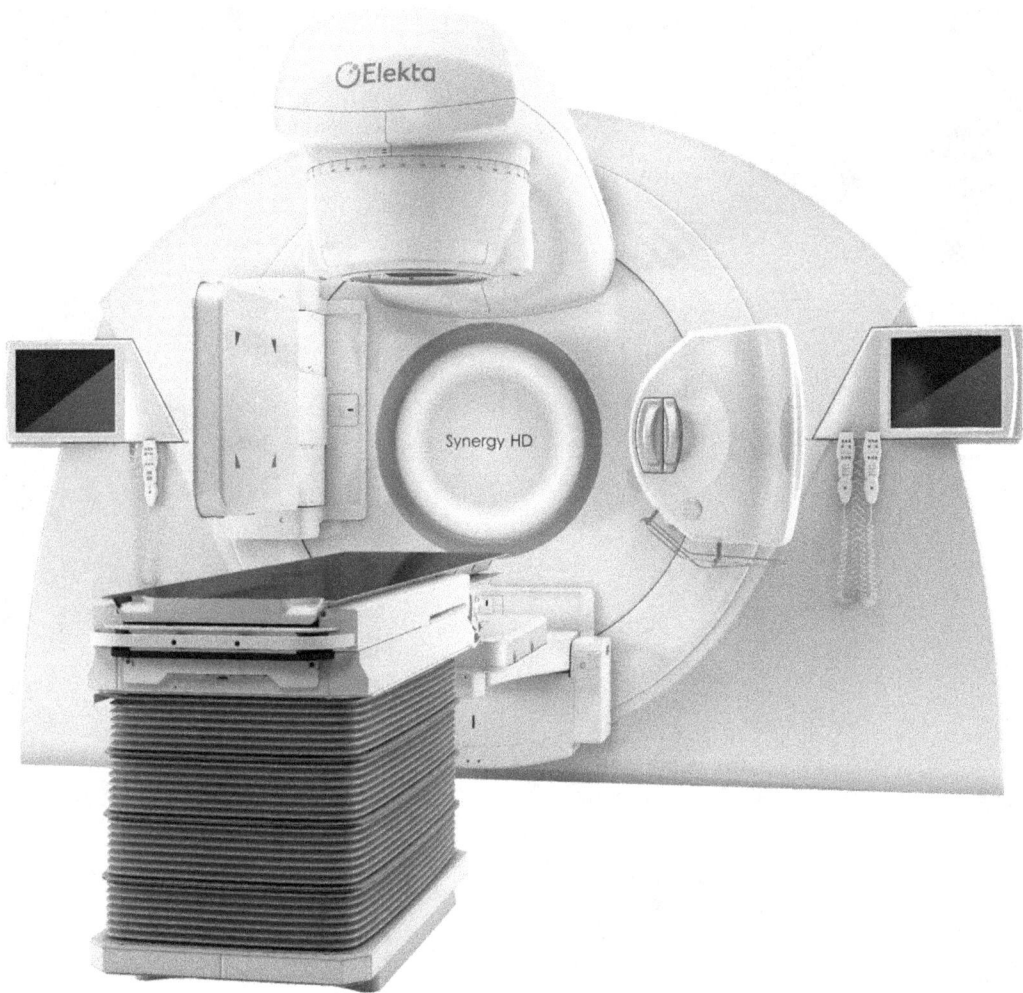

FIGURE 2.12
Elekta Synergy – left-angled front view. (Image courtesy of Elekta. Source: https://www.elekta.com/company/gallery.)

such as sterilization of medical devices. Around the world, thousands of hospitals use particle accelerators and dosimetry equipment designed, produced, maintained and upgraded by IBA. Their work is made of four core activities: Industrial Solutions, Radio-Pharma Solutions, Proton Therapy and Dosimetry.

According to the IBA 2020 Annual Report, there were more than 100,000 patients treated on IBA PT equipment, they sold 60 PT centers and their proton therapy market share was 42%!

IBA is the world leader in proton therapy with IBA customers having treated more than half of the proton therapy patients treated on commercial systems. The company has been leading proton therapy development for the last 30 years and has built the largest user community worldwide. IBA offers the highest uptime rates and can install a system in less than 12 months (Figs. 2.15 and 2.16).

The technological roadmap of IBA is focused on three areas: Motion Management, Arc Therapy and FLASH Irradiation. Motion management tools are needed to ensure accurate treatment delivery by managing the challenges caused by tumor motion.

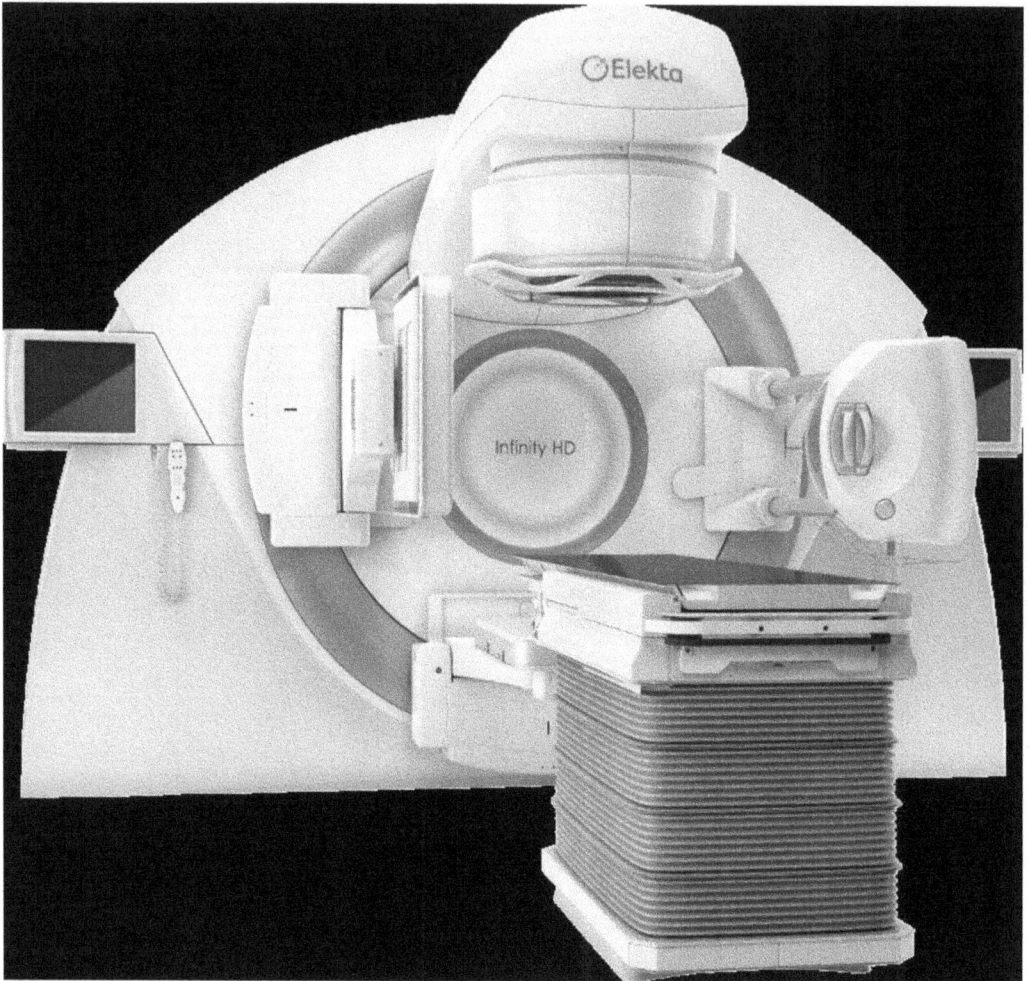

FIGURE 2.13
Infinity HD – Right Angled Front on View. (Image courtesy of Elekta. Source: https://www.elekta.com/company/gallery.)

With motion management, a proton therapy clinic will be able to treat more patients with more confidence.

Proton arc therapy has the possibility for further improvement of the quality of the treatment. This technological evolution will offer patients numerous advantages: (i) Potentially enhanced dose conformity at the tumor level and a potential reduction of the total dose received by the patient. (ii) Simplified treatment planning and delivery without performing the multiple field adjustments. (iii) Less time in the treatment room and a maximized patient throughput thanks to an optimized workflow.

FLASH irradiation refers to the fast and powerful treatment that delivers a high dose of radiation at an ultra-high dose rate. A novel technique could potentially shorten treatment time from 6 to 8 weeks to less than a week 6. It also has the potential to significantly reduce side effects in patients and as such, FLASH therapy has the potential to

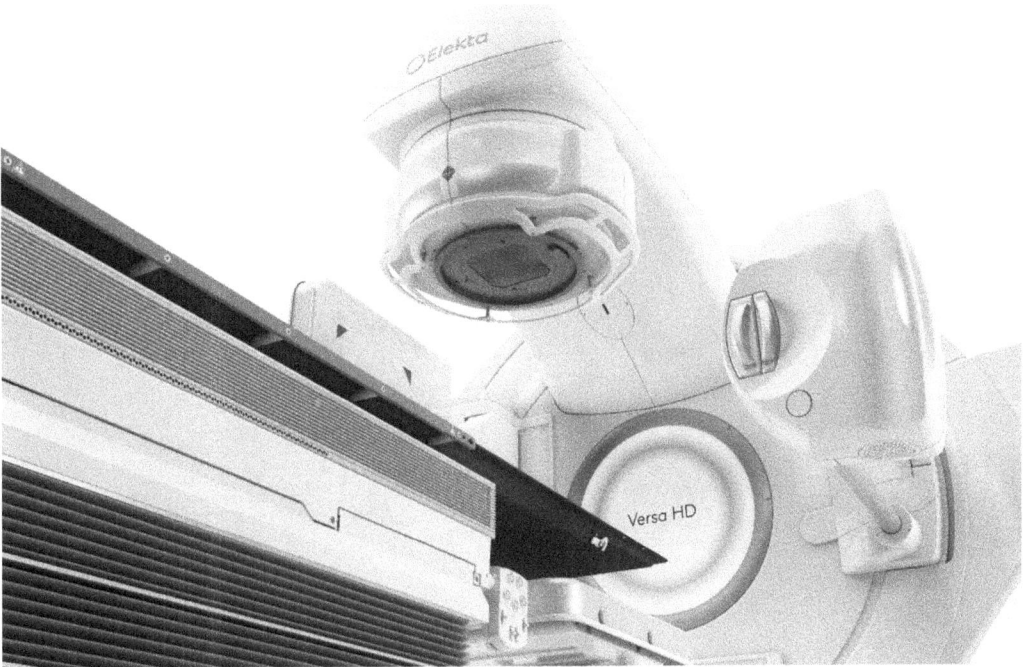

FIGURE 2.14
Versa HD – Image from below. (Image courtesy of Elekta. Source: https://www.elekta.com/company/gallery.)

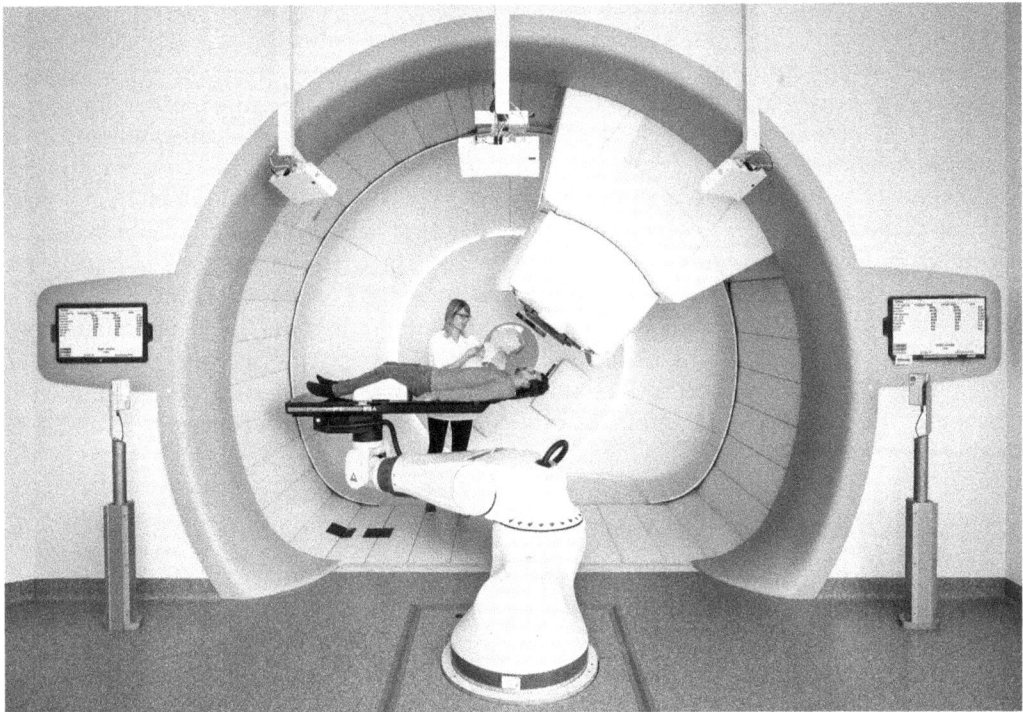

FIGURE 2.15
IBA's Proteus® ONE. (Courtesy IBA.)

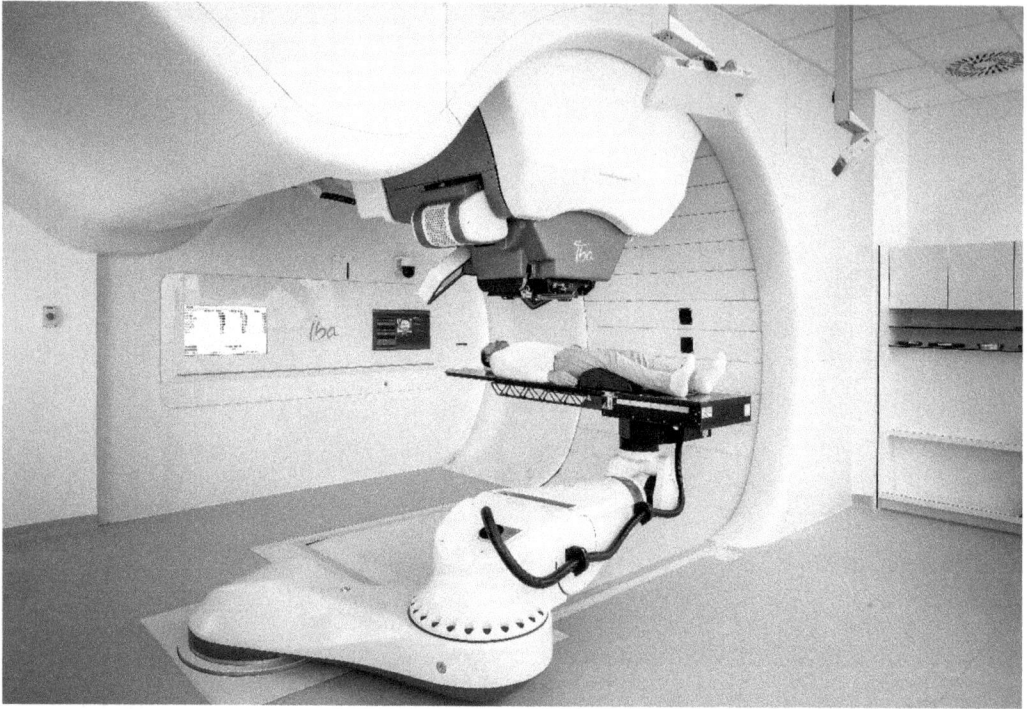

FIGURE 2.16
IBA's Proteus® PLUS. (Courtesy IBA.)

dramatically change the landscape of radiotherapy and patient cancer care, making it more effective and more accessible than conventional radiotherapy.

As a world leader in proton therapy, IBA is developing numerous partnerships with the aim of making its solutions more competitive, accessible for its clients and beneficial for their patients. The main area in which these partnerships are active concerns the integration of software and imaging solutions. A cutting-edge treatment like proton therapy demands great precision in the alignment of patients and constant monitoring of the development of the tumor during the treatment. The current technologies enable IBA to provide an adaptive treatment in which the tumor is constantly monitored and the treatment adapted accordingly. By intelligently integrating the most advanced capabilities of partners such as Elekta, Philips and RaySearch, IBA can leverage their IT developments in imagery and conventional radiotherapy to offer the most sophisticated treatment. This software integration results in unequaled performances in terms of quality of treatment functionality, workflow and automation.

2.7.5 SUMITOMO Heavy Industries, Ltd (SHI), Japan

The cyclotron systems for PET, available from the SHI Group, provide radioactive isotopes, which are required for the manufacture of these tracer RIs, in a safe and stable manner.

Since delivery of the AVF cyclotron, a large scale accelerator for nuclear physics experiments, to the Research Center for Nuclear Physics at Osaka University in 1972, the SHI Group has been developing and manufacturing numerous accelerators, such as large scale ring cyclotrons or linear accelerators, as well as ultra-compact synchrotron radiation

(SR) rings, etc., thereby contributing towards the development of accelerator technologies in Japan. The HM series of products of compact cyclotrons for PET was developed based on such a proprietary accelerator technology. Since the first unit was delivered to Kyoto University Hospital in 1981, the SHI Group has been taking the lead in the proliferation and development of cyclotrons for PET in Japan (The SHI Group has the top market share in Japan, with a delivery record exceeding 100 facilities.)

The SHI Group is the only Japanese manufacturer delivering cyclotrons for PET and providing support for PET diagnostics in Japan, from development through to after sales service, backed by an infallible framework (A manufacturing location was established in Niihama City, in Ehime Prefecture, with service sites established at eight locations nationwide from Hokkaido to Kyushu in order to provide speedy service).

Here, we shall mention two of the SHI Group product lines: CYPRIS HM-20 (see Fig. 2.17) and CYPRIS HM-12 (see Fig. 2.18).

CYPRIS HM-20 applications are in clinical, research and large-scale facilities.

Accelerator energy levels are for protons 20 MeV and for deuterons 10 MeV. Integrated shield around the accelerator is optional. The characteristics of this machine are:

- Flexible target configuration with max. 8 targets (4 targets × 2 ports). Simultaneous irradiation is available.
- Excellent production capacity with high current beam and targets for clinical, R&D and delivery.
- Significant space saving with self-shielding (option).

FIGURE 2.17
CYPRIS HM-20. (Courtesy of Sumitomo Heavy Industries, Ltd.)

FIGURE 2.18
CYPRIS HM-12. (Courtesy of Sumitomo Heavy Industries, Ltd.)

- Easy, safe and less radiation exposure for maintenance and service with vertical beam orientation.
- Capacity for metal and halogen isotope production (^{68}Ga, ^{64}Cu, ^{89}Zr, ^{123}I, ^{124}I, etc.).

CYPRIS HM-12 applications are mainly clinical (tumor, brain and circulatory).

Accelerator energy levels are for protons 12 MeV and for deuterons 6 MeV. Integrated shield is optional. The characteristics of this machine are:

- Flexible target configuration with max. 8 targets (4 targets × 2 ports). Simultaneous irradiation is available.
- High production capacity with high current beam and targets for clinical and R&D.
- Significant space saving with self-shielding (option).
- Easy, safe and less radiation exposure for maintenance and service with vertical beam orientation.
- Capacity for metal and halogen isotope production (^{68}Ga, ^{64}Cu, ^{89}Zr, ^{123}I, ^{124}I, etc.).

The SHI Group developed cyclotrons for proton beam therapy by utilizing the accelerator technology that it has been developing since the early 1970s. In 1997, the SHI Group delivered the first proton beam therapy cancer treatment system to the National Cancer Center Hospital East. In 1998, the SHI Group commenced operations as Japan's first proton beam therapy cancer treatment facility to be installed in the hospital (and the second in the world). The system continues to offer stable operation to this day. Proton beam therapy systems are comprised of an accelerator (cyclotron), an energy selection system, a beam transport system and rotating gantry irradiation equipment (gantry, irradiation nozzle and patient positioning equipment).

The SHI Group has the networks of the facilities where their systems are delivered in Japan (four facilities), Taiwan (two facilities) and South Korea (one facility) and provides operations of cyclotrons, overall system maintenance and management and after-sales service through the provision of information between facilities.

2.7.6 TeamBest®, USA

TeamBest® is a multinational medical company founded in 1977 in Springfield, Virginia, USA. TeamBest®, through Best Cyclotron Systems, Inc. (BCSI), Vancouver, Canada, offers radioisotopes and production capabilities for nuclear medicine and radiotherapy with its range of cyclotron systems. There are now five general energy domains for BCSI cyclotron systems: 15, 25, 30, 35 and 70 MeV. The applications for these accelerator systems are different, and the configurations will reflect that difference, (Tables 2.9 and (2.10).

- The BCSI 15p is a 15 MeV proton cyclotron. 15 MeV has been selected to offer a significant ^{18}F production yield with the investment efficiency of small power costs and low initial cost. As the user moves to other PET radioisotopes, the energy offers adequate production yields of radioisotopes such as ^{18}F and ^{64}Cu with the added benefit of optimized specific activity of the product. The cyclotron system has a small footprint and can be either installed in its own shielded room with or without local shielding, (reduces radiation) or can be shielded with integrated (self) shielding. Production targets for ^{18}F, ^{11}C, ^{13}N, ^{15}O and ^{64}Cu are

TABLE 2.9

Best cyclotrons

Best Cyclotrons	1–3 MeV	Deuterons for materials analysis
	70–150 MeV	For Proton Therapy
	3–90 MeV	High current proton beams for neutron production and delivery
Best 15p Cyclotron	15 MeV	Proton only, capable of high current up to 1000 µA, for medical radioisotopes
Best 20u/25p Cyclotrons	20, 25–15 MeV	Proton only, capable of high current up to 1000 µA, for medical radioisotopes
Best 30u/35p Cyclotrons	30, 35–15 MeV	Proton only, capable of high current up to 1000 µA, for medical radioisotopes
Best 70p Cyclotron	70–35 MeV	Proton only, capable of high current up to 1000 µA, for medical radioisotopes
Best 150p Cyclotron	From 70 MeV up to 150 MeV	For Medical Treatments of Benign and Malignant Tumors for Neurological, Eye, Head/Neck, Pediatric, Lung Cancers, Vascular/Cardiac/Stenosis/Ablation, etc.

TABLE 2.10

The capability of producing radiopharmaceuticals on demand

Isotope	Dose	Target	Reaction
^{11}C	40 mCi	Water enriched in ^{18}O	^{18}O+^{1}H^{+} → n+^{18}F
^{18}F	100 mCi	Nitrogen gas	^{14}N+^{1}H^{+} → ^{4}He^{++}+^{11}C

available. These systems are designed for individual hospital or small regional pharmacy use.

- The BCSI 25p is a 25 MeV proton cyclotron. It is designed to provide additional proton energy to make some specific single photon (SPECT) radioisotopes available for users. Most notable is ^{123}I, where the majority of production yield is covered by the 25 MeV energy beam. A group producing a specific radiopharmaceutical can, with the Best 25p, produce their own input material to maintain independence and cost-effective manufacturing. The key feature is a unique design for each application that streamlines cost and production efficiency.

- The Best 30 UPGRADEABLE (B30u) has all the features of the B25p but may be upgraded to a B35p when the user requires additional capability. The key upgrade feature is that the maximum energy of the B30u as delivered is 30 MeV. This provides about twice the amount of radioisotope that the B25p delivers. A production facility based on the B30u is designed to allow modifications for the upgrade to the B35p.

- The BCSI 35p is a 35 MeV proton cyclotron. It is designed to produce radioisotopes using the (p,2n) or the (p,3n) reaction. There is a broad range of single-photon emitters that are used in nuclear diagnostic imaging and therapy that are accessible in this energy range. TeamBest will partner with the end user to create a facility that will satisfy the end user's requirements and provide some of TeamBest's radioisotope supply requirements. Both solid and gas target systems can be added to the BCSI 35p system.

- The BCSI 70p is a 70 MeV proton cyclotron. The energy provides access to radionuclides produced by (p,xn) reactions and is a research accelerator as well as a radioisotope production cyclotron. TeamBest will collaborate with the end user to create a facility that will satisfy the end user's requirements and provide some of TeamBest's radioisotope supply requirements, together with the opportunity for joint research projects. Both solid and gas target systems can be added to the BCSI 70p system (Figs. 2.19–2.23).

TeamBest®, through Best ABT Molecular Imaging, Inc., Knoxville, Tennessee (http://www.bestabt.com/) is marketing the BG-75 Biomarker Generator, called "dose on demand", which produces unit doses of molecular imaging drugs for PET at the point of use, see Fig. 2.24. It encompasses an entire PET production lab in a 5.5 m × 5.5 m (30.25 m^2) room. It includes a self-shielded micro-cyclotron for producing the positron-emitting isotopes of ^{18}F and ^{11}C and microchemistry system for labeling specific molecules with the positron-emitting isotopes. The uniqueness of this system is that it is much smaller than conventional PET cyclotrons, easier to install and simple to operate.

The ABT self-shielded cyclotron operates at 7.5 MeV and 6 µA. It produces three or more unit doses of fludeoxyglucose (FDG) doses per run of 60 minutes, with synthesis accomplished in additional 60 minutes.

2.8 AMIT Project

AMIT stands for Advanced Molecular Imaging Technologies. The project started as a three-year project with partners 10 companies, 14 research laboratories and other

FIGURE 2.19
The BCSI 15p: a 15 MeV proton cyclotron. (Courtesy of Best Group of Companies.)

FIGURE 2.20
The BCSI 25p: a 25 MeV proton cyclotron. (Courtesy of Best Group of Companies.)

collaborators funded by the Spanish Ministry of Science and Innovation. The target was development of the core technology for molecular imaging in Medicine and Biomedicine with special focus in the human brain and in particular in mental diseases. The main step was the work package on development of a compact cyclotron for ^{11}C and ^{18}F production under the CIEMAT (Spanish Centre for Energy, Environment and Technology). The motivation was to develop a Compact cyclotron able to produce short half-life isotopes for sintering PET radiotracers, including:

- Extending the production of radioisotopes to hospitals and institutes which are not prepared for hosting conventional facilities.
- Disposing a back-up system for producing selected radiotracers at prices that can compete in specific cases with those of standard production centers.

FIGURE 2.21
The Best 30p (Upgradeable). (Courtesy of Best Group of Companies.)

FIGURE 2.22
The Best 35p Cyclotron. (Courtesy of Best Group of Companies.)

FIGURE 2.23
The BCSI 70p: a 70 MeV proton cyclotron. (Courtesy of Best Group of Companies.)

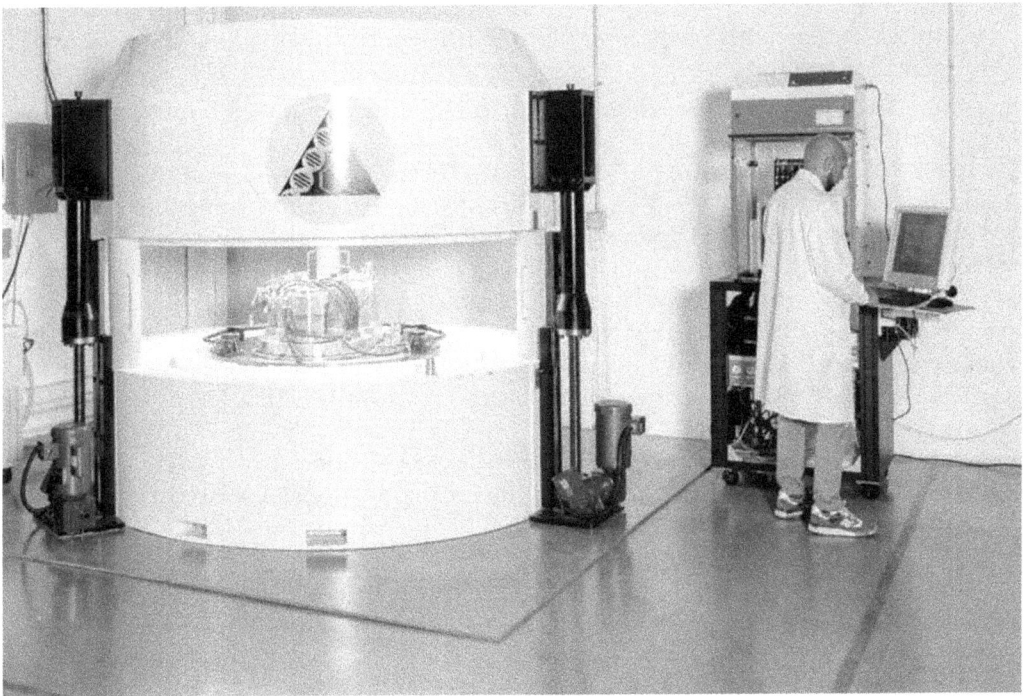

FIGURE 2.24
ABT self-shielded cyclotron with synthesis system for FDG production. (Courtesy of Best Group of Companies.)

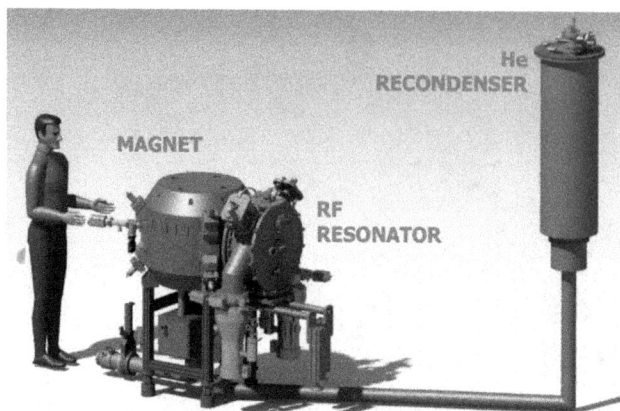

FIGURE 2.25
Schematic view of superconducting cyclotron dedicated to PET radioisotopes production (following Munilla 2016).

The AMIT continued by following phases:

- 2014–2017 The OPTIMHAC program (Spanish National funding): Beam optimization.
- 2014–2017 CIEMAT Funds.
- 2016–2019 The CIENTO program (Spanish Nacional funding): Cryogenics and operational optimization.
- 2017–2021 The ARIES program (H2020): Integration as a research infrastructure.

For details see Munilla (2016) and Portela (2017). The physical size of their cyclotron (beam requirements: E > 8.5 MeV I > 10μA) including all the components and the cryogenic supply system is shown by schematic drawing in Fig. 2.25.

The International Conference on Cyclotrons and their Applications has a long history, dating back to the 1950s. The conference takes place every 3 years with the most recent events being held: 18th in Giardini Naxos, Italy (2007), 21st in Zurich/Switzerland (2016), Vancouver/Canada (2013) and Lanzhou/China (2010). The 22nd conference in the series was held in Cape Town, South Africa and hosted by iThemba LABS. The conference took place from 22–27 September 2019.

In addition, we should mention an event of potential interest to readers, Academia-Industry Matching Event on Superconductivity for Accelerators for Medical Applications. 24–25 November 2016, CIEMAT, held in Madrid, Spain.

2.9 Biomaterials

Biomaterials play an important role in medicine today, especially by restoring function and facilitating healing for people after injury or disease. Origin of biomaterials may be natural or synthetic, they are both used in medical applications to support, enhance, or replace damaged tissue or biological functions. Biomaterials engineering and devices for human applications are described in details in the book by Wise et al. (2000a and 2000b). A review of the role of biomaterials in biomedical field is also described in articles by Yadav and Gangwar (2018), Yaqub and Min-Hua (2019) and Zhao et al. (2019).

The present demand for lifelong implants and bone replacement is characterized by biocompatibility, bioactivity and mechanical properties of biomaterials, without the immune rejection shows a great challenge. Presently stainless steel, titanium alloys, Co–Cr alloys, zirconia, alumina, ultra-high molecular weight polyethylene, polymethyl methacrylate, etc. are some of the biomaterials that satisfy the requirements of physical and chemical properties, mechanical strength and biocompatibility for biomedical applications. Significant progress has been made to improve the function of artificial joints, but currently the major efforts have been devoted to minimize the wear and to increase the lifespan of the implants or prosthesis during its functionality in the human body (see Baum 2021).

The role of accelerators is best known in their use for surface hardening. When used for surface hardening, the ions of the doping agent material are fired at the target material and only penetrate the material a very short distance, essentially remaining at the surface of the material. Ions are chosen that will complement the atomic structure of the target material, making it stronger. This process is used to create hard surfaces for materials that are used for example in objects like artificial heart valves (Oveissi et al. 2020) and other prosthetic implants.

In the case of artificial heart valves, there is an urgent need for developing alternative heart valves to overcome long-term stability and biocompatibility issues that are associated with artificial heart valves. The main challenge faced by material engineers is to create new, self-healing materials that can offer adequate mechanical properties for highly dynamic biological environments. While various self-healing materials have been developed, none so far has been used for the fabrication of heart valves.

Design technology and manufacturing process of orthopedic implants using electron beam melting (EBM) technology, which is a type of 3D printing, is the latest method of implant production in the world, where each piece is made directly according to the 3D computer model. In this method, the raw materials are melted and thin layers of materials complete the layer by layer of the desired part. Because each person's bones are different in shape and require unique surgery to repair them when they break, making any plaque or prosthesis by following the patient's bone line makes orthopedic surgery more successful. In this way, implant surgery has been able to help a lot to increase human life expectancy and rejuvenate organs. Studies have shown that the EBM technology method has great potential in orthopedics. Making custom plates and plates based on the EBM method has resulted in a significant reduction in the high cost of orthopedic procedures (MONIB 2021).

The cellulose membrane (CM) is a major component of plant cell walls and is both a chemically and mechanically stable synthetic polymer with many applications for use in tissue engineering. However, due to its dissolution difficulty, there are no known physiologically relevant or pharmaceutically clinical applications for this polymer. Thus, research is underway on controlled and adjusted forms of cellulose depolymerization. To advance the study of applying CM for tissue engineering, Eo et al. (2016) have suggested new possibilities for electron beam (E-beam) treatment of CM. Treatment of CM with an E-beam can modify physical, chemical, molecular and biological properties, so it can be studied continuously to improve its usefulness and to enhance value. The authors review clinical applications of CM, cellulose-binding domains, cellulose crosslinking proteins, conventional hydrolysis of cellulose and depolymerization with radiation and focus their experiences with depolymerization of E-beam irradiated CM.

We find of interest the recent paper by Seyfoori et al. (2021) which briefly reviews the progress in the areas of biomaterial use for pulmonary tissue regeneration and integration with current bioengineered platforms including engineered tissues, organoids and organs-on-a-chip platforms for viral respiratory disease studies. The authors also present a brief overview of the opportunities by organ-on-a-chip systems for studying COVID-19 and subsequent drug development.

2.10 Trace Elements

Trace element concentrations in body fluids and tissues hold much promise as a clinical test. Nuclear analytical methods offer a very interesting approach to these problems because of their ability to detect simultaneously several trace elements in very small samples (biopsy, hair, blood, etc.).

The specific aims of such investigations are to determine the clinical usefulness of multiple trace element concentration measurements for the definitive relationships to disease activity and efficiency of therapy; to explore further the preliminary observations that changes in trace element levels precede other symptoms or disease activity; to determine trace element concentrations in normal vs disease tissues.

In the years coming, the presence of traces of heavy elements in the human organism is going to be of great interest in the detection and probably in the treatment of human diseases. Nuclear analytical techniques have already become a very effective means of detecting trace elements in biomedical samples and sensitivities from 0.1 to 1.0 ppm wet weight can now easily be obtained. These sensitivities can generally be achieved in ~15 min, using as little as 100 μg of material,

Trace element analysis of biopsy samples represents a difficult analytical problem because of the small amount of material available for analysis (wet weight: from 100 μg to 2–3 mg) and the lack of adequate reference materials for micro-analytical investigations. Valković et al. (1993) have evaluated the capabilities of X-ray fluorescence in both standard and total reflection geometry, proton-induced X-ray emission and neutron activation analysis for multi-elemental analysis of biopsy samples. Normal mucosa and mucosa with ulcerative colitis (a pre-cancer stage) were chosen for the described study. It has been suggested that Zn, Se and Ca could be involved in the colon cancer development. Analysis of biopsies is of invaluable importance as an ultimate diagnostic tool, especially in the case of tumors. The recognition of the role played by trace elements in tumor development and growth has resulted in the need for multi-elemental analysis of biopsy material. The Se and Ca supplementation seems to reduce the colon cancer incidence, while a Zn deficiency could have a protective effect.

Intestinal mucosa samples were collected during endoscopy examination (colonoscopy), colonic segments were taken by using a steel pliers and then, immediately frozen (at −70°C) in polystyrene beakers. Afterward, the samples were freeze-dried to constant weight with a mean drying time of 24 h. For both, standard XRF and PIXE, biopsy samples were prepared as sandwich between Mylar and Formvar foils. Typical spectra are shown on Fig. 2.26, top: XRF spectrum, bottom: PIXE spectrum.

It is evident that the concentrations of the following elements could be determined: S, Cl, K, CaTi, Mn, Fe, Cu, Zn, Br, Rb and Pb. Compared with XRF, 3.0 MeV PIXE has better sensitivities for lighter elements and not so good sensitivities for other elements of interest, for example, Se.

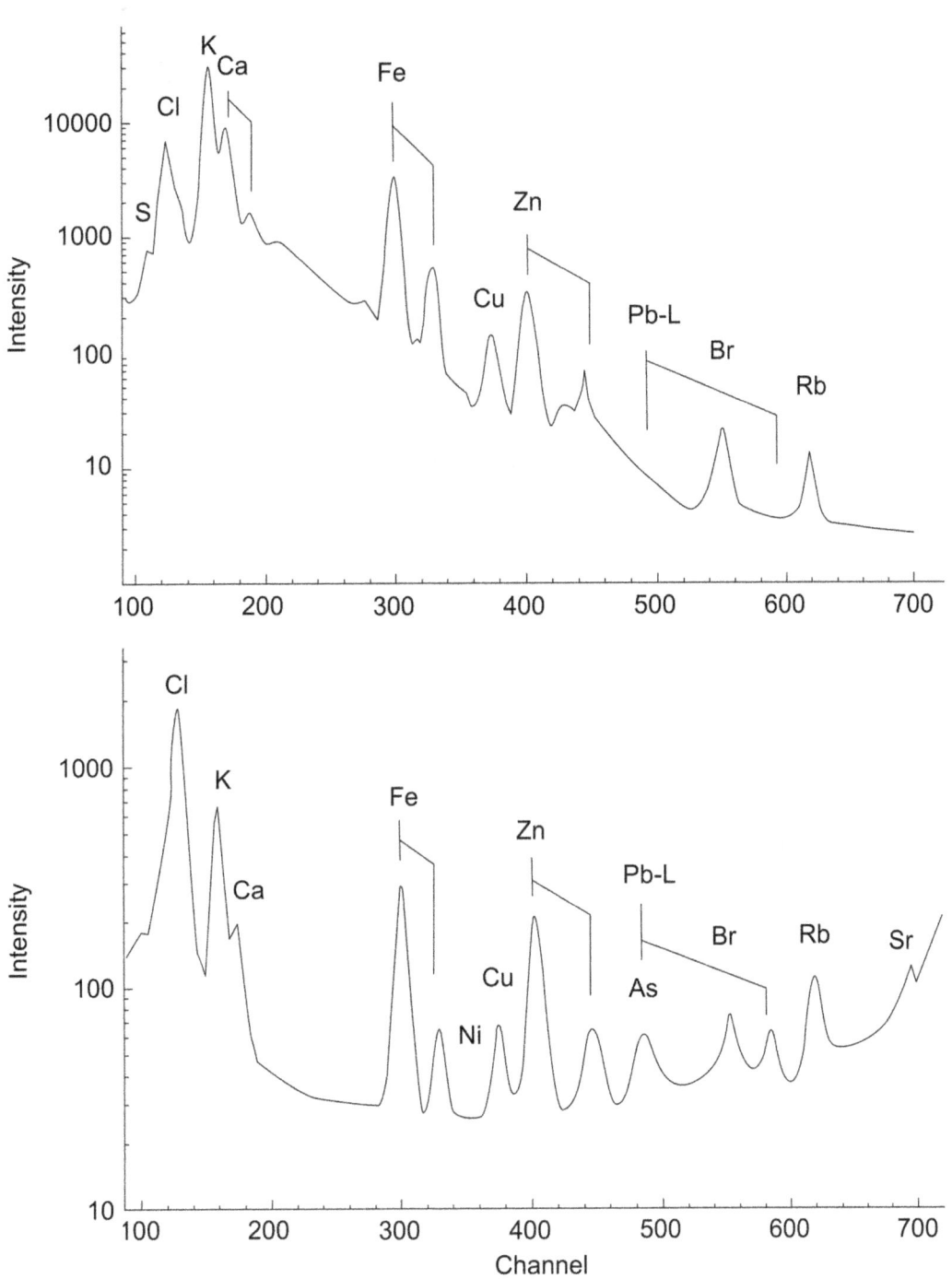

FIGURE 2.26
X-ray emission spectroscopy of biopsy sample, top: XRF, bottom: PIXE.

2.10.1 PIXE Analysis of Biological Material

Nuclear analytical techniques have been successfully applied to biological material since their introduction, see Valković (1980). Applications of nuclear microprobes in the life sciences have been done sometime later by Boss et al. (1983, 1985), Llabador and Moretto (1998) and others. X-ray emission spectroscopy and PIXE in particular has been applied to a variety of problems.

Here, we shall discuss to some extent applications in the biomedical field. The accumulation of some metals in some organs in several diseases has been known for some time. Much attention has been paid to the accumulation of early measurable trace metals in organs during the growth of tumors. This has been suggested as a diagnostic tool. Since Cu, Fe and Zn concentrations in blood serum (normal values: 1–2 ppm) hold much promise as a clinical test, several methods have been developed for determining their concentrations.

Charged-particle-induced X-ray emission spectroscopy offers a very interesting approach to these problems because of its ability to detect simultaneously several trace elements. In addition, the technique has the advantage that only very small quantities of serum are needed in target preparation; good results can be obtained with a drop of blood serum on an appropriate backing.

The specific aims of such investigations should be: to determine the clinical usefulness of multiple trace element concentration measurements for the definitive relationships to disease activity and efficiency of therapy; to explore further the preliminary observations that changes in trace element levels precede other symptoms or disease activity; to determine trace element concentrations in normal vs. diseased tissues.

Concentrations of the essential trace elements in body fluids are rather low. For example, blood samples contain a large number of trace elements, however; only Cu, Fe, Zn and Sr are found in concentrations higher than 1 ppm in healthy adults. These elements can be easily determined by X-ray emission spectroscopy using samples of some 60 µl dried on an appropriate backing material. However, many elements of biochemical interest (some of them holding a promise to be used as a clinical test in pathological conditions) occur in much lower concentrations. The same is true of the other body fluids as well. An example of essential trace elements that occur at low concentrations in body fluids is selenium (human blood serum concentration 30–100 ppb). It has been shown that with a suitable preconcentration procedure (acid digestion) both PIXE and X-ray tube induced XRF (T-XRF) could be used in the analysis of Se in human blood serum. Much promise is given by the use of SR-induced XRF (SR-XRF) in total reflection geometry (Phillips et al. 1986). Because of the small source size and natural collimation of SR exiting the storage ring, X-rays can be directed with high precision at small glancing angles to the specimen and are particularly effective for use in the total reflection geometry. In addition, the high degree of polarization, high intensity and continuous tenability of the beam energy are contributing factors for obtaining good analytical sensitivity. Such an arrangement is described by Spal et al. (1984). An MDL of 8 ppb was determined for selenium in human blood serum and in proposed NBS-SRM 1598 bovine serum. The results show that this method is sufficiently sensitive for the analysis of Se in blood serum.

Let us mention that micro-PIXE allows the measurement of elemental profiles in individual blood cells. For example, Johansson and Lindh (1984) have observed altered elemental profiles in erythrocytes and neutrophil granulocytes in patients with chronic lymphatic leukemia, chronic and acute myeloid leukemia. In another report, Johansson et al. (1987) observed that the elemental profiles of erythrocytes and neutrophil granulocytes in breast cancer patients displayed significant alteration in some essential and nonessential elements.

Other body fluids have been also measured by nuclear analytical techniques. For example, Maeda et al. (1990) have reported PIXE analysis of human spermatozoa free from contaminations of seminal plasma. Trace element concentrations in urine have been widely used to monitor occupational exposure and intoxication with heavy metals, such as chromium.

Bone is a dynamic structure, constantly remodeling in response to changing mechanical and environmental factors. This is particularly evident in the mineral component encrusting the collagenous framework. The mineral is principally in the form of calcium apatite, but calcium can exchange with strontium, both during the cellular processes of mineralization and resorption and by passive exchange with the deposited crystals. Mineralization is generally characterized by densitometry, but because of the differences in absorption cross-sections of calcium and strontium, it can be misleading in studies of composition. In their work, Bradley et al. (2007) have used X-ray diffraction to identify calcium and strontium apatite and X-ray fluorescence to quantify strontium and calcium distribution. The beam characteristics available from SR, this has enabled us to obtain microscopic resolution on thin sections of bone and cartilage from the equine metacarpophalangeal joint.

The authors investigated the distribution of minerals in the bone-cartilage interface and within individual trabeculae. In trabecular bone, the ratio of strontium to calcium concentration was typically 0.0035 ± 0.0020, and higher by a factor of ~3 at the periphery than in the center of a trabeculum (possibly reflecting the more rapid turnover of mineral in the surface layer). In the dense subchondral bone, the ratio was similar, approximately doubling in the calcified cartilage.

Bradley et al. (2007) also explored the changes in mineralization associated with the development of osteoarthritis. They analyzed lesions showing cartilage thinning and changes in the trabecular organization and density of the underlying bone. At the center of the lesion, the ratio of strontium to calcium was much lower than that in normal tissue, although the calcified cartilage still showed a higher ratio than the underlying bone. In the superficially normal tissue around the lesion, the calcified cartilage returned to a normal ratio much more rapidly than the underlying bone. These data demonstrate the complex relationship between changes in cartilage and the underlying bone.

Kaabar et al. (2010) have studied the bone-cartilage interface. μ-PIXE analysis has been employed in investigating and quantifying the distribution of a number of essential elements in thin human diseased articular cartilage sections affected by osteoarthritis. Various cations Ca, P and Zn have been reported to play an important role both in the normal growth and in remodeling of articular cartilage and subchondral bone as well as in the degenerative and inflammatory processes associated with the disease. They also act as co-factors of a class of enzymes known as metalloproteinase that is believed to be active during the initiation. Progress and remodeling processes associated with osteoarthritis. Other important enzymes such as alkaline phosphatase are associated with cartilage mineralization. SR X-ray fluorescence for mapping of elemental distributions in bone and cartilage has also been employed by the present group and others. Small-Angle X-ray Scattering was carried out on decalcified human articular cartilage to explore the structural and organizational changes of collagen networks in diseased articular cartilage.

2.10.2 AMS Analysis of Biological Material

AMS has revolutionized high-sensitivity isotope detection in biomedical research because it allows the direct determination of the amount of isotope in a sample rather than measuring its decay, and thus the quantitative analysis of the fate of the radiolabeled probes under the given conditions. AMS was first used in the early 1990s for the analysis of biological samples containing enriched ^{14}C for toxicology and cancer research. Today, the biomedical applications of AMS range from *in vitro* to *in vivo* studies, including the studies of (i) toxicant and drug metabolism, (ii) neuroscience, (iii) pharmacokinetics and (iv) nutrition and metabolism of endogenous molecules such as vitamins. In addition, a new drug development concept that relies on the ultra-sensitivity of AMS, known as human micro dosing, is being used to obtain early human metabolism information of candidate drugs (Hah 2009).

For a practical AMS measurement, biological samples containing 0.2–5 mg of carbon must be converted to solid carbon (graphite or fullerene) using a two-step process. In a quartz tube, and using excess copper oxide (CuO), the sample's biological carbon is oxidized to CO_2. The CO_2 is then reduced to solid carbon by both reduction with titanium hydride and zinc powder and catalyzation with either iron or cobalt (Ognibene et al. 2003).

AMS has been adopted as a powerful bio-analytical method for human studies in the areas of pharmacology and toxicology. The exquisite sensitivity (10^{-18} mol) of AMS has facilitated studies of toxins and drugs at environmentally and physiologically relevant concentrations in humans. Such studies include risk assessment of environmental toxicants, drug candidate selection, absolute bioavailability determination, and more recently, assessment of drug-target binding as a biomarker of response to chemotherapy (Enright et al. 2016). Combining AMS with complementary capabilities such as high-performance liquid chromatography (HPLC) can maximize data within a single experiment and provide additional insight when assessing drugs and toxins, such as metabolic profiling.

AMS can bridge this gap and provide a sensitive and accurate method to measure low-level doses that can range from environmentally relevant concentrations of toxicants to sub-therapeutic and micro-dosing levels of drugs. Micro dosing is the administration of a low sub-pharmacological dose (up to 100 µg or 1% of the therapeutic dose, whichever is smaller) to study the effects of a drug or toxin at levels that are unlikely to produce an observable effect, but high enough for assessment of cellular and systemic response, see Combes et al. (2003) and Garner (2010). AMS allows for direct testing of toxicants in humans, allowing for a more accurate risk assessment.

Combining AMS analysis with other highly sensitive techniques such as PET can provide complementary information and maximize clinical data within the same study including: long-term pharmacokinetic information, cellular binding and metabolite profiling. These measurements may provide critical information when assessing new drug candidates or potential toxicity of environmental materials and chemicals (Enright et al. 2016).

Another example is presented in the work by Klein 2013) who used the 1 MV multi-element AMS system for biomedical applications at the Netherlands Organization for Applied Scientific Research (TNO) for analyses of ^{14}C, ^{41}Ca and ^{129}I. The TNO performs clinical research programs for pharmaceutical and innovative foods industry to obtain early pharmacokinetic data and to provide anti-osteoporotic efficacy data of new treatments. The AMS system analyzes carbon, iodine and calcium samples for this purpose. The measurements on blank samples indicate background levels in the low 10^{-12} for calcium and iodine, making the system well suited for these biomedical applications. Carbon blanks have been measured at low 10^{-16}. For unattended, around-the-clock analysis, the system features the 200 sample version of the HVE SO-110 hybrid ion source and user-friendly control software.

2.10.3 Live Cell Imaging

Ion micro beams originally were designed to apply a defined number of ions to one or more cells in a cell population. In contrast to radiobiological experiments with a broad ion beam where the number of ion traversals of a cell varies over the population of cells due to Poisson distribution, application of defined numbers of ions minimizes cell-to-cell dose variation. In the beginning of biological research using ion microbeams, the focus was placed on the impact of very small doses, down to one particle traversal within a whole cell population (Matsumoto et al. 2009). The possibility of targeting single cells with a defined number of ions, in combination with the huge progress of molecular and cellular biological techniques in the last two decades, opened a much broader field of biological/ biomedical microbeam research

The irradiation setup at the ion microprobe SNAKE, attached to 14 MV Munich tandem accelerator, is used to irradiate living cells with single energetic ions. The ir- radiation accuracy of 0.55 µm and respectively 0.40 µm allows irradiating substructures of the cell nucleus. By the choice of ion atomic number and energy, the irradiation can be performed with a damage density adjustable over more than three orders of mag- nitude. Immunofluorescence detection techniques show the distribution of proteins involved in the repair of DNA double-strand breaks. In one of the first experiments, the kinetics of the appearance of irradiation-induced foci in living HeLa cells was ex- amined. In other experiments, a new effect was detected which concerned the inter- action between irradiation events performed at different time points within the same cell nucleus (Hauptner et al. 2006).

The live-cell imaging facility at SNAKE is now able to routinely perform ambitious experiments. In addition to recording time series after single ion irradiation in defined patterns (Hable et al. 2012), experiments with selective high dose irradiation of cells within a reasonable time span can be performed. This raises the possibility to define the minimal dose needed for DNA repair proteins to accumulate in visible foci. Further, cellular substructures like nucleoli or different chromatin states (heterochromatin vs. euchromatin) can be targeted and irradiated with a defined number of ions, e.g., to in- vestigate differential response mechanisms. Now, the setup is modified to perform three- colour LCI experiments. This enables the study the behaviour of three different proteins tagged with different fluorescent markers.

Ion microbeams are important tools in radiobiological research. Still, the worldwide number of ion microbeam facilities where biological experiments can be performed is limited. Even fewer facilities combine ion micro irradiation with live-cell imaging to allow microscopic observation of cellular response reactions starting very fast after ir- radiation and continuing for many hours. At SNAKE in Munich 14 MV tandem accel- erator, a large variety of biological experiments is performed on a regular basis. Drexler et al. (2015) have presented the experimental developments and ongoing research projects at the SNAKE setup with specific emphasis on live-cell imaging experiments. An over- view of the technical details of the setup is given, including examples of suitable biolo- gical samples. By ion beam focusing to sub micrometer beam spot size and single ion detection, it is possible to target subcellular structures with defined numbers of ions. Focusing on high numbers of ions to single spots allows studying the influence of high local damage density on the recruitment of damage response proteins.

Lim et al. (2021) reported a means by which atomic and molecular secondary ions, including cholesterol and fatty acids, can be sputtered through single-layer graphene to enable secondary ion mass spectrometry imaging of untreated wet cell membranes in solution at subcellular spatial resolution. They could observe the intrinsic molecular distribution of lipids, such as cholesterol, phosphoethanolamine and various fatty acids, in untreated wet cell membranes without any labeling. They showed that graphene-covered cells prepared on a wet substrate with a cell culture medium reservoir are alive and that their cellular membranes do not disintegrate during SIMS imaging in an ultra-high vacuum environment. Ab initio molecular dynamics calculations and ion dose-dependence studies suggest that sputtering through single-layer graphene occurs through a transient hole generated in the graphene layer. Cholesterol imaging shows that methyl-β-cyclodextrin preferentially extracts cholesterol molecules from the cholesterol-enriched regions in cell membranes (Lim et al. 2021).

The risk assessment for low doses of high LET radiation has been challenged by a growing body of experimental evidence showing that non-irradiated bystander cells can receive signals from irradiated cells to elicit a variety of cellular responses. These may be significant for radiation protection but also for radiation therapy using heavy ions. Charged particle microbeams for radiobiological application provide a unique means to address these issues by allowing the precise irradiation of single cells with a counted numbers of ions.

Voss et al. (2008) are focused specifically on heavy-ion microbeam facilities currently in use for biological purposes, describing their technical features and biological results. Typically, ion species up to argon are used for targeted biological irradiation at the vertically collimated microbeam at JAEA (Takasaki, Japan). At the SNAKE microprobe in Munich, mostly oxygen ions have been used in a horizontal focused beam line for cell targeting. At GSI (Darmstadt), a horizontal microprobe with a focused beam for defined targeting using ion species up to uranium is operational. The visualization of DNA da-mage response proteins relocalizing to defined sites of ion traversal has been accom-plished at the three heavy ion microbeam facilities described above and is used to study mechanistic aspects of heavy ion effects. However, bystander studies have constituted the focus of biological applications. While for cell inactivation and effects on cell cycle pro-gression a response of non-targeted cells has been described at JAEA and GSI, respec-tively, in part controversial results have been obtained for the induction of DNA damage measured by double-strand formation or at the cytogenetic level. The results emphasize the influence of the cellular environment, and standardization of experimental conditions for cellular studies at different facilities as well as the investigation of bystander effects in the tissue will be the aims of future research. At present, the most important conclusion of radiobiology studies at heavy-ion microbeams is that bystander responses are not ac-centuated for increasing ionizing density radiation (Voss et al. 2008).

References

Allonia, D., Prataa, M., and Smilgysc, B. 2019. Experimental and Monte Carlo characterization of radionuclidic impurities originated from proton irradiation of $[^{18}O]H_2O$ in a modern medical cyclotron. *Appl. Radiat. Isot.* 146: 84–89.

Anand, S. S., Singh, H., and Dash, A. K. 2009. Clinical Applications of PET and PET-CT. *Med. J. Armed Forces India* 65(4): 353–358.

Baum, H. 2021. New materials for body implants & prostheses. https://www.uc.edu›content›dam›docs›OLLI.

Bensaleh, S., Bezak, E., and Borg, M. 2009. Review of MammoSite brachytherapy: advantages, disadvantages and clinicaloutcomes. *Acta Oncol.* 48(4): 487–494.

Bestcyclotron Systems. http://www.bestcyclotron.com/products.html.

Bos, A. J. J., v.d. Stap, C. C. A. H., Lenglet, W. J. M., Vis, R. D., and Valković, V. 1983. Measurements of trace element concentration profiles across the diameter of human hair with micro-PIXE. *IEEE Trans. Nucl. Sci.* NS-30(2): 1249–1251.

Bos, A. J. J., v.d. Stap, C. C. A. H., Valković, V., Vis, R. D., and Verheul, H. 1985. Incorporation routes of elements into human hair, implications for hair analysis used for monitoring. *Sci. Total Environ.* 42: 157–169.

Bradley, D. A., Muthuvelu, P., Ellis, R. E., et al. 2007. Characterisation of mineralisation of bone and cartilage: X-ray diffraction and Ca and Sr KC(X-ray fluorescence microscopy. *Nucl. Instrum. Methods Phys. Res. B* 263: 1–6.

Brady, Z., Taylor, M. L., Maynes, M., et al. 2008. The clinical application of PET/CT: A contemporary review. *Aust. Phys. Eng. Sci. Med.* 31(2): 90–109.

Combes, R. D., Berridge, T., Connelly, J., Eve, M. D., Garner, R. C., Toon, S., and Wilcox, P. 2003. Early microdose drug studies in human volunteerscan minimise animal testing: Proceedings of a workshop organised by Volunteersin Research and Testing. *Eur. J. Pharm. Sci.* 19(1): 1–11.

Costouros, N. G. and Hawkins, R. A. 2009. Cardiac and neurological PET-CT applications. Applied Radiology August 5, 2009.

Damato, B., Kacperek, A., Errington, D., and Heimann, H. 2013. Proton beam radiotherapy of uveal melanoma. *Saudi J. Ophthalmol.* 27(3): 151–157.

Dehnel, M. P., Trudel, A., Duh, T. S., and Siewart, T. 2005. Beamline developments in commercial cyclotron facilities. *Nucl. Instrum. Methods Phys. Res. B* 241: 655–659.

Drexler, G. A., Siebenwirth, C., Drexler, S. E., et al. 2015. Live cell imaging at the Munich ion microbeam SNAKE – a status report. *Radiat. Oncol.* 10: 42 (8 pages). 10.1186/s13014-015-0350-7

Durante, M. 2019. Proton beam therapy in Europe: more centres need more research. *Br. J. Cancer* 120: 777–778.

Durante, M. and Friedl, A. A. 2011. New challenges in radiobiology research with microbeams. *Radiat. Environ. Biophys.* 50(3): 335–338.

Enright, H. A., Malfatti, M. A., Zimmermann, M., Ognibene, T., Henderson, P., and Turteltaub, K. W. 2016. The use of accelerator mass spectrometry in human health and molecular toxicology. *Chem. Res. Toxicol.* 29(12): 1976–1986.

Eo, M. Y., Fan, H., Cho, Y. J., Kim, S. M., and Lee, S. K. 2016. Cellulose membrane as a biomaterial: from hydrolysis to depolymerization with electron beam. *Biomater. Res.* 20: 16 (13 pp).

Finn, R., Capitelli, P., Sheh, Y., Lom, C., Graham, M., and Germain, J. St. 2005. Engineering refinements to overcome default nuclide regulatory constraints. *Nucl. Instrum. Methods Phys. Res. B* 241: 665–669.

Garner, R. C. 2010. Practical experience of using human microdosing with AMS analysis to obtain early human drug metabolism and PK data. *Bioanalysis* 2(3): 429–440.

Gragoudas, E. S., Goitein, M., Koehler, A. M., et al. 1977. Proton irradiation of small choroidal malignant melanomas. *Am. J. Ophthalmol.* 83: 665–673.

Gragoudas, E. S. 2006. Proton beam irradiation of uveal melanomas: The First 30 Years The Weisenfeld Lecture. *Investig. Ophthalmol. Vis. Sci.* 47: 4666–4673.

Hable, V., Drexler, G. A., Brüning, T., et al. 2012. Recruitment kinetics of DNA repair proteins Mdc1 and Rad52 but not 53BP1 depend on damage complexity. *PLoS One.* 7: e41943.

Hah, S. S. 2009. Recent advances in biomedical applications of accelerator mass spectrometry. *J. Biomed. Sci.* 16: Art. no. 54. https://doi.org/10.1186/1423-0127-16-54.

Hauptner, A., Cremer, T., Deutsch, M., et al. 2006. Irradiation of living cells with single ions at the ion microprobe SNAKE. *Acta Phys. Polonica A* 109(3): 273–278.

IAEA. 2008. Cyclotron Produced Radionuclides; Principles and Practice. Technical Reports Series No. 465. International Atomic Energy Agency, Vienna, Austria.

IAEA. 2015. Quality Management in Nuclear Medicine Practices, 2nd edition. Human health Series No. 33. International Atomic Energy Agency, Vienna, Austria.

IAEA. 2015b. The Transition from 2-D Brachytherapy to 3-D High Dose Rate Brachytherapy. IAEA Human Health Reports No. 12. International Atomic Energy Agency, Vienna, Austria.

IAEA. 2015c. Implementation of High Dose Rate Brachytherapy in Limited Resource Settings, IAEA Human Health Series No. 30. International Atomic Energy Agency, Vienna, Austria.

IAEA. 2017. The Transition from 2-D Brachytherapy to 3-D High Dose Rate Brachytherapy: Training Material. Training Course Series (CD-ROM) No. 61. IAEA. International Atomic Energy Agency, Vienna, Austria.

IAEA. 2020a. Accelerator Knowledge Portal. https://nucleus.iaea.org/sites/accelerators/Pages/Cyclotron.aspx, https://nucleus.iaea.org/sites/accelerators/knowledgerepository/Lists/OnlineResources/AllItems.aspx

IAEA. 2020b. Nuclear Medicine Resources Manual, International Atomic Energy Agency, Vienna, Austria.

IAEA. 2021. Production, Quality Control and Clinical Applications of Radiosynovectomy Agents. IAEA Radioisotopes and Radiopharmaceuticals Reports No. 3. International Atomic Energy Agency, Vienna, Austria.

Inoue M. 1998. Status of accelerators in Japan. Contribution to the 1st Asian Particle Accelerator Conference (APAC 98). Proc. The first Asian Particle Accelerator Conference APAC98 (1999): 23–27.

Jadvar, H. 2021. Expanding Applications for PET and PET/CT Scanning. Medscape, March 10, 2021. https://www.medscape.org/viewarticle/553465.

Johansson, E. and Lindh, U. 1984. Leukemia – its manifestation in the microelement profile in individual blood cells as determined in the nuclear microprobe. *Nucl. Instrum. Methods B* 3(1–3): 637–640.

Johansson, E., Lindh, U., Johansson, H., and Sundstrom, C. 1987. Micro-PIXE analysis of macro- and trace elements in blood cells and tumors of patients with breast cancer. *Nucl. Instrum. Methods Phys. Res. B* 22(1–3): 179–183.

Johnson, R. R., Erdman, K., Gyles, W., Burbee, J., Manegoda, A., Sabaiduc, V., Kovac, B., VanLier, E., Wong, J., Watt, R., Wilson, J., and Zyuzin, A. 2005. Radioisotope production targets and modules. *Nucl. Instrum. Methods Phys. Res. B* 241: 670–675.

Kaabar, W., Gundogdu, O., Laklouk, A., et al. 2010. μ-PIXE and SAXS studies at the bone-cartilage interface. *Appl. Radiat. Isot.* 68: 730–734.

Kim, E. E., Lee, M.-C., Inoue, T., and Wong, W.-H. (Eds.). 2013. *Clinical PET and PET/CT Principles and Applications*. Springer-Verlag, New York.

Kim, T. W., Choi, E., Park, J., et al. 2018. Clinical outcomes of proton beam therapy for choroidal melanoma at a single institute in Korea. *Cancer Res. Treat.* 50(2): 335–344.

Kitson, S. L., Cuccurullo, V., Ciarmiello, A., Salvo, D., and Mansi. L. 2009. Clinical applications of positron emission tomography (PET) imaging in medicine: oncology, brain diseases and cardiology. *Current Radiopharm.* 2(4): 224–253.

Klein, M., Vaes, W. H. J., Fabriek, B., et al. 2013. The 1 MV multi-element AMS system for biomedical applications at the Netherlands Organization for Applied Scientific Research (TNO). *Nucl. Instrum. Methods Phys. Res. B* 294: 14–17.

Košťál, M., Losa, E., Schulc, M., et al. 2019. The methodology of characterization of neutron leakage field from PET production cyclotron for experimental purposes. *Nucl. Instrum. Methods Phys. Res. A* 942: 162374, (8 pp).

Koukourakis, G., Kelekis, N., Armonis, V., and Kouloulias, V. 2009. Adv. Urol. vol. 2009: article ID 327945 (11 pp).

Kreiner, A. J., Baldo, M., Bergueiro, J. R., et al. 2014. Accelerator-based BNCT. *Appl. Radiat. Isot.* 88: 185–189.

Lane, A. M., Kim, I. K. S., and Gragoudas, E. S. 2015. Long-term risk of melanoma-related mortality for patients with uveal melanoma treated with proton beam therapy. *JAMA Ophthalmol.* 133(7): 792–796.

Lim, H., Lee, S. Y., Park, Y., Jin, H., Seo, D., Jang, Y. H., and Moon, D. W. 2021. Mass spectrometry imaging of untreated wet cell membranes in solution using single-layer graphene. *Nat. Methods* 18: 316–320.

Llabador, Y. and Moretto, P. 1998. *Applications of Nuclear Microprobes in the Life Sciences*. World Scientific Publishing Co., Singapore.

Mahajan, A. and Cook, G. 2017. Clinical Applications of PET/CT in Oncology. In: Khalil M. (eds) *Basic Science of PET Imaging*. Springer, Cham. 10.1007/978-3-319-40070-9_18.

Maeda, K., Sasa, Y., Kusuyama, H., Yoshida, K., and Uda, M. 1990. PIXE analysis of human spermatozoa isolated from seminal plasma. *Nucl. Instrum. Methods Phys. Res. B* 49: 228–230.

Matsumoto, H., Tomita, M., Otsuka, K., and Hatashitam M. 2009. A new paradigm in radioadaptive response developing from microbeam research. *J Radiat Res.* 50: A67–A79.

Merrill, F. E. 2015. Flash proton radiography. *Rev. Accel. Sci. Technol.* 8: 165–180. 10.1142/S179362 6815300091.

Michalec, B., Swakoń, J., Sowa, U., et al. 2010. Proton radiotherapy facility for ocular turnors at the IFJ PAN in Krak6w Poland. *Appl. Radiat. Isot.* 68: 738–742.

Mishra, K. K., and Daftari, I. K. 2016. Proton therapy for the management of uveal melanoma and other ocular tumors. *Chin. Clin. Oncol.* 5(4): 50 (7 pp).

MONIB. 2021. The latest methods of designing and manufacturing orthopedic implants. https://monib-health.com/en/post/65-orthopedic-implant-manufacturing-process.

Müller, C., et al. 2012. A unique matched quadruplet of terbium radioisotopes for PET and SPECT and for α- and β- radionuclide therapy: an in vivo proof-of-concept study with a new receptor-targeted folate derivative. *J. Nucl. Med.* 53(12): 1951–1959.

Munilla, J. 2016. Compact Accelerators for Radio Isotope Production: The AMIT Project. https://indico.cern.ch/event/659942/contributions/2691992/attachments/1525348/2384898/AMIT_ARIES.pdf).

Myers, S., Degiovanni, A., and Farr, J. B. 2019. Future prospects for particle therapy accelerators. *Rev. Accel. Sci. Technol.* (Eds: Chao, A. W., and Chou, W) 10(1): 49–92.

Nye, J. A., Dick, D. W., Avila-Rodriguez, M. A., and Nickles, R. J. 2005. Radiohalogen targetry at the University of Wisconsin. *Nucl. Instrum. Methods Phys. Res. B* 241: 693–696.

Ognibene, T. J., Bench, G., Vogel, J. S., Peaslee, G. F., and Murov, S. 2003. A high-throughput method for the conversion of CO_2 obtained from biochemical samples to graphite in septa-sealed vials for quantification of 14C via accelerator mass spectrometry. *Anal. Chem.* 75: 2192–2196.

Oveissi, F., Naficy, S., Lee, A., Winlaw, D. S., and Dehghani, F. 2020. Materials and manufacturing perspectives in engineering heart valves: a review. *Mater. Today Bio* 5: 100038. (20 pp).

Phillips, J. C., Baldwin, K. J., Lehnert, W. F., Le Grand, A. D., and Prewitt, C. T. 1986. SUNY X21 beamline at NSLS: A multi-use port on a dedicated high brightness synchrotron radiation source. *Nucl. Instrum. Methods Phys. Res. A*, 246(1–3): 182–189

Portela, P. C. 2017. Compact Accelerators for Radio Isotope Production: The AMIT Project. (https://indico.cern.ch/event/659942/contributions/2691992/attachments/1525348/2384898/ AMIT_ARIES.pdf).

PR NewsWire. 2017. Global Radiotherapy Technologies and Markets 2016–2021 – Major Players are Varian, Elekta, IBA-Worldwide Proton Therapy and Accuray. Jul 24, 2017.

Sadeghi, S., Mirzaei, M., Rahimi, M., and Jalilian, A. R. 2014. Development of ^{111}In-labeled porphyrins for SPECT imaging. *Asia Ocean. J. Nucl. Med. Biol.* 2(2): 95–103.

Sakurai, H., Ishikawa, H., and Okumura, T. 2016. Proton beam therapy in Japan: current and future status. *Jpn. J. Clin. Oncol.* 46(10): 885–892.

Schaeffer, P., et al. 2015. Direct Production of ^{99m}Tc via [100]Mo(p,2n) on Small Medical Cyclotrons. *Phys. Procedia* 66: 383–395.

Schillo, M., Geisler, A., Hobl, A., et al. 2001. Compact superconducting 250 MeV proton cyclotron for the PSI PROSCAN proton therapy project. *AIP Conf. Proc.* 600: 37–39.

Schmor, P. 2011. Review of cyclotrons for the production of radioactive isotopes for medical and industrial applications. *Rev. Accel. Sci. Technol.* 4: 103–116.

Schueller, M. J., Ferrieri. R. A., and Schlyer, D. J. 2005. An automated system for oxygen-18 water recovery and fluorine-18 delivery. *Nucl. Instrum. Methods Phys. Res. B* 241: 660–664.

Seyfoori, A., Amereh, M., Dabiri, S. M. H., Askari, E., Walsh, T., and Akbari, M. 2021. The role of biomaterials and three dimensional (3D) in vitro tissue models in fighting against COVID-19. *Biomater. Sci.* 9: 1217–1226.

Skowronek, J. 2017. Current status of brachytherapy in cancer treatment – short overview. *J. Contemp. Brachyther.* 9(6): 581–589.

Spal, R. Dobbyn, R. C., Burdette, H. E., Long, G. G., Boettinger, W. J., and Kuriyama, M. 1984. NBS materials science beamlines at NSLS. *Nucl. Instrum. Methods* 222(1–2): 189–192.

Stewart, A. J. and Viswanathan, A. N. 2006. Current controversies in high-dose-rate versus low-dose-rate brachytherapy for cervical cancer. *Cancer* 107(5): 908–915.

Streeter Jr, O. E., Vicini, F. A., Keisch, M., et al. 2003. MammoSite radiation therapy system. *Breast* 12(6): 491–496.

Urano, S., Wakamoto, I., and Yamakawa, T. 2003. Electron beam sterilization system. *Mitsubishi Heavy Indus. Tech. Rev.* 40(5) (5 pp).

Valkovic, V. 1980. *Analysis of Biological Materials for Trace Elements Using X-Ray Emission Spectroscopy.* CRC Press, Boca Raton, Florida, USA.

Valković, V. Bernasconi, G., Haselberger, N., Makarewicz. M., Ogris, R., Moschini, G., Bogdanović, I., Jakšić, M., and Valković, O. 1993. Multi-element analysis of biopsy samples. *Nucl. Instrum. Methods Nucl. Res. B* 75: 155–159.

Valkovic, V., and Moschini, G. 1992. Application of charged-particle beams in science and technology. *Riv. Nuovo Cim.* 15: 1–73.

van der Meulen, N. P., Hasler, R., Talip, Z., et al. 2020. Developments toward the implementation of 44Sc production at a medical cyclotron. *Molecules* 25: 4706. 10.3390/molecules25204706.

Voss, K. O., Fournier, C., and Taucher-Scholz; G. 2008. Heavy ion microprobes: a unique tool for bystander research and other radiobiological applications. *New J. Phys.* 10: 075011.

Weber, D. C., Langendijk, J. A., Grau, C., and Thariat, J. 2020. Proton therapy and the European Particle Therapy Network: the past, present and future. *Cancer Radiother.* 24(6): 687–690.

Wise, D. L., Trantolo, D. J., Lewandrowski, K.-U., Gresser, J. D., Cattaneo, M. V., and Yaszemski, M. J. (Eds.). 2000a. *Biomaterials Engineering and Devices: Human Applications. Volume 2. Orthopedic, Dental, and Bone Graft Applications.* Springer Nature, Switzerland.

Wise, D. L., Trantolo, D. J., Lewandrowski, K.-U., Gresser, J. D., Cattaneo, M. V., and Yaszemski, M. J. (Eds.). 2000b. Biomaterials Engineering and Devices: Human Applications. Volume 1. Fundamentals and Vascular and Carrier Applications. Humana, bought by Springer in 2006, Switzerland.

Yadav, S. and Gangwar, S. 2018. An overview on recent progresses and future perspective of biomaterials. *IOP Conf. Ser. Mater. Sci. Eng.* 404: 012013. (5 pp).

Yaqub, K. M. and Min-Hua, C. 2019. A review on role of biomaterials in biomedical field. *Int. J. Bio-Pharma Res.* 8(9): 2788–2793.

Zhao, X., Cui, K., and Li, Z. 2019. The role of biomaterials in stem cell-based regenerative medicine. *Future Med. Chem.* 11(14): 10.4155/fmc-2018-0347.

Additional reading

Barnabé, S., Brar, S. K., Tyagi, R. D., Beauchesne, I., and Surampalli, R. Y. 2009. Pre-treatment and bioconversion of wastewater sludge to value-added products--fate of endocrine disrupting compounds. *Sci. Total Environ.* 407(5): 1471–1488.

Cirilli, M. 2021. CERN's impact on medical technology. *CERN Courier* 61(4): 23–27.

Melvin, S. D. and Leusch, F. D. 2016. Removal of trace organic contaminants from domestic wastewater: A meta-analysis comparison of sewage treatment technologies. Environ Int. 92–93

3

Ion Beam Analysis: Analytical Applications

3.1 Introductory Remarks

Ion beams are used to analyse the elemental composition in many different fields (materials science, metallurgy, geology, biology, medicine, archaeology, art, etc.). The main techniques used are: Rutherford Backscattering (RBS), Proton-Induced X-ray Emission (PIXE), Charged Particle Activation Analysis (CPAA) or Nuclear Reaction Analysis (NRA), Secondary Ionization Mass Spectrometry (SIMS), Particle Desorption Mass Spectrometry (PDMS) and Extended X-ray Absorption Fine Structure (EXAFS) using synchrotron radiation.

When a sample is bombarded with a beam of charged particles a number of processes take place, some of them are indicated in Fig. 3.1.

All of these processes can be used to obtain analytical information about sample investigated, provided that one understands the underlying physics (Fig. 3.2).

Progress in accelerator-based nuclear analytical techniques (NAT) is associated with developments in areas of

- Beam focusing and construction of microprobes.
- Sample preparation methods.
- Simultaneous use of several detection methods.
- Hardware and software for data collection and analysis.

The number of software packages is available either in the form of freeware, software or shareware. On its Accelerator Knowledge Portal, IAEA (2020) provides the list of available codes. Here we list some of the more important:

- NRC: BEAMnrc: Software tool to model radiation beams. The main application for BEAMnrc is modeling the treatment head of radiotherapy linear particle accelerators (linacs) used by medical physicists to treat cancer. Due to its flexible, modular design and companion utilities, this software can also be used for a vast range of applications, including the simulation of research and industrial linac beams, x-ray emitters, radiation dose delivery to a patient, radiation shielding, and more. Available as freeware from https://www.nrc-cnrc.gc.ca/eng/solutions/advisory/beam_index.html. Application field – radiotherapy.
- V. Palonen et al.: CAR4AMS: A program for Bayesian analysis of Accelerator Mass Spectrometry (AMS). The program performs an analysis of a whole AMS experiment and outputs a Markov chain, points of which are distributed as the probability of the isotope concentrations of the samples. Histograms of the points hence represent the

DOI: 10.1201/9781003033684-3

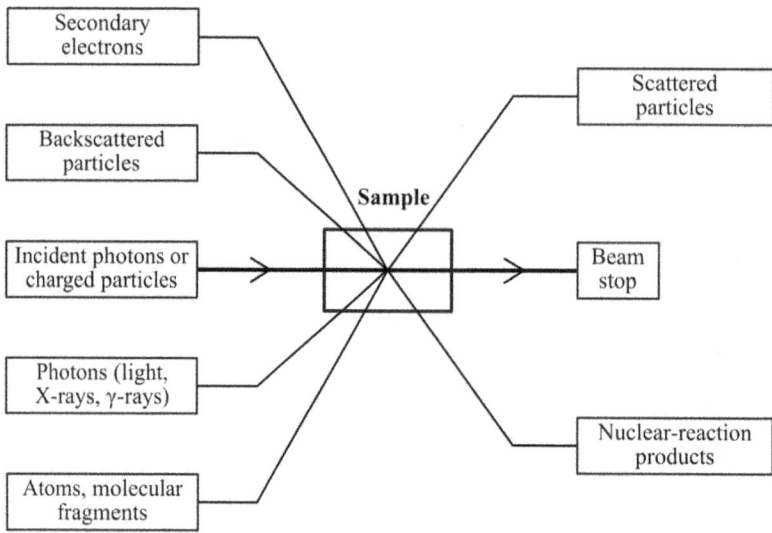

FIGURE 3.1
Schematic presentation of processes taking place during sample irradiation by the charged particle beam.

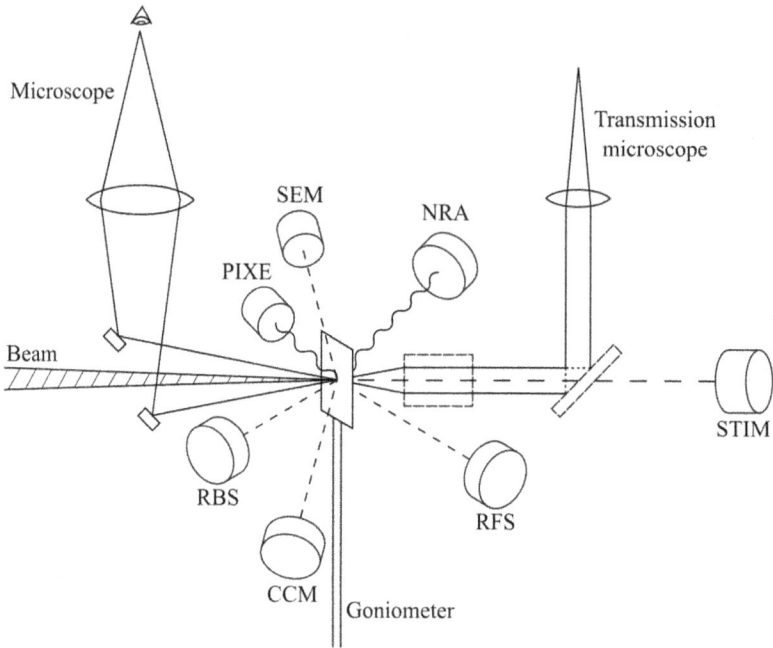

FIGURE 3.2
PIXE – Particle Induced X-Ray Emission. RBS – Rutherford Backscattering. RFS – Rutherford Forward Scattering. NRA – Nuclear Reaction Analysis. CCM – Channeling Contrast Microscopy. STIM – Scanning Transmission Microscopy. SEM – Secondary Electron Microscopy.

probability density functions of the isotope concentrations. Installation and use of the program are described, along with the format of the input and output files. Available as a freeware from http://beam.acclab.helsinki.fi/~vpalonen/car4ams/.

- F. Schiettekatte: CORTEO: Corteo is a Monte Carlo (MC) simulation program (freeware) intended to make possible MC simulations of ion beam analysis (IBA) spectra within a few seconds. It computes the trajectory of ions in materials, based on binary collision, central potential and random phase approximations. Computations are accelerated by extracting the scattering angle components from tables indexed using the binary representation of reduced energy and impact parameters. It also uses the "main collision" approximation: after computing the trajectory of the incident ion to a certain depth, CORTEO generates a scattered ion or a recoil in the direction of the detector. The trajectory of this ion is computed until it emerges out of the target, and its intersection with the detector is checked. Implemented techniques include RBS, ERD, RBS-TOF, ERD-TOF and coincidence of identical ions.

   Link http://www.lps.umontreal.ca/~schiette/index.php?n=Recherche.Corteo.

- N. Barradas, with C. Jeynes and R. Webb: DATAFURNACE: The IBA DataFurnace is a fitting code, not a simulation code (although it has a simulator, of course). It has a core code to do the physics called NDF and written in Fortran, and a graphical user interface (GUI) code called WiNDF and written in Visual Basic. It is designed to facilitate accurate and automatic analysis of large batches of complex samples. The fits obtained are generally "perfect": the purpose is to extract all the information from the spectra (well, as much information as possible!). Channeling is not supported. This software is available at www.surreyibc.ac.uk/ndf.

- E. Szilágyi: DEPTH: Performs calculation of energy and depth resolution of RBS and ERD analysis techniques. This freeware is available at http://www.kfki.hu/~ionhp/doc/prog/wdepth.htm.

- C. Ryan: GeoPIXE: This is quantitative micro-PIXE imaging and analysis software. The GUI interface of GeoPIXE™ allows interactive spectrum fitting, list-mode sorting and quantitative image projection using Dynamic Analysis (or simple energy windows, or regions of interest), and quantitative analysis of arbitrary regions and line projections of images (all elements simultaneously). The windows are linked and communicate with each other to provide an efficient interactive PIXE/SXRF analysis and imaging environment. Version 6 of GeoPIXE™ available at http://www.nmp.csiro.au/GeoPIXE.html.

- J. L. Campbell: GUPIXWIN: GUPIXWIN is a versatile software package for fitting PIXE spectra from thin, thick, intermediate and layered specimens. It extracts peak intensities and converts these to concentrations via the H-value standardization method. X-ray excitation may be via protons, deuterons, ^{3}He or ^{4}He ions; X-ray spectrometry may be via Si(Li) or Ge detectors. Interactive or batch processing may be selected. In either its earlier GUPIX form or in the newer GUPIXWIN form, this software has been supplied to about 140 IBA laboratories in 30 countries. GUPIX has been tested extensively using various standard reference materials and inter-comparisons. The Windows version GUPIXWIN became available in version 1.1 on 1 July 2005. It retains most of the features of the original GUPIX and offers a marked increase in user convenience. Because of the very large volume of feedback and suggestions, further development was done to incorporate the needs of users, leading to Version 1.3. Development of V2.0 was completed in early 2007, providing an elegant new output manager together with batch mode for both single-detector and dual-detector operation. V2.1 was released early in 2008 and included an extension of proton energies to 5 MeV, plus various improvements that were suggested at the PIXE2007 conference in Mexico. Link for this software is http://pixe.physics.uoguelph.ca/gupix/main/.

- C. Pascual-Izarra: HOTSTOP: HOTSTOP is scientific software that allows to extract experimental ion stopping force curves from RBS spectra (as well as to estimate uncertainties) It uses Simulated Annealing and Bayesian Inference algorithms. Freeware is available at http://hotstop.sourceforge.net/.

- C. P. Isarra: LIBCPIXE: Open-source library for Particle Induced X-ray Emission (PIXE) simulation. Freeware is available at http://cpixe.sourceforge.net/.

- K. Arstila: MCERD: MCERD is a MC simulation program, which is used for simulations of ERD (Elastic Recoil Detection) measurements. This freeware is available at https://www.jyu.fi/fysiikka/en/research/accelerator/abasedmat/software.

- N. P. Barradas: NDF: NDF is a DOS code dedicated to the analysis of RBS, ERDA, PIXE, non-resonant NRA and NDP data for any ion, any target, any geometry and the number of spectra. The free version of NDF is a simulator only. It does not implement fitting or Bayesian inference. The full version of NDF includes local search fitting, simulated annealing fitting, Bayesian inference, as well as an excellent GUI that can be found in http://www.ionbeamcentre.co.uk/ndf/. The general philosophy of NDF is to put accuracy before calculation speed. In some cases, this leads to calculations that can be orders of slower magnitude, i.e., a few seconds in modern PCs. Faster calculations can be made by turning the appropriate options on or off. NDF is a DOS program that reads input files and creates output files. No graphical interface or output is supplied. The users must use their own graphics package to visually inspect the fits and depth profiles obtained. Link for this software is http://www.itn.pt/facilities/lfi/ndf/uk_lfi_ndf.htm.

- G. Grime: OMDAQ: A complete software and hardware solution for collecting and processing data from a nuclear microprobe facility. Combines the functionality of OMDAQ98, XYZ and Dan32, Runs under Windows XP with PCI interface cards, Works with the OM-1000e 8-channel ADC interface. Completely re-written multi-thread code with many new features, Integration of sample stage XYZ and video controls, GUPIX interface and RBS simulation. Link for this software is http://www.microbeams.co.uk/download.html.

- B. Ramsey: OXCAL: A program designed for the analysis of chronological information from ^{14}C dating experiments. The capabilities of the program can be divided into two main categories:

  1. The calculation of probable age ranges for scientifically dated samples (through radiocarbon calibration, sapwood estimates, etc.)

  2. The analysis of groups of events that are related either through generic groupings or through stratigraphic relationships.

  This freeware is available at link https://c14.arch.ox.ac.uk/oxcal/OxCal.html.

- IAEA: QXAS: Quantitative X-ray analysis system. Link for this software is http://qxas.software.informer.com/.

- M. Thompson: RUMP: RBS analysis package, providing comprehensive analysis and simulation of RBS and ERD spectra. Freeware is available at http://www.genplot.com./.

- A. Gurbich: SIGMACALC: The SigmaCalc software has been developed in order to provide the IBA scientist with a tool for computing the differential cross sections required for an analytical work. The reliability of the calculated cross-sections was proved by comparisons with posterior measurements and benchmark experiments. SigmaCalc is based on the already published and some new results of the data evaluation. Different theoretical models (optical, S- and R-matrix) with optimized parameters are used for calculations. No expertise in nuclear physics is needed in order to perform the calculations of the required cross-sections with SigmaCalc.

  A theoretical evaluation of the cross-sections grounded on appropriate physics seems to be the only way to resolve the problem of nuclear data for IBA. Though experiments undertaken to determine the required cross-sections are often reported and many data are accumulated in databases, all these cross-sections should be evaluated prior to their widespread use. The reasons are as follows. The analysis of the available experimental information revealed numerous discrepancies in the reported cross sections values, which are far beyond quoted experimental errors. In addition, due to cross-sections dependence on a scattering angle the available data are valid only in the case of a scattering geometry very close to the geometry used in cross sections measurements. It was shown in numerous papers that the evaluation of the cross-sections by combining a large number of different data sets in the framework of the theoretical model makes it possible to calculate excitation functions for analytical purposes for any scattering angle, with reliability exceeding that of any individual measurement. This freeware is available at the link http://www.surreyibc.ac.uk/sigmacalc/.

- M. Mayer: SIMNRA: SIMNRA is a Microsoft Windows program for the simulation of backscattering spectra for IBA with MeV ions. SIMNRA is mainly intended for the simulation of non-RBS, nuclear reactions and elastic recoil detection analysis (ERDA). More than 300 different non-Rutherford and nuclear reactions cross-sections for incident protons, deuterons, ^{3}He, ^{4}He and Li-ions are included. SIMNRA can calculate any ion-target combination including incident heavy ions and any geometry including transmission geometry and arbitrary foils in front of the detector. SIMNRA is not free. It is a shareware program, which can be used for a trial period of 30 days without fee. If you want to use SIMNRA after this period, you have to register. Link for this software is http://home.rzg.mpg.de/~mam/.

- J. F. Ziegler: SRIM: SRIM is a collection of software packages, which calculate many features of the transport of ions in matter. Typical applications include:

  1. Ion Stopping and Range in Targets: Most aspects of the energy loss of ions in the matter are calculated in SRIM, the Stopping and Range of Ions in Matter. SRIM includes quick calculations, which produce tables of stopping powers, range and straggling distributions for any ion at any energy in any elemental target. More calculations that are elaborate include targets with complex multi-layer configurations.

  2. Ion Implantation: Ion beams are used to modify samples by injecting atoms to change the target chemical and electronic properties. The ion beam also causes damage to solid targets by atom displacement. Most of the kinetic effects associated with the physics of this kind of interaction are found in the SRIM package.

3. Sputtering: The ion beam may knock out target atoms, a process called ion sputtering. The calculation of sputtering, by any ion at any energy, is included in the SRIM package.

4. Ion Transmission: Ion beams can be followed through mixed gas/solid target layers, such as occurs in ionization chambers or in energy degrader blocks used to reduce ion beam energies.

5. Ion Beam Therapy: Ion beams are widely used in medical therapy, especially in radiation oncology.

Typical applications are included. Freeware is available at http://www.srim.org/.

Intercomparison of PIXE spectrometry software packages has been published in (IAEA 2003). The participating programs are: Geopixe (Ryan et al. 1990), Gupix (Maxwell et al. 1989), Pixan (Clayton 1986), Pixeklm (Szabo and Borbely-Kiss 1993), Sapix (Sera and Futatsugawa 1996), Winaxil (Maxwell et al. 1989) and Witshex (Lipworth et al. 1993). The results of the intercomparison indicated that most of the programs performed reasonably well with respect to peak areas. Some disagreements existed, for example in cases of low precision (intensity) peaks overlapping with or close to high precision (intensity) ones, and when K and L lines overlap.

NAT are extremely useful tools for the analysis of materials and are able to support and provide significant contributions to many of the research and application areas. While NAT are often complementary to non-nuclear techniques, NAT can provide information that could not be obtained by alternative methods. International Atomic Energy Agency (IAEA) has published a number of reports dealing with some of the problems in this field. Let us mention only some:

- Nuclear techniques for cultural heritage research (IAEA 2011).
- Hands-on training courses using research reactors and accelerators (IAEA 2014).
- Utilization of accelerator-based real time methods in investigation of materials with high technological importance (IAEA 2015).
- Development of a reference database for particle-induced gamma ray emission (PIGE) spectroscopy (IAEA 2017).
- Accelerator simulation and theoretical modeling of radiation effects in structural materials (IAEA 2018).
- Improvement of the reliability and accuracy of heavy IBA (IAEA 2019).

3.2 Ion Sources

Here, we shall discuss ion sources used in particle accelerator systems dedicated to IBA techniques. Key performance and characteristics of some ion sources are discussed: emittance, brightness, gas consumption, sample consumption efficiency, lifetime, etc. (see: Podaru and Mous 2016). Different analysis techniques have different requirements on the ion source performance, so it is important to understand the factors influencing their performance.

Parameters that characterize ion source performance include ion beam emittance, ion beam brightness, plasma hash, energy spread and gas/solid target consumption efficiency.

The ion beam emittance ε is defined as the area S of the two-dimensional phase space divided by π. $\varepsilon = S/\pi$ and also $\varepsilon_x = x_o \times x'_o$ and $\varepsilon_y = y_o \times y'_o$, where x_0 and y_0 are the beam radii in horizontal (x) and vertical (y) planes at a waist, while x'_o and y'_o are the beam half-angle divergences in the x and the y planes. The energy-normalized emittance is obtained by multiplying the emittance ε with the square root of ion beam energy. Degradation of normalized beam emittance from the ion source to the target is a measure of the quality of the beam transport system (Podaru and Mous 2016).

The ion beam brightness (B) is defined as the ion beam current, I, that can be transported through two apertures of areas, A_o (object area) and A_c (collimator area), separated by a drift L, at a given ion beam energy, E,

$$B = (I \times L^2)/(A_o \times A_c \times E) \tag{3.1}$$

To allow comparison of the performance of the microprobes at various facilities, Szymansky and Jamieson (1997) defined the normal brightness at a half-angle the highest ion beam brightness. High beam brightness is advantageous for microprobe applications. It is clearly desirable to employ the brightest possible source in order to focus the smallest possible probe size on the specimen, with the highest possible beam current.

The ion beam energy spread can negatively influence the optical properties of high-quality lenses, i.e., nuclear microprobe lenses. For tandem accelerators, the ion beam energy spread ΔE is dependent on mainly three factors: the initial beam energy spread from the ion source, the energy spread introduced by the interaction with the stripper gas and the terminal voltage ripple. The beam energy spread introduced by the ion source is dependent on many plasma parameters, which ultimately transfer kinetic energy to the ions. Ion sources with ion beam energy spread below 10 eV are considered good for IBA experiments. This value should be compared to the terminal voltage ripple for all-solid-state particle accelerators.

Single-ended particle accelerators benefit from the use of direct positive extraction ion sources located in the high voltage terminal. Positive ion sources are more prolific in producing high-intensity ion beams.

Tandem accelerators require negative ion beams for the injection. In contrast to single-ended particle accelerators, the ion sources are easily accessible and the injection configuration allows multiple ion sources to be applied with ease. Tandem accelerators have the advantage that the ion source maintenance can be done much faster than in single-ended particle accelerators. Negative ion sources for tandem injectors include gas sources: von Ardenne type, multicusp type or RF type and cesium sputter type ion sources using energetic Cs^+ beam of particles that strike a solid surface with target material.

A high brightness proton injector for the Tandetron accelerator developed by High Voltage Engineering Europe (HVEE) is described by Pelicon et al. (2014). The multicusp ion source has been tuned to deliver at the entrance of the Tandetron™ accelerator H^- ion beams with a measured brightness of 17.1 A m^{-2} rad^{-2} eV^{-1} at 170 µA, equivalent to an energy normalized beam emittance of 0.767 π mm mrad $MeV^{1/2}$. The high brightness of the ion source enables the reduction of object slit aperture and the reduction of acceptance angle at the nuclear microprobe, resulting in a reduced beam size at selected beam intensity, which significantly improves the probe resolution for micro-PIXE applications.

Klein and Mous (2017) describe the cesium sputter ion source designed to fulfill the stringent requirements of AMS. It has a storage capacity of up to 200 samples for unattended operation and accepts solid as well as gaseous CO_2 samples. The samples are stored in a separate vacuum chamber and transported upon use into the hot central part of the ion source, thereby minimizing cross-talk between the samples.

RF ion sources (only positive ion extraction) coupled to charge exchange canals (vapors of Li, Na, Rb) can provide negative H, He, C, O ion beams for IBA.

A series of ion sources can produce direct negative extraction of ion beams of, e.g., H, C, NH_x, O. For example, duoplasmatron ion source can operate with direct negative extraction for ion beams of H (in off-axis beam extraction mode) or is used for direct positive extraction for He ion beams.

Quax et al. (2010) at HVEE developed a high-current light-ion injector for tandem accelerators in response to an increasing demand for high current MeV ion beams of H, D and He. High current MeV H and He are used, among others, in deep level implantation and charge carrier lifetime control in semiconductor power devices, as well as material damage studies. MeV H and D serve neutron production (Welton et al. 2004) and neutron reference fields while high-current MeV He is a candidate as a diagnostic tool in fusion reactors (Sasao et al. 2006).

3.3 Proton Induced X-Ray Emission

The fundamental principles of the PIXE method lays on the use of ion beam to induce characteristic X-ray emission. The energy of the incident beam is usually in the range of 1–5 MeV per nucleon, its intensity varying between 0.1 and 100 nA. It is a widely used IBA technique for determination of element concentrations above Z = 12 in matrix composed of light elements.

A major advantage of the PIXE method is that it permits analysis of very small absolute masses of an element. This is particularly true of the measurements made with the use of proton micro-beams. Another advantage is the relatively short time required for the irradiation of a single specimen (Figs. 3.3 and 3.4).

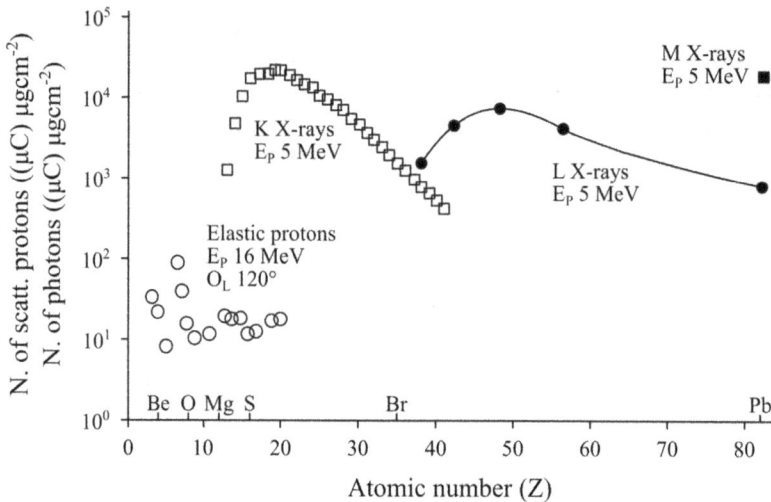

FIGURE 3.3
Elemental sensitivity for PIXE and ESA as a function of target atomic number.

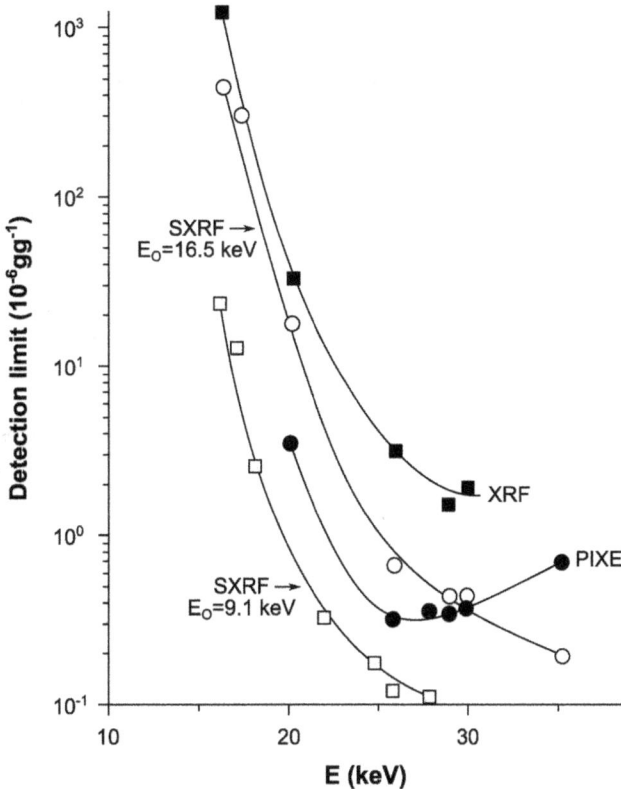

FIGURE 3.4
Detection limits comparison of XRF, PIXE and SXRF methods.

Some of the typical PIXE applications are:

- Air pollution analysis and multi-elemental analysis of environmental samples.
- Medical and biological applications, tissue sections.
- Determination of trace elements in geological sections.
- Solid state and material analysis.
- Measuring the composition and thickness uniformity of layers.
- Agriculture studies.
- Trace element mapping, when used in conjunction with the microprobe.

For some applications, an external beam setup is required. Such a setup is usually located on the same beamline as the in-vacuum setup (see Fig. 3.11). The focused beam is usually brought into the air through $d \leq 10$ μm Kapton window. Proton beam intensity should be such that no damage to the Kapton foil is made. Such a system is described in details by Lövestam and Swietlicki (1989).

Some PIXE programs worldwide is shown in Table 3.1.

In PIXE measurements, samples have to be introduced into the vacuum (scattering chamber) in order to perform measurements. This is a limitation, especially for liquid samples, large samples, etc. In order to avoid these difficulties external beam PIXE (see

TABLE 3.1

Some PIXE programs worldwide

Laboratory	Ion energy (MeV)	Primary applications
AUSTRALIA		
Canberra	p: 3.0	Mineral
Lucas Heights	p: 3.0	Archeology
Melbourne	p: 2.5	Biological, material science
AUSTRIA		
Linz	p, α: 0.1–0.8	Materials science
BANGLADESH		
Dacca	p, α: 2.5	Fluids
BELGIUM		
Geel	p, α: 1–3	Aerosols
Gent	p: 2.4	Medical
Liege	p, α: 3, 18	Minerals
Namur	p, α: 0.5–3.2	Biological
BRAZIL		
Rio de Janeiro	p: 3.0	Different
Sao Paolo	p	Aerosols
CANADA		
Chalk River	p: 0.5	Minerals science
Guelph	p: 2.0	Biological
Manitoba	p: 20–50	Aerosols
Montreal	p	Medical
CHILE		
Santiago	p: 6	Aerosols
CHINA		
Fudan	p: 3.0	Archeology
Shanghai	p: 3.5	Water
TAIWAN		
Lungan	p: 3.0	Materials science
DENMARK		
Aarhus	p: 0.5–5	Materials science
Copenhagen	p: 2.0–3.0	Aerosols
FINLAND		
Helsinki	p: 2.0	Aerosols
Turku	p, d, α: 5–21	Biological
FRANCE		
Saclay	p, α: 4–22, 10–30	Materials science
Strasbourg	p: 4.0, heavy ions	Materials science
GERMANY		
Bochum	p: 1.5–4	Biological
Bonn	α: 30	Materials science
Frankfurt		Aerosols
Marburg	p: 2–4	Aerosols
GREECE		
Athens	p: 2.0	Medical

TABLE 3.1 (Continued)

Some PIXE programs worldwide

Laboratory	Ion energy (MeV)	Primary applications
HUNGARY		
Budapest	p, d, α: 1.4–4	Different
Debrecen	p, α: 0.8–5, heavy ions	Different
INDIA		
Bhudaneswar		Forensic
Kanpur	p: 2.0	Fluids
IRAN		
Teheran	p: 0.5	Water
ISRAEL		
Weizmann	Heavy ions: 34	Materials science
ITALY		
Catania	p, α: 1.0–2.5	Aerosol
Milan	p: 2.8	Aerosol
Padova		Biological, different
JAPAN		
Nagoya	p: 0.15–2	Different
Osaka	p, α: 1.0, 20	Material science
Sendai	p: 3.5	Different
Tokyo	α: 30	Biological
Tsukuba	p, tandem	Material science
Yokohama	p: 2.5	Fluids
NETHERLANDS		
Amsterdam	p: 3.0	Biological
Eindhoven	p: 3.0	Fluids
NEW ZEALAND		
Lower Hutt	p: 2.5	Biological
POLAND		
Cracow	p: 2.9	Biological
Warsaw	p:2 -3	Minerals
PORTUGAL		
Sacavem	p: 0.2–2	Different
RUMANIA		
Bucharest	p: 4	Biological
SOUTH AFRICA		
Johannesburg	p: 3	Aerosols
Pelindaba	p: 2.4	Aerosols
Faure	p: 1.5–3.5	Biological
SWEDEN		
Lund	p: 2.5	Aerosols
Uppsala	α: 1.8	Art
UK		
Birmingham	p: 1–3	Biological
Hammersmith	p: 2.8	Medical

(Continued)

TABLE 3.1 (Continued)

Some PIXE programs worldwide

Laboratory	Ion energy (MeV)	Primary applications
Harwell	p: 3.0	Materials science
Manchester	p: 2.0	Materials science
Surrey	p, α: 0.3–2	Different
USA		
Bell Labs	α: 2–4	Materials science
Brgham Young	p: 2	Aerosols
Brooklyn	p: 3	Different
Colorado	p: 4.3	Biological
Davis (Univ. Calif)	α: 18, 4.5	Aerosols
Duke	p: 3	Biological
Florida State	p: 4	Aerosols
Florida Univ.	p: 3.8	Biological
Genesero, NY	p: 0.4–2.0	Biological
IBM, NY	p, α <2.5	Materials science
Ill. Inst. Technology	p	Aerosols
Kentucky	p: 2.4	Minerals
Mississippi	p: 3	Gasses
Naval Research Lab.	p: 3	Materials science

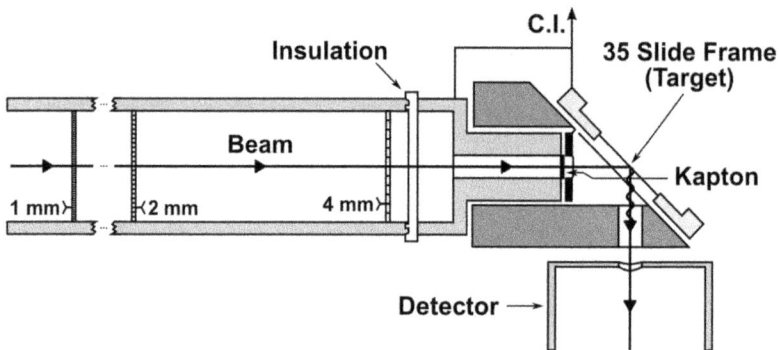

FIGURE 3.5
External beam PIXE set-up.

Fig. 3.5) is being used in many laboratories. Early work on the external beam PIXE method and its applications have been reviewed by Williams (1984).

The use of external beams with the PIGE method has been still rather limited. By applying both PIXE and PIGE methods, a very good overall picture of the elemental composition of a sample may be obtained. Some examples of such applications are described in papers by Räisänen (1986) and Boni et al. (1989).

3.3.1 PIXE/RBS/PIGE Scattering Chamber

In PIXE measurements, samples have to be introduced into vacuum (scattering chamber) in order to perform measurements. This is a limitation, especially for liquid samples, large samples, etc. In order to avoid these difficulties, external beam PIXE is being used in many laboratories.

Such scattering chamber, shown in Fig. 3.6, has been constructed and manufactured for many years in Laboratory for Nuclear Microanalysis, Ruder Boskovic Institute, Zagreb, Croatia. Its features are as follows:

- The beam is defined to a desired diameter by a set of 4 collimators and a diffuser foil.
- X-ray detector is positioned 50 mm from the sample at an angle of 135° (optional 90°).
- Three filters, interchangeable from outside are positioned between the sample and the detector.
- Particle detector is positioned at a fixed angle (150° or 165° specified by user).
- γ-ray detector (up to 7.5 cm in diameter) can be positioned at an angle of 135° or 45°.
- Sample changer with 16 positions is remotely controlled by TTL pulses.
- Current is measured:
  - Directly from the sample (thick targets) or via a Faraday cup (thin targets) or both for intermediate targets.
  - Indirectly, by counting backscattered particles from a gold foil inside the collimator assembly.
- Cover flange has a transparent window to enable control of sample positioning.
- Beam diffuser can be interchanged with a quartz beam viewer.
- Additional free side ports (CF 50) enable easy mounting of other auxiliaries (vacuum
- Gauges, detector connections, etc.).
- Chamber is made of Al alloy and anodized.

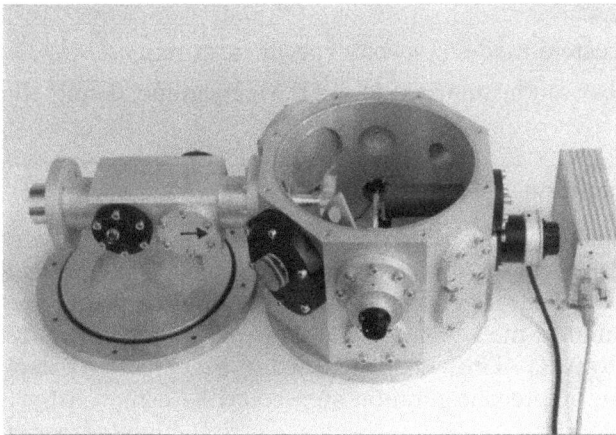

FIGURE 3.6
PIXE/RBS/PIGE scattering chamber.

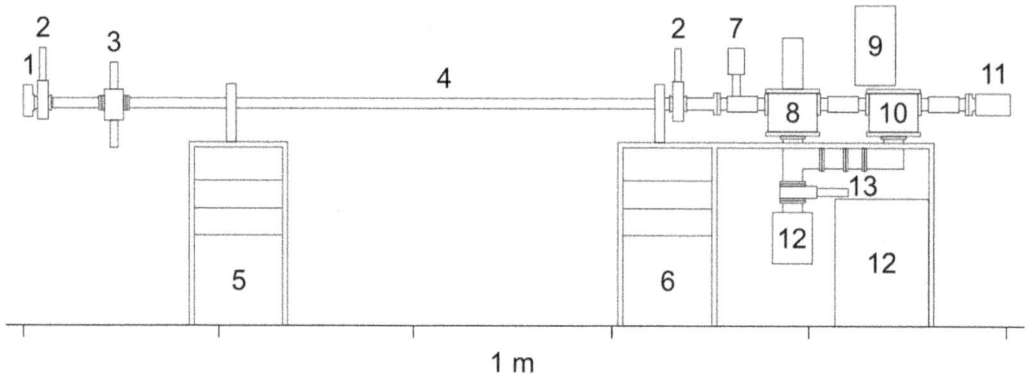

FIGURE 3.7
Beam line with PIXE and RBS set-up.

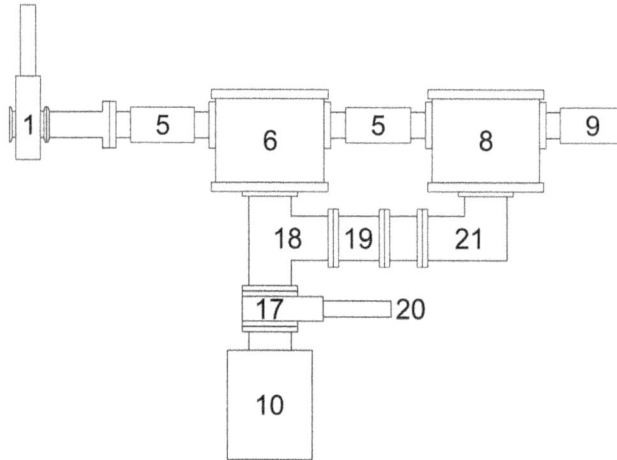

FIGURE 3.8
Details of beam line with PIXE and RBS scattering chambers.

- Chamber is custom made to specify specific user requests.
- Such a chamber can be attached to the PIXE beamline, details shown in Figs. 3.7 and 3.8.

3.3.2 Target Preparation

Samples for PIXE analysis could be prepared as thin or thick targets. Steps in thin target and thick target preparations of biological material are schematically presented in Figs. 3.9 and 3.10, respectively. Detailed description of target preparation procedures for different materials could be found in the IAEA Manual on Sample Preparation Techniques in Trace Element Analysis by X-Ray Emission Spectroscopy, prepared by Valković (1983).

Matrix reduction, or preconcentration step could be achieved by freeze drying (liophylization), low-temperature ashing (oxygen excited), high-temperature ashing, or wet oxidation procedures. Preparation of homogenous solution could be accomplished by (i) ash from low-temperature ashing being dissolved in hydrochloric acid or in nitric acid,

Thin Target PIXE

```
┌─────────────────────────┐
│   Biological material   │
└─────────────────────────┘
              │
              ▼
┌─────────────────────────┐
│    Matrix reduction     │
└─────────────────────────┘
              │
              ▼
┌─────────────────────────┐
│  Homogenous solution or │
│   finely divided powder │
└─────────────────────────┘
              │
              ▼
┌─────────────────────────┐
│    Internal standard    │
│        spiking          │
└─────────────────────────┘
              │
              ▼
┌─────────────────────────┐
│     Transfer onto       │
│        backing          │
└─────────────────────────┘
```

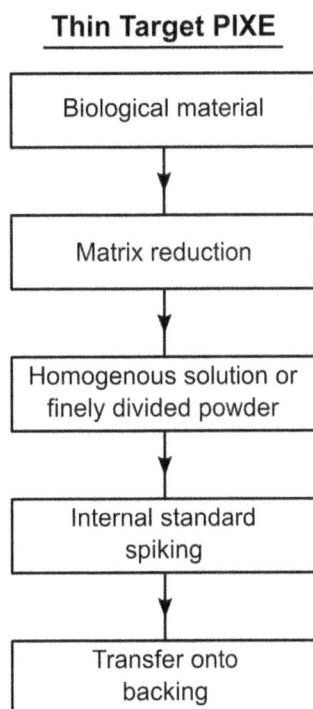

FIGURE 3.9
Thin target PIXE biological material sample preparation.

(ii) acid digestion: acid is boiled of the sample charred by use of flame burner; procedure should be repeated until clear solution is obtained.

PIXE has been applied to the study of numerous problems. For example, Bellagamba et al. (1993) have applied PIXE to the study of trace element behaviour in the coal combustion cycle. In their work, the elemental composition of parent coal and of different combustion products were measured and an insight into partitioning of major, minor and trace elements upon combustion and abatement by control devices in power plants was obtained. The characterization of the pathways of the different elements through the combustion cycle, including combustion itself and fly ash abatement, is essential for the understanding, quantification and improvement of the fly ash precipitation mechanism.

In the work by Bellagamba et al. (1993), PIXE has been combined with PIGE to characterize South African coal and its fly ash collected at the inlet and outlet of the electrostatic precipitators of a power plant. Enrichment and penetration factors for the major and trace elements have been determined thus pointing out the different behaviours of elements according to the chemical and physical properties of fly ash particles to which they are associated, as well as their dependence on the plant operating conditions.

3.4 Proton-Induced Gamma-Ray Emission

The physics behind this method lies on the emission of gamma rays due to the excitation of nuclei by a proton beam. Many nuclear reactions lead to gamma emission. Typical PIGE reactions are: radioactive capture (p,γ), inelastic scattering (p,p') and other nuclear reactions like resonances and direct nuclear reactions $(p, particle + \gamma)$.

Thick Target PIXE

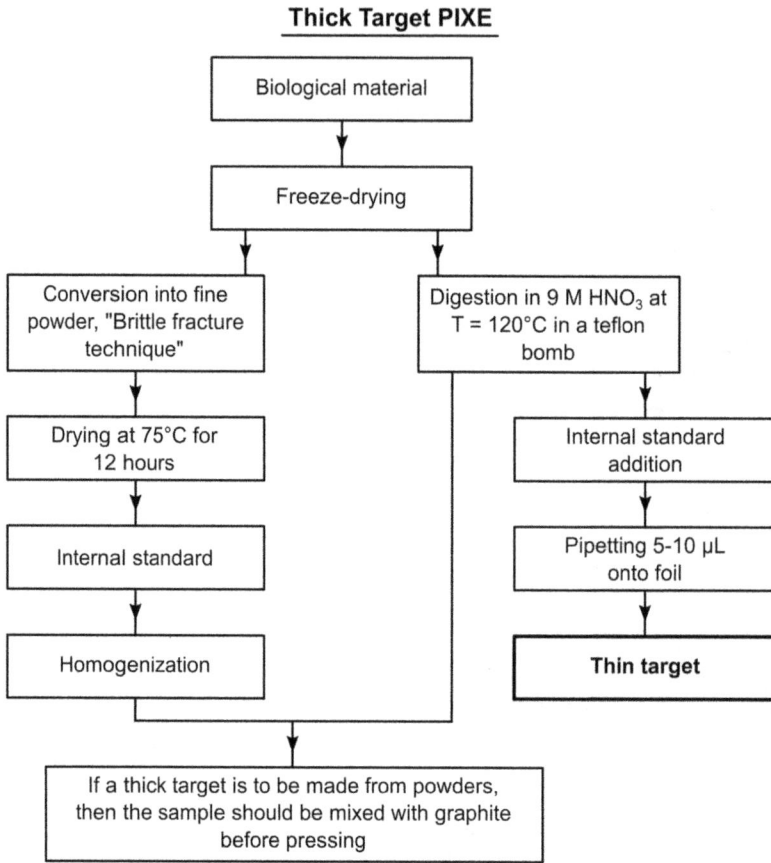

FIGURE 3.10
Thick target PIXE biological material sample preparation.

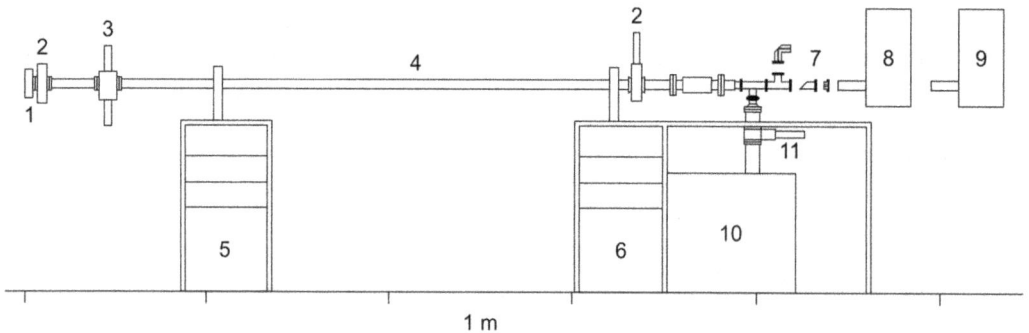

FIGURE 3.11
Beam line with PIGE, PIXE and in-air PIXE setup.

PIGE method is extremely useful for the detection of low Z elements. It is a widely used method to complement the PIXE analyses. It is also very useful for depth profiling (spatial resolution below 10 nm) in addition to the fact that it is a non-destructive profiling technique. It is a widely used method for the determination of carbon concentrations (Fig. 11).

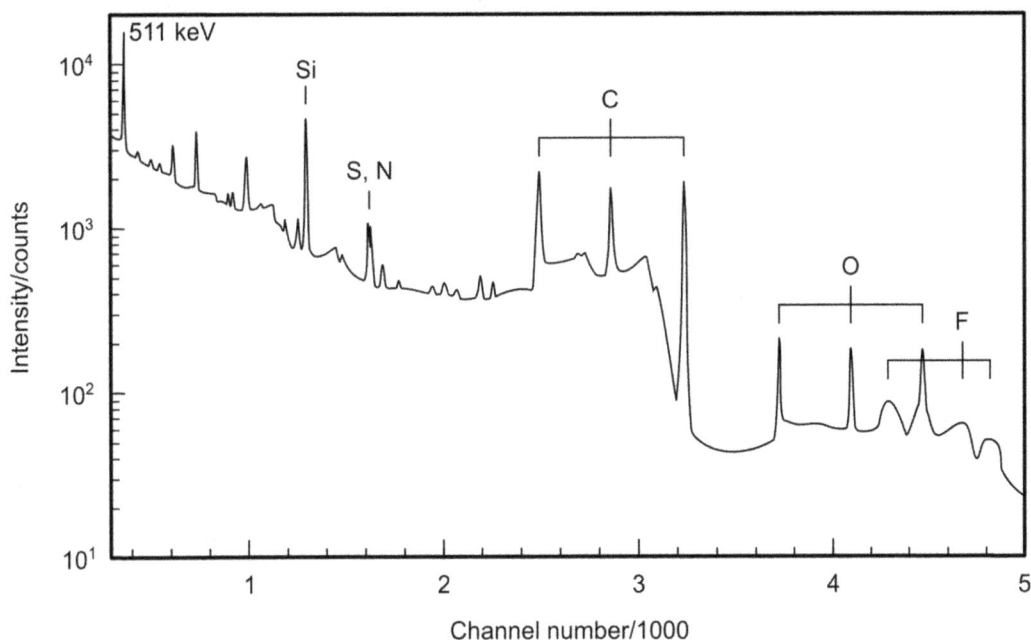

FIGURE 3.12
Gamma ray spectrum resulting from 6.5 MeV proton bombardment of coal SRM: NIST 1632.

Caridi et al. (1992) have described the use of PIGE for the detection of fluorine in coal and coal fly ash samples for the study of fluorine enrichment phenomena in the coal combustion cycle. During their study, pulverized South African coal (Amcoal) samples were collected twice a day during the investigation period. Ash sampling was carried out at the inlet and outlet units of a 3 MWe pilot electrostatic precipitator (ESP) installed in a slip stream derived from the flue gas duct of a power plant. Size fractionated fly ash samples were collected at the ESP inlet unit using an Anderson high-capacity stack sampler, with two impaction stages, a cyclone and a thimble backup filter (Fig. 3.12).

Caridi et al. (1992) used a thick-PIGE method for the analysis of coal samples. Coal samples were powdered and pelletized and then bombarded with 6.5 MeV protons for a known recorded charge with 10–20 nA beam current. The $^{19}F(p,\alpha_2\gamma)^{16}O$ reaction ($E_\gamma = 6129$ keV) was used. The quality control was performed by the analysis of NIST reference materials 1633a and 1635. Fly ash, prepared as thin samples has been analyzed using PIXE setup. Valuable information on fluorine enrichment from parent coal to fly ash and through the abatement process have been obtained for specific operational modes of the electrostatic precipitator.

3.5 Nuclear Reaction Analysis

This method is mainly used for the detection of light elements in a variety of matrices. It is based on detection and analyses of reaction products in the final state. Measurements are usually done with ions accelerated to the energy of 0.1–4.0 MeV. In addition the

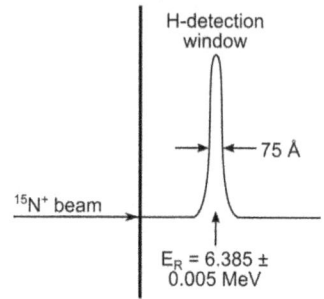

FIGURE 3.13

Schematic presentation of nuclear-resonant reaction $^1H(^{15}N,\alpha\gamma)^{12}C$ used for profiling hydrogen concentration in material.

determination of concentration levels, the method is also used for the measurement of concentration profiles.

An example of the measurements of ^1H profile implanted in Ta is presented in the report by Weiser and Kalbitzer (1985). They studied the quantum mechanical diffusion of ^1H in Ta at low temperatures using $^1H(^{15}N,\alpha\gamma)^{12}C$ reaction. The technique depends on the resonant nuclear reaction between protons and ^{15}N, which yields an α-particle and a characteristic 4.43 γ-ray. This resonant reaction occurs at the center-of-mass energy of 402 keV, has a width of 0.9 keV, and a peak cross-section of 200 mb. The cross-section of resonance is more than 2 orders of magnitude smaller. To utilize this reaction as a probe for hydrogen, one bombards the sample with ^{15}N (see Fig. 3.13) and measures the yield of the characteristic 4.43 MeV γ-rays using a NaI scintillation counter.

If there is hydrogen on the surface of the sample and the ^{15}N has an energy of 402 keV in the center of mass (6.385 MeV in the laboratory), the yield of 4.43 MeV γ-rays is proportional to the amount of the hydrogen on the surface. To determine the concentration of hydrogen inside the sample, the ^{15}N energy is raised so that it is above the resonant energy for hydrogen on the surface, but as the ^{15}N slows down inside the sample, it reaches the resonant energy at a certain depth inside the surface.

Reactions used determination of surface concentrations of some light elements are:

- Hydrogen isotopes: $^1H(^{11}B,\alpha)2\alpha$, $^2H(^3He,p)^4He$.
- Lithium: $^7Li(p,\alpha)^4He$.
- Beryllium: $^9Be(\alpha,n)^{12}C$, $^9Be(d,p)^{10}Be$.
- Boron isotopes: $^{10}B(p,\alpha)^7Be$, $^{11}B(p,\alpha)^8Be$.
- Carbon: $^{12}C(d,p)^{13}C$.
- Nitrogen: $^{14}N(d,p)^{15}N$, $^{14}N(d,\alpha)^{12}C$.
- Oxygen isotopes: $^{16}O(d,p\gamma)^{17}O$, $^{16}O(d,p)^{17}O$, $^{18}O(p,\alpha)^{15}N$.
- Fluorine: $^{19}F(p,\alpha\gamma)^{16}O$, $^{19}F(p,p\gamma)^{19}F$.

In a very broad sense, NRA term is applied to the situation when a sample is bombarded and a reaction product is measured. Often the differentiation is made based upon the particle being detected, only if a particle being detected is a product of nuclear reaction, like (p,d), (p,α) or similar, one speaks about NRA.

NRA can be used in the above-mentioned reactions especially for the determination of light element concentration levels or concentration profiles. Very convenient are nuclear resonant reactions, an example being already mentioned H profiles determination by the

TABLE 3.2

Some nuclear reactions used in surface analysis

Impurity nucleus	Nuclear reaction	Bombarding ion energy/keV	Emitted particle energy/MeV
^{16}O	$^{16}O(d,p)^{17}O$	830	1.52
^{14}N	$^{14}N(d,\alpha)^{12}C$	1300	6.76
^{12}C	$^{12}C(d,p)^{13}C$	1000	3.01
^{27}Al	$^{27}Al(p,\gamma)^{28}Si$	992	
^{11}B	$^{11}B(p,\alpha)^{8}Be$	700	4.07
^{2}H	$^{2}H(d,p)^{3}H$	550	2.45

use of $^{1}H(^{15}N,\alpha\gamma)^{12}C$ reaction. Characteristics of some of the nuclear reactions very often used in the surface analysis are presented in Table 3.2.

One difference between this technique and RBS is that in this case the technique is not suited for the identification of unknown surface impurities since the choice of the beam and its energy depend on some prior knowledge of the surface impurity.

Several factors have to be considered in choosing the nuclear process to be exploited in concentration measurements. These include isotope abundance, nuclear-level structure of isotopes, nuclear reaction Q-values, existence of resonances, cross-section magnitudes, etc. Additionally, the choice of reaction is also determined by the characteristics of the available accelerator. Accelerators able to accelerate protons, deuterons, tritons, ^{3}He or ^{4}He particles up to energies of the order of 10 MeV are the most easily accessible. They include the large number of electrostatic accelerators (Van de Graaff and Cockcroft-Walton type) as well as cyclotrons.

Next, we shall discuss light elements (A < 20), their nuclear level data have been compiled by several groups. The best-known energy level compilations for light nuclei (A = 5–12) are those published by Ajzenberg-Selove and Lauritsen (1968, 1974) and by Ajzenberg-Selove (1986, 1987, 1990, 1991) for A = 11–20, see also Tilley et al. (2004) and TUNL 2018.

For the overview of ion beam techniques for the analysis of light elements in thin films, including depth profiling see IAEA (2004).

Next, we shall discuss individual light elements (hydrogen – fluorine as well as some heavier elements of interest). The material presented follows the article on the application of charged-particle beams in science and technology published as a special issue of Il Nuovo Cimento authored by myself and my colleague and friend prof. Guiliano Moschini (1940–2016), see Valković and Moschini (1992).

3.5.1 Hydrogen

Hydrogen in nature occurs in the form of two stable isotopes (^{1}H, abundance 99.085% and ^{2}H, abundance 0.015%). The radioactive isotope ^{3}H ($\tau_{1/2}$ = 12.26 years) also occurs in nature produced by cosmic-ray bombardment and as a byproduct of different human activities.

The $^{2}H(^{3}He, p)$ ^{4}He reaction has the distinction of having a particularly high Q-value (18.35 MeV) and a relatively large cross-section (695 mb for E_d = 450 keV). Using this reaction, Picraux and Vook (1974) carried out an elegant channeling lattice location study of ^{2}H in tungsten.

The reaction ^{2}H(t,n)^{4}He has similar properties. The reaction Q-value is positive (Q = 17.59 MeV) and the cross-section shows a strong peak at 0.107 MeV with σ_{peak} = 5.0 barns. This resonance corresponds to a level in ^{5}He at excitation energy of E_x = 16.76 MeV. The width of the observed resonance is Γ_{lab} = 135 keV.

The presence of hydrogen in various materials and the ways in which it affects their properties has been the subject of attention of the scientific community for decades. A major area developed in this context in the second half of the 20th century was hydrogen in metals (Alefeld and Völkl 1978, Fukai 2005). Of special interest is hydrogen in steel. For example, in order to reduce the fuel consumption, it is necessary to reduce the weight of the vehicles. Therefore, the new generation, high-strength steels are largely integrated in the modern car bodies. In recent time, aluminized boron steel, a fully martensitic grade coated with Al-Si alloy, experiences a rapid growth in the anti-intrusion applications of the automotive structures, e.g., bumpers or doors, due to its excellent mechanical properties after hot stamping (Hein and Wilsius 2008). The practice of using aluminized boron steel, in particular the absorption/desorption of diffusible hydrogen in aluminized boron steel is discussed in details by Georges et al. (2013).

A reaction that can be used for the hydrogen analysis is ^{1}H(^{11}B,α)^{8}Be. For example, Ligeon and Giuvarch (1974) used this reaction to determine the concentration of hydrogen implanted into silicon. They have measured the profile of implanted H over a depth of 400 nm; a depth resolution of about 40 nm was achieved. Alpha particles were detected at an angle of 90° with respect to the beam direction. This method allows the detection of surface layers of the order of 5×10^{13} atoms of hydrogen per cm^2. With a 1.793 MeV ^{11}B beam one reaches a 163 keV resonance in the (p,α) reaction. Provided experimental facilities exist to handle such high-mass beams, the potential of such an analytical method for routine hydrogen examination is very large. However, some measurements of light elements have also been performed by the detection of particles scattered at forward angles. When light elements are being measured, both nuclei can escape from the target material, and both can be detected. For particles of equal mass, the conservation laws require that the angle between two emerging particles be 90° at the target. This fact has been used by Cohen et al. (1972) for the determination of hydrogen by proton–proton elastic scattering. The two outgoing protons were detected by two detectors subtending an angle of 90° at the target. This method is highly specific to hydrogen and sensitivity below 1 ppm could be achieved.

The hydrogen concentration in the sample principle could be determined simultaneously with the determination of other elements. The experimental arrangement for such an analysis is slightly more complicated, and it includes detectors and electronics necessary for the determination of hydrogen concentration in the sample by the detection of pop coincidences simultaneously with the determination of heavy elements by X-ray emission and light elements by backward scattering of protons. This represents an "all-element" analysis method if the sample is prepared in the form of a thin target.

Schweitzer et al. (2005) applied this technique using the ^{1}H(^{15}N;α,γ)^{12}C reaction to study the time dependence of the chemical reactions involved in the curing of cement. By using the Dynamitron Tandem accelerator at the Ruhr Universitat, Bochum, Germany, they have been able to achieve a few nanometer spatial resolutions at the surface of cement grains and to study the hydrogen distributions to a depth of about 2 μm. By applying a technique for stopping the chemical reactions at arbitrary times, the time dependence of the chemical reactions involving specific components of cement could be investigated. In addition, the effects of additives on chemical reactions have been studied.

By using resonant ^{1}H(^{15}N,$\alpha\gamma$)^{12}C reaction, Wilde et al. (2016) have measured surface H coverages with a sensitivity in the order of ~10^{13} cm^{-2} (~1% of a typical atomic monolayer density) and H volume concentrations with a detection limit of ~10^{18} cm^{-3} (~100 at. ppm). The near-surface depth resolution was 2–5 nm for surface-normal ^{15}N ion incidence onto the target and could be enhanced to values below 1 nm for very flat targets by adopting a surface-grazing incidence geometry. The method is versatile and readily applied to any high vacuum compatible homogeneous material with a smooth surface (no pores).

Resonant nuclear reactions are a powerful tool for the determination of the amount and profile of hydrogen in thin layers of material. Usually, this tool requires the use of a standard of well-known composition. The work by Reinhardt et al. (2016) deals with standard-less hydrogen depth profiling. This approach requires precise nuclear data, e.g., on the widely used ^{1}H(^{15}N,$\alpha\gamma$) ^{12}C reaction, resonant at 6.4 MeV ^{15}N beam energy. Here, the strongly anisotropic angular distribution of the emitted γ-rays from this resonance has been remeasured, resolving a previous discrepancy. Coefficients of (0.38 ± 0.04) and (0.80 ± 0.04) have been deduced for the second and fourth-order Legendre polynomials, respectively. In addition, the resonance strength has been re-evaluated to (25.0 ± 1.5) eV, 10% higher than previously reported. A simple working formula for the hydrogen concentration was given for cases with known γ-ray detection efficiency.

Weiss (2021) recently published a critical review of methods used in the analysis of hydrogen in inorganic materials and coatings; he has reviewed the currently used bulk analysis and depth profiling methods for hydrogen in inorganic materials and inorganic coatings. Bulk analysis of hydrogen is based on fusion of macroscopic samples in an inert gas and the detection of the thereby released gaseous H_2 using inert gas fusion and thermal desorption spectroscopy. They offer excellent accuracy and sensitivity. Depth profiling methods involve glow discharge optical emission spectroscopy and mass spectrometry, laser-induced breakdown spectroscopy, SIMS, NRA and ERDA. The methodology, calibration procedures, analytical performance and major application areas are described in the article. The synergies and the complementarity of various methods of hydrogen analysis are described. The special value of the paper is the critical evaluation of the listed literature.

3.5.2 Helium

When a thin sample containing helium is bombarded by low-energy beam of α-particles, the conservation laws require that the angle between two emerging α-particles be 90°. By the detection of both outgoing α-particles in coincidence, the concentration of the ^{4}He implanted into different materials can be determined.

Another elastic-scattering process often used is p-α scattering. Measurements of the mean depth and profile of implanted helium distributions in copper foils of various thickness by proton backscattering (Ep = 2.5 MeV) are described by Blewer (1974). Distributions of ions implanted at bombarding energies of (50–150) keV agree to within 10 nm with calculated range for helium in Cu. At room temperature, the shape of the distribution is approximately Gaussian. The isotopes ^{3}He and ^{4}He do not have low-lying excited states. ^{4}He has several highly excited states ($E_x > 20$ MeV) that decay by particle emission and are of no use in analytical application. The α-particle is a very tightly bound nucleus and a lot of energy is required to break it. As a result, there are no reactions useful for the determination of the concentration of ^{4}He. The break-up of incoming deuterons is the reaction with the least negative Q-value.

However, this process is not selective. Cross-sections for the elastic scattering of ^{4}He+t and of ^{4}He+^{3}He have resonances at bombarding energies around 5.2 MeV that might find some potential use. The ^{3}He isotope, which constitutes only 0.00013% of naturally occurring helium, can be used in some cases for the determination of helium concentration. For example, density profile of implanted ^{3}He atom in a nickel absorber were determined by detecting the protons from the ^{3}He(d,p)^{4}He reaction (Hufschmidt et al. 1975). The density distributions in an absolute scale were calculated from the proton energy spectra by means of computer programs. For such applications the resonance in the ^{3}He(d,p)^{4}He reaction at E_r = 450 keV, with Γ ~ 450 keV, is used. It corresponds to the J^Π = 3/2$^+$, 16.66 MeV excited state in ^{5}Li.

Szakács et al. (2008) have discussed the determination of migration of ion-implanted helium in silica by proton backscattering spectrometry. Understanding the processes caused by ion implantation of light ions in dielectric materials such as silica is important for developing the diagnostic systems used in fusion and fission environments. Such a tool is require to study the process by which the ion-implanted helium is able to escape from SiO$_2$ films.

In the work by Pentecoste et al. (2017) the first stages of defect formation in W due to the accumulation of helium (He) atoms inside the crystal lattice was investigated. NRA technique has been used to quantify the He implanted amount in W. This has been done by performing implantation with ^{3}He isotope and bombarding the implanted material with 900 keV deuterium beam which induces the reaction: ^{3}He(d,p)^{4}He. Protons were detected and the intensity of the corresponding signal is compared to that obtained on a reference sample containing 10^{16} He cm^{-2}. This analysis has been carried out in a dedicated apparatus DIADDHEM at the CEMHTI laboratory, Orléans, France (Sauvage et al. 2004).

3.5.3 Lithium

Natural lithium is made up of two isotopes, ^{6}Li (7.42%) and ^{7}Li (92.58%). Both can be used in analytical applications through a variety of reactions. A possible way of prompt activation is by using inelastic-scattering process. Target nuclei are excited by energy transfer from projectile and left in an excited state. After a very short time, the excited-state-ground-state transition occurs and a γ-ray is determined by the energy difference between the two states involved in transitions. The resonant structures in the cross-section of the ^{7}Li(d,p)^{8}Li, ^{7}Li(d,n)^{8}Be, ^{7}Li(d,α)^{5}He and ^{7}Li(d,γ)^{9}Be reactions occur at several energies: 360 keV, 777 keV, 1030 keV, 2000 keV, etc. These resonances correspond to ^{9}Be states. High-energy γ-rays for the identification of the presence of ^{7}Li are available from the ^{7}Li(α,γ)^{11}B reaction, which has narrow resonances at 401 keV, 819 keV and 958 keV. The most intense γ-transitions for these resonances are:

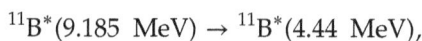

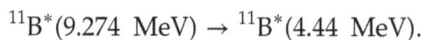

$$^{11}B^*(8.919 \text{ MeV}) \rightarrow {}^{11}B\text{g.s.,}$$

$$^{11}B^*(9.185 \text{ MeV}) \rightarrow {}^{11}B^*(4.44 \text{ MeV}),$$

$$^{11}B^*(9.274 \text{ MeV}) \rightarrow {}^{11}B^*(4.44 \text{ MeV}).$$

Lithium concentrations have been determined using the ^{6}Li(d,α)^{4}He and ^{7}Li(p,α)^{4}He reactions by several investigators. The advantage of both reactions is in the fact that for the bombarding energies of (1.0–7–1.5) MeV outgoing α-particles have rather high energies

(8–7–9) MeV. Dearnaley (1973) has used proton and α-particle beams from 6 MeV Van de Graaff accelerator to determine quantitatively the composition of complex corrosion films produced by the exposure of 316 stainless-steel specimens to a variety of alkaline solutions (LiOH, NaOH). Analysis was performed by a combination of backscattering and nuclear-reaction techniques. The $^7Li(p,\alpha)$ reaction was used. This reaction exhibits a strong, broad resonance peaking at a proton energy of 3 MeV (see also Hartley 1975). The specimens were bombarded with protons at this energy and for calibration purposes; a $LiNbO_3$ crystal was included in the study. The lithium–iron ratio was determined by comparing the α-yield to the scattered proton yield from Fe and Nb, respectively, and correcting for the differences in the Rutherford scattering cross-section.

Their series of experiments have demonstrated that ion backscattering, coupled with nuclear-reaction analysis (and coupled with more conventional techniques of surface analysis) offers a very powerful means of examining corrosion films. Another reaction of interest is the $^7Li(p,\gamma)^8Be$ reaction. This reaction requires only 0.5 MeV protons, has a relatively large resonance cross-section of 6 mb and produces a readily detected 17.6 MeV γ-ray.

3.5.4 Beryllium

The simplest way to characterize the ^9Be nucleus is by the small binding energy of the neutron (only about 1.6 MeV). Many charged-particle-induced reactions have positive Q-values. However, because of low-lying $^8Be + n$ threshold, only transitions to the ground state should be considered. Transitions to excited states will result in a background whose structure is determined mainly by phase space. Both the $^9Be(\gamma,n)^8Be$ and $^9Be(p,n)^9B$ reactions have resonant structures; however, they are of little use in concentration profile measurements. The use of these reactions requires the determination of the neutron energy by means of time-of-flight (TOF) measurements. There is a number of resonances in the ^9Be $(p,\gamma)^{10}B$ reaction together resulting in intensive γ-transitions. These gamma lines may be used as a signature of ^9Be presence within the target bombarded by protons.

3.5.5 Boron

A large number on nuclear reactions involving ^{10}B and ^{11}B may be used in the measurements of boron concentrations. The abundances of ^{10}B (19.79%) and ^{11}B (80.22%) favor the use of the ^{11}B isotope.

For the determination of boron, usually the following two nuclear reactions are used:

$$^{11}B(p,\,\alpha)^8Be \quad E_p \sim 700 \text{ keV}$$

$$^{10}B(n,\,\alpha)^7Li \quad E_n \sim \text{thermal}$$

The $^{11}B(p,\alpha)^8Be$ reaction has been employed for boron detection, mainly for channeling experiments on ion-implanted silicon (North and Gibson 1970; Akasaka and Horie 1973). Applications of this reaction have included the analysis of boron in steels. The amount of B detected was in the range 1–100) ppm (Oliver et al. 1975).

3.5.6 Carbon

Carbon has two stable isotopes: ^{12}C (abundance 98.89%) and ^{13}C (abundance 1.11%). Both isotopes have been used for the determination of carbon concentrations in different specimens. The low-lying excited states are well separated: (0–4.439–7.653–9.638) MeV. Gamma transitions between them may be used as a signature of the presence of carbon. Elastic scattering of protons shows two sharp peaks (E_p = 0.46 and 1.7 MeV) corresponding to states in ^{13}N. Other reactions on ^{12}C used for the determination of carbon include:

$$^{12}C(p, \gamma)^{13}N$$

$$^{12}C(d, p)^{13}C$$

$$^{12}C(He, p)^{15}N$$

$$^{12}C(t, p)^{14}C.$$

The reaction $^{12}C(p,\gamma)^{13}N$ exhibits two low-energy resonances that have been often used. For example, let us mention the work by Lorenzen (1974) in which a method is described for measuring the depth distribution of carbon in steel surfaces by mean of the $^{12}C(p,\gamma)^{13}N$ reaction. At proton energies of 0.457 MeV and 1.698 MeV, the excitation function for this reaction exhibits sharp resonances that have been used to activate the carbon in thin layers beneath the surface. By means of irradiations at selected proton energies, this technique facilitates the measurement of the depth distribution of the carbon content. It has been found that this method can be used to reveal carbon concentration profiles down to a depth of 20 μm with a resolution between 0.26 and 1.7 μm for the two resonances. The detection limit obtained for carbon is 0.1 % and the shape of the depth distribution is reproducible within 15%. The relative $^{12}C/^{13}C$ concentration can be measured using (p,γ) reactions. Close et al. (1973) have described an investigation of the $^{12}C(p,\gamma)^{13}N$ and $^{13}C(p,\gamma)^{14}N$ reactions and their applicability for assaying materials for ^{13}C enrichments. The system was calibrated by measuring the ratio of the yield from the $^{12}C(p,\gamma)^{13}N$ reaction to the yield from $^{13}C(p,\gamma)$ reaction as a function of the ^{13}C enrichment.

The $^{12}C(d,p)^{13}C$ reaction has the advantage of good sensitivity at relatively low deuteron energies. The cross-section has a broad resonance at about 1.3 MeV ($d\sigma/d\Omega$ = 56 mb/sr for θ_{lab} = 135°). Many (d,p) reactions have Q-values in excess of that for $^{12}C(d,p)$; however, the most likely interferences at bombarding energy of 1.3 MeV are protons from $^{16}O(d,p)$ and $^{14}N(d,p)$ reactions. Pierce et al. (1974) have made extensive use of the $^{12}C(d,p)^{13}C$ reaction at 1.3 MeV deuteron energy to examine carbon diffusion profiles in steel. They point out that great care must be exercised to avoid spurious counts from hydrocarbon contaminant films deposited during analysis. The effects of contamination can be minimized either by implanting carbon sufficiently deep to ensure clear separation of the counts from a surface peak or by using a cold trap very close to the specimen surface. The sensitivity to carbon of the $^{12}C(d, p)^{13}C$ reaction was found to be 75–100 ppm for steels (Pierce et al. 1974). A major problem in an experiment involving the detection of low levels of C is the background resulting from hydrocarbons in the vacuum system being deposited on the target surface by the incident beam. Surface levels of approximately 10^{16} atoms of C/cm^2 can be expected. Concentration of carbon in steel has also been studied by $^{12}C(d, p)^{13}C$ and $^{13}C(d, p)^{14}N$ reactions at 1.1 MeV (Feldman et al.

1973). They have determined the lattice site location of carbon in steel by combining measurements of the above reactions and channeling techniques.

Barrandon and Seltz (1973) have described a method for the microanalysis of carbon on the surface of metal samples by the detection of protons from the (t, p) reaction at bombarding energies below 2 MeV. For the reaction $^{12}C(t, p)^{14}C$, Q = 4.641 MeV, the cross-sections for optimal bombarding energies E_t = 1.930 and 1.740 MeV are $d\sigma/d\Omega$ (θ = 160°) = 12.36 mb/sr and $d\sigma/d\Omega$ (θ = 150°) = 4.05 mb/sr, respectively. A large number of γ-rays can be produced in the process of the bombardment of ^{13}C nucleus by differently charged particles. Inelastic scattering can generate some characteristic γ-rays. The $^{13}C(p, \gamma)^{14}N$ reaction has several narrow low-energy resonances corresponding to ^{14}N excited states. A large number of γ-rays is also produced in $^{13}C(d,n)^{14}N$ and $^{13}C(d,p)^{14}C$ reactions. The γ-rays produced are due to the transitions involving excited states of ^{14}N and ^{14}C nuclei.

3.5.7 Nitrogen

Nitrogen concentration measurements can be performed using nuclear reactions on two isotopes: either ^{14}N or ^{15}N. Isotope ^{14}N is more abundant in natural nitrogen having an abundance of 99.63%. Many γ-ray transitions are allowed, and the de-excitation of the ^{14}N nucleus gives rise to a complicated γ-ray spectrum. Deuteron-induced reactions are characterized by high Q-values. Both (p, γ) and (α,γ) reactions on ^{14}N show several low-energy resonances. Resonances in the (α, γ) reaction occur for E_α = 0.559, 1.140 and 1.395 MeV, for α-particle energies under 1.5 MeV.

The abundance of the ^{15}N isotope is only 0.37%. However, many reactions have convenient Q-values and cross-section energy dependence to be of use in elemental analysis. Alpha particles from proton bombardment can be detected with or without coincidence requirements on simultaneous γ-rays. Transitions to $^{12}C^*$ (4.43 MeV) in the $^{15}N(p,\alpha)^{12}C$ reaction are accompanied with 4.43 MeV γ-ray. The cross-section for α-γ coincidences shows several low-energy resonances. Several deuteron-induced reactions have positive Q-values; among them the (d, α) reaction has the highest, with Q = 7.687 MeV. The $^{15}N(\alpha,\gamma)^{19}F$ reaction has two sharp resonances for energies below 1.5 MeV: one for E_α = 0.850 MeV and another for E_α = 1.385 MeV.

As seen, several reactions can be used to determine nitrogen concentrations. However, the most popular are $^{14}N(d,p)^{15}N$ and $^{14}N(d, \alpha)^{12}O$. Although deuteron-induced reactions produce large amounts of background (mainly in the form of neutrons and γ-rays), the Q-values for deuteron-induced reactions are frequently very large (for $^{14}N(d, p)^{15}N$, Q = +8.61 MeV).

As a result, the emitted protons are usually well separated from the elastic low-energy part of the spectrum and the energy loss through the absorber foils is small. Oliver et al. (1975) and Oliver and Pierce (1974) have used deuteron-induced reactions to determine the distribution of nitrogen in metal samples over the range of 0.05–2% using E_d = 1.0–3.0 MeV. Protons and ex-particles emitted from the $^{14}N(d, p)^{15}N$ and $^{14}N(d,\alpha)^{12}O$ reactions were measured as 3 MeV deuteron beams of a few micrometers in diameter (from Van de Graaff accelerator) were scanned across the sample surface. The obtained results are found to be in good agreement with the results of the chemical analysis.

3.5.8 Oxygen

Although oxygen occurs in nature in the form of three stable isotopes: ^{16}O, ^{17}O and ^{18}O, the ^{17}O isotope can generally be ignored because of its small abundance (0.037%). There are no excited states in the ^{16}O nucleus under 6 MeV excitation energy. As a result, inelastic-scattering processes are not possible for low bombarding energies. This fact allows the use of (p, and p) (α, and α) elastic-scattering processes in the measurements of oxygen concentrations. Several deuteron-induced reactions on ^{16}O have positive Q-values.

The ^{18}O isotope (abundance 0.204%) has several low-lying excited states. Gamma-transitions between them are allowed. Deuteron-induced reactions: (d, p), (d,n) and (d, α) on ^{18}O have all positive Q-values. The $^{18}O(p, α)$ reaction is also characterized by a positive Q-value. The cross-section for the $^{18}O(p,α)^{15}N$ reaction shows several resonances, the most intense one occurring at E_p = 0.85 MeV.

Oxygen concentrations have been measured using the following reactions: $^{18}O(p,α)^{15}N$, $^{18}O(d, α)^{16}N$, $^{16}O(d, α)^{14}N$, $^{18}O(d, p)^{19}O$, $^{16}O(t, p)^{18}O$, $^{16}O(d, n)^{17}F$ and $^{16}O(d, p)^{17}O$. A narrow resonance in the $^{18}O(p, α)^{15}N$ reaction at 1.763 MeV (width, Γ = 4.0 keV) may be used to determine concentration profiles by exciting the resonance at specific depths below the surface and calculating the appropriate scattering centers at that depth (see Barnes et al. 1973).

Calvert et al. (1974) to look at oxygen diffusion in steels used the same resonance. These authors have developed a deconvolution technique to extract the original profile. Another resonance in the (p, α) reaction on ^{18}O often used is the one at 639 keV. Whitton et al. (1971) used the 639 keV resonance in the $^{18}O(p, α)^{15}N$ reaction to measure the depth distribution of oxygen implanted into gallium phosphide at implantation energies of 20, 40 and 60 keV. Nield et al. (1972) used the same resonance in measuring the oxide thickness on chromium and oxygen diffusion constructed in TiO_2 and Cr_2O_3. Lindstrom and Heuer (1974) have constructed a magnetic spectrometer for studies of the distribution of ^{18}O within a solid sample using the $^{18}O(p, α)^{15}N$ reaction. This approach allows probing deeper concentration profiles than does the absorber foil technique.

Accurate determination of metal/oxygen ratios and oxide thicknesses can be accomplished also by using $^{16}O(α, α)^{16}O$ elastic scattering. For this purpose, it is necessary to choose regions over which the cross-section is varying smoothly or is constant. For α-particle beams there are two such bombarding energy intervals:

(1.8–2.2) MeV, where the cross-section obeys the Rutherford law, and (2.6–2.9) MeV, where the cross-section remains constant for detection of alphas near θ = 165°.

Dearnaley at al. (1975) have determined oxygen concentrations by elastic p-^{16}O scattering above 4 MeV bombarding energy. Over a wide range of scattering angles, the energy differential cross-section remains constant and is approximately 80 times the Rutherford cross-section. The penetration of these energetic protons is such that it was possible to use thin foil specimens (3–6) μm and achieve a complete separation of the peaks in the energy spectrum due to scattering from metal or oxygen nuclei and to distinguish between the oxygen on the front and back of the foil. In principle, it is feasible to make a direct comparison of the oxygen absorption of the implanted face and the unimplanted rear face.

Oxygen content in glass and some bulk chemicals was also determined by the (d,pγ) reaction. The $^{16}O(d, pγ)^{17}O$ reaction has large resonances at 0.950 and 1.300 MeV that permit the detection of oxygen at the 5 ppm level (Mandler and Semmler 1974). When using deuteron beams for analysis, a boron shield is used around the detector to minimize the effects of neutrons generated by the (d, n) reaction. In the work reported by Lorenzen and Brune (1975), oxygen in zircaloy surfaces has been determined by means of charged-particle activation analysis employing the following two reactions:

(a) $^{16}O(d, n)^{17}F^{(\beta^+)} \rightarrow {}^{17}O$ Q = −1.63 MeV

(b) $^{16}O(d; p, \gamma)^{17}O$ Q = +1.05 MeV

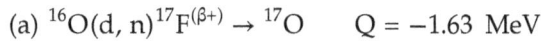

The detection limits for oxygen in such surfaces have been investigated by measuring the promptly emitted 0.87 MeV γ-rays, reaction (b) and also the 511 keV annihilation radiation that arises from β^+ decay of ^{17}F, reaction (a). At a deuteron energy of 3 MeV, a detection limit of 0.7×10^{-7} g/cm^2 was obtained from reaction (a).

Another example is the work reported by Markowitz and Mahoney (1962) in which ^{3}He-induced reactions were applied to activation analysis for oxygen. Analyses were performed by the detection of the radioactive isotope ^{18}F produced by the $^{16}O(^3He,p)^{18}F$ reaction. This isotope is a positron emitter with a 110 min half-life. ^{3}He-induced reactions have positive Q-values. However, the variation of the reaction cross-section with energy complicated the problem of measuring the quantity of an element in the sample.

Barrandon and Seltz (1973) described a method for microanalysis of oxygen on the surface of metal samples by detection of protons from (t, p) reactions at bombarding energies below 2 MeV. Using (t, p) reactions, simultaneous analysis of oxygen and carbon surface layers is possible with a sensitivity of 5×10^{-2} µg/cm^2. The oxygen concentration is determined using the $^{16}O(t,p)^{18}O$ reaction, Q = 3.730 MeV, at bombarding energies of 1.930 MeV or 1.870 MeV.

3.5.9 Fluorine

Fluorine occurs in nature in the form of only one stable isotope: ^{19}F. Its nucleus has many low-lying excited states; there are eight under 4.0 MeV excitation energy. This level structure gives rise to several low-energy y-rays. Several reactions on ^{19}F have positive Q-values, (p,α) and (p,γ) being the most useful. For example, some experimentalists have measured fluorine contamination on and below the surface of zircaloy (a common material for the cladding of fuel elements in nuclear reactors) by means of a resonance at 1375 keV in the ^{19}F (p,αγ) ^{16}O reaction. The same resonance and reaction are used for the determination of the fluorine and hydrogen distributions in lunar samples (Oliver and Pierce 1974). The fluorine distribution was obtained using a proton beam. Due to the sharp resonant nature of this reaction (5 keV wide), the γ-ray counting rate is proportional to the fluorine content at a particular depth. This depth can be determined by the choice of the incident proton energy, since higher-energy protons penetrate deeper before being slowed down to the resonant energy. An analogous situation holds for the measurement of hydrogen distributions, in which a ^{19}F beam is used to induce the reaction on the hydrogen contained in the lunar sample. In either case, the γ-ray counting rate is measured at a number of incident-beam energies near the depth profile using proportionalities between beam energy and depth and between counting rate and concentration of hydrogen and fluorine. The hydrogen profile was measured in the layer (0–0.5) µm, with a depth resolution of 0.05 µm. An alternative method of detecting fluorine is by using the $^{19}F(p,\alpha)^{16}O$ reaction at higher bombarding energies (1.2–2.6) MeV. The activation analysis can be done by the ^{3}He-induced reaction $^{19}F(^3He,\alpha)^{18}F$, $\tau_{1/2}$ = 110 min. The (p,αγ) reaction has been used by several investigators for the determination of the fluorine content in teeth and for measurements of the fluorine concentration profile.

3.5.10 Other Elements

In principle, the concentration of any element can be determined by performing some nuclear reaction and the subsequent detection of reaction products. The sensitivity for a particular element will depend on the nuclear reaction used, the cross-section for the process and interferences from nuclear reactions on other elements within the sample. The use of charged particle-induced reactions is limited only to light nuclei. Heavier nuclei have more complicated energy level diagrams. The level density is much higher than for light nuclei. Furthermore, heavier nuclei (larger Z) have higher Coulomb barriers for both incoming and outgoing charged particles. These factors play an important role in analytical applications.

Proton, deuteron, ^{3}He- and ^{4}He-induced reactions with resulting outgoing charged particles and photons can be used in analytical applications. For example, the sodium content in the surface layer of glass was determined using (p,γ), $(p,p\gamma)$ and $(p,\alpha\gamma)$ reactions (Mandler and Semmler 1974). Sulfur was detected nondestructively in thin film in a copper-nickel alloy in the presence of carbon and oxygen by the $^{32}S(d,p)^{33}S$ reaction (Wolicki and Knudson 1967). Ricci and Hahn (1967) have discussed some ^{3}He-induced reactions. Besides the use of the $^{14}Ne(^3He, d)^{15}O$ and $^{19}F(^3He,\alpha)^{18}F$ reactions, they also discuss the use of the $^{35}Cl(p,d)^{34m}Cl$ reaction, in which the isomeric state of ^{34}Cl is formed. ^{34}Cl has a half-life of $\tau_{1/2} = 32.4$ min. In addition, chlorine can also be detected using the $^{37}Cl(d,p)^{38}Cl$ reaction.

By definition, nuclides that have the same A and Z but different states of excitation are called isomers. Spin and parity quantum numbers of the excited state and lower states may be such that electromagnetic transitions are slowed down to a point that the formed excited state has a very long lifetime. The lifetime of an isomeric state (or metastable state) can be anywhere from 10^{-10} s to 10^6 years. The lower limit on the lifetime of the state to be considered an isomeric state is sometimes arbitrarily set on 0.1 s. On the other hand, Kantele and Tannila (1968) list all isomeric states with $\tau_{1/2} \sim 10^{-10}$ s. Their table is useful when an unidentified radioisotope is encountered. They list the energies of the most intense gamma-rays associated with the given half-life. Other types of radiations are mentioned only if no gamma-rays are emitted.

There are no isomers for very light nuclei. This is because to be metastable an excited level must differ from lower energy levels by several units of angular momentum. Even at higher A, nuclear isomeric states are not spread uniformly among all nuclei but are preferentially concentrated in "islands" of nuclei with Z or N just below the magic numbers (for even A). Isomers with both N and Z even are very rare. The shell model of the nucleus can satisfactorily explain all these facts. The lifetime of excited states of light nuclei is generally rather short; it is said that they emit prompt γ-rays. Detection of prompt γ-rays allows the determination of concentrations of many elements. Mandler and Semmler (1974) have calculated sensitivities for several elements.

In their work Caridi et al. 1992 described the application of proton resonant scattering at $E_p = 4.6$ MeV to the composition determination of zirconium oxide thin films, reactively evaporated at different oxygen partial pressures using either zirconium metal or oxygen deficient ZrO_{2-x} as starting materials. The results are correlated with the optical and structural properties of ZrO_{2-x} films.

Electron beam evaporated zirconium oxide films are suitable materials for optical coatings operating at visible and infrared wavelengths. The optical properties of these films are related to the microstructure and the crystallographic phase composition, which depend on the oxygen partial pressure in the evaporation chamber and on the substrate temperature. These deposition process parameters are particularly critical in the case of reactive evaporation from zirconium metal, a method that seems very effective for the mass production of optical coatings.

3.5.11 Stopping Power

When a monoenergetic beam of particles with atomic number Z_1 and mass number M_1, with initial energy E_0, crosses a monoelemental thin foil of thickness Δx and areal density N_t, composed of an element with atomic number Z_2 and mass M_2 the transmitted beam energy will be $E_f = E_0 - \Delta E$, where ΔE is the energy loss of the beam crossing the foil. The stopping power of the medium, S, is defined as the energy loss per path length:

$$S \equiv -dE/dx = -\lim_{\Delta x \to 0} \Delta E/\Delta x \qquad (3.2)$$

The stopping power is, thus, a property of the medium; the energy loss is a property of the beam. The correct terminology should state the stopping power of a given element (or compound) for a given ion species or, alternatively, the stopping power for a given ion in a given element (or compound), see IAEA 2019.

The units of the stopping power (S) are, for instance, eV/nm or MeV/μm. S is also called the linear stopping power. The mass stopping power, with units such as MeV/(mg/cm^2), is defined as the ratio between the linear stopping power and the density (ρ) of the material:

$$S/\rho = -1/\rho \ dE/dx = 1/\rho \ S \qquad (3.3)$$

Most IBA techniques, such as ERDA or RBS, are not sensitive to actual thickness or depth, but to areal density or, in other words, to the amount of matter crossed, which is commonly given in units of atoms/cm^2, μg/cm^2 or mg/cm^2. Thus, the stopping cross-section, ε, is often the most useful quantity:

$$\varepsilon = -1/N \ dE/dx = -1/N \ S \qquad (3.4)$$

where N is the atomic number surface density of the medium. The stopping cross-section is, like the stopping power, a property of the medium. Commonly used units are, for instance, eV/(10^{15} atoms/cm^2), eV/(μg/cm^2) or keV/(mg/cm^2).

As a beam of particles crosses a material, each ion undergoes a different number of collisions and follows a slightly different path. After crossing a certain thickness, an initially monoenergetic beam has a certain energy distribution. The energy loss defined above is the average of this energy distribution (sometimes the median or the most probable energy loss is taken). This phenomenon of the spreading of the energy distribution of the beam is called "straggling".

The statistical fluctuations in the number of collisions lead to so-called energy loss straggling. Multiple scattering – that is, the combined effect of very many small angle scattering events – leads to a change in the angular distribution of the beam, which contributes to an increase in its energy distribution. Geometric straggling includes the effects of a finite-sized beam, beam spot and detector. Other effects can be important in special circumstances. Energy straggling results in the depth resolution degrading with depth. The result is a broadening of any interface signals in the spectra obtained and a distortion of the spectral shape. Any interface, intermixing and diffusion studies with RBS or ERDA (or other IBA techniques) must take energy straggling into account as accurately as possible, or the results will be incorrect. State of the art calculations of depth resolution as a function of depth are made by the computer code DEPTH, which is based on the models developed by Amsel et al. (2003). The databases for straggling are, however, much sparser and less accurate than for stopping power. The theoretical models are, therefore, more difficult to test against data, and no statistical analysis of their accuracy has been presented so far.

3.5.12 Concentration Profile Measurements

Measurement of depth profiles can be accomplished by the use of nuclear reactions with a sharp resonance in the cross-section at a given bombarding energy. This technique takes advantage of the fact that resonance reactions occur only at specific energy while the energy of a charged particle beam progressively decreases as it penetrates the sample. In such a way, information on depth variation of element concentration can be determined. The precision with which the depth of a resonance can be located depends mainly on the straggling, and on the width of resonance, the initial energy distribution of the beam and the energy resolution of the system.

Information on range and range straggling of heavy ions are important in semiconductor device fabrication by ion implantation. The projected range and range straggling have been measured for many ions in different materials mainly using backscattering techniques with several MeV $^4\text{He}^+$ ions. For compounds, one relies on the applicability of Bragg's rule on additivity of stopping powers.

The depth at which a given resonance occurs is equal to the difference between the total range $R(E_o)$ and the range of a particle with the initial energy equal to E_R:

$$dR(E_0) = R(E_0) - R(E_R) \tag{3.5}$$

This can be calculated from energy range tables for a given material and given bombarding particle.

The apparent width of a resonance in a nuclear reaction and the yield of particles or γ-rays from the resonance are influenced by the thickness of the target. Assuming that the stopping cross-section of material and the energy loss ΔE of the particle beam in the target are independent of the beam energy in the vicinity of resonance, the yield will be given by

$$Y = 1/\varepsilon \int_{E-\Delta E}^{E} \sigma(E)\,dE \tag{3.6}$$

E and ε are related by $\Delta E = Nt\varepsilon$, where N is the number of nuclei/cm^3 in the target, while t is the target thickness in cm. Nt is then a surface density measuring the number of atoms per cm^2. When (J (E) is given by a Breit-Wigner relation, the yield Y can be calculated:

$$Y = \sigma_R \Gamma / 2\varepsilon [\text{tg}^{-1}(E - E_R/\tfrac{1}{2}\Gamma) - \text{tg}^{-1}(E - E_R - \Delta E/\tfrac{1}{2}\Gamma)] \tag{3.7}$$

where $\sigma_R = \sigma(E_R)$. For a given ΔE the above equation has a maximum at $E = E_R + \Delta E$ given by

$$Y_{\text{max}} = (\sigma_R \Gamma / \varepsilon)\,\text{tg}^{-1}(\Delta E / \Gamma) \tag{3.8}$$

For an "infinitely thick" target the maximum yield of "thick target step" is

$$Y_{max}(\infty) = \frac{1}{2}\pi\sigma_R\Gamma/\varepsilon \qquad (3.9)$$

The observed width Γ' Is related to the natural width Γ of the resonance by

$$\Gamma' = (\Gamma^2 + \Delta E^2)^{1/2} \qquad (3.10)$$

The ratios $Y_{max}(\Delta E)/Y_{max}(\infty)$ and Γ'/Γ are presented graphically in tables prepared by Marion and Young (1968).

As an example, we shall briefly discuss the problem of profiling hydrogen in materials. The presence of hydrogen in some materials is known to have dramatic effects on the properties of these materials. For example, the presence of hydrogen in steel can cause "hydrogen embrittlement", and the bombardment of many metals with high fluxes of low-energy hydrogen can result in the formation of blisters and the spalling away of macroscopic pieces of material. Many of these phenomena are poorly understood because there has not been available a technique with which to measure the concentration of hydrogen as a function of depth in the material. Even such fundamental quantities as the range of penetration of low-energy hydrogen into materials and the mobility of hydrogen in solids have not been carefully studied because of the lack of an analytical method.

Two reactions are often used for hydrogen profiling: $^{19}F+^1H$ and $^{15}N+^1H$. The latter reaction is of more interest, in spite of its lower yield, because it gives a depth resolution approximately four times better (smaller); it can be easily used to a depth of over 3 μm, while the $^1H+^{19}F$ can be easily used only to about one-half the energy compared with the $^1H+^{15}N$ reaction.

The technique depends on the resonant nuclear reaction between protons and ^{15}N that yields an α-particle and a characteristic 4.43 MeV γ-ray. This resonant reaction occurs at the centre-of-mass energy of 402 keV, has a width of 0.9 keV, and a peak cross-section of 200 mb. The cross-section of resonance is more than 2 orders of magnitude smaller. To utilize this reaction as a probe for hydrogen, one bombards the sample with ^{15}N and measures the yield of the characteristic 4.43 MeV γ-rays using a NaI scintillation counter.

If there is hydrogen on the surface of the sample and the ^{15}N has an energy of 402 keV in the center of mass (i.e., 6.385 MeV in the laboratory), the yield of 4.43 MeV γ-rays is proportional to the amount of hydrogen on the surface. To determine the concentration of hydrogen inside the sample, the ^{15}N energy is raised so that it is above the resonant energy for hydrogen on the surface, but as the ^{15}N slows down inside the sample, it reaches the resonant energy at a certain depth inside the surface.

Several authors have also described the use of PIXE for depth profiling. This is illustrated in Fig. 3.13; let p be the surface density of contaminant (g/cm^2), while N_0 represents the total number of contaminant atoms. The X-ray production cross-section σ(E) for the proton bombardment of impurity atoms depends on the proton energy at a given depth x. K X-rays are emitted isotropically, therefore the fraction of X-rays detected within a Si (Li) detector's solid angle dΩ is $(d\Omega/4\pi)\exp[-\mu x \sec\theta]$. The coefficient μ is the mass absorption coefficient for the impurity's X-rays in the target under investigation. The total detected yield Y, the number of X-rays per proton, is given by

$$Y = N_0(d\Omega/4\pi)\varepsilon_0 \int_0^1 \sigma[E(x)] \exp[-\mu x \sec\theta]dx \qquad (3.11)$$

where ε_0 is the detector efficiency for the impurity's X-rays. For very thin impurity films bombarded with N protons the above formula reduces simply to

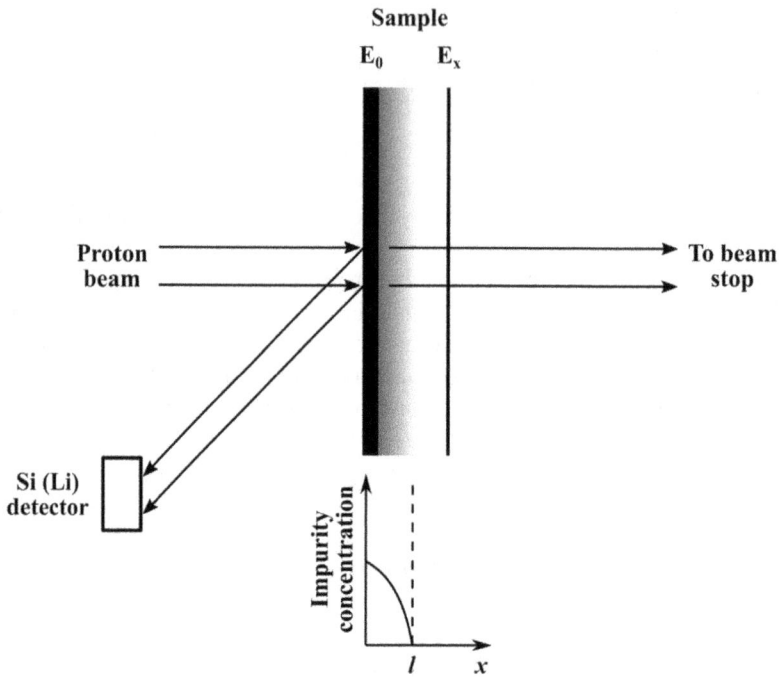

FIGURE 3.14
Concentration profile determination can be accomplished by the measurement of the yield of proton-induced characteristic X-ray emission.

$$Y \approx N_0 (d\Omega/4\pi)\varepsilon_0 N\sigma(E_p)\cdot l \tag{3.12}$$

This has been derived assuming no change in proton energy and no change in the X-ray production cross-section as a function of depth (Fig. 3.14).

Another tool for the measurement of element concentration depth profiles is to use PIXE measurements at different bombarding energies. If the sample is bombarded with monoenergetic protons of energy E_1 under an angle θ, the X-ray yield from an infinitesimal volume at depth x of the element with $n(x)$ atoms per volume, is given by

$$dI = Q/e \; n(x)\sigma[E \; (E_1, x/\cos \theta)] \; dx \tag{3.13}$$

where Q/e represents the proton flux, and $\sigma[E(E_1, x/\cos\theta)]$ the cross-section for the proton energy $E(E_1, x/\cos\theta)$ at the depth $x/\cos\theta$. The fraction of X-rays emitted into the angle ϕ at which the detector with efficiency α and solid angle Ω is positioned is then given by

$$dI = Q\alpha\Omega/4\pi e \; n(x)\sigma[E \; (E_1, x/\cos \theta)] \; \exp[-\mu x/\cos \phi] \; dx \tag{3.14}$$

where μ is the absorption coefficient of the considered X-rays. If we integrate the above equation over the whole proton range R, the expression of the total yield is obtained:

$$I = k \int_0^{R\cos\theta} c(x) \; \sigma[E \; (E_1, x/\cos \theta)] \; \exp[-\mu x/\cos \phi] \; dx \tag{3.15}$$

where $C(x) = n(x)W/N_0$, W is the atomic weight of the element, N_0 Avogadro's number and $k = Q\alpha\Omega N_0/4\pi eW$. For $C(x) = 1$ the above equation presents the efficiency Y of the PIXE for the thick-target measurements. From this equation, it can be seen that the $c(x)$ profile can, in principle, be obtained by varying the absorption of cross-section term. This can be done, for example, by varying the proton energy or tilting the target. The basic rule for deconvolution of an unknown profile is to consider $c(x)$ as

$$c(x) = \sum_j c_j f_j(x) \tag{3.16}$$

where c_j are the components of the f basis. The f's most often used are steplike functions with edges determined by the proton ranges used. Equation can now be transformed into the matrix equation:

$$I_i = \sum_j A_{ij} c_j \tag{3.17}$$

Suffix i corresponds to the i-th energy used. The matrix A_{ij} is, in fact, the contribution to the X-ray yield of each slab with unit concentration of the element of interest:

$$A_{ij} = k \int_{X_{j-1}}^{X_j} \sigma[E \ (E_i, x/\cos\theta)] \ \exp[-\mu x/\cos\phi] dx \tag{3.18}$$

The A_{ij} values can be easily computed by using tabulated values from the literature or by measuring the yields of homogeneous samples. To determine the c_j values from the measured I_i results, the matrix inversion procedure must be applied. Because of statistical errors in measuring I_i and errors in A_{ij}, erratic fluctuations in c are produced. A smoothing procedure, which consists in changing the A_{ij} or I_i values, is the only way to get reasonable results for c_j.

Here, we shall mention two examples of ERDA measurements. In the work reported by Ishigami et al. (2005) ERDA measurements in reflection geometry have been performed on titanium deuteride (TiD_x) set in air of 1 atm and titanium hydride (TiH_x) set in a mixture of hydrogen gas of 0.5 atm and deuterium gas of 0.5 atm. A ^{4}He beam with m energy of 15 MeV is extracted through a molybdenum foil with a thickness of 5 μm into the gases. Spectra from deuterium and hydrogen in these samples are obtained by subtraction of spectra of titanium from spectra of the TiD_x and TiH_x, respectively. Depth profiles of the deuterium and hydrogen atoms are calculated from the obtained spectra. The atomic ratios of deuterium and hydrogen to titanium in the samples are 0.83 and 0.79 up to a depth of 8 μm, respectively.

The TOF-ERDA method is one of the IBA methods that are capable of analyzing light elements in a sample with excellent depth resolution. In this method, simultaneous measurements of recoil ion energy and time of flight are performed, and ion mass is evaluated. The energy of recoil ions is calculated from TOF, which gives better energy resolution than conventional silicon semiconductor detectors (SSDs). TOF-ERDA is expected to be particularly applicable for the analysis of light elements in thin films. In his review article, Yasuda (2020) described the principle of TOF-ERDA measurement and details of the measurement equipment along with the performance of the instrumentation, including depth resolution and measurement sensitivity.

In another report, Döbeli et al. (2005) have discussed the use of heavy ion ERDA (HIERDA). Previously, HIERDA has been implemented mainly at accelerator facilities with beam energy between 10 and 100 MeV. This is an energy region where gas ionization chambers allow for good resolution and particle identification. Using an ionization chamber with a very thin-patterned Si entrance window heavy ions with an energy of a few hundred keV can be detected. By combining this detector with a TOF spectrometer equipped with an ultra-thin diamond-like-carbon foil it is now possible to perform HIERDA with projectiles (e.g., iodine) of energies down to 1.5 MeV with good resolution. Beams of this type can be produced even with the smallest accelerators used in IBA. Consequences concerning the applicability of ERDA as well as scattering cross-sections, beam damage and multiple scattering will be discussed.

3.6 Rutherford Backscattering

RBS has proved to be very useful in characterization of heavy element distribution implanted into a light element matrix. The method can be improved by the use of recoil detection analysis, with measurement both particle mass and its energy. Fig. 3.15 shows the schematic

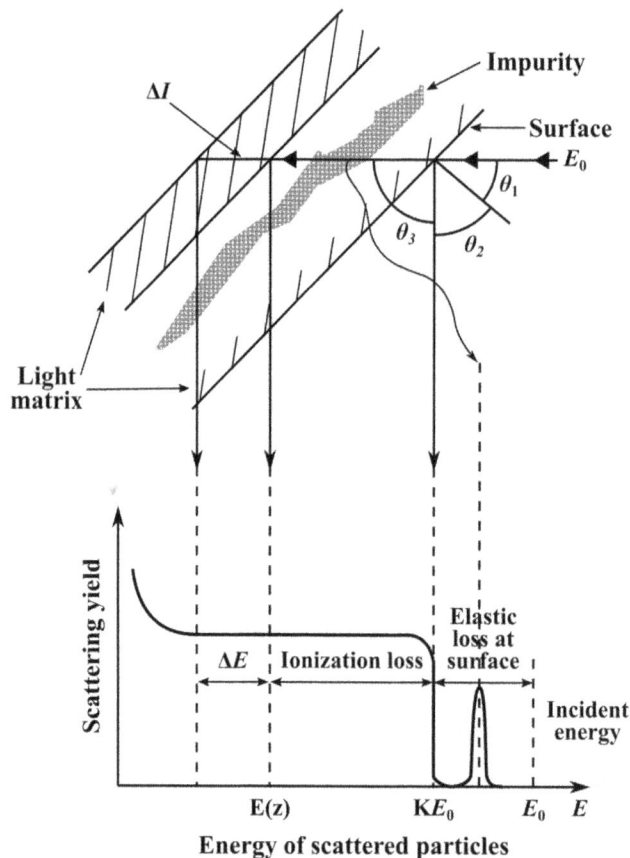

FIGURE 3.15
Schematic of elastic backscattering for an idealized sample of light material containing a heavy impurity.

of elastic backscattering for an idealized sample of light material containing a heavy impurity. Particles scattered from the surface region of the matrix (z = 0) will have an energy

$$E_0(0) = k^2 E_1 \tag{3.19}$$

where E_1 is the incident energy and k^2 is the kinematic factor given by

$$k^2 = [M_1 \cos \beta + (M_2^2 - M_1^2 \sin^2 \beta)^{1/2}] / [M_1 + M_2] \tag{3.20}$$

where M_1 is the mass of the incident ion and M_2 that of the scattering atom. From Eqs. (3.19) and (3.20) it follows that each surface impurity has its own characteristic value of k^2 uniquely defined by its mass M_2 and hence will yield a unique value of $E_0(0)$.

For the case of high-energy ions, the analysis can be put on a quantitative basis by using the Rutherford scattering cross section

$$\sigma = [Z_1 Z_2 e^2 / E]^2 \csc^4(\beta/2) \tag{3.21}$$

where Z_1 and Z_2 are the atomic numbers of the incident and target atoms, respectively. Using this equation the scattering yield Y_x from a particular species x with density N_x is given by

$$Y_x = [Z_1 Z_2 e^2 \csc^2(\beta/2) / E]^2 N_x \Delta\Omega \tag{3.22}$$

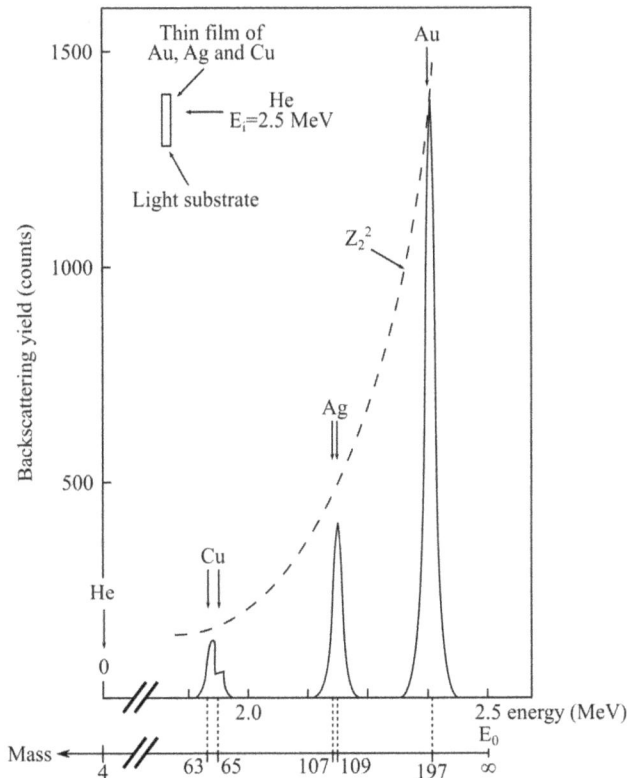

FIGURE 3.16
The Rutherford cross-section dependence on the atomic number of the scattered ions.

where $\Delta\omega$ is the solid angle of the detector. N_x can be calculated from the measured value of Y_x.

Fig. 3.16 shows the Rutherford cross-section dependence on the atomic number of the scattered ions.

In order to illustrate the type of information, one can obtain by measurement of backscattered particles Figs. 3.17 and 3.18 are presented.

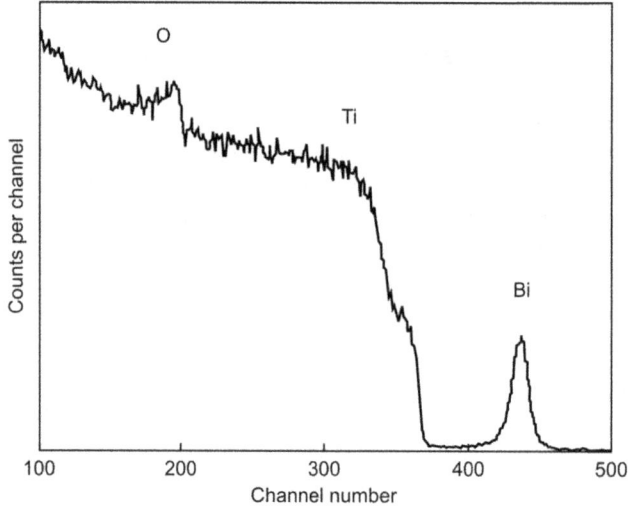

FIGURE 3.17
Energy spectrum of 1.5 MeV ^{4}He ions scattered at 164° from titanium target implanted with bismuth ions and oxidized.

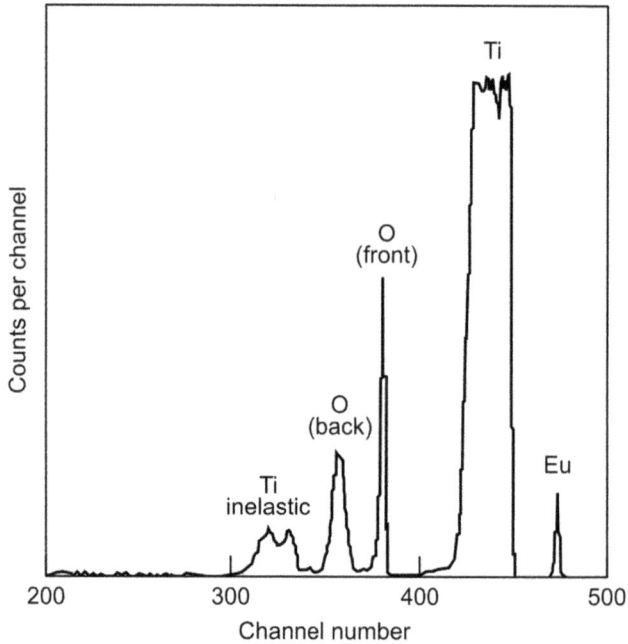

FIGURE 3.18
Energy spectrum of 4 MeV protons scattered at 164° from a 5 μm thick foil of titanium, implanted with europium ions and subsequently oxidized.

An important parameter to be discussed in the analytical applications of RBS is mass resolution. Clearly, if two peaks corresponding to scatter from nuclei x and y lie close together, they can only be resolved and measured if their energy separation is greater than the used detector resolution, ΔE:

$$E_1\left(k_x^2 - k_y^2\right) > \Delta E \tag{3.23}$$

where E_1 is the incident ion bombarding energy, while the energy of the particle scattered from the surface of the sample is given by Eqs. (3.19) and (3.20).

From these formulae following conclusions can be obtained:

- For a fixed incident ion (M_1) the mass resolution decreases as M_2 increases.
- The mass resolution for heavier elements improves as the mass M_1 of the analyzing beam increases.
- The peak separation increases linearly with the beam energy E_1.

Kinematics limits the use of backscattering to light and medium-heavy nuclei. Although the resolution increases as M_2 decreases the sensitivity of the method is influenced by the fact that the lower the charge, Z, the lower the Rutherford cross-section. The use of heavier mass ions is limited by the fact that the radiation damage is more important and also the backscattering occurs as long as $M_2 > M_1$. When $M_1 > M_2$ both particles are scattered in forward laboratory angles.

In the case when the impurity atoms are distributed uniformly throughout the host solid material, the energy scale may be converted into a depth scale (Morgan 1975; Bogh 1973):

$$E_2(z) = E_1 k^2 - z[S(E_1)k^2 + S(E_2)/\cos\,\theta]. \tag{3.24}$$

where S(E) represents the stopping power or rate of energy loss. Thus by knowing the stopping power of target material, the depth at which a particular event took place may be calculated. This is the basis for the determination of implantation profiles by RBS and is most readily suited to heavy impurities within a light substrate since under these conditions the backscattering peak is separated from the continuum.

^{4}He ions have few nuclear reaction resonances with target nuclei in the energy region 0.4–4.0 MeV and many investigators using nuclear microanalysis techniques have employed ^{4}He beams.

The stopping cross-section in a material is defined as

$$E = dE/d\rho_A = (1/N)\,(dE/dx) \tag{3.25}$$

where $dE/d\rho_A$ is the energy loss per unit area density (atoms/cm^2), N is the number of atoms per cm^3 and (dE/dx) is the energy loss per unit path length. Some authors used a term "stopping power" for quantities dE/dx or ρdE/dx, where ρ is the target density.

Backscattering techniques allow the determination of elemental composition as a function of depth with typical values of depth resolution ranging from 10 to 30 nm. The technique is ideally suited to measurements in thin-film system of thicknesses of several hundred nm and have provided a basis for examination of thin-film interactions. Backscattering has been used for measurements of film thickness, contaminant concentration, interface barrier and compound formation kinetics.

Next, we shall describe some of the numerous applications reported in the literature. Most authors exposed specimens to be analyzed to monoenergetic beams of light ions, protons or α-particles at currents of ~10–30 nA and with energies typically less than about 3 MeV. A high degree of collimation is not required, although it is important to have well-defined beam for channeling experiments.

Myers et al. (1973) have studied the implantation and diffusion of Cu in a single crystal of Be. Backscattering of 2 MeV He^+ ions was used to determine the concentration vs. depth profile to ~2 μm with a resolution of ~30 nm. A solid-state detector was positioned at $\theta = 170°$.

Alexander et al. (1973) used channeling and backscattering of ^{14}N and ^4He ions were used to determine the location of Br implanted into Fe single crystals. The $^{14}N^+$ energy was 3.0 MeV. With ^{14}N ions a greater angular resolution can be obtained for the same beam divergence; this is due to the fact that critical angles for channeling are larger for ^{14}N than ^4He of the same energy. For both types of ions, it was necessary to correct the counts in the Br window for background, which was typically around 20% of the total counts.

Bayley and Townsend (1973) measured helium backscattering on a silica glass sample implanted with 300 keV Kr ions, flux 3×10^{16} ions/cm^2; the results revealed a distribution peaking at 726 nm. After annealing at 850 °C for 1 h this peak was not detectable anymore, They have also noticed that implantation produces refractive index changes which are due mainly to radiation damage.

Van der Weg et al. (1973) studied the contamination of silicon surfaces by proton backscattering and PIXE spectroscopy using 100–140 keV protons. Backscattered protons were observed at 135° with surface barrier detectors, while X-rays were measured at forward angles using two detectors: proportional counter (gas: argone–methane mixture) and Si(Li) detector. Atoms of elements C, O and I were observed on the surface. During the bombardment proton energy spectra were taken for random incidence and under channeling conditions. Yields obtained with X-ray generation were higher than those obtained from backscattering data.

Blewer (1974) has performed analysis of near-surface helium and hydrogen isotope depth profiles by a proton backscattering technique. The method used has been demonstrated as a means to characterize the concentration, initial position, and subsequent migration behaviour of all atom species. The projected range of helium implanted at 50 keV in Nb, V, Ti and Cu has been measured and, for the latter two metals, has been found to agree with theoretical calculations within 10 nm. Deuterium has also been detected and profiled (in Ti) using this technique. Detection sensitivity has been demonstrated at the 7 at % level for D in Ti and at the 0.5 at % level for ^4He in Cu. In addition, surface and bulk distributions of carbon, oxygen and ^3He have been profiled and, in principle, the depth distributions of tritium, ^6Li, ^7Li, Be, B, N and F are also simultaneously resolvable if contained within the foil samples. The technique has been used to investigate the effect of radiation damage and in situ annealing on implanted helium profiles in copper.

Luomajärvi et al. (1985) investigated oxygen detection by proton backscattering. The oxygen detection sensitivity of 2.5 MeV proton backscattering is shown to exceed that of ^4He backscattering by even a factor of about 15 depending on the matrix. The needed proton elastic scattering cross-sections of oxygen for $\theta_{lab} = 170°$ have been measured in the energy range $E_{lab} = 770–2480$ keV relative to Ti and Sn elastic scattering cross sections using thin TiO_2 and SnO_2 samples. The angular dependence of the cross-section was measured at energies $E_{lab} = 1790, 1990, 2191$ and 2382 keV for backscattering angles. The experimental cross-sections were found to be 1.1–5.7 times the pure Coulomb cross-section.

Theoretical calculations for the scattering cross-sections were performed and their in-applicability to experimental purposes was demonstrated.

Franco et al. (2005) have analyzed $Si(100)/Al_2O_3$ 500 nm/(CoFeB 2–7 nm/Al_2O_3 0.7–1.5 nm)$_{x4}$/Ta 2 nm multilayers. Layer thickness, and roughness, as well as interdiffusion on annealing, are important parameters to be determined. This poses an extreme challenge to any analysis technique. The authors have shown that the technique of choice could be RBS with an inexpensive Si surface barrier detector, complemented by elastic recoil detection analysis and X-ray diffraction experiments. The authors showed that extremely demanding nanolayered systems could be successfully analyzed with conventional RBS using an inexpensive and easy to handle detector. 1 and 2 MeV He* beams were used. The main requirement was that grazing angle in both incidence and exit had to be used. For the thinnest sample, this angle was as small as 6° between beam and the sample surface. By complementing the RBS results with ERDA and XRD, the authors could provide the answers for the most important compositional and structural questions on these samples.

In another report, Wielunski et al. (2005) studied ultra-thin films of aluminum oxide and hafnium oxide are currently being explored as high-k gate dielectrics for next-generation CMOS and related devices. Among the many methods to produce such films, atomic layer deposition (ALD) appears very promising as it enables deposition of ultra-thin layers on Si and other substrates with monolayer control. For device applications, it is critical to be able to measure and control the total oxide film thickness, as well as that of the SiO_2 that often forms at the high-k/Si interface. The authors showed how Hf, O, AI and Si from ultra-thin oxides could be detected and separated from the Si substrate scattering spectrum. In order to separate the Si substrate signal from the amorphous ultra-thin oxide O, Al and Si signals, ion channeling was applied. While the separation of Al and Si signals from thin oxide layers in typical 2 MeV He RBS is difficult due to the small mass difference, the authors have demonstrated that separating these signals is possible using higher energy scattering, 2.7 MeV.

In the work reported by Martinez-Martinez et al. (2005) RBS with ^{4}He energies from 2 to 6 MeV has been used to study the properties of amorphous photo-luminescent Al_2O_3:Ce, Mn films grown by spray pyrolysis on Coming 7059 glass substrates. The source solutions were $AlCl_3$, $CeCl_3$ and $MnCl_2$ dissolved in deionized water. Different molar concentrations (Ce 10%; Mn 1%, 3%, 5%, 7% and 10%) were investigated under the same deposition conditions at a substrate temperature of 300 °C. The RBS spectra show a homogeneous depth profile of both Ce and Mn within the films, and the measured quantities are consistent with the original solution concentrations. An important amount of Cl, which plays a significant role in luminescent properties, was detected in both the doped and undoped samples.

Laricchiuta et al. (2019) demonstrated that RBS could be extended from a metrology concept applied to blanket films toward a method to analyze confined nanostructures. By a combination of measurements on an ensemble of devices and extensive simulations, it is feasible to quantify the composition of InGaAs nanostructures (16–50 nm) embedded periodically in SiO_2 matrix. The methodology is based on measuring multiple fins simultaneously while using the geometrical shape of the structures, obtained from a transmission electron microscopy analysis, as input for a multitude of trajectory calculations. In this way, the authors were able to reproduce the RBS spectra and to demonstrate the sensitivity of the RBS spectra to the quantitative elemental composition of the nanostructures and to variations of their shape and mean areal coverage down to one nanometer. Thus, the authors establish RBS as a viable quantitative characterization technique to probe the composition and structure of periodic arrays of nanostructures.

RBS in channeling mode (RBS/C) is an efficient technique for characterizing crystallographic defects. Recently, Jin et al. (2020) used a RBS/C simulation code based on the binary collision approximation called RBS simulation in arbitrary defective crystals and successfully applied it to predict the RBS/C spectra from different damaged materials, whose structures were generated in high-dose ion irradiation atomistic simulations. In their paper, they introduced new developments improving the flexibility of the developed software and its applicability to different types of materials. More precisely, they modified the algorithm describing the slowdown process of backscattered ions, added fitting parameters in the collision partner search routine, modified the routine taking into account target atom thermal vibrations and provided new descriptions of the ion beam divergence. For example, the effect of the modifications on simulated RBS/C spectra was shown for $\langle 011 \rangle$-oriented UO_2 crystal analyzed with a 3.085 MeV He^{2+} ion beam. Some of these changes proved necessary to achieve satisfying agreement between simulations and experimental data. Similar observation was made for $\langle 001 \rangle$-oriented Si and $\langle 001 \rangle$-oriented GaAs crystals analyzed with a 1.4 MeV He^+ ion beam. In these simulations, the modifications have also resulted in good agreement with experiment (Jin et al. 2020).

3.7 Channeling

Evidence for the support of ion channeling through thin crystals were experimentally established in the early 1960s, see for example Robinson and Oen (1963), Dearnaley (1964), Kornelsen et al. (1964) and others. In the studies of penetration of ions through thin crystals, it was found that transmitted ion current had maxima at positions when ions were incident along symmetry axis. The effects of ion channeling are also evident in reduction of cross-sections for Coulomb scattering, nuclear reaction and production of characteristic x-rays. This effect is due to steering of ions away from the rows of crystal nuclei. This will not happen for atoms located in interstitial positions centered on a channel direction; for details see textbooks Mayer and Rimini 1977, and Feldman et al. 1982.

The first requirement for experiments aiming to determine the lattice location of impurity atoms is to orient the specimen so that incoming atoms enter along a major channeling axis. After that the energy spectra of backscattered particles or reaction products are measured for two low index directions and along randomly oriented direction (see Sattler and Dearnaley 1967, Gemmell 1974). A channeling effect has been also observed in silicon detectors with α particles (Cipolla and Taroni 1967).

The most used channeling method for atom location is RBS technique. The limitation of this technique is that it can be used only for study of heavy ions implanted in the lattice of lighter atoms. There are many published reports and books on the study of ions implanted into semiconductors. Mayer et al. (1970) treat fundamental aspects of ion channeling in the early book. Few years later several practical books on the subject appeared including Crowder (1973), Wilson and Brewer (1973) and Dearnaley et al. (1973). There are more books published on this problem, see for example Carter and Grant (1976), Ryssel and Ruge (1986); however detailed information on all aspects of ion implantation can be found in the series of Ion Beam Modification of Materials (IBMM) conferences and Ion Implantation Technology (IIT) conferences as well as in the proceedings of numerous other conferences.

Typical applications of the technique include:

- Assessment of degree of crystalline perfection in annealing treatments.
- Measurement of damage or damage profile from ion implantation.
- Location of lattice position of impurities.
- Measurement of thickness of amorphous layers on single-crystal substrates.
- Assessment of degree of epitaxy in grown structures.
- Measurement of light element (C, O, N) surface contaminants.

Fig. 3.19 shows random alignment and axially aligned RBS spectra for a structure containing an amorphous layer and a substitutional impurity. The amorphous layer produces no reduction in the yield. The yield variation of the impurity enables the degree of substitutionality to be measured.

International Conference on IBMM (held biannually since 1971) is the world's leading platform for ion beam experts and educators to exchange and report their most recent significant findings in the ion beam community. The scope of the conference ranges from fundamental radiation materials science to industrial applications. The Proceedings of the

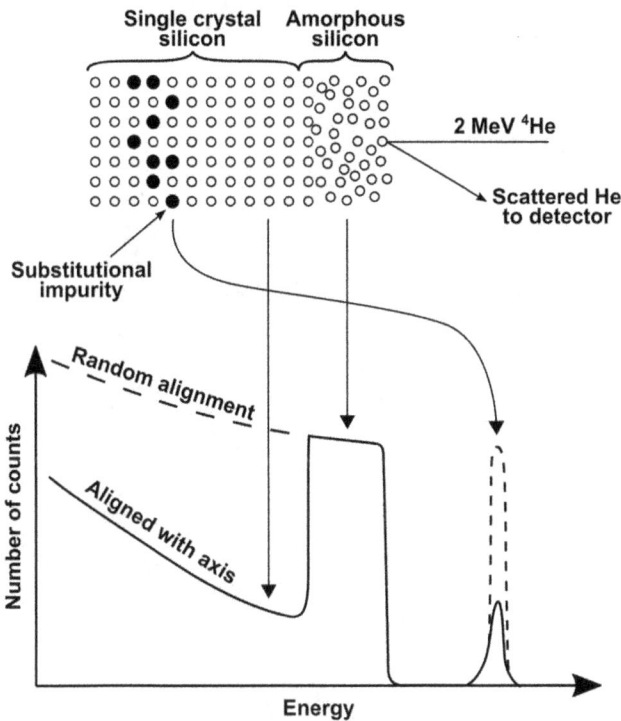

FIGURE 3.19
Schematic presentation backscattered He energy spectra for random (dotted line) and axially aligned (solid line) from silicon-containing amorphous surface layer and substitutional impurity.

last 21st International Conference on IBMM (IBMM 2018) is published as a special issue of journal Nuclear Instruments and Methods in Physics Research Section B: Beam Interactions with Materials and Atoms.

The 22nd International Conference on Ion Implantation Technology, IIT 2018 was held 16–21 September in Würzburg, Germany. The IIT conferences are an open forum for discussion of major challenges in current and emerging technologies related to the tools and processes for ion implantation, annealing of semiconductors, and non-semiconductors, implanted devices, metrology of implanted layers and devices, as well as methods related to ion implantation. The proceedings of conferences are published as IEEE conference proceedings.

Fig. 3.20 shows a 18SDH-2 6 MV Pelletron system with 6 beamlines and four NEC supplied sources with a possible expansion of up to three additional sources. It is a versatile system designed to perform various methods including AMS and IBA. Installed in Japan in 2014 to replace a 12 MV accelerator system that was severely damaged during an earthquake.

FIGURE 3.20
A picture of a versatile 18SDH-2 6 MV Pelletron system with 6 different beamlines and four ion sources with a possible expansion of up to three additional sources. (Courtesy of National Electrostatics Corp.)

3.8 Microprobe Techniques

Scanning by spot-size ion beams has transformed the PIXE technique from an analytical tool into an imaging device to map the distribution of elements – an actual nuclear "microscope" (Fig. 3.21).

Already in 1990 many laboratories have operational microprobes and used them in the field of biomedicine (see Lindh 1990) and Table 3.3.

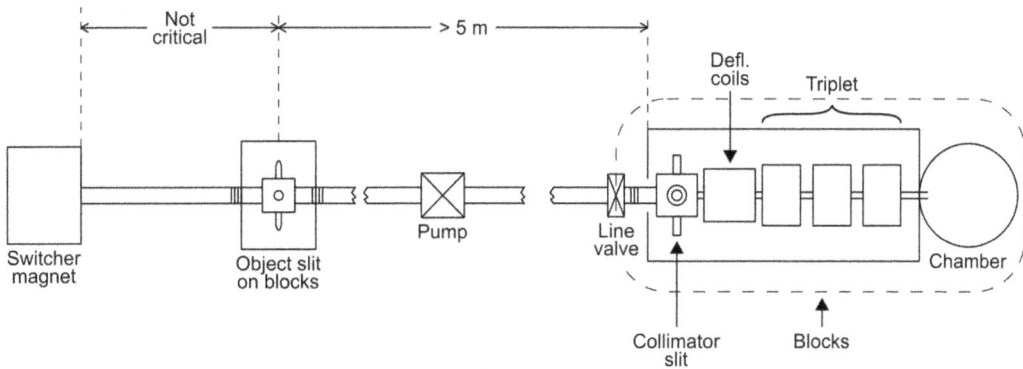

FIGURE 3.21
Sketch of microbeam layout.

TABLE 3.3

Nuclear microprobes used in biomedicine

Location	Size of beam spot	Applications
SUNY, Albany, USA	$1.3 \times 1.3 \ \mu m^2$	Biology, Trace elements
Vrije Univ., Amsterdam, NL	$5 \times 5 \ \mu m^2$	Biology 40%
Univ. Birmingham, UK	$5 \times 5 \ \mu m^2$	Biological
Univ. Bochum, Germany	$10 \ \mu m$	Medicine 50%, Biology 20%
Bordeaux-Gradignan, France	$1.8 \times 1.8 \ \mu m^2$	Biology
Eindhoven, Univ. of Tech. NL	$10 \times 10 \ \mu m^2$	Biology 80%
Univ. Oregon, Eugene, USA	$8 \ \mu m$	Biology 70%
Faure, South Africa	$4 \times 4 \ \mu m^2$	Biology
MPI, Heidelberg, Germany	$3 \times 2 \ \mu m^2$	Biology 30%
Queen's Univ., Kingston, Canada	$10 \ \mu m$	Biology 30%
Lund Univ. of Tech. Sweden	$5 \times 5 \ \mu m^2$	Biomedicine 50%
Manitoba, Canada	$5 \times 5 \ \mu m^2$	Medical, Biophysics
Univ. Melbourne, Australia	$1 \ \mu m$	Biomedicine 50%
Univ. Oxford, UK	$\leq 1 \times 1 \ \mu m^2$	Biology 50%, Medicine 30%
Z. Kernforsch., Rossendorf, Germany	$6 \ \mu m$	Biology 10%
Schonland Res. C., South Africa	$<1 \mu m$	Biology 20%
Waseda Univ., Tokyo, Japan		Biology
Univ. Uppsala	$2.9 \times 2.9 \ \mu m^2$	Biomedicine
Institute RB, Zagreb, Croatia		Different

Some of the microprobe facilities are described in details in the scientific and technical literature. One of them is Zagreb nuclear microprobe facility (Jakšić et al. 1993). The microbeam system is positioned at one of the beam lines of the EN tandem accelerator. The system consists of the Oxford design magnetic quadrupole doublet and the scattering chamber constructed at the Institute. The scattering chamber is of octagonal design. At the 135° port the 80 mm² Si(Li) detector with 12.5 μm Be window was positioned. The sample xyz manipulator situated at the 90° port is motorized and can be remotely controlled. The sample holder is designed to enable a high solid angle for the Si(Li) detector (up to 0.4 sr). Beam focusing and sample positioning are controlled through the vertical large working distance microscope which views the sample by the hollow mirror positioned in front of the sample. On the video monitor, a microscope view field is simultaneously seen. The mirror can be remotely removed. A backscattering particle detector is also mounted in the chamber. The target chamber is pumped using oil diffusion pumps.

HVEE has developed an ultra-stable 3.5 MV Sigletron accelerator for Nano probe applications (Mous et al. 1997) shown in Fig. 3.22. Because the energy stability and ripple of the beam, and beam brightness are of vital importance for the performance of a nanoprobe, special care has been taken in optimizing these parameters. The system consists of an RF source that is directly mounted on the accelerator tube, a switching magnet to bend

FIGURE 3.22
Singletron, 3.5 MV In-line for micro-beam applications. (Courtesy HVEE.)

the beam into a chamber for standard analysis purposes and an analysis magnet that directs the beam into the nanoprobe. The accelerator has been installed at the University of Leipzig, Germany. For one of the main applications, the system is connected to a nano-beamline to achieve submicron resolution (Butz et al. 2000).

Numerous are applications of proton microprobe to the analysis of different samples and mapping of element concentrations on μm-scale. Let us mention some examples. In the work of Caridi et al. (1993) proton microprobe has been applied to the study of electrostatic precipitation of single fly ash particles. Altogether 69 fly ash particles (23 collected at the inlet and 46 collected at the outlet), with the average diameter of about 1 μm, from the electrostatic precipitator have been analyzed. From the measured concentration values of 18 elements, two main groups of fly ash particles have been identified in the inlet and four groups in the outlet in additional to some "anomalous" particles not belonging to any of the above groups. These data could be used to evaluate the importance of chemical composition effect on the electrostatic precipitation.

Coal combustion is still a major source of particulate emissions. Electric precipitators show a minimum in collection efficiency for particles in the 0.1–1.0 μm size, which have longer atmospheric residence time and worse effects on health and visibility. The concentration of trace species in fly-ash increases with decreased particle size.

Another application of proton microprobe is illustrated by measurements or elemental concentrations across the radial cut of human hair (Figs. 3.23 and 3.24).

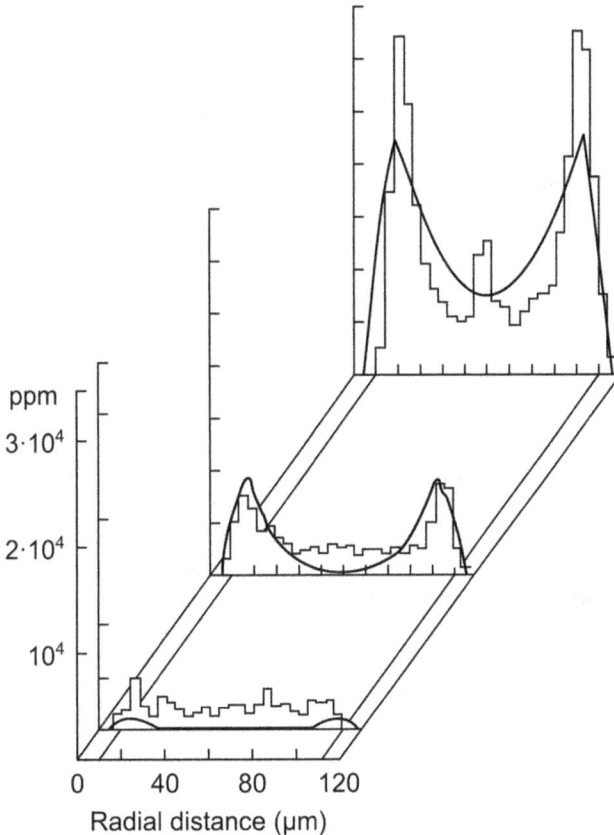

FIGURE 3.23
Radial distribution of Ca in human hair.

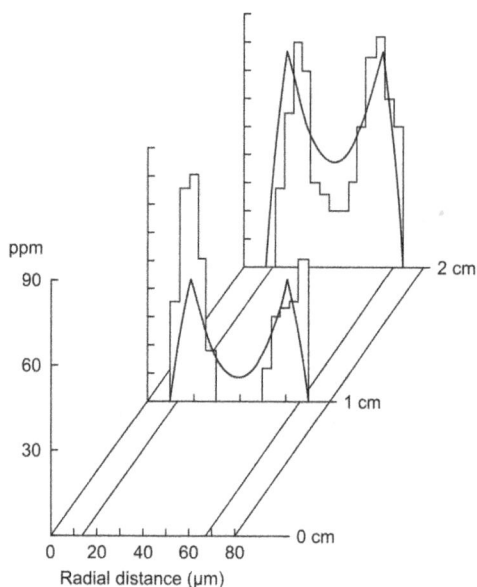

FIGURE 3.24
Radial distribution of Pb in human hair.

An external beam microprobe facility, based on a quadrupole doublet supplied by Oxford Microbeam Ltd, has been installed on a new beamline at the 3 MV single-ended Van de Graaff accelerator in Florence. The goal was to obtain a beam with a spot size on target of 10–20 μm and a current in the order of at least 1 nA, in order to allow PIXE, PIGE and RBS elemental analysis in air or in a helium atmosphere. The beam was extracted from the vacuum lines through a 0.1 μm thick Si_3N_4 window to minimize lateral straggling (Giuntini et al. 2007; Massi et al. 2002).

3.9 Scanning Transmission Ion Microscopy, STIM

The principle of STIM method is based on the following: as the ion beam passes through the specimen into a detector, each individual ion gives in its energy loss a quantitative measurement of the localized specimen mass per unit area.

In order to avoid the errors of counting statistics (though not those of energy straggling) and the efficiency (almost 100%) demands the use of very low beam currents hence decreasing all aberrations. The best resolutions achievable are about 50 nm (Bench and Legge 1989). STIM can also be used for thick specimens. STIM is equally applicable to material analysis, and here it may be combined with channeling (CSTIM) to provide a virtually non-damaging but highly sensitive analytical tool for thin crystals (Cholewa et al. 1993). With the removal of dechanneled beam, CSTIM has been used as a very sensitive tool for mapping crystal damage.

In nuclear elastic collisions, the incoming ion can be forward scattered (θ < 90) or backscattered (θ > 90; RBS). By detecting forward-scattered ions at angles θ ~5–10, one can obtain an image that reflects elemental information about the sample. In this technique, called forward STIM (FSTIM), the scattering cross-section increases as Z^2 and therefore there is a higher probability of scattering from high-Z heavy elements compared with the low-mass elements.

When STIM is applied to 3D tomography, it provides a means of quantitatively imaging the entire volume of microscopic objects without resorting to difficult and destructive serial sectioning. There are many such objects which are too small for the spatial resolution of x-ray tomography and too large (thick) for the requirements of electron tomography.

Increasing interest in the use of nanoparticles to elucidate the function of nanometer-sized assemblies of macromolecules and organelles within cells, and to develop biomedical applications such as drug delivery, labeling, diagnostic sensing, and heat treatment of cancer cells has prompted investigations into novel techniques that can image nanoparticles within whole cells and tissue at high resolution. Using fast ions focused on nano dimensions, Chen et al. (2013) show that gold nanoparticles (NPs), AuNPs, inside whole cells can be imaged at high resolution, and the precise location of the particles and the number of particles can be quantified. High-resolution density information of the cell can be generated using STIM, enhanced contrast for AuNPs can be achieved using forward scattering transmission ion microscopy, and depth information can be generated from elastically backscattered ions (RBS). These techniques with appropriate associated instrumentation allow the whole-cell three-dimensional imaging at <10 nm resolution (Chen et al. 2013) (Fig. 3.25).

Emmrich et al. (2021) have designed a dark-field STIM detector for the helium ion microscope. The detection principle is based on a secondary electron conversion holder with an exchangeable aperture strip allowing its acceptance angle to be tuned from 3 to 98 mrad. They investigated the contrast mechanism and performance using freestanding nanometer-thin carbon membranes. The results demonstrated that the detector could be optimized either for the most efficient signal collection or for maximum image contrast.

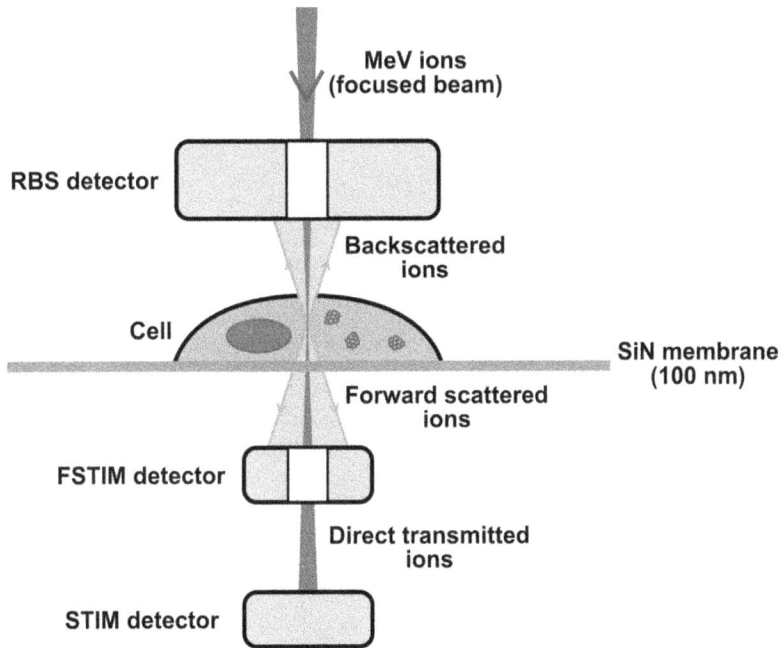

FIGURE 3.25
Schematic diagram of the experimental setup for cell imaging using fast ions.

The designed setup allows for the imaging of thin low-density materials that otherwise provide little signal or contrast and for a clear end-point detection in the fabrication of nano-pores. In addition, the detector was able to determine the thickness of membranes with sub-nanometer precision by quantitatively evaluating the image signal and comparing the results with MC simulations.

Nano-pores fabricated from glass micro-capillaries are used in applications ranging from scanning ion conductance microscopy to single-molecule detection. Still, evaluating the tip by a noninvasive means remains challenging. For instance, electron microscopy characterization techniques can charge, heat and contaminate the glass surface and typically require conductive coatings that influence the final tip geometry. On contrary, electrical characterization by the means of ion current through the capillary lumen provides only indirect geometrical details of the tips. Zweifel et al. (2016) have shown that helium STIM provides a nondestructive and precise determination of glass nano-capillary tip geometries. This enables the reproducible fabrication of axially asymmetric blunt, bullet and hourglass-shaped tips with opening diameters from 20 to 400 nm by laser-assisted pulling. Accordingly, this allows for an evaluation of how tip shape, pore diameter and opening angle affects ionic current rectification behaviour and the translocation of single molecules. Zweifel et al. (2016) have shown that current drops and translocation dwell times are dominated by the pore diameter and opening angles regardless of nano-capillary tip shape.

Eichhoorn et al. (2015) applied STIM to measure sputter yields of thin Kovar foil. The results have been found in very good agreement with values determined by the weight loss method, demonstrating STIM as a feasible alternative measurement technique for sputter yield estimation of thin material samples. Measurements have been carried out under normal xenon ion incidence for ion energies in the range between 100 eV and 1000 eV. In addition, sputter yields of Kovar bulk samples are reported. The data might be interesting for ion beam applications such as solar electric propulsion, in which materials with low sputter yields are preferred to ensure a long operational lifetime of the system components.

For more reading about basics and applications of nano-lithography see contributions to Nanolito 2021.

3.10 Accelerator Mass Spectrometry

The electric and magnetic fields, used for the analysis of ions, provide only information about two parameters E/Q and M/Q, where:

> E = kinetic energy of the ion,
> M = mass of the ion,
> Q = charge of the ion.

There are four ways in which the parameters E/Q and M/Q may be determined:

i. Magnetic selection $(Bxr)^2 = 2(M/Q) \times (E/Q)$

ii. Electrostatic selection $Exr = 2(E/Q)$

iii. Cyclotron selection $1/f = (\pi/B) \times (M/Q)$

iv. Velocity selection $v^2 = 2(E/Q) \times (M/Q)$

where B is the magnetic field, r is the radius of the ion path, E is the electric field, f is the cyclotron frequency, v is the ion velocity. Low-resolution measurements that separate neighboring isotopes as final output only the parameters M/Q and E/Q and ambiguities can arise if M and Q have common factors. It is for this reason that some flexibility in the choice of Q is desirable. If it is possible to measure the energy of ion also, then it is possible to determine Q and so determine the mass from the ratio M/Q. The use of energy, mass and charge signatures, at the energies such that charge state 3^+ or higher is dominant, is the basis for the AMS of almost all stable isotopes.

AMS was introduced in 1977 by Muller, who suggested that a cyclotron could be used for detecting ^{14}C, ^{10}Be, and other long-lived radioisotopes (Muller 1977), and independently by the Rochester group (Purser et al. 1977), who demonstrated that ^{14}C could be separated from the isobar ^{14}N by relying on the instability of negative ion ^{14}N.

AMS is an analytical technique that uses an ion accelerator and its beam transport system as an ultrasensitive mass spectrometer. AMS method is one of the most important accelerator-based IBA techniques. This technique has made it possible to separate isotopes from isobars, with the highest possible resolution of 10^{-15}– 10^{-16} atoms per atom. AMS measurements are being made at more than 30 accelerator laboratories around the world, and half of these are dedicated to AMS measurements of long-lived radioisotopes. One of the most popular applications of the AMS is radiocarbon dating. Areas of applications include Archaeology, Geology, Groundwater analysis, Metabolite profiling, Microdosing, Atmospheric sciences, Oceanography, Nuclear waste management, Toxicology, Cosmogenic study, Astrophysics, Pharmacokinetics, and many more. The main AMS area of applications are:

- Biomedical sciences.
- Geosciences.
- Non-proliferation safeguarding and monitoring.
- Nutritional science.
- Oceanographic and atmospheric sciences (as important parts of geoscience).
- Agricultural applications.
- Industrial and material sciences applications.
- Environmental applications.

For detection of radioisotopes with long life times, >1 year, it may often be more advantageous to use AMS-counting techniques rather than traditional decay counting. This is especially true for measurements where efficiency is a criterion, as for small samples, or if high precision is required. While atom counting has a counting rate that is essentially independent of decay lifetime and sample size, the decay counting rates are comparable only if the isotopic half-life is less than one year for a sample size of order of 1 mg. Of course, if sufficient material is available, the decay-counting rate can always be improved by using more material.

Accelerator mass spectrometers detect atoms of specific elements according to their atomic weights. The process of AMS begins by processing sample material, often as little as 50 micrograms, and placing it into the ion source. The sample material is converted to a negative ion and accelerated to extraordinarily high kinetic energies. The ion beam passes through a gas or foil stripper to remove electrons, and the resulting ion beam passes through mass and electrostatic filters to remove isobars. AMS can detect $^{14}C/^{12}C$ ratios below 3×10^{-16}. Carbon AMS is by far the most common, although there is growing worldwide interest in studying beryllium, aluminum, calcium, iodine, chlorine and the actinides.

3.10.1 Experimental Arrangements

Figs. 3.26 and 3.27 show the two possible arrangements of various components for a typical AMS system. Tandem electrostatic accelerators are generally believed to be the best accelerators for most AMS measurements. Although, there have been some important results with cyclotrons, and dedicated small cyclotrons have been developed for AMS.

A system designed for AMS measurements of carbon, beryllium and aluminum is shown in Fig. 3.26. It includes a Multi-Cathode SNICS (MC-SNICS) source (back left), 90 degree magnet (front left) electrostatic analyzer magnet (ESA) (front right), 45 degree magnet for enhanced beryllium background (back right), and AMS detector (back center). Dual-source configurations are possible.

Fig. 3.27 illustrates another possibility when a larger target room space is available. All components of the system are indicated in the figure.

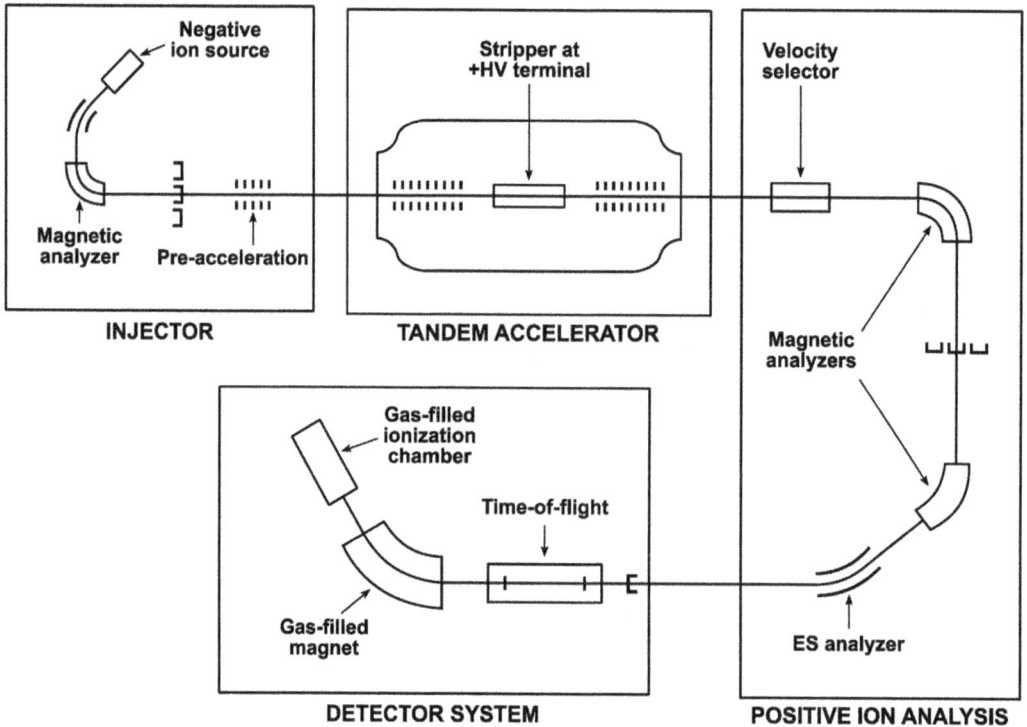

FIGURE 3.26
Components of AMS system based on the use of Tandem electrostatic accelerator.

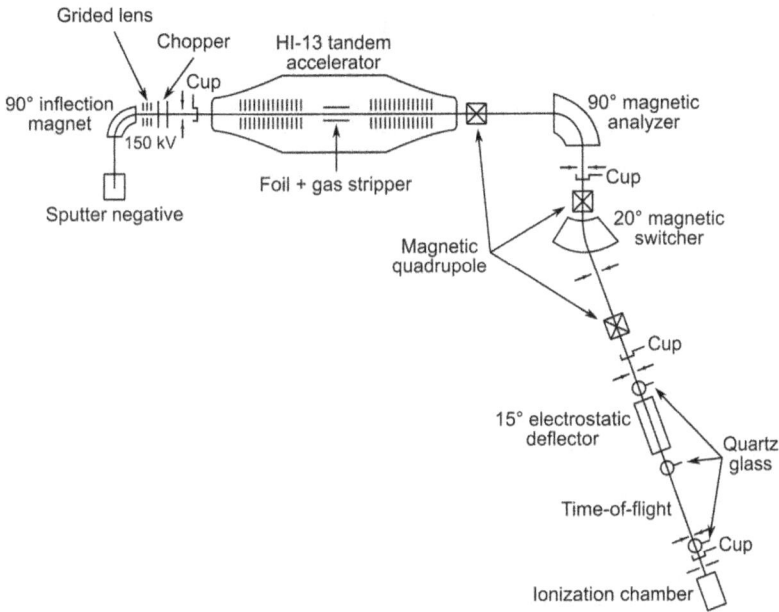

FIGURE 3.27
Components of AMS system based on the use of Tandem electrostatic accelerator.

A system designed for carbon only AMS is shown in Fig. 3.28. It includes a MC-SNICS source (left), followed by a 90° analyzing magnet, offset Faraday cup, 250 kV acceleration tube, and the 250 kV high voltage deck (right). Within the deck is a molecular dissociator, 90° magnet, 90° electrostatic spherical analyzer, and AMS detector. Installed in Lithuania in 2015.

In 1994, High Voltage Engineering Europa B.V. has introduced the new generation Tandetron Accelerator Mass Spectrometer for direct determination of isotopic concentration ratio's from sub-milligram samples. Their model 4130-AMS spectrometer was suitable for single-atom detection of ^{3}H, ^{10}Be, ^{14}C, ^{26}Al and, optionally ^{36}Cl and ^{129}I. When measuring recent ^{14}C samples statistical precisions of ~0.3% could be achieved in times less than one hour. Full computer control allowed data collection and analysis for several days without operator intervention (Fig. 3.29).

The dedicated ^{14}C AMS system made by HVEE is shown in Fig. 3.28 and described in details in Klein et al. (2019). A unique feature of the design is a magnetic charge state selector that is incorporated in the high voltage terminal, which reduces the primary source of background that originates from the injection of ^{13}CH$^-$ to negligible levels. As a result, background levels in the low 10^{-16} regimes are achievable, thus supporting the most stringent ^{14}C dating applications. Another feature of the system is the incorporation of several slit feedback loops for stabilization of the position of the ion beam throughout the system, which avoids drift and ensures stable, long-term operation. This system is also described by Scognamiglio et al. (2021) together with a multi-element HVEE AMS system (Fig. 3.30).

Sequential injection or bouncing has a number of properties, which can lead to a reduction of the analysis accuracy if no appropriate measures are taken. Klein et al. (2004) at HVEE developed a special injection system in order to eliminate these shortcomings. In a sequential injection system, the isotopes are injected one after the other. The vacuum chamber of the injection magnet is electrically isolated. A voltage applied to the chamber changes the energy of the ions that pass through the magnetic field. The voltage determines

FIGURE 3.28
A picture of a 1.5SDH XCAMS 500 kV compact Pelletron system. (Courtesy of National Electrostatics Corp.)

FIGURE 3.29
A picture of a Single Stage AMS (SSAMS) 250 kV open-air system (with safety cage removed). (Courtesy of National Electrostatics Corp.)

FIGURE 3.30
Tandem, 200 kV, AMS ^{14}C. (Courtesy HVEE.)

the mass of the injected ions. The cyclic change of this voltage for the injection of a sequence of isotopes is often referred to as bouncing. Data acquisition is usually stopped during injection of the stable isotopes, which eliminates the background originating from their injection. Sequential injection provides the flexibility to examine virtually all isotopes of the periodic table. The accuracy of the timing of the injections well as the switching characteristics of the bouncer voltage power supply influence the achievable precision of the analysis. This is of special importance for ^{14}C dating where the highest precision is required. To push the performance of sequential injection AMS to its limits, HVEE has optimized the bouncer set-up and its electronics, as described in (Klein et al. 2004).

NEC manufactures AMS systems in different energies, depending on the isotopes to be studied:

- 250 kV systems are used for measuring carbon – ^{14}C.
- 500 kV systems are used for measuring carbon – ^{14}C, beryllium – ^{10}Be, aluminum – ^{26}Al, and iodine – ^{129}I.
- 1 MV systems provide better isobaric separation for the above isotopes and are used for measuring calcium – ^{41}Ca, and the actinides.
- 3 MV systems provide further isobaric separation for the above isotopes.
- 6 MV systems are recommended for measuring chlorine – ^{36}Cl.

These long-lived radioisotopes can now be measured in small (mg) natural samples having isotopic abundances in the range 10^{-12}–10^{-15} and as few as 10^5 atoms.

Isotope ratios are obtained by alternatively selecting each stable isotope and measuring its beam current in a removable or offset Faraday cup and then by measuring the radioisotope counting rate in the detector. Standards are periodically measured for normalization, and blanks to measure the background.

The Accelerator Mass Spectrometers could be found in laboratories shown in Table 3.4. Included are also institutions listed in European Accelerator Mass Spectrometry Facilities by Macková et al. (2016).

TABLE 3.4

Accelerator mass spectrometers

Laboratory/Institution	Country	Accelerator	Main application
ANSTO, Sydney	Australia	2 MV tandem	^{14}C
Aust. Nat. Univ. Canberra	Australia	14 UD Pelletron	^{36}C
VERA, Univ. of Vienna	Austria	3 MV tandem	^{36}Cl, ^{129}I, ^{10}Be, ^{14}C, ^{26}Al, ^{41}Ca
RICHBrussels	Belgium	Micadas (Mini Carbon Dating System)	^{14}C
Chalk River	Canada	MP	^{14}C
McMaster	Canada	FN	^{10}Be, ^{14}C
AEL-AMS Lab. Univ. of Ottawa	Canada	HVE3 MV tandem	^{3}H, ^{10}Be, ^{14}C, ^{26}Al, ^{36}Cl, ^{129}I, ^{236}U.
Toronto	Canada	Tandetron	^{14}C, ^{26}Al, ^{129}I
Beijing	China	HI-13 TD	^{10}Be
Xi'an AMS Center	China	HVE Tandem 3 MV	^{10}Be, ^{14}C, ^{26}Al, ^{129}I
Institute of Earth Environment CAS, Xian	China	Tandetron	Research and applications
Aarhus	Denmark	EN tandem	^{14}C, ^{32}Si, $^{22,24}Na$
Ion beam analysis laboratory, University of Helsinki	Finland	TAMIA 5 MV HV tandem	^{14}C
CEREGEU niversity of Aix Marseille	France	Tandetron	cosmogenic nuclides
CEREGE Aix en Provence	France	5 MV ASTER accelerator	Cosmogenic Nucleides
LMC14 at LSCESaclay	France	3 MV NEC Pelletron	^{14}C
ARTEMIS Gif-sur-Yvette	France	Tandetron	^{10}Be, ^{14}C, ^{26}Al
Christian-Albrechts-University, Kiel	Germany	Tandetron	^{14}C
Darmstadt	Germany	Linac	^{41}Ca
University of Köln	Germany	tandem voltage 6 MV	^{36}Cl, ^{10}Be, ^{14}C, ^{26}Al.
AMS Maier/Leibnity Lab., Munich	Germany	MP+linac	^{36}Cl, ^{41}Ca, ^{53}Mn
DREAMS, HZDR, Drezden	Germany	Tandetron 6 MV	^{7}Be, ^{10}Be, ^{26}Al, ^{36}Cl, ^{41}Ca, ^{129}I.
Max-Planck-Institut für Bio-Geo Chemie, Jena	Germany	Tandetron	Analysis of different samples by isotope ratio measurements
Darmstadt	Germany	Linac	^{41}Ca
GAMS Accelerator Mass Spectrometry Group, Garching	Gremany	14 MV MP tandem	^{26}Al, ^{36}Cl, ^{41}Ca, ^{53}Mn, ^{59}Ni, ^{60}Fe, ^{79}Se, ^{93}Zr, ^{97}Tc, ^{99}Tc, ^{238}U, ^{244}Pu and others

TABLE 3.4 (Continued)

Accelerator mass spectrometers

Laboratory/Institution	Country	Accelerator	Main application
Erlangen	Germany	EN	^{10}Be, ^{14}C
ATOMKI Debrecen	Hungary	MICADAS ^{14}C system	^{14}C
IUAC New Delhi	India	15UD Pelletron	trace elements
AURiS Ahmedabad	India	1MV accelerator	^{14}C, ^{10}Be and ^{26}Al.
Rehevot	Israel	14U pelletron tandem	^{36}Cl, ^{41}Ca, ^{129}I
CIRCE, INNOVA, Napoli	Italy	3MV-Tandem	^{14}C
CEDAD Univ. of Salento, Lecce	Italy	HVE 3 MV Tandetron	^{14}C by AMS
Legnaro, Padova	Italy	XTU	^{10}Be, ^{36}Cl,
INFN-HH Napoli	Italy	HVE 3MV tandetron	^{14}C
Nagoya	Japan	Tandetron	^{14}C
Osaka	Japan	AVF cyclotron	^{41}Ca
Tokyo	Japan	4MV-TD	^{10}Be, ^{14}C, ^{26}Al
NIES-TERRA Tsukuba	Japan	Tandem 15SDH-2	^{14}C, ^{10}Be, ^{26}Al, ^{36}Cl, ^{41}Ca, ^{129}I, etc.
LEMA, UNAM Mexico City	Mexico	Tandetron	Radiocarbon analyses/ dating
Lower Hutt	New Zeeland	EN	^{14}C
Poznan Radiocarbon Laboratory	Poland	1.5 SDH-Pelletron	^{14}C
KIGAM Daejeon	Republic of Korea	Tandetron	All types of geological samples
RoAMS Bucharest - Magurele	Romania	Tandetron	Research and applications
BINP AMS Novosibirsk	Russia	Tandem 1 MeV Made by BINP	^{14}C
CNA University of Sevilla	Spain	1MV Tandetron	^{14}C
Upsala University	Sweden	EN Tandem	^{14}C
Lund University	Sweden	3 MV Pelletron	^{14}C
ETH Zurich Paul Scherrer Ins.	Switzerland	EN	^{10}Be, ^{14}C, ^{26}Al, ^{36}Cl, ^{32}Si
LARA AMS Laboratory, University of Bern	Switzerland	MICADAS (MIni CArbon DAting System)	^{14}C
Taiwan Univ.Taipei	Taiwan, R.O.C.	HVE 1.0 MV Tandetron-4110	^{10}Be, ^{14}C, ^{26}Al.
Universiuty of Groningen	The Netherlands	Tandetron 2.5 MV	^{10}Be, ^{36}Cl,
Utrecht	The Netherlands	EN	^{10}Be, ^{14}C
Argonne	USA	FN+linac	^{41}Ca, ^{44}Ti, ^{26}Al, ^{59}Ni, ^{60}Fe
CAIS at Univ. of Georgia, Athens	GA, USA	NEC Pelletron 1.5 SDH-1	^{14}C
Arizona	USA	Tandetron	^{14}C, ^{10}Be

(Continued)

TABLE 3.4 (Continued)

Accelerator mass spectrometers

Laboratory/Institution	Country	Accelerator	Main application
Trace Element AMS, Naval Research Lab.	DC, USA ...	3MV Pelletron tandem	trace elements
Beta Analytic Lab. Miami	FL, USA	Accelerator (?)	^{14}C
CAMS at LLNL, Livermore	CA, USA	10-MV FN tandem, NEC 1-MV tandem	Different
Pennsylvania State Univ	PA, USA	FN	^{10}Be, ^{26}Al, ^{41}Ca
Xceleron, Inc. Germantown	MD, USA	-	^{14}C
Mega SIMSUCLA	CA, USA	NEC3SDH-I-2	Isotopic analysis of solar wind: N, O
University of Rochester	NY, USA	MP tandem Van de Graaff accelerator	^{10}Be, ^{14}C, ^{36}Cl, ^{41}Ca, and ^{129}I
Perdue Univ. West Lafayette	IN, USA	FN tandemNEC pelletron	^{10}Be, ^{14}C, ^{26}Al, ^{36}Cl, ^{41}Ca, ^{129}I
NOSAMS, Woods Hole Ocean. Inst.	MA. USA	3MV tandetron, 500kV pelletron	^{14}C
Rutgers	USA	FN	^{10}Be
KCC-AMS Irvine, Univ of Calif.	CA, USA	0.5MV NEC 1.5 SDH-1	^{14}C
Direct AMS (D-AMS) Labs, Seattle	WA, USA	FN	^{14}C
Accium BioSciences, Seattle	WA, USA	NEC 1.5 5DH-1 Pelletron	^{14}C
Rochester	USA	MP	^{36}Cl, ^{129}I, ^{10}Be, ^{14}C, ^{26}Al, ^{41}Ca
NRL	USA	cyclotron	^{3}H
NERC, SUERC, Univ. Glasgow	Scotland, UK	250 kV	^{14}C
		NEC 5 MV AMS	^{0}Be, ^{14}C, ^{26}Al, ^{36}Cl, ^{41}Ca, ^{129}I
Queen's Univ. Belfast	Northern Ireland, UK	NEC compact model 0.5 AMS	^{14}C
University of Oxford	U.K.	Tandetron	^{14}C
14 Chrono Queen's University, Belfast	UK	MICADAS (MIni CArbon DAting System)	^{14}C
University of Bristol	UK	NEC compact model 0.5 AMS	^{14}C
University of Oxford	UK	Tandetron	^{14}C

Some examples:

- VERA, Vienna, Austria is The Vienna Environmental Research Accelerator Project (Kutschera et al. 1997). The Isotope Research and Nuclear Physics group operates the Vienna Environmental Research Accelerator (VERA), established in 1996. Based on a 3 MV tandem accelerator, it was one of the first accelerators especially dedicated and designed for universal AMS across the nuclear chart and was continuously extended to reach new demands. VERA can measure all standard AMS isotopes (10Be, 14C, 26Al, 36Cl, 41Ca, 129I), and has an outstanding reputation for the development and improvement of AMS methods and methodologies. The system is especially well suited for the determination of actinides (236U, 233U, Np, Pu, Am). The recently commissioned ILIAMS (Ion-Laser Interaction Mass Spectrometry) beamline is expected to extend the scope of AMS applications beyond the established isotopes. AMS measurements on the complete set of established isotopes are performed for many national and international partners with a recent emphasis on actinides.

- While a minor fraction of the measurements is carried out on a commercial basis (mainly radiocarbon dating), most samples originate from collaborative research projects. About 5–10% of the regular working hours are presently used for student training. Research at VERA contributes to the faculty research area "Physics and the Environment", see https://www.ionbeamcenters.eu/RADIATE-project-partners/vienna/.

- CologneAMS is the Centre for AMS at the University of Cologne, see Klein et al. (2011). It operates a dedicated AMS system designed to measure all standard cosmogenic nuclides (^{10}Be, ^{14}C, ^{26}Al, ^{36}Cl, ^{41}Ca, ^{129}I) and which uses a 6 MV Tandetron™ accelerator equipped with an all solid-state power supply, foil and gas stripper. The system also enables a sensitive detection of heavy ions up to ^{239}U and

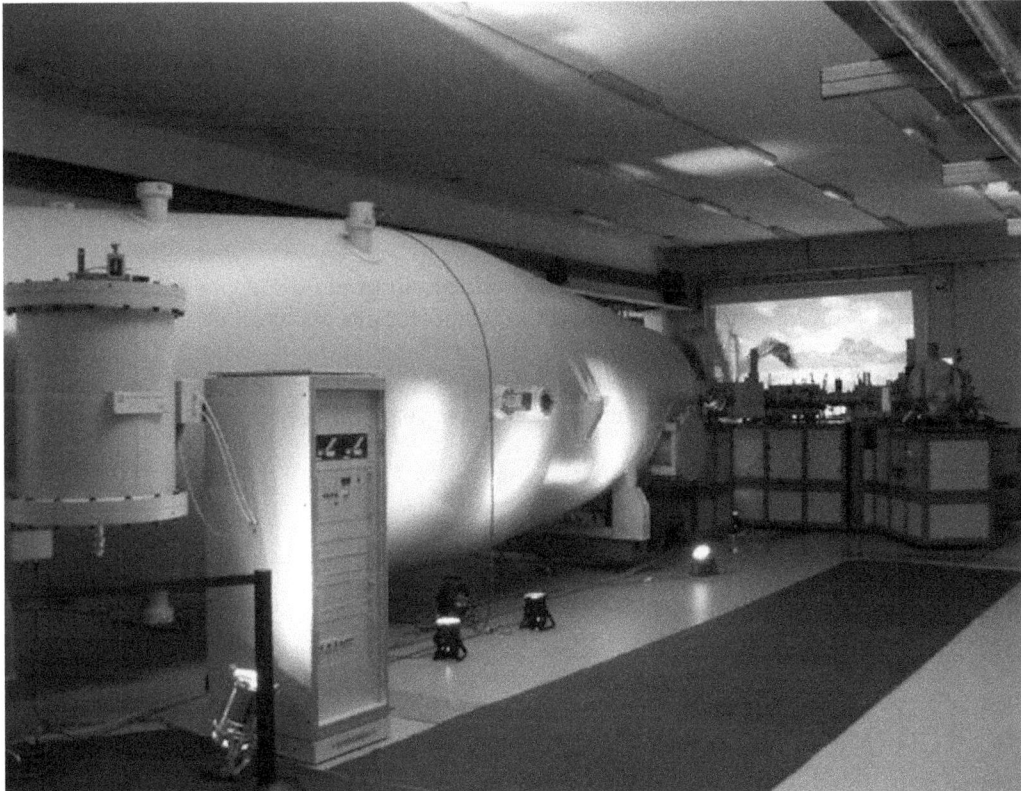

FIGURE 3.31
HVEE Tandetron 6.0 MV, AME ME+IBA. (Courtesy HVEE.)

^{244}Pu. The high-energy mass-spectrometer consists of a 90° magnet with a radius of 2 m and a mass-energy product of 351 AMU MeV to allow the detection of ^{244}Pu^{5+} up to the maximum terminal voltage of 6 MV. An electrostatic energy analyzer and a switching magnet that can transport the rare isotope beam into various beamlines follow this magnet. The switching magnet forms a third analyzing element that is needed especially for the sensitive detection of heavy elements. Two beamlines are equipped with their own detection system. One of these lines is used for suppression of isobaric background in the case of the analysis of e.g., ^{36}Cl. This is accomplished by an absorber foil which generates a Z-dependent energy loss in combination with a momentum/charge-state selection via a 120° magnet that features up to 30 mrad acceptance for efficient beam transport (Fig. 3.31).

- A novel tabletop AMS system with overall dimensions of only 2.5–3 m^2 has been built and tested (Synal et al. 2007). Their mini radiocarbon dating System (MICADAS) is based on a vacuum insulated acceleration unit that uses a commercially available 200 kV power supply to generate acceleration fields in a tandem configuration. At the high-energy end, ions in charge state 1$^+$ are selected and interfering molecules of mass 14 amu are destroyed in multiple collisions. The new system is the prototype of a new generation of radiocarbon spectrometers that fulfill the requirements for radiocarbon dating applications as well as for the less demanding ^{14}C/^{12}C isotopic ratio measurements as needed, e.g., in biomedical applications.

There are many overviews of AMS accomplishments in different fields let us mention some: Synal (2013), Kutschera (2016) an Aerts-Bijma et al. (2021).

3.10.2 AMS and Cosmocronology

Some of the commonly measured long-lived cosmogenic isotopes are shown in Table 3.5.

3.10.2.1 ^{10}Be

^{10}Be has a half-life of 1.5×10^6 years and is produced in upper atmosphere by cosmic-ray induced spallation of atmospheric N and O. This isotope is attached to aerosols and it is deposited in snow and ice in the top soil or it enters lakes and seas, where is accumulated in sediments. The mean global fallout is about 10^6 atoms cm^{-2} per year. The mean residence time of ^{10}Be in the atmosphere is about 1–2 years, in oceans is much larger about several hundred years.

TABLE 3.5

Long-lived cosmogenic isotopes detected by AMS

Isotope	Half-life (years)	Stable isobar	Interfering Isotopes	Chemical form	Sample Size (mg)	AMS Detection limit[a]	Range of terrestrial conc.[a]
^{10}Be	1.5×10^6	^{10}B	^{9}Be	BeO	0.2	7×10^{-15}	$10^{-8} - 10^{-14}$
^{14}C	5.7×10^3	^{14}N	^{12,13}C	C	0.25	0.3×10^{-15}	$10^{-12} - 10^{-16}$
^{26}Al	7.2×10^5	^{26}Mg	^{27}Al	Al$_2$O$_3$	3	10×10^{-15}	$\sim 10^{-14}$
^{36}Cl	3.1×10^5	^{26}S, ^{36}Ar	^{35,37}Cl	AgCl	2	0.2×10^{-15}	$10^{-12} - 10^{-17}$
^{41}Ca	1.3×10^5	^{41}K	^{40,42}Ca	CaF$_2$	10	500×10^{-15}	$10^{-15} - 10^{-16}$
^{129}I	15.9×10^6	^{129}Xe	^{127}I	AgI	2	100×10^{-15}	$\sim 10^{-16}$

[a] Compared to the stable isotope of the same element.

Some of the problems which have been studied by the ^{10}Be measurements are:

- Subduction of tectonic plates from ^{10}Be measurements in volcanic lavas (see, for example, Morris et al. 1990).
- Growth of manganese nodules from measurements of ^{10}Be profiles (see, for example, Guichard et al. 1978; Somayajulu 2000).
- Origin of tektites (see Serefiddin et al. 2007; Rochette et al. 2018).
- Record of geomagnetic reversals in lake sediments (Just et al. 2019; Simon et al. 2019).
- Ice core record of solar modulations and climatic variations (Steinhilber et al. 2012).
- Origin of meteorites (see for example Sossi et al. 2017).
- Variability of galactic cosmic ray flux from ^{10}Be profiles in ocean sediment cores (see for example, McHargue and Donahue 2005).

3.10.2.2 ^{14}C

Three isotopes of carbon are present in nature; ^{12}C, ^{13}C and ^{14}C. ^{12}C accounts for ~99.8% of all carbon atoms, ^{13}C accounts for ~1% of carbon atoms while ^{14}C represents only 1 ppb (one part per billion) of natural carbon. Carbon isotope ^{14}C is radioactive and has a half-life of 5730 years. Because this decay is constant it can be used as a "clock" to measure elapsed time assuming the starting amount is known. A unique characteristic of ^{14}C is that it is constantly formed in the upper atmosphere where neutrons from cosmic rays knock a proton from ^{14}N atoms. These newly formed ^{14}C atoms rapidly oxidize to form ^{14}CO$_2$ that is chemically indistinguishable from ^{12}CO$_2$ and ^{13}CO$_2$. Photosynthesis incorporates ^{14}C into plants and therefore animals that eat the plants. ^{14}C enters the dissolved inorganic carbon pool in the oceans, lakes and rivers. From there it is incorporated into shell, corals and other marine organisms. When a plant or animal dies it no longer exchanges CO$_2$ with the atmosphere. This starts the radioactive decay "clock". ^{14}C decays by emitting an electron, which converts a neutron to a proton, converting it back to its original ^{14}N form.

Radiocarbon dating has been well-established field for more than a half century. The potential applications of this method are limited by the large amount of carbon needed for measurement. The AMS allows one to reduce sample size from typical a few grams to as little as 50 μg. This has opened many new possibilities of AMS applications.

Applications in archaeology and geoscience require high-precision measurement of ^{14}C concentration in CO$_2$ solid samples. Statistical error of 1% made in such a measurement corresponds to an error in the radiocarbon age of 80 years independently of the age of the sample. Achieving sufficient statistical precision for ^{14}C is not a problem for AMS, except for very small (0.5 mg) or very old (>20,000 years) samples. The ^{14}C counting rates for modern samples (standards) are of the order of 20 c/s, so that a small statistical error (part of 1%) can be obtained in less than an hour.

3.10.2.3 ^{26}Al

The Al isotope ^{26}Al is produced in the atmosphere by spallation of ^{40}Ar by cosmic rays. Only very low ^{26}Al/^{27}Al ratios (10^{-14}) are observed in terrestrial samples of interest. Much higher ^{26}Al/^{27}Al ratios (10^{-11} or higher) are found in extraterrestrial material.

3.10.2.4 ^{36}Cl

Radioisotope ^{36}Cl is produced by the following three processes:

1. Spallation mainly of ^{40}Ar.
2. Neutron activation induced by neutrons from spallation process by the reaction ^{36}Ar(n,p)^{36}Cl.
3. Slow neutron activation through the reaction ^{35}Cl(n,γ)^{36}Cl.

The first two reactions occur mainly in the atmosphere; the third process occurs in the lithosphere where the U and Th decay series produce direct fission neutrons and (α,n) reactions non-light elements.

3.10.2.5 ^{41}Ca

Radioisotope ^{41}Ca is produced by the ^{40}Ca(n,γ) ^{41}Ca reaction with ^{40}Ca/^{41}Ca ratio estimated to be about 10^{-14}.

3.10.2.6 ^{129}I

This radioisotope is highly mobile.

3.10.3 AMS in Geophysics, Geology and Mineralogy

The most common applications of AMS in earth sciences are surface-exposure dating and erosion rates surface-exposure dating (Gosse and Phillips 2001). In the simplest case (i.e., no erosion or inherited nuclides), a rock that has previously been shielded from cosmic rays is suddenly exposed at the surface. As time passes, the rock accumulates an inventory of cosmogenic nuclides through in situ production; the number of such nuclides is a measure of how long the rock has been exposed to cosmic rays. This method thus dates the event that exposed the rock.

Radioisotope ^{36}Cl has a wide range of applications in geoscience. Of particular interest is its geochemistry because chlorine has a very high electron affinity, and occurs in natural circumstances as the chloride anion. The fact that chloride is one of the most hydrophilic substances and passes through various systems with only minimal interaction makes it especially suitable for hydrological tracing and ground water dating. The isotopic abundance of ^{36}Cl/Cl in natural chloride samples extracted from the relevant media such as polar ice, ground water sediment and rocks varies between 10^{-10} and 10^{-16}.

Establishing glacial chronologies is an important part of paleoclimate research. Radiocarbon methods can be used in cases where organic matter can be associated with glacial events (Clark et al. 1995). The development of surface-exposure methods has made dating possible for older events (Fabel and Harbor 1999); when erosion rates are exceptionally low, surface-exposure methods can date events back to several million years. Such dating helps, for example, in determining the sequence of glacial events in a particular area (Gosse et al. 1995).

Radioisotope ^{129}I is highly mobile in its inorganic form and therefore appears to be potentially powerful tool for studying the timing of geological and hydrological processes. Table 3.6 (according to McMahon 1992) shows the mass corresponding to 1 MBq for a number of important radionuclides. The long-lived radionuclides like for example

TABLE 3.6

The mass of 1 MBq for some radionuclides of different half-lives

Radionuclide	Half-live (years)	Mass of 1 MBq (g)
^{232}Th	1.4×10^{10}	2.5×10^{-7}
^{238}U	4.5×10^{9}	8.1×10^{-8}
^{129}I	1.7×10^{7}	1.6×10^{-10}
^{99}Tc	2.1×10^{5}	1.6×10^{-12}
^{239}Pu	2.4×10^{4}	4.3×10^{-13}
^{14}C	2.5×10^{3}	6.1×10^{-15}
^{137}Cs	30	3.1×10^{-16}

^{232}Th can easily be measured with non-radiometric methods. The relatively short-lived radionuclides should be measured by radiometric methods such as AMS.

Surface-exposure dating is making an important contribution toward understanding tectonic events over the past few million years. Establishing chronologies for movements along faults allows slip rates to be determined and earthquakes to be dated. Two ways to establish such chronologies involve surface exposure dating of fault scarps (Mitchell et al. 2001) and of alluvial sediments that have been affected by fault motion (Zehfuss et al. 2001).

It should be mentioned that ^{10}Be and ^{36}Cl concentrations in the ice cores contain information concerning the history of the Earth's magnetic field, the solar cycle, and the Earth's climate regimes. A key challenge in this research is unraveling all the various effects that are embodied in the nuclide signal. One technical aspect of this research is that a careful analysis of a long ice core usually requires many AMS measurements; progress in this field is therefore related to increased efficiency of AMS measurements of ^{10}Be and ^{36}Cl (Fifield 1999).

3.11 QA/QC Procedures: Reference Materials

A quality assurance (QA) program is an essential part of the sound analytical protocol. It must include the use of well-designed methods, adequate instrumentation, trained staff, provision of representative samples adequate sample preparation method, use of reliable reference materials, proficiency testing and more.

Quality control in terms of accuracy can be best improved by participation in inter- and intra-laboratory testing programs. In addition, to assess the accuracy of his method the analyst can test it with certified reference materials (CRM). The use of CRMs permits interlinking of the results obtained by the laboratory with the outside world, in order to achieve the maximum accuracy possible at the time. Maier (1990) summarizes requirements for the production of CRM and they include requirements on homogeneity, stability, similarity with real material, and accuracy and uncertainty of certified values.

CRMs are generally produced and certified for their chemical composition by major national standards organizations such as the National Institute of Standards and Technology, USA; the National Institute of Environmental Studies, Japan, the Bundesanstalt fur Materialprufung, Germany, the Laboratory of the Government Chemist, the UK. These national entities have been joined by regional organization, e.g., Community Bureau of Reference, Belgium and the IAEA on an international scale. Through the standards organizations and the Agency's Analytical Quality Control Services (AQCS) program virtually all chemical and radio-analytical laboratories are in the position to use procedures for QA. The demand for these procedures and the supporting CRMs is still rapidly developing. The evolution of laboratory accreditation programs in industrialized countries points the way that soon analytical results will not be accepted from other than "accredited" laboratories. It is obvious that for global trade and industrial development, for the protection of human and environmental health, and for the science of chemical metrology all laboratories strive to obtain compatibility in their measurement system. It prevents disagreement among parties and helps promote credibility and product quality. When compatibility is based on the international measurement system, it can provide evidence of compliance to environmental regulations, appropriate patient diagnosis, adequate nutrition, proper manufacturing and material processing.

In the system of standards, we have seen a rapid increase in the availability and use of natural matrix CRMs. From the first biological standard that was made available in 1956 (Kenworthy et al. 1956), the analytical community nowadays has several hundreds of different vegetation, animal and human tissue, food, sediment, and soil samples. These CRMs are being used with most analytical techniques, however, direct applications of many existing CRMs in micro-analytical procedures for trace element determinations such as EDXRF and PIXE or solid sampling atomic absorption spectrometry (SS-AAS) is often difficult or sometimes impossible. The requirements for these methods need detailed attention.

Most of the natural matrix CRMs are being produced from bulk or composite, batches of the material collected in nature or in a processing facility. They are mostly solid or coarse grain products, which from their very nature are inhomogeneous; only on rare occasions, the initial material is a fine grain powder, such as flour, milk powder, and some sediments. Hence, the producer of CRMs must convert the natural materials to products that allow portioning into thousands of subsamples that are of equal composition. Consequently, the original material is transformed to the product with rather conventional means of size reduction such as cutting, grinding, and milling as well as blending and mixing to improve the homogeneity and the representativeness of the individual subsample. The conventional processes in general achieve that 200 mg test portions are representative for the lot with an uncertainty of <1% relative. This is obviously not sufficient for the techniques that analyze much smaller samples.

Homogeneity is always a relative term. In powdered materials, once blended and defined in their physical characteristics, it depends on the distribution of the constituents of interest in each particle and on the distribution of the various particles within the powder, thus being usually different for each constituent, and increasing as the amount of the material sampled increases.

The homogeneity is a primary requisite of any powdered CRM and can be quantitatively expressed through the estimate of the "degree of inhomogeneity", i.e., the scatter of the constituents of interest about their mean value from portion to portion of a sample or from sample to sample, which in turn can be expressed as a relative standard deviation.

CRMs are usually supplied in bottles. The "degree of inhomogeneity" can therefore be twofold: the reproducibility between bottles and the reproducibility within each bottle.

Overall random error is

$$s = \left(s_a^2 + s_h^2\right)^{+1/2} \tag{3.26}$$

where

s = measured standard deviation,

$s_a{}^2$ = variance due to random errors of analysis (or variance due to analysis of inhomogeneous test sample),

$s_h{}^2$ = variance due to random sample inhomogeneity.

The value of s_a is accessible to the direct measurement only by instrumental methods that allow repeatable measurements of one identical sample under identical conditions (could be done by XRF or INAA).

In addition

$$s_h = \left(s_b^2 + s_k^2\right)^{1/2} \tag{3.27}$$

where $s_b{}^2$ is variance related to the inhomogeneity between bottles, while $s_k{}^2$ is variance related to the inhomogeneity within the bottle. Investigations of the degree of inhomogeneity (s_h) have eliminated potential reference materials. In addition, it was shown that the degree of inhomogeneity increases for most of the materials for decreasing concentrations. This difficulty arises especially for geological and environmental materials, since some of the elements are concentrated mainly in one mineral of the matrix such as zirconium in zircons or native gold in a rock.

The lAEA practice for homogeneity testing involves determinations of concentrations of some of the elements by the available techniques. Usually, elements like Ca, Fe, Mg, Mn, Ca and P are determined by ICP-OES, Cu, Fe, Mg and K by AAS and Br, Na, and Cl by NAA on several samples (>10) taken from one bottle and the results were compared with those obtained on sub-samples taken from several bottles (>10) chosen at random. If the results did not differ significantly, by applying the *F*-test, it can be found that, for example, a material could be considered homogenous for a sample size of greater or equal to 300 mg at a given confidence level (usually 0.05).

IBA has some very specific problems. The list of problems includes:

- Precise energy calibration is one of the prerequisites for depth profiling by the backscattering techniques.
- Contaminant layers growing on the targets during irradiation.

These problems make quantification without CRMs very difficult. In addition, ion beams see a very small mass of the target. In spite of these difficulties, some materials are available for calibration purposes in IBA. Examples are evaporated standard layers of metals (Al, Ti, V, Cr, Cu, N, Pd, Ag, Pt, Au) deposited on light substrates (mainly vitreous carbon) available from the Central Bureau for Nuclear Measurements (Wätjen et al. 1990). Due to the unique in vacuum weighing procedure, the thickness of the layers in term of areal density can be traced back directly to the SI unit of mass (20 µg to 7 mg). Analytical results of high accuracy can also be achieved by employing internal standard layers evaporated directly onto the unknown samples with accurately determined densities. Other useful standards are prepared specifically for SIMS, by low energy (30–120 keV) implantation into Si or any other electrically conductive material. A certificate of implanted dose (10^{15} ions/cm^2) and energy usually accompanies each standard.

Some of the reference materials may from their nature be suitable for the use with the ion beam analytical techniques. The materials that are very fine powders and are obtained in their final form through a rather constant homogeneous process. Among these are fly ash samples, certain sediments, spray dried milk powder, and flours. Other materials require state-of-the-art size reduction and homogenization techniques. Recently two techniques have been successfully used to produce CRMs from which small test portions can be taken: (1) jet milling of dried biological materials and (2) cryogenic homogenization of fresh or dry materials. Following the principles of homogenous non-contaminating and production, Versieck et al. (1988) have described the certification of freeze-dried human serum for 14 trace elements; PIXE results were as accurate and precise as those of other techniques applied due to extreme homogeneity of the tested material.

The AQCS program was established by the IAEA to support and assist those laboratories that are involved in the analysis of nuclear, environmental (including marine) and biological materials (see Fajgelj et al. 1997). This assistance includes:

1. The organization of intercomparison studies to establish the levels of organic compounds, inorganic compounds and nuclides (stable and radioactive) in the above materials.
2. The sale and distribution of these materials.
3. Training and consulting with fellows and scientific visitors from those same laboratories.
4. Co-ordinated research and co-operation in the production and certification of CRMs.

Laboratories involved in the analysis of samples with complex matrices must somehow determine the precision (reproducibility) and accuracy (estimate of the maximum deviation from the true value) of their result. The precision of the results can easily be determined by repeated analysis of the same material. The determination of accuracy, however, in most cases requires more detailed procedures involving one or more of the following:

- An analyte is determined by a number of different methods and analytes in an attempt to remove any biases introduced by the methods.
- Analytes are determined in samples that include a reference material that represents the sample matrix as closely as possible.
- The laboratory may regularly participate in intralaboratory studies. The agreement between its reported result and the recommended or certifies value determined by some statistical averaging procedure, implies that the method provides accurate results.
- As most laboratories do not have the time or manpower to determine their accuracy using the first option, the Agency provides a most valuable service in offering laboratories the option of determining their accuracy via the latter two options with minimal or no cost. Information on these services is presented below.
  1. Intercomparison Runs: Participation in intercomparison runs is carried out on a voluntary and cost-free basis for the participants. Once all the results are received and evaluated, a report containing the results and statistical evaluation is issued and a copy send to each participant.

2. Reference Materials: The IAEA has a number of reference materials (RMs) available for sale and is regularly organizing intercomparison runs to re-plenish deleted RMs or initiate new RMs. These RMs can be used as secondary standards for quality control in the laboratory, can be used to check analytical methods and instruments and to verify new procedures or can be used for training new personnel.

Although in principle, one can achieve the required accuracy of analytical data without the use of CRMs such an approach would require a tremendous investment in personnel, apparatus, various analytical techniques, control bodies, and additionally would also need, for each type of determination, a substantial participation in intercomparisons with many other laboratories of the same high quality. This is possible, but the investment in time and expertise to produce high quality accurately CRM gives an indication of the complexity of the task required. Accuracy of results is more easily obtained with CRMs, therefore CRMs should be an integral part of QA/QC system practiced in the analytical laboratory. To satisfy the needs of the micro analytical techniques, a new generation of CRMs should be produced that satisfies homogeneity requirements down to less than 1 mg of material.

References

Aerts-Bijma, A. T., Paul, D., Dee, M. W., Palstra, S. W. L., and Meijer. H. A. J. 2021. An independent assessment of uncertainty for radiocarbon analysis with the new generation high-yield accelerator mass spectrometers. *Radiocarbon* 63(1): 1–22.

Ajzenberg-Selove, F. 1986. Energy levels of light nuclei A = 16 – 17. *Nucl. Phys.* A460: 1–110.

Ajzenberg-Selove, F. 1987. Energy levels of light nuclei A = 18 – 20. *Nucl. Phys.* A475: 1–75.

Ajzenberg-Selove, F. 1990. Energy levels of light nuclei A = 11 – 12. *Nucl. Phys* A506: 1–158.

Ajzenberg-Selove, F. 1991. Energy levels of light nuclei A = 13 – 15. *Nucl. Phys.* A523(1): 1–196.

Ajzenberg-Selove, F. and Lauritsen, T. 1968. Energy levels of light nuclei A = 11 – 12. *Nucl. Phys.* A114: 1–142.

Ajzenberg-Selove, F. and Lauritsen, T. 1974. Energy levels of light nuclei A = 5 – 10. *Nucl. Phys.* A227: 1–243.

Akasaka, Y. and Horie, K. 1973. *Channelling Analysis and Electrical Behaviour of Boron Implanted Silicon. In Ion Implantation in Semiconductors and Other Materials.* Crowder, B. L. (Ed.) Plenum Press, New York. pp. 147–157.

Alefeld, G., and Völkl, J. (Eds.). 1978. *Hydrogen in Metals II: Application-Oriented Properties.* Springer: Berlin/Heidelberg, Germany, 1978.

Alexander, R. B., Collaghan, P. T., and Poate, J. M. 1973. An exacting test of the chanelling technique for atom location: Br implanted into Fe, pp. 477–490. In Crowder, B. L. (Ed.) *Ion Implantation in Semiconductors and Other Materials.* Plenum Press, New York.

Amsel, G., Battistig, G., and L'Hoir, A. 2003. Small angle multiple scattering of fast ions, physics, stochastic theory and numerical calculations. *Nucl. Instrum. Methods Phys. Res. B* 201: 325–388.

Barnes, D. G., Calvert, J. M., kay, K. A., and Lees, D. G. 1973. The role of oxygen transport in oxidation of Fe-Cr alloys. *Philos. Mag.* 28: 1303.

Barrandon, J. N. and Seltz R. 1973. Methode de dosage simultané du carbone et de l'oxigene à la surface des metaux par la reaction (t,p) à basse énergie. *Nucl. Instrum. Methods* 111: 595–603.

Baxter, M. S. and Walton, A. 1970a. Glasgow University Radiocarbon Measurements III. *Radiocarbon* 12: 496–502.

Baxter, M.S. and Walton, A. 1970b. Radiocarbon DATING OF Mortars. *Nature* 225: 937–938.

Bayley, A. R., and Townsend, P. D. 1973. Refractive Index Profilesproduced in Silicaglass by Ion Implantation, pp. 575–584. In Crowder, B. L. (Ed.) *Ion Implantation in Semiconductors and Other Materials*. Plenum Press, New York.

Bellagamba, B., Caridi, A., Cereda, E., Braga Marcazzan, G. M., and Valković, V. 1993. PIXE applications to the study of trace element behaviour in coal combustion. *Nucl. Instrum. Methods Phys. Res.* B75: 222–229.

Bench, G. S. and Legge, J. F. 1989. High resolution STIM. *Nucl. Instrum. Methods Phys. Res. B* 40–41 (1, 2): 655–658.

Blewer, R. S. 1974. Proton backscattering as a technique for light ion surface interaction studies in CTR materials investigations. *J. Nucl. Mater.* 53: 268–275.

Blewer, R. S. (1974) Depth Distribution and Migration of Implanted Helium in Nickel Foils Using Proton Backscattering, pp 557–572. In Picraux, S. T., EarNisse, E. P. and Vook, F. L. (Eds.) *Applications of Ion Beams to Metals*. Plenum Press, New York.

Bogh, E. 1973. *Channeling: Theory, Observation and Applications*. In Morgan, D. V. Ed., Wiley Interscience, New York. 435.

Boni, C., Caruso, E., Cereda, E., Braga Marcazzan, G. M., and Redaelli, P. 1989. A PIXE-PIGE system for the quantitative elemental analysis of thin samples. *Nucl. Instrum. Methods B*, 40–41: 620–623.

Butz, T., Flagmeyer, R.-H., Heitmann, J. et al. 2000. The Leipzig high-energy ion nanoprobe: A report on first results. *Nucl. Instrum. Methods Phys. Res. B* 161–163: 323–327.

Calvert, J. M., Derry, D. J., and Lees, D. G. 1974. Oxygen diffusion studies using nuclear reactions. *J. Phys. D Appl. Phys.* 7: 940–953.

Caridi, A., Cereda, E., Fazinić, S., Jakšić, M., Braga Marcazzan, G. M., Valković, O., and Valković, V. 1992. Fluorine enrichment phenomena in coal combustion process. *Nucl. Instrum. Methods Nucl. Res. B* 66: 298–301.

Caridi, A., Cereda, E., Fazinić, S., Jakšić, M., Braga Marcazzan, G. M., Scagliotti, M., Valković, V. 1992. Proton resonant scattering for oxigen stoichometry of reactively evaporated ZrO_{2-x} Films. *Nucl. Instrum. Methods Nucl. Res. B* 64: 774–777.

Caridi, A., Cereda, E., Grime, G. W., Jaksic, M., Braga Marcazzan, G. M., Valković, V., and Watt, F. 1993. Application of proton microprobe analysis to the study of electrostatic precipitation of single fly ash particles. *Nucl. Instrum. Methods Phys. Res.* B77: 524–529.

Carter, C. and Grant, W. A. 1976. *Ion Implantation of Semiconductors*. Edward Arnold Ltd., London.

Chen, X., Chen, C.-B., Udalagama, C. N. B., et al. 2013. High-Resolution 3D imaging and quantification of gold nanoparticles in a whole cell using scanning transmission ion microscopy. *Biophys. J.l* 104: 1419–1425.

Cholewa, M., Saint, A., Legge, G. J. F., Jamieson, D. N., and Nishijima, T. 1993. Channeling STIM and its applications. *Nucl. Instrum. Methods Phys. Res.* B77(1–4): 184–187.

Cipolla, F. and Taroni, A. 1967. A channeling effect in solid state detectors. *Nucl. Instrum. Methods* 54(2): 331–332.

Clark, D. H., Bierman, P. R., and Larsen, P. 1995. Improving in situ cosmogenic chronometers.*Q. Res.* 44: 367–377.

Clayton, E. 1986. PIXAN, the Lucas Heights PIXE Analysis Computer Package. Report AEC/M113.

Close, D. A., Malanify, J. J. and Umbarger, C. J. 1973. Isotopic carbon analysis using low energy protons. *Nucl. Instrum. Methods* 113: 561–571.

Cohen, B. L., Fink, C. L., and Degnan, J. H. 1972. Nondestructive analysis for trace amounts of hydrogen. *J. Appl. Phys.* 43: 19–25.

Crowder, B., (Ed.). 1973. Ion Implantation in Semiconductors and Other Materials. The IBM research Symposia Series. Springer

Dearnaley, G. 1964. The channelling of ions through silicon detectors. *IEEE Trans. Nucl. Sci.* 11(3): 249–253.

Dearnaley, G. 1973. Ion Penetration and Channeling. In Dearnaley, G. et al. (Eds.) *Ion implantation*. North Holand Pub. Co., Amsterdam. 3.

Dearnaley, G., Freeman, J. H., Nelson, R. S., and Stephen, J. 1973. *Ion Implantation*. North-Holland, Amsterdam

Dearnaley, G., Hartley, N. E. W., Turner, J. F., Garnsey, R., and Woolsey, I. S. 1975. Ion–beam analysis of corrosion films on 316 steel. *J. Vac. Sci. Technol.* 12: 449.

Delibrias, G. and Labeyrie, J. 1964. Dating of old mortars by the carbon-14 method. *Nature* 201: 742–743.

Delibrias, G., Guillier, M. T., and Labeyrie, J. 1964. Saclay natural radiocarbon measurements I. *Radiocarbon* 6: 233–250.

Döbeli, M., Kottler, C., Glaus, F., and Suter, M. 2005. ERDA at the low energy limit. *Nucl. Instrum. Methods Phys. Res. B* 241: 428–435.

Eichhorn, C., Manova, D., Feder, R. et al. 2015. Sputter yield measurements of thin foils using scanning transmission ion microscopy. *Eur. Phys. J. D* 69: 19 (7 pp).

Elmore, D. and Phillips, F. D. 1987. Accelerator mass spectrometryfor measurements of long-lived radioisotopes. *Science* 236: 543–550.

Emmrich, D., Wolff, A., Meyerbröker, N., et al. 2021. Scanning transmission helium ion microscopy on carbon nanomembranes. *Beilstein J. Nanotechnol.* 12: 222–231.

Enright, H. A., Malfatti, M. A., Zimmermann, M., Ognibene, T., Henderson, P., and Turteltaub, K. W. 2016. The use of accelerator mass spectrometry in human health and molecular toxicology. *Chem Res Toxicol.* 29(12): 1976–1986.

Fabel, D. and Harbor, J. 1999. The use of in situ produced cosmogenic nuclides in glaciology and glacial geomorphology. *Ann. Glaciol.* 28: 103–110.

Fajgelj, A., Parr, R. M., Dekner, R., Danesi, P. R., Valkovic, V., and Vera Ruiz, H. 1997. The IAEA's Analytical Quality Control Services (AQCS) programme on intercomparaison runs and reference materials. Proc. Internat. Symp. On Harmonization of health related environmental measurements using nuclear and isotopic techniques, Hyderabad, India, 4–7 November 1996 International Atomic Energy Agency, Vienna, Austria Report IAEA-SM-344/3. (pp 175–187).

Fazinić, S., Jakšić, M., Kukec, L., Valković, O., and Valković. V. 1992. Light element analysis using PIXE and PIGE Spectroscopies. *Nucl. Instrum. Methods Nucl. Res. B* 66: 273–276.

Feldman, L. C., Poate, L. M., Ermanis, F., and Schwartz, B. 1973. The combined use of he backscattering and He induced X-Rays in the study of anodically grown oxide films on GaAs. *Thin Solid Films* 19: 81–89.

Feldman, L. C., Mayer, J. W. and Picraux, S. T. 1982. *Material Analysis by Ion Channelling*. Academic Press, New York.

Fifield, L.K. 1999. Accelerator mass spectrometry and its applications. *Rep. Prog. Phys.* 62: 1223–1274.

Folk, R.L. and Valastro, S. Jr. 1979. Dating of Lime Mortar by 14C: Radiocarbon Dating. Proceedings of the ninth International Conference, eds.Berger, R. and Suess, H. E. p. 721– 732. California Press.

Franco, N., Gouveia, J. A. A., Alves, E., Cardoso, S., Freitas, P. P., and Barradas, N. P. 2005. Analysis of nanolayered samples with a ^{4}He beam. *Nucl. Instrum. Methods Phys. Res. B* 241: 361–364.

Fukai, Y. 2005. *The Metal-Hydrogen System*. Springer: Berlin/Heidelberg, Germany.

Gemmell, D. S. 1974. Channeling and related effects in the motion of charged particles through crystals. *Rev. Mod. Phys.* 46: 129–227.

Georges, C., Sturel, T., Drillet, P., and Mataigne, J. M. 2013. Absorption/desorption of diffusible hydrogen in aluminized boron steel. *ISIJ Int.* 53(8): 1295–1304.

Giuntini, L., Massi, M., and Calusi, S. 2007. The external scanning proton microprobe of Firenze: A comprehensive description. *Nucl. Instrum. Methods A* 576: 266–273.

Gosse, J. C. and Phillips, F. M., 2001. Terrestrial cosmogenic nuclides: theory and applications.*Q. Sci. Rev.* 20: 1475–1560.

Gosse, J. C., Klein, J., Evenson, E. B., Lawn, B., and Middleton, R. 1995. Beryllium-10 dating of the duration and retreat of the last Pinedale glacial sequence. *Science* 268: 1329–1333.

Gove, H. E., Litherland, A. E., and Elmore, D. 1987. Proceedings of the IV International symposium on accelerator mass spectrometry, Niagara-on-the-Lake, Ontario, Canada. *Nucl. Instrum. Methods Phys. Res.* B29: 1.

Guichard, F., Reyss, J. and Yokoyama, Y. 1978. Growth rate of manganese nodule measured with 10Be and 26Al. *Nature* 272: 155–156.

Hah, S. S. 2009. Recent advances in biomedical applications of accelerator mass spectrometry. *J. Biomed. Sci.* 16: article number 54. 10.1186/1423-0127-16-54.

Hartley, N. E. W. 1975. Rutherford Backscattering, Nuclear Reactions and X-ray Methods for Surface Composition Analysis. AERE Report 8048.

Hein, P. and Wilsius, J. 2008. Status and innovation trends in hot stamping of Usibor 1500P. *Steel Res. Int.* 79: 85–91. https://www.ionbeamcenters.eu/RADIATE-project-partners/vienna/

Hufschmidt, M., Heintze, V., Möller, W. and Kamke, D. 1975. The density profile of implanted ^{3}He measured by means of the ^{3}He(d,p)^{4}He reaction. *Nucl. Instrum. Methods* 124: 573–577.

IAEA. 1995. IAEA Meeting on Accelerator Mass Spectrometry. Zagreb, Croatia, April 19–21. 1995. https://inis.iaea.org/collection/NCLCollectionStore/_Public/35/066/35066061.pdf.

IAEA. 2003. Intercomparison of PIXE spectrometry software packages. IAEA-TECDOC-1342 (65 pp). IAEA, Vienna, Austria.

IAEA. 2004. Ion beam techniques for the analysis of light elements In thin films, including depth profiling. IAEA-TECDOC-1409. (126 pp). IAEA, Vienna, Austria.

IAEA. 2011. Nuclear techniques for cultural heritage research. IAEA Radiation Technology Series No. 2. STI/PUB/1501 (205 pp). IAEA, Vienna, Austria.

IAEA. 2014. Hands-on training courses using research reactors and accelerators. Training Course Series No. 57. IAEA-TCS-57 (103 pp). IAEA, Vienna, Austria.

IAEA. 2015. Utilization of accelerator based real time methods in investigation of materials with high technological importance. IAEA Radiation Technology Reports No. 4. STI/PUB/1649 (104 pp). IAEA, Vienna, Austria.

IAEA. 2017. Development of a reference database for particle induced gamma ray emission (pige) spectroscopy. IAEA-TECDOC-1822. IAEA, Vienna, Austria.

IAEA. 2018. Accelerator simulation and theoretical modelling of radiation effects in structural materials. IAEA Nuclear Energy Series No. NF-T-2.2. STI/PUB/1732 (116 pp). IAEA, Vienna, Austria.

IAEA. 2019. Improvement of the reliability and accuracy of heavy ion beam analysis. Technical reports series no. 485. International Atomic Energy Agency, Vienna, Austria.

IAEA. 2020. Accelerator Knowledge Portal. International Atomic Energy Agency, Vienna, Austria. https://www.iaea.org/resources/databases/accelerator-knowledge-portal. https://nucleus.iaea.org/sites/accelerators/Pages/default.aspx

IBMM. 2018. Proceedings of the 21st International Conference on Ion Beam Modification of Materials. *Nucl. Instrum. Methods Phys. Res. B* 465: 47–52.

Ishigami, R., Ito, Y. Yasuda, K., and Hatori, S. 2005. ERDA with 15 MeV ^{4}He ions under atmospheric pressure. *Nucl. Instrum. Methods Phys. Res. B* 241: 423–427.

Jakšić, M., Kukec, L., and Valković, V. 1993. The Zagreb nuclear microprobe facility. *Nucl. Instrum. Methods Phys. Res.* B77: 49–51.

Jin, X., Crocombette, J.-P., Djurabekova, F., Zhang, S., Nordlund, K., Garrido, F., and Debelle, A. 2020. New developments in the simulation of Rutherford backscattering spectrometry in channeling mode using arbitrary atom structures. *Model. Simul. Mater. Sci. Eng.* 28(7): 075005.

Just, J., Sagnotti, I., Nowaczyk, N. R., Francke, A., and Wagner, B. 2019. Recordings of Fast paleomagnetic reversals in a 1.2 Ma Greigite-Rich Sediment Archive From Lake Ohrid, Balkans. *J. Geophys. Res. Solid Earth* 124: 12,445–12,464.

Kantele, J. and Tannila, O. 1968. Isomeric state ($\tau_{1/2} \geq 10^{-10}$ s) listed according to increasing half-life. *Nucl. Data* A4: 359–395.

Kenworthy, A. L., Miller, E. J., and Mathis, W. T. 1956. Nutrient-element analysis of fruit tree leaf samples by several laboratories. *Proc. Am. Soc. Hortic. Sci.* 67: 16–21.

Klein, M., Mous, D. J. W., and Gottdang, A. 2004. Fast and accurate sequential injection AMS with gated Faraday cup current measurement. *Radiocarbon* 46(1): 77–82.

Klein, M. G., Dewald, A., Gottdang, A., Heinze, S., and Mous, D. J. W. 2011. A new HVE 6 MV AMS system at the University of Cologne. 2011. *Nucl. Instrum. Methods Phys. Res. B* 269: 3167–3170.

Klein, M. and Mous, D. J. W. 2017. Technical improvements and performance of the HVE AMS sputter ion source SO-110. *Nucl. Instrum. Methods Phys. Res. B* 406: 210–213.

Klein, M., Podaru, N. C., and Mous, D. J. W. 2019. A novel dedicated HVE ^{14}C AMS system, *Radiocarbon* 61: 1441–1448.

Kornelsen, E. V., Brown, F., Davies, J. A., Domeij, B., and Piercy, G. R. 1964. Penetration of heavy ions of keV energies into monocrystalline tungsten. *Phys. Rev.* 136 (3A): A849.

Kutschera, W., Collon, P., Friedmann, H., Golser, R., Hille, P., Priller, A., Rom, W., Steier, P., Tagesen, S., Wallner, A., Wild, E., and Winkler, G. 1997. VERA: A new AMS facility in Vienna. *Nuclear Instum. Methods Phys. Res. B* 123 (1–4), pp. 47–50.

Kutschera, W. 2016. Accelerator mass spectrometry: state of the art and perspectives. *Adv. Phys. X* 1(4): 570–595.

Laricchiuta, G., Vandervorst, W., Vickridge, I., Mayer, M., and Meersschaut, J. 2019. Rutherford backscattering spectrometry analysis of InGaAs nanostructures. *J. Vac. Sci. Technol. A* 37: 020601.

Ligeon, E. and Giuvarch, A. 1974. A new utilization of ^{11}B beams: hydrogen analysis by H(^{11}B,α)2α nuclear reaction. *Radiat. Effects* 22: 101–105.

Lindh, U. 1990. Micron and submicron nuclear probes in biomedicine. *Nucl. Instrum. Methods Phys. Res.* B49: 451–464.

Lindstrom, W. W. and Heuer, A. H. 1974. A magnetic spectrometer for nuclear microanalysis of concentration distributions. *Nucl. Instrum. Methods* 116: 145–155.

Lipworth, A. D., Annegarn, H. J., and Kneen, M. A. 1993. Advanced, enhanced HEX program for PIXE. *Nucl. Instr. and Meth.* B75: 127–130.

Lorenzen, J. 1974. Depth distribution studies of carbon in steel surfaces by means of the nuclear reaction ^{12}C(p,γ)^{13}N. *Nucl. Instrum. Methods* 121: 467–475.

Lorenzen, J. and Brune, D. 1975. Determination of oxygen in zircaloy surfaces by means of charged particle activation analysis. *Nucl. Instrum. Methods* 123: 379–384.

Lövestam, N. E. G. and Swietlicki, E. 1989. An external beam setup for Lund proton microprobe. *Nucl. Instrum. Methods Phys. Res.* B43: 104–111.

Luomajärvi, M., Rauhala, E. and Hautala, M. 1985. Oxygen detection by non-Rutherford proton backscattering below 2.5 MeV. *Nucl. Instrum. Methods Phys. Res. B* 9(3): 255–258.

Macková, A., MacGregor, D., Azaiez, F., Nyberg, J., and Piasetzky, E. (Eds.). 2016. Nuclear Physics for Cultural Heritage. Appendix A: European facilities using nuclear techniques to study cultural heritage. Published by Nuclear Physics Division of the European Physical Society.

Maier, E. A. 1990. Environmental analysis: is accuracy possible without certified reference materials? *Anal. Proc.* 27: 269–270.

Malone, G., Valastro, S. Jr., and Vazela, A. G. 1980. Carbon-14 chronology of mortar from excavation in the Medieval Church of Saint-B'enigne, Dijon, France. *J. Field Archaeol.* 7: 329–343.

Mandler, J. W. and Semmler, B. A. 1974. Trace element content and depth distribution using proton and deuteron interactions. In *Proceedings of 3rd International Conference on Applications of Small Accelerators*. Eds. Duggan, J. L. and Morgan, I. L. Denton, Texas: CONF-741040-P1. pp 173–183.

Marion, J. B. and Young, F. C. 1968. *Nuclear Reaction Analysis: Graphs and Tables*. North Holland Pub. Co., Amsterdam.

Markowitz, S. S. and Mahoney, J. D. 1962. Activation analysis for oxygen and other elements by helium-3-induced nuclear reactions. *Anal. Chem.* 34: 329–335.

Martinez-Martinez. Rickards, R., Garcia-Hipolito, M., Trejo-Luna, R., Martmez-Sanchez, E., Alvarez-Fregoso, O., Ramos-Brito, F., and Falcony, C. 2005. RBS characterization of Al_2O_3 films doped with Ce and Mn. *Nucl. Instrum. Methods Phys. Res. B* 241: 450–453.

Massi, M., Giuntini, L., Chiari, M., Gelli, N., and Mandò, P. A. 2002. The external beam microprobe facility in Florence: Set-up and performance. *Nucl. Instrum. Methods Phys. Res. B* 190: 276–282.

Maxwell, J. A., Campbell, J. L., and Teesdale, W. J. 1989. The Guelph PIXE software package. *Nucl. Instrum. Methods* B43: 218–230.

Mayer, J. W., Erikson, L., and Davies, J. A. 1970. *Ion implantation in semiconductors: Silicon and Germanium*. Academic Press, New York.

Mayer, J. W. and Rimini, E. 1977. *Ion Beam Handbook for Material Analysis*. Academic Press, New York.

McHargue, I. R. and Donahue, D. J. 2005. Effects of climate and the cosmic-ray flux on the 10Be content of marine sediments. *Earth Planet. Sci. Lett.* 232(1): 193–207.

McMahon, A. W. 1992. An intercomparison of non-radiometric methods for the measurement of low levels of radionuclides. *Appl. Radiat. Isot.* 43: 289–303.

Mitchell, S. G., Matmon, A., Bierman, P. R., Enzel, Y., Caffee, M., and Rizzo, D. 2001. Displacement history of a limestone fault scarp, northern Israel, from cosmogenic ^{36}Cl. *J. Geophys. Res.* 106: 4247–4264.

Morgan, D. V. 1975. Recent advances in surface studies: ion beam analysis. *Contemp. Phys.* 16: 221–241.

Morris, J., Leeman, W., and Tera, F. 1990. The subducted component in island arc lavas: constraints from Be isotopes and B–Be systematics. *Nature* 344: 31–36.

Mous, D. J. W., Haitsma, R. G., Butz, T. et al. 1997. The novel ultrastable HVEE 3.5 MV Singletron''' accelerator for nanoprobe applications. *Nucl. Instrum. Methods Phys. Res. B* 130: 31–36.

Muller, R. A. 1977. Radioisotope dating with a cyclotron. *Science* 196: 489–494.

Myers, S. M., Beerhold, W. and Picraux, S. T. 1973. Implantation and diffusion of Cu in Be. In Crowder, B. L. (Ed.). *Ion Implantation in Semiconductors and Other Materials*. Plenum Press, New York. pp. 455–464.

Nanolito. 2021. Summer School in Basics and Applications of Nanolithography. June 29–30 & July 1st 2021, Univesity of Salamanca, Spain.

Nield, D. J., Wise, P. J., and Barnes, D. G. 1972. Measurement of ^{18}O concentration profiles using resonant nuclear reactions. *J. Phys. D* 5(12): 2292–2299.

North, J. E. and Gibson, W. M. 1970. Channeling study of boron-implanted silicon. *Appl. Phys. Lett.* 16: 126–129.

Ognibene, T. J., Bench, G., Vogel, J. S., Peaslee, G. F., and Murov, S. 2003. A high-throughput method for the conversion of CO_2 obtained from biochemical samples to graphite in septa-sealed vials for quantification of ^{14}C via accelerator mass spectrometry. *Anal. Chem.* 75: 2192–2196.

Oliver, C. and Pierce, T. B. 1974. The determination of boron in steels by measurement of the alpha-particles emitted during charged-particle irradiation. *Radiochem. Radioanal. Lett.* 17: 335–342.

Oliver, C., McMillan, J. W. and Pierce, T. B. 1975. The use of the nuclear microprobe for the examination of nitrogen distributions in metal samples. *Nucl. Instrum. Methods* 124: 289–297.

Pelicon, P., Podaru, N. C., Vavpetič, P., et al. 2014. A high brightness proton injector for the Tandetron accelerator at Jožef Stefan Institute. *Nucl. Instrum. Methods Phys. Res. B* B332: 229–233.

Pentecoste, L., Thomann, A. L., and Brault, P. et al. 2017. Substrate temperature and ionkinetic energy effects on first steps of He+ implantation intungsten: experiments and simulations. *Acta Materialia* 141: 47–58.

Picraux, S. T. and Vook, F. L. 1974. Lattice Location Studies of 2D and 3He in W, pp. 407–422. In Picraux et al. (Ed.). *Applications of Ion Beams to Metals*. Plenum Press, New York.

Pierce, T. B., McMillan, J. W., Peck, P. F., and Jones, I. G. 1974. An examination of carbon diffusion profiles in steel specimens by means of the nuclear microprobe. *Nucl. Instrum. Methods* 118: 115–124.

Podaru, N. C. and Mous, D. J. W. 2016. Recent developments and upgrades in ion source technology and ion beam systems at HVE. *Nucl. Instrum. Methods Phys. Res. B* 371: 137–141.

Purser, R. A., et al. 1977. An attempt to detect stable N⁻ ions from a sputter ion source and some implication of the results for the design of tandems for sensitive carbon analysis. *Rev. Phys. Appl.* 12 (10): 1487–1492.

Quax, G. W. W., Gottdang, A., and Mous, D. J. W. 2010. A high-current light-ion injector for tandem accelerators. *Rev. Sci. Instrum.* 81, 02A701 (3 pp).

Räisänen, J. 1986. Detection limits of external beam PIXE in the analysis of thick samples for elements with Z > 26. *X-Ray Spectrom.* 15: 159–166.

Reinhardt, T. P., Akhmadaliev, S., Bemmerer, D., Stöckel, K., and Wagner, L. 2016. Absolute hydrogen depth profiling using the resonant ^{1}H(^{15}N,$\alpha\gamma$)^{12}C nuclear reaction. arXiv.org > nucl-ex > arXiv:1605.02526v1. 13 pp.

Ricci, E. and Hahn, R. L. 1967. Sensitivities for activation analysis of fifteen light elements with 18-MeV helium-3 particles. *Anal. Chem.* 39(7): 794–797.

Rochette, P., Braucher, R., Folco, L., Horng, C. S., Aumaître, G., Bourlès, D. L., and Keddadouche, K. 2018. ^{10}Be in Australasian microtektites compared to tektites: size and geographic controls. *Geology* 46(9): 803–806.

Robinson, M. T. and Oen, O. S. 1963. The channeling of energetic atoms in crystal lattices. *Appl. Phys. Lett.* 2 (2): 30–32.

Ryan, C. G., Cousens, D. R., Sie, S. H., and Griffin, W. L. 1990. Quantitative analysis of PIXE spectra in geoscience applications. *Nucl. Instrum. Methods* B49: 271–276.

Ryssel, H. and Ruge, I. 1986. *Ion Implantation.* Wiley, Chichester.

Sasao, M., Shinto, K., Isobe, M., et al. 2006. Confined alpha particle diagnostic system using an energetic He^0 beam for ITER. *Rev. Sci. Instrum.* 77, 10F130 (3 pp).

Sattler, A. R. and Dearnaley, G. 1967. Channeling in diamond-type and zinc-blende lattices: comparative effects in channeling of protons and deuterons in Ge, GaAs, and Si. *Phys. Rev.* 161: 244–252. Erratum: Phys. Rev. 165 (1968) 750.

Sauvage, T., Erramli, H., Guilbert, S., et al. 2004. Profile measurements of helium implanted in UO$_2$ sintered pellets by using the ^{3}He(d,α)^{1}H nuclear reaction analysis technique. *J. Nucl. Mater.* 327(2–3): 159–164.

Schweitzer, J. S., Livingston, R. A., Claus Rolfs, C., et al. 2005. Nanoscale studies of cement chemistry with ^{15}N resonance reaction analysis. *Nucl. Instrum. Methods Phys. Res. Sect. B Beam Interact. Mater. Atoms.* 241(1–4): 441–445.

Scognamiglio, G., Klein, M., van de Hoef, F. L., Podaru, N. C., and Mous, D. J. W. 2021. Low-energy ^{14}C and multi-element HVE AMS systems. *Nucl. Instrum. Methods Phys. Res. B* 492: 29–33.

Sera, K. and Futatsugawa, S. 1996. Personal computer aided data handling and analysis for PIXE. *Nucl. Instrum. Methods* B109/110: 99–104.

Serefiddin, F., Herzog, G. F., and Christian Koeberl, C. 2007. Beryllium-10 concentrations of tektites from the Ivory Coast and from Central Europe: evidence for near-surface residence of precursor materials. *Geochim. Cosmochim. Acta* 71: 1574–1582.

Simon, Q., Suganuma, Y., Okada, M., Haneda, Y., ASTER Team. 2019. High-resolution ^{10}Be and paleomagnetic recording of the last polarity reversal in the Chiba composite section: age and dynamics of the Matuyama–Brunhes transition. *Earth Planet. Sci. Lett.* 519: 92–100.

Somayajulu, B. L. K. 2000. Growth rates of oceanic manganese nodules: implications to their genesis, palaeo-earth environment and resource potential. *Curr. Sci.* 78(3): 300–308.

Sonninen, E., Erametsa, P. and Jungner, H. 1989. Dating of mortar and bricks: an example from Finland: Archaeometry, Proceedings of the 25th International Symposium, Maniatis, Y. Ed. Amsterdam-Oxford-New York-Tokyo, p. 99–107.

Sossi, P. A., Moynier, F., Chaussidon, M., Villeneuve, J., Chizu Kato, C., and Gounelle, M. 2017. Early Solar System irradiation quantified by linked vanadium and beryllium isotope variations in meteorites. arXiv:1705.01808v1 [astro-ph.EP].

Steinhilber, F., Abreu, J. A. Beer, J., et al. 2012. 9,400 years of cosmic radiation and solar activity from ice cores and tree rings. *PNAS* 109(16): 5967–5971.

Synal, H. A., Stocker, M., and Suter, M. 2007. MICADAS: a new compact radiocarbon AMS system. *Nucl. Instrum. Methods Phys. Res. Methods B* 259: 7–13.

Synal, H.-A. 2013. Developments in accelerator mass spectrometry. *Int. J. Mass Spectrom.* 349–350: 192–202.

Szabo, G. and Borbely-Kiss, I., 1993. PIXYKLM computer package for PIXE analyses. *Nucl. Instrum. Methods* B75: 123–126.

Szakács, G., Szilágyi, E., Pászti, F., and Kótai, E. 2008. Determination of migration of ion-implanted helium in silica by proton backscattering spectrometry. *Nucl. Instrum. Methods in Phys. Res. Sect. B* 266(8): 1382–1385.

Szymansky, R. and Jamieson, D. N. 1997. Ion source brightness and nuclear microprobe applications. *Nucl. Instrum. Methods Phys. Res. B* 130(1–4): 80–85.

Tilley, D. R., Kelley, J. H., Godwin, J. L., Millener, D. J., Purcell, J. E., Sheu, C. G., and Weller H. R. 2004. Energy levels of light nuclei A = 8, 9, 10. *Nucl. Phys. A* 745: 155–362.

TUNL 2018. Nuclear Data Evaluation Project. Energy Levels of Light Nuclei, A = 3–20. http://www.tunl.duke.edu/nucldata/figures.shtml. Last modified: 05 March 2018.

Valković, V. 1983. Sample Preparation Techniques in Trace Element Analysis by X-Ray Emission Spectroscopy. IAEA Manual, IAEA-TECDOC-300. International Atomic Energy Agency, Vienna, Austria.

Valković, V. and Moschini. G. 1992. Application of charged particle beams in science and technology. *La Revista del Nuovo Cimento* 15(3): 1–73.

Valković, V. 1992. Human hair analysis by PIXE. *Int. J. PIXE* 2: 1–18.

Valković, O., Jakšić, M., Caridi, A., Cereda, E., Haque, A. M. I., Cherubini, R., Moschini, G., Menapace, E., Markowicz, A., Makarewicz, M., and Valković. V. 1992. X-Ray and gamma spectroscopy of coal and coal ash samples. *Nucl. Instrum. Methods Nucl. Res. B* 66: 479–484.

Valković, V., Cereda, E., Menapace, E., Moschini, G., and Valković, O. 1992. Environmental and human health impact of coal burning power stations. *Int. J. of PIXE* 2: 575–592.

Valković, V., Zeisler, R., Bernasconi, G., and Danesi, P. R. 1992. Reference materials for micro-analytical nuclear techniques. *Int. J. PIXE* 2: 651–664.

Valkovic, O., Jaksic, M., Fazinic, S., Valkovic, V., Moschini, G., and Menapace, E. 1995. Quality control of PIXE and PIGE nuclear analytical techniques in geological and environmental applications. *Nucl. Instrum. Methods Phys. Res.* B99: 372–375.

Van der Weg, V. F., Kool, W. H., Roosendaal, H. E., and Saris, F. W. 1973. Silicon surface studies by means of proton backscattering and proton induced X-ray emission. *Radiat. Eff.* 17: 245-

Van Strydonck, M., Dupas, M., and Dauchot-Dehon, M. 1983. Radiocarbon Dating of Old Mortars. *PACT J.* 8: 337–343.

Van Strydonck, M., Dupas, M., Dauchot-Dehon, M., Pachiaaudi, C., and Maréchal, J. 1986. The influence of contaminating (fossil) carbonate and the variations of kl3C in mortar dating. *Radiocarbon* 28(2A): 702–710.

Versieck, J., Van Ballenberghe, L., de Kesel, A., et al. 1988. Certification of a second-generation biological reference material (freeze-dried human serum) for trace element determinations. *Anal. Chim. Acta* 204: 63–75.

Wätjen, U., Schroyen, D., Bombelka, E., and Rietveld, P. 1990. *Nucl. Instrum. Methods Phys. Res.* B50: 172–176.

Weiss, Z. 2021. Analysis of hydrogen in inorganic materials and coatings: a critical review. *Hydrogen* 2: 225–245.

Weiser, M. and Kalbitzer, S. 1985. Observation of quantum mechanical diffusion of 1H in ta at low temperatures. *Zeitschriftfür Physikalische Chemie Neue Folge 143: 183–106.* München, Germany: R. Oldenbourg Verlag.

Welton, R. F., Stockli, M. P., Murray, S. N., and Keller, R. 2004. The status of the spallation neutron source ion source. *Rev. Sci. Instrum.* 75(5): 1793–1795.

Whitton, J. L., Mitchell, I. V., and Winterbon, K. B. 1971. *Can. J. Phys.* 49: 1225–1232.

Wielunski, L. E., Chabal, Y., Paunescu, M. et al. 2005. Ion backscattering study of ultra-thin oxides: Al_2O_3 and $AlHfO_x$ films on Si. *Nucl. Instrum. Methods Phys. Res. B* 241: 377–381.

Wilde, M., Ohno, S., Ogura, S., Fukutani, K., and Matsuzaki, H. 2016. Quantification of hydrogen concentrations in surface and interface layers and bulk materials through depth profiling with nuclear reaction analysis. *J. Visual. Exp.* 109: 53452.

Williams, E. T. 1984. PIXE analysis with external beams: systems and applications. *Nucl. Instrum. Methods B* 3(1–3): 211–219.

Wilson, R. G. and Brewer, G. R. 1973. *Ion Beams with Application to Ion Implantation.* Wiley Interscience, New York.

Wölfli, W., Polach, H. A., and Andersen, H. H. (Editors). 1984. Proceedings of the III International Symposium on Accelerator Mass Spectrometry, Zürich, Switzerland, in Nucl. Instrum. Methods Phys. Res. B5:91.

Wolicki, E. A. and Knudson, A. R. 1967. Nuclear detection method for sulfur in thin films. *Int. J. Appl. Radiat. Isot.* 18(6): 429–433.

Yasuda, K. 2020. Time-of-flight ERDA for depth profiling of light elements. *Quantum Beam Sci.* 4(4): 40 (15 pp).

Zehfuss, P. H., Bierman, P. R., Gillespie, A. R., Burke, R. M., and Caffee, M. W. 2001. Slip rates on the Fish Springs Fault, Owens Valley, California, deduced from cosmogenic 10Be and 26Al and soil development on fan surfaces.*Geol. Soc. Am. Bullet.* 113: 241–255.

Zweifel, L. P., Shorubalko, I., and Lim. R. Z. H. 2016. Helium scanning transmission ion microscopy and electrical characterization of glass nano-capillaries with reproducible tip geometries. *ACS Nano* 10(2): 1918–1925.

Additional reading

Alfassi, Z. B., Ed. 1990. *Activation Analysis, Vol I and II.* CRC Press, Boca Raton, Florida, USA.

Alfassi, Z. B. and Peisach, M. 1991. *Elemental Analysis by Particle Accelerators.* CRC Press, Boca Raton, Florida. ISBN: 978-0849360312.

Black, C., Chevallier, O. P., and Elliott, C. T. 2016. The current and potential applications of Ambient Mass Spectrometry in detecting food fraud. *TrAC Trends Anal. Chem.* 82: 268–278.

Borderie, B., Barrrandon, J. N., Delaunay, B., and Basutsu, M. 1979. Use of coulomb excitation by heavy ions (35Cl,55 MeV) for analytical purposes: possibilities and quantitative analysis. *Nucl. Instrum. Methods* 163: 441–451.

Bosch, F., El Goresy, A., Martin, B. Povh, B., Nobiling, R., Scwalm, D., and Traxel, K. 1978. The Heidelberg Proton microprobe. *Nucl. Instrum. Methods* 149: 665–668.

Calligaro, T., Coquinot, Y., Pichon, L., and Moignard, B. 2011. Advances in elemental imaging of rocks using the AGLAE external microbeam. *Nucl. Instrum. Methods Phys. Res. B* 269: 2364–2372.

Ebrahim, G. H. 2019. The application of stereo scanning transmission ion microscopy (stereo-STIM) imaging in biological specimen. *Nucl. Instrum. Methods Phys. Res. B* 450: 127–130.

Hobbs, C. P., McMillan, J. W., and Palmer, D. W. 1988. The effects of surface topography in nuclear microprobe Rutherford backscattering analysis. *Nucl. Instrum. Methods Phys. Res. B* 30: 342–348.

Johansson, S. A. E., Campbell, J. L. and Malmqvist, K. G., Eds. 1995. *Particle-Induced X-Ray Emission Spectrometry (PIXE).* John Wiley & Sons, Inc., New York, USA.

Mackova, A., Groetzschel, R., Eichhorn, F., Nekvindova, P., and Spirkova, J. 2004. Characterization of $Er:LiNbO_3$ and $APE:Er:LiNbO_3$ by RBS–channeling and XRD techniques. *Surf. Interface Anal.* 36: 949–951.

Macková, A., and Pratt, A. 2014. Ion/Neutral Probe Techniques. In Handbook of Spectroscopy: Second Edition (Eds. Gauglitz, G., Moore, D. S.). Wiley-VCH, vol. 2: 741–779.

Munnik, F., Sjoland A., and Watjen, U. 2000. Using microprobe distribution maps to determine homogeneity and correlation between elements. *Nucl. Instrum. Methods Phys. Res. B* 161–163: 348–353.

Murr, L. E. 1991. *Electron and Ion Microscopy and Microanalysis: Principles and Applications.* Marcel Dekker, New York. ISBN: 978-0824785567.

Nowicki, L. R., Ratajczak, R., Stonert, A. Turos, A., Baranowski, J. M., Banasik, R., and Pakula, K. 2000. Characterization of InGaN/GaN heterostructures by means of RBS/channeling. *Nucl. Instrum. Methods Phys. Res. B* 161-163: 539–543.

Stori, E.M., de Souza, C. T., Amaral, L., et al. 2013. Use of STIM for morphological studies of microstructured polymer foils. *Nucl. Instrum. Methods Phys. Res. B* 306: 99–103.

Tirira, J., Serruys, Y. and Trocellier, P. 2006. *Forward Recoil Spectrometry.* Plenum Press, New York. ISBN: 978-1-4613-0353-4.

Turteltaub, K. W., Feiton, J. S., Gledhill, B. L., Vogel, J. S., Southon, J. R., Caffee, M. W., Finkel, R. C., Nelserv D. E., Proctor, I. P., and Davis, J. C. 1990. Accelerator mass spectrometry in biomedical dosimetry: relationship between low-level exposure and covalent binding of heterocyclic amine-carcinogens to DNA. *Proc. Natl. Acad. Sci. USA* 87: 5288–5292.

Volfinger, M. and Robert, J. L. 1994. Particle-induced gamma-ray emission spectrometry applied to the determination of light elements in individual grains of granite minerals. *J. Radioanal. Nucl. Chem.* 185: 273–291.

Yasuda, K., Batchuluun, C., Ishigami, R., and Hibi, S. 2010. Development of a TOF-ERDA measurement system for analysis of light elements using a He beam. *Nucl. Instrum. Methods Phys. Res. Sect. B Beam Interact. Mater. Atoms* 268(11): 2023–2027.

4

Industrial Applications

4.1 Introductory Remarks

Accelerators and their products are used in almost all branches of high technology and modern medicine. Some typical applications of low-energy accelerators – most of them being cyclotrons, electrostatic generators (Van de Graaff or similar), and linear accelerators ("linacs") – are briefly described below. The term "industrial accelerators" includes all accelerators producing charged particle beams except those for medical therapy and physics research. The category does not include devices generating internal beams like cathode ray tubes, X-ray tubes, RF tubes and electron microscopes or lithography systems. The category covers approximately ½ of all accelerators being sold, with vendors list changing constantly with time.

However, although the practical uses of a new kind of accelerator are usually explored soon after its invention, its widespread adoption as an industrial tool can take decades. For example, William Shockley first proposed ion implantation of semiconductor materials, now the largest industrial application of accelerators, in the 1950s. However, it was not a widely accepted industrial technique until the 1970s (see Rubin and Poate 2003).

Such lengthy acceptance cycles are common because, in the beam business, the users are primarily concerned with cost-effectiveness. They are less interested in the details of the accelerator's technology; to them, the accelerator is a black box. Most industrial applications have evolved from science programs using research accelerators at universities or national laboratories. Although originally developed for physics research they have found a variety of uses in manufacturing and commerce.

An example to the contrary is the Rhodotron®, a high-power electron accelerator specifically developed for the sterilization of medical products in an industrial setting. These accelerators, about 3 m wide, can produce up to 700 kW of beam power at electron energies of 7 MeV meeting the rigid requirements of routine production. Before a system can be widely accepted by industry, it must first be developed into a reliable production tool and then tested in an industrial setting for a number of years.

The performance requirements demanded by the industry have led also to significant advances in accelerator technology. Industrial accelerators include both electron-beam and ion-beam systems spanning essentially all of the acceleration methods developed for research: electrostatic systems, RF linacs, betatrons, cyclotrons and synchrotrons. Industrial accelerators could be grouped into:

    i. Direct Voltage: In this group of machines voltage gradient is used to accelerate charged particles (electrons or ions). Members of this group are Van de Graaff type accelerators. They use a charge-carrying belt or "chain". Energies range

DOI: 10.1201/9781003033684-4

from 1 to 15 MeV at currents from a few nA to a few mA. Dynamitron and Cockcroft Walton generator type accelerators are basically voltage multiplier circuits at energies to up to 5 MeV and currents up to 100 mA. Inductive Core Transformer type accelerators are based on a transformer charging circuit with energies to 3 MeV at currents to 50 mA.

ii. RF Linacs: This group of accelerators uses RF-generated voltage to accelerate "bunches" of charged particles. Electron linacs are characterized by standing wave cavities from 0.8 to 9 GHz. They produce beams of energies from 1 to 16 MeV at beam power to 50 kW. Ion linacs are also members of this group; they all use RFQs at 100 to 600 MHz to produce energies from 1 to 70 MeV at beam currents up to mA.

iii. Circular: This group of accelerators is using a magnetic field to maintain a circular orbit. Cyclotrons accelerate ions to energies from 10 to 70 MeV at beam currents to several mA. Betatrons accelerate electrons to energies up to 15 MeV at a few kW beam power. Rhodotrons accelerate electrons to energies from 5 to 10 MeV at beam power up to 700 kW. Synchrotrons accelerate electrons to energies up to 3 GeV and ions to energies up to 300 MeV/amu.

Hamm and Hamm (2011, 2012) described in some detail the various types of accelerators employed in a number of industrial applications. They considered all three broad categories of accelerators: direct voltage; linear accelerators with RF voltage; and cyclic accelerators using magnetic fields, with or without RF voltage. Direct-voltage accelerators (also known as DC accelerators) make up the category most widely used in industry for both electron and ion beams. Many use a relatively low DC voltage (300 kV or less) applied across a short gap, with the external power supply connected to the accelerator through a high voltage cable. Among them are ion-implantation accelerators; small, deuteron-beam fusion-neutron generators for the oil industry and electron-beam systems for materials processing. The high-voltage power supply is an integral part of higher-energy DC systems such as Van de Graaff accelerators, which use charge-carrying belts; Cockcroft–Walton generators, which use voltage multipliers; and inductive-core transformers (ICTs) (see Hamm and Hamm 2011; Hamm 2012).

Cyclic accelerators are not yet as widely used in industry as linacs and DC machines are. Cyclotrons are the oldest and most commonly used form of cyclic accelerator. Because mass-dependent relativistic effects make it impractical for them to accelerate electrons to useful energies, they are used only to accelerate ions. A magnetic field confines the ions in an outwardly spiraling path as they are accelerated by the RF voltage. High-energy cyclotrons (with ion energies above 200 MeV) are used to treat cancer, but the lower-energy ones are used in the industry primarily for the production of radionuclides. Because that requires higher beam currents than could be produced by early research machines, it has led to significant advances in cyclotron technology. Among such advances is the acceleration of negative ions for better beam extraction and the use of large-gap, strong-focusing magnets to minimize beam losses.

The betatron is the next oldest technology in the cyclic category. It was developed to accelerate electrons to high energies using only a time-varying magnetic field. Many betatrons have been built for nondestructive testing and medical applications, but they have now been replaced for those applications by compact modern electron linac structures. Nowadays, betatrons are used only to create high-energy X-rays for industrial radiography. The Rhodotron® is a newer cyclic accelerator, designed to accelerate high

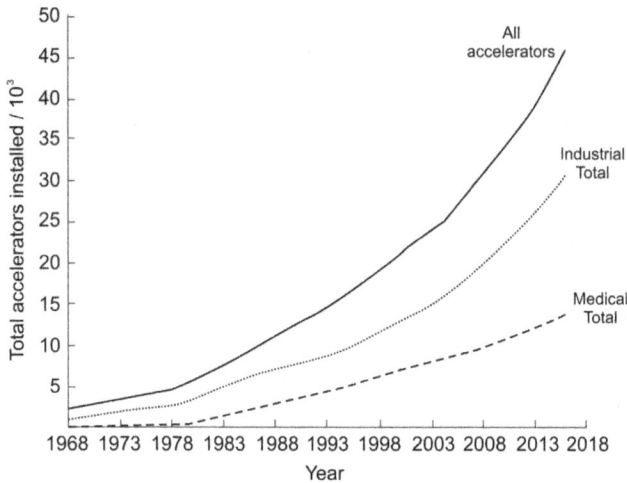

FIGURE 4.1
The number of all accelerators built and installed for various applications worldwide over 50 years (period 1968–2018).

currents of electrons using a coaxial resonator to achieve an energy boost of 1 MeV per pass. Bending magnets arrayed around the circumference of the cavity constrain the beam to make repeated passes through the accelerating cavity. The Rhodotron® was specifically developed for industrial applications (Fig. 4.1).

The use of accelerators in material modification and subsequent analysis has been a fast-growing area of accelerator-based technologies. Numerous applications of ion implantation technology have been transferred from the research laboratory to the industry (Fig. 4.1). It is an accepted view that a new industrial revolution is in progress by the establishment of advanced material processing and machining technologies that can create new materials down to an atomic and molecular level. This is being accomplished by accelerator-based technologies through the development of high-energy ion beams that are focused, clustered and wide-ranging. Modifications and analyses of different materials employing ion beams in the MeV range are being done with an increasing number of charged particle accelerators. Some of the tasks performed include ion implantation and processing; synthesis of thin films and surface modifications; fabrication of biomaterials; study of corrosion-erosion phenomena; concentration profile measurements; and diffusion phenomena studies. Some companies already have high-energy ion implantation systems running, designated to accelerate singly or doubly charged heavy ions such as boron, nitrogen, phosphorus, arsenic and antimony. By controlling the beam's energy, the systems can be used to treat materials at desired surface depths. In this way, for example, extremely high-quality multi-layers or modified surface layers having numerous functions can be formed on ordinary materials.

Another noteworthy application is ion projection lithography. The microelectronics industry requires the development of lithography capabilities below 0.3 µm for advanced silicon devices and capabilities below 0.1 µm for hetero-structure and quantum coupled devices. Ion projection lithography may meet these strict requirements, surmounting the limits of optical and X-ray lithographic methods. Industrially developed countries are aware of the potential of accelerator-based technologies. For example, Germany has 23 electrostatic accelerators, nine of them being tandems with experimental set-ups for hydrogen profile determination, RBS, ion implementation, channeling, microprobe and AMS. Many of these facilities devote more than 50% of their operating time to applied research. In addition, there are 16 cyclotrons, some designed exclusively for isotope production, at least three with PET capabilities, and 11 synchrotron and linear accelerators mainly used for heavy-ion acceleration. In Japan, medical applications of accelerators alone include 13 cyclotrons with PET capabilities, heavy-ion accelerators, and more than 500 linear accelerators used for therapeutic applications.

The role of industrial accelerators and many of their applications are described in several very nice and useful reports. First, let us mention Volume 4 of the Reviews of Accelerator Science and Technology, Accelerator Applications in Industry and the Environment (Chao and Chou 2011). Then, there are two nice articles by Robert and Marry Hamm "The Beam Business: Accelerators in Industry" in 2011 and "Industrial Accelerators and Their Applications" in 2012.

4.2 Radiation Processing by Electron Beams

Effects of electron beam (EB) interaction with materials can be divided into two groups: thermal processes and non-thermal processes. Thermal processes result in heat production and must be used in a vacuum environment. They include evaporation, melting, welding/joining, hardening, micro-structuring and similar. The non-thermal processes can be subdivided into chemical reactions and biocidal effects; both could be performed in the atmospheric environment. The chemical reaction processes include curing, cross-linking, drying print-inks and surface modification/grafting. The non-thermal processes producing biocidal effects could be applied to disinfection of animal feed, seed treatment, sterilization of products, sterile packaging and inactivation of pharmaceutical waste.

EB irradiation accelerators could be subdivided into groups according to EB energy, which in turn define the area of application:

i. 100–300 keV: Single gap, self-shielded beam system without beam scanning. Beam currents from 10 to 2000 mA, capable of treating 1–3 m wide material. Such machines are used for curing thin film coatings and cross-linking laminates and single-strand wires.

ii. 450–1000 keV: Larger dc systems with scanned beams and self-shielding. Beam currents from 25 to 250 mA, treating 0.5–2 m wide material. Mainly used for cross-linking, curing and polymerization processes in the tire, rubber and plastics industry.

iii. 1–5 MeV: Scanned beam DC systems capable of 25–200 kW beam power, scanned beam width up to ~2 m. Used for cross-linking and polymerization of thicker materials, and for sterilization of medical products.

iv. 5–10 MeV: High energy scanned beam systems capable of 25–700 kW beam power. Used for medical product sterilization and cross-linking and polymerization of thicker materials. They are also used as X-ray generators for food irradiation, wastewater remediation and gemstone colour enhancement for topaz and diamonds.

Radiation processing has been widely accepted for use in many areas of the global economy. Sterilization, polymer cross-linking (tapes, tubes, cables), tire component curing, the conservation of art objects, the irradiation of selected food items are well-established technologies. Both types of irradiators, gamma sources and electron accelerators, are used in these processes.

Among the industrial applications, radiation processing by EBs is the most widely used technology as a part of the production line. Radiation processing is characterized by the following advantages over conventional processing: temperature independence: low temperature capability; ease of controlling process; capability of sterilization.

Here, we shall present one accelerator from one company: Rhodotron® from IBA-industrial, shown in Fig. 4.2. The IBA Rhodotron® E-beam particle accelerator has gone through a major development program. Pulsing the beam of the second generation

FIGURE 4.2
Second Generation Rhodotron®, E-beam accelerator, power on demand. (Courtesy IBA.)

Rhodotron® is one of the main enablers bringing multiple additional advantages to the previous generation. Flexibility is also one of its key features since multiple applications are possible through multiple beamlines, i.e., multiple energies. More than 45 Rhodotron®s are installed and supported by IBA worldwide. The manufacturer lists the following areas of applications:

- Sterilization of medical devices and pharmaceuticals.
- Phytosanitary treatment of imported/exported produce.
- Cold pasteurization of a wide variety of foodstuff.
- Biohazard reduction of contaminated materials.
- Curing of advanced and wood-plastic composites.
- Improving polymers by crosslinking.
- Modification of melt flow index.
- Controlled scissioning of PTFE.
- Wood pulp.
- The coloration of glass and gemstones.
- Doping of semiconductors.
- Air, land and sea port cargo screening.
- Sterile Insect Technique (SIT).

The operating principle of most E-beam accelerators is that electrons gain energy when they cross a region where an electric field exists. With Rhodotron® technology, a resonating electric field is generated within a toroid-shaped accelerating cavity and electrons travel back and forth through different diameter crossings within the cavity. Hence, the Rhodotron® is a recirculating accelerator. Every pass through the central pillar accelerates electrons by increments.

The accelerator is available in five models, as shown in Table 4.1.

Rhodotron® can be configured with multiple beamlines to provide alternate beam voltages or technologies (E-beam and X-ray). This can be done using the same accelerator, the same conveyor and the same irradiation room, reducing investment costs and facility size. The 270° bending magnet allows the change of the beam direction and the possibility

TABLE 4.1

Five models of Rhodotron® product line

MODEL	TT 50	TT 100	TT 200	TT 300	TT 1000
Energy/MeV	2–10	2.5–10	2–10	2–10	2–7
Max. power at 10 MeV/kW	20	40	100	245	560
Maximum current/mA	2	4	10	35	80
Diameter/m	1.3	1.6	3	3	3 + 1
Height/m	Adjustable	1.7	2.4	2.4	2.4
MeV/Pass	0.8–1	0.833	1	1	1.166
Number of passes	10	12	10	10	6
Modular design	No	No	Yes	Yes	Yes
Beam profile	Pulsed	CW	Pulsed/CW	Pulsed/CW	Pulsed/CW

of considering various irradiation configurations. Adding an X-ray target to the Rhodotron® beamline allows treating much thicker products such as medical devices on pallets or products having high densities.

Industrial application of radiation processing is growing in many countries. Japan is probably a leading country; it has several hundred units of EB accelerators used for industrial purposes and R&D. Commercial production of cross-linked polyethylene for insulation of wire and cable was first achieved by using radiation processing in the USA in the 1950s. Since then, research and development activities have brought about successful industrial applications. These products have unique properties. In many cases, radiation processing provides distinct advantages over conventional processes in terms of product properties, economies, wide processing temperature range and environmental protection. Industrial products of polymeric materials using radiation processes include: cross-linked wire and cable (heat resistant); foamed polyethylene; heat shrinkable tubing and sheets; curing of surface coating for wood panel, paper, roof tile, steel plate, gypsum tile, overcoat of printing and floppy disks; adhesive tapes; wood-plastic composites abrasion resistant, water resistant); polymer flocculant (high molecular weight); automobile tires (cross-linking); teflon powder (by decomposition of used Teflon); contact lenses; water absorbents (for disposable diapers, etc.); deodorant *fiber*; cross-linked polyurethane (cable for antilock brake sensor); cross-linked nylon; battery separators (AAc grafted polyethylene).

R&D is being carried out aiming at the preparation of advanced materials, such as new drug delivery systems, biocompatible materials, super high-temperature resistant silicon carbide fiber, etc.

Because of the interaction between electrons and generated radicals, microorganisms are dead by a process of DNA chain cleavage. The amount of interaction between EB and object is called the absorbed dose, defined as the energy absorbed per unit mass ($[J/kg]$ = $[Gy]$). The logarithmic value of the survival fraction is proportional to the absorbed dose. The absorbed dose required to reduce the survival fraction to $1/10$ is called the D value, and a specific value is set for each microorganism. Therefore, the required absorbed dose increases in proportion to the target reduction level. Depth and does depend greatly on the acceleration energy, and hence adequate setting according to the individual object is necessary (Fig. 4.3).

EB materials processing is a mature business with large growth in developing countries. Applications of electron guns date as far as the early 1900s see Richardson (1901), who later (in 1928) received the Nobel Prize in Physics for the theoretical description and equation for thermionic emission. Understanding and applications of these processes led to the invention of the vacuum diode by J.A. Fleming (1905), and the development of the vacuum tube triode by L. De Forest (1907) using the third terminal to provide current control and amplification, leading to the subsequent electronic revolution. Nowadays, EB material processing is critical to automotive production, used in refractory metals and dissimilar metals industry, for precision cutting and drilling. For these activities EB energy from 60 to 200 keV and beam power from 6 to 200 kW are required.

Cross-linking of materials is the largest application of EB irradiators. Table 4.2 shows the cross-linking applications by type of material and its use.

Intense heat shrinks the heavy-duty tubing over a power cable. Electron accelerators are used in producing the tubing. As an example let us mention The Trans-Alaska Pipeline that runs through 800 rugged miles of mountains, forests and rivers. Wrapped around much of this sturdy pipe, a durable coating known as heat shrink has held the line against corrosion for more than 30 years. The same material protects the tangles of wires in the

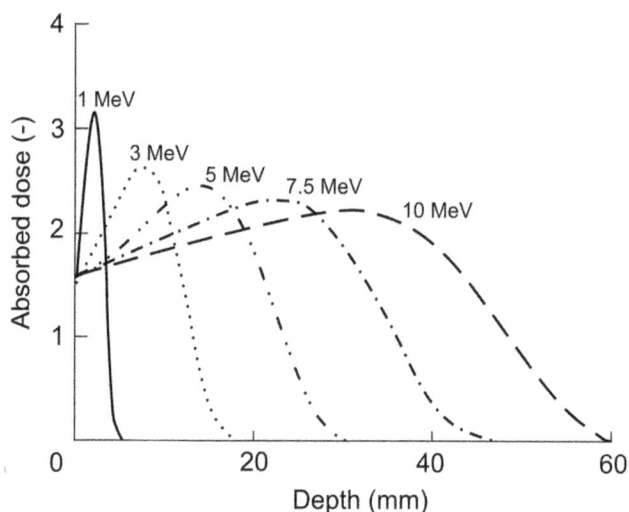

FIGURE 4.3
Relation between the depth and delivered dose.

TABLE 4.2

Cross-linking applications

No.	Material type	Material use
1	Cross-linked polyethylene, PE and PVC	Heat and chemical-resistant wire insulation
2	Cross-linked foam polyethylene	Insulation, packing and flotation material
3	Cross-linked rubber sheet	High quality automobile tires
4	Cross-linked polyurethane	Cable insulation
5	Cross-inked nylon	Heat and chemical resistant auto parts
6	Heat resistant SiC fibers	Metal and ceramic composites
7	Vulcanized rubber latex	Surgical gloves and finger cots
8	Cross-linked hydrogel	Wound dressings
9	Acrylic acid grafted PE film	Battery separators
10	Grafted polyethylene fiber	Deodorants
11	Curing of paints and inks	Surface coating and printing

electronics we use every day. Heat shrink arrives from the factory in colourful, rubbery sheets or tubes. When heated, instead of melting, the polymer-based product shrinks to form a tight seal. Specific grades of this material are designed to prevent wear, resist burning, keep water out and keep electricity in. Heat shrink owes its incredible capabilities to treatment with an EB from a particle accelerator.

Here is an example; Tyco Electronics' Menlo Park, California, plant sends millions of feet of heat-shrink tubing each week to factories all over the United States. Inside the plant, vats of white pellets melt. Colour is added if desired, and the polymer flows through an extruder into the cool water, where it solidifies into tubes. The tubing threads through an opening in a six-foot-thick wall and passes through the beam of a small electron accelerator. Peering through slanted, water-filled portholes, you see a blue shimmer over the tubing. That is the electrons ripping air molecules apart. Inside the polymer, the radiation shears off hydrogen atoms, leaving the equivalent of sticky ends that bond together and make the polymer stronger. However, the material's chemical structure has changed. Its molecules will remember their positions even after being stretched apart. To finish up, machinery heats the material, and then quickly cools it while blowing it up with air like a balloon. The tubing "freezes" at up to four times its original diameter. Tyco Electronics produces plenty of custom and standard tubes with different properties, colours, thicknesses, diameters and lengths. Workers can easily slide the expanded tubing over cables that need protection or colour-coding. When they heat the tubing again, it shrinks back to its original size. Heat shrink shields wires inside airplanes, buses, alarm clocks and computers.

The global heat-shrink tubing market is dominated by a few major players that have a wide regional presence. The leading players in the heat-shrink tubing market are TE Connectivity (Switzerland), 3M (USA), Sumitomo Electric (Japan), ABB (Switzerland), HellermannTyton (West Sussex), Alpha Wire (USA), Woer (China), Qualtek (USA), Panduit (USA), Zeus (USA), Guanghai Materials (China), Thermosleeve (USA), Insultab (USA), Dasheng Heat Shrinkable Material (China) and Changchun Heat Shrinkable Materials (China). The global heat-shrink tubing market is projected to reach 2.3×10^9 US $ by 2024 from an estimated 1.8×10^9 US $ in 2019, at a compound annual growth rate (CAGR) of 5.3% during the forecast period (Research and Markets 2019).

It should be mentioned that sterilization of single-use disposable medical products – surgical gowns, surgical gloves, syringes and sutures – is a growing application. The largest potential application is food and waste irradiation.

Fossil fuel reserves would be exhausted at the current rate of consumption in the near future. Therefore, an alternative renewable energy technique for producing alcohol by fermenting an alcohol spawn of grain has been developed in many countries. Cellulose is not easy to hydrolyze and lignin and hemi-cellulose are bad in accessibility, a manufacturing process is lengthy and not efficient. Therefore, there is an urgent need for a manufacturing process that mass production is possible. Three kinds of industrial electron accelerators (400 keV 12 mA Cockcroft Walton type, 2 MeV 50 mA ELV type, and 10 MeV 3 mA Linac Type) and their irradiation facilities including shielding blocks have been designed, fabricated and under construction by the KAPRA (Korea Accelerator and Plasma Research Association). In this paper, Lee et al. (2008) reported a cellulose modification study for producing biofuel using EB irradiation and its applications; see also (Henniges et al. 2013).

Some gemstones are exposed to radiation to enhance or change their colour. Topaz is the most commonly treated stone. Typically orange topaz becomes blue after it has been exposed. Diamonds and other precious gems may also be treated with radiation. This process of irradiation can make the gems slightly radioactive. That is why the US Nuclear Regulatory Commission, NRC, regulates the initial distribution of these gemstones. The NRC requires the stones to be set aside, typically for a couple of months, to allow any radioactivity to decay. Under the NRC license, a distributor must conduct radiological surveys before the stones can be put on the market (NRC 2021).

Idris et al. (2014) evaluated the gemstone colour enhancement by using different kinds of precious and non-precious gemstones. By using the irradiation technique, selected gemstones are exposed to the highly ionizing radiation EB to knock off electrons to generate colour centers culminating in the introduction of deeper colour. The colour centers may be stable or unstable depending on the nature of colour center produced. The colour change of irradiated stones was measured by Hunter Lab colour measurement. At 50 kGy, Topaz shows changes colour from colourless to golden. Meanwhile, pearl shows changes from a pale colour to grey. Kunzite and amethyst show colour changes from colourless to green and pale colour to purple. Measurement by gamma survey meter confirmed that irradiation treatment with 3 MeV EB machine does not render any activation that activates the gems to become radioactive.

For more information on heat treatment and irradiation of gem stones, in particular topaz, see (Manutchehr-Danai 2009). Topaz, which is mined clear, becomes sky blue when treated with high-energy EB. The colour of quartz, tourmaline, aquamarine and emerald gemstones can be similarly enhanced when treated with Gamma radiation.

4.2.1 Electron Beam Modification of Semiconductor Devices

Fig. 4.4 illustrates the radiation paths for electron, ion and photon beams in an irradiated semiconductor material. Heavy charged particles (ion beams) lose energy in small steps. Their well-defined short penetration range depends on ion energy and the target material properties. Ion beams may cause frequent atomic displacements, whereas electrons can generate only some. Gamma rays and X-rays can generate rare atom displacements through the Compton effect.

EB modification of semiconductor devices is based on the decrease of the lifetime of minority carriers after radiation treatment. To obtain suitable conditions for this process a number of defects are created in silicon crystals by high-energy electrons. EB processed fast switching power thyristors and diodes are used in a growing number of applications for which the long-term stability, the efficiency and energy savings semiconductor components are important and assured through competitive pricing. Power interruptions at data processing centers and at hospitals with life-support equipment are examples of the need for such power thyristors and diode devices, along with many other applications in metallurgy, mining, transportation, household uses and other areas (Zimek 2001).

Pavlov et al. (2016) have described the accelerator-based EB technologies for modification of bipolar semiconductor devices. Devices of different classes with a wide range of operating currents (from a few mA to tens kA) and voltages (from a few volts to 8 kV) were processed on a large scale including power diodes and thyristors, high-frequency bipolar and IGBT transistors, fast recovery diodes, pulsed switching diodes, precise temperature compensated Zener diodes (in general more than 50 device types), produced by different enterprises. The necessary changes in electrical parameters and characteristics of devices caused by the formation in the device structures of electrically active and

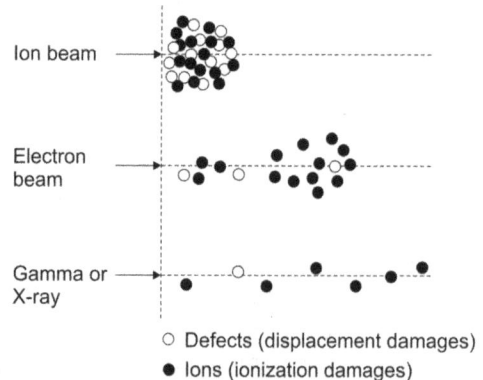

FIGURE 4.4
Radiation paths of electron and ion beams and gamma rays and X-rays in an irradiated semiconductor material.

○ Defects (displacement damages)
● Ions (ionization damages)

stable in the operating temperature range sub-nanoscale recombination centers. Technologies implemented in the air with high efficiency and controllability, and are an alternative to diffusion doping of Au or Pt, γ-ray, proton and low-Z ion irradiation.

The main features and advantages of EB technology for the modification of semiconductor devices are as follows:

- Possibility of processing in air.
- Electron flux density adjustment by beam current and distance from the output window due to scattering in air.
- Providing a large processing area due to operation in the scanning mode.
- Very ample opportunities for the design and construction of under-beam equipment, including beam control systems and a variety of holders, automated sample delivery systems, freestanding vacuum/environment chamber with a beam input window for special investigations.
- Simultaneous pass-through processing of several semiconductor wafers arranged one behind the other due to deep penetrating ability of accelerated electrons (range in Si about 1–2 cm).
- Samples heating by beam and temperature control if necessary.
- Small footprint and favorable cost of equipment, easy to use and maintain.

Liu et al. (2018) have studied the electron-beam radiation-induced degradation of silicon nitride and its impact on semiconductor failure analysis by TEM. They found that high-dose electron-beam radiation at 200 kV led to rapid degradation of silicon nitride process layers, i.e., thin downing of nanostructured silicon nitride, inter-diffusion of O and N, the formation of bubble-like defects and segregation of N at neighboring interfaces. Further detailed analysis revealed that radiation-induced modification in the microstructure and chemical composition of silicon nitride layers could be ascribed to the electron radiation induced knock-on damage and ionization damage. The radiation-enhanced diffusion (RED) accounted for the continuous thin down of the nitride process layer and the formation of bubble-like defects in thick nitride spacer process layers. The work well demonstrated the electron-beam sensitivity of nanostructured silicon nitride materials in the semiconductor devices, and thus may give useful information about electron-dose control during TEM failure analysis of the semiconductor devices containing nanostructured silicon nitride process layers (Liu et al. 2018).

The effects of electron-beam irradiation on the organic semiconductor rubrene and its application as a dosimeter were discussed by Kim et al. (2016). They found that electron-beam irradiation induces n-doping of rubrene. Authors fabricated rubrene thin-film transistors with pristine and irradiated rubrene and found that the decrease in transistor properties originated from the irradiation of rubrene and that the threshold voltages were shifted to the opposite directions as the irradiated layers. With this experience, they were able to fabricate a highly sensitive and air-stable electron dosimeter based on a rubrene transistor.

Finally, Costa et al. (2020) demonstrated that the EB irradiation of materials, typically used in characterization measurements, could be also employed for advanced fabrication, modification and functionalization of composites. They developed irradiation equipment using an EB irradiation source to be applied in materials modification. Using this equipment, the formation of a thick Ag film on the Ag_3PO_4 semiconductor was carried out by EB irradiation for the first time. This was confirmed by various experimental techniques (X-ray diffraction (XRD), field-emission scanning electron microscopy, Raman spectroscopy and X-ray photoelectron spectroscopy) and *ab initio* molecular dynamics simulations. Their calculations demonstrate that, at the earlier stages, metallic Ag growth is initiated preferentially at the (110) surface, with the reduction of surface Ag cations forming metallic Ag clusters. As the (100) and (111) surfaces have smaller numbers of exposed Ag cations, the reductions on these surfaces are slower and are accompanied by the formation of O_2 molecules.

4.2.2 Applications to Coal Burning Power Stations

4.2.2.1 Cleaning Flue Gases by Electron Beams

The use of coal for power production has a serious side effect that of atmosphere pollution by exhaust flue gases. Consequently, lots of effort has gone into finding the processes, which can improve this situation. One of the radiation processes which was successfully demonstrated in many laboratories and pilot plant facilities is the reduction of SO_2 and NO_x pollutants from flue gases emitted during fuel combustion in boilers for electrical power and heat production (Zimek 1995). The full-scale industrial implementation of an EB process for flue gases treatment would require accelerator modules with a beam power of over 500 kW and electron energy in the range 1–1.5 MeV. The 500 MW power plant may require 5–8 MW of EB power deposited in the flue gas.

The amount of emission of SO_2 is still increasing and was estimated to 80 million tons in 2020 damaging the environment. In China, for example, coal is the major primary energy source and 26 million tons of SO_2 is emitted in a year 37% of which is from power station. The EB technology to remove SO_2 and NO_x simultaneously was first developed by Japanese group (Machi et al. 1977), being followed by research groups in Poland, USA, Germany, China, Bulgaria, Brazil and Malaysia (Chmielewski et al. 2004; Kawamura et al. 1980; Park et al. 2019; Zwolińska et al. 2020; EPA 2015; Shemwell et al. 2002). It should be noted that the IAEA has also played an important role for the development of this technology (see IAEA 1987).

Distinct advantages of this EB process are: simultaneous removals of SO_2 and NO_x, dry process without waste water treatment, simple system and smaller space requirement, no expensive denitrization catalyst, valuable by-products of fertilizer, applicability for high SO_2 content flue gases (Fig. 4.5).

In Poland, the INCT and Dolna Odra Power Co. have finally achieved this industrial operation of the plant to clean 270,000 Nm^3/h from power station with the strong support of IAEA and the Japanese Government. Four accelerators of 800 keV, 300 kW of the Nisshin High Voltage are used with 2 power supplies. In Changdu, China a flue gas cleaning plant, constructed by Ebara Co. and State Power Cooperation, has successfully been operated for almost 3 years. The plant has a capacity of 300,000 Nm^3/h and is installed with 2 accelerators of 800 kV, 400 mA. The removal rate is more than 80% by irradiation of 3.2 kGy for flue gases contain SO_2 of 500 to 2,400 ppm. Another industrial plant has been completed in China for Hangzhou Power Company by Ebara Co. and is now under operation. The plant has a treatment capacity of 300,000 Nm^3/h using 2 units of EB machine of 800 kV, 400 mA to remove more than 75% of SO_2 and 40% of NO_x when the initial concentration of SO_2 and NO_x is 969 ppm and 200 ppm, respectively.

Haigfeng Wu of Tsinghua University reported at the meeting in JAERI in December 2002 (Wu 2002) that 4 power plants have submitted the plan to the Government to install EB plant for cleaning flue gases. Many other power plant managements are actively

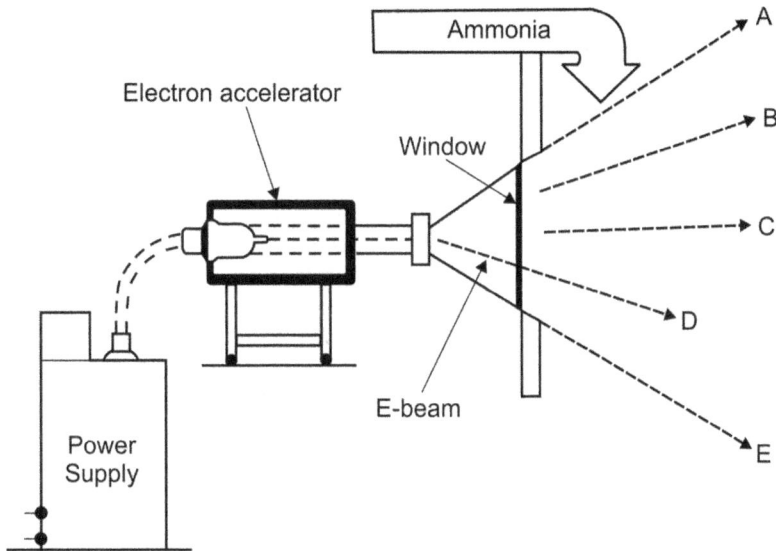

FIGURE 4.5
Reaction mechanisms involved in electron beam removal of SOx and NOx from exits gasses: A – Ammonia-Flue gas interaction, B – Production of radicals, C – Removal of SO_2 and NOx, D – Production of H_2SO_4 and HNO_3, E – Production of ammonium sulfate and ammonium nitrate.

evaluating this technology because in a few years the regulation for SO_2 emission will be stricter. The proposed plant has the capacity of 10,000 m^3/h and demonstrated to remove 92–99% of SO_2 and 30–33% of NO_x at the lower dose of 1 kGy.

The IAEA Project in Bulgaria of the pilot plant construction and operation for cleaning flue gas of coal-burning power plant containing as high as 5,000 ppm of SO_2 was implemented by the support of the JAERI. The plant capacity was 10,000 m^3/h using 2 accelerators of 800 kV donated by the JAERI. The JAERI Takasaki has achieved the efficient removal of dioxins from the flue gas of municipal waste incineration plant by using a pilot plant of 1,000 m^3/h capacity. The accelerator used is 300 kV, 40 mA and dose given to the flue gas is up to 14 kGy. The plant was installed at a waste incineration plant near Takasaki. More than 90% of dioxin (poly-chlorinated- di-benzo-paradioxins and poly-chlorirated-di-benzo- furans) was removed at 14 kGy.

Irradiation is also a quick and credible method of disinfection of municipal sewage sludge. Sludge disinfected with electron accelerator can be used as soil fertilizer immediately after treatment provided the heavy metals content in the treated sludge is under the limits acceptable for agricultural application.

Marine transport is a newly considered emission source. For example, two stroke Diesel motor (up to 81 MW) with 6–14 pistons (each 1820 dm^3) consumes 250 tons of heavy oil per day. The exhaust gases contain 13% O_2, 5.2% CO_2, 5.35% H_2O, 1500 ppm NO_x, 600 ppm SO_x, 60 ppm CO, 180 ppm HC and 129 mg/m^3 particulate matter. The first use of accelerator technology on a sea going ship for exhaust gas cleaning is described in the HERTIS (2021) – Hybrid Exhaust-gas-cleaning Retrofit Technology for International Shipping project. The HERTIS project will develop novel, hybrid technology based on the concept of combining two methods to clean up exhaust gases: irradiation by an EB accelerator and subsequent purification by improved wet-scrubbing technology. This innovative, hybrid, exhaust gas-cleaning retrofit technology will simultaneously provide a solution for the SO_x,

NO_x and PM emissions challenge in a single technological system, which will cost less than operating on low-sulfur fuel or deploying a conventional scrubber.

The applicability of the EB flue gas treatment technology for purification of marine diesel exhaust gases containing high SO_2 and NO_x concentration gases was studied by Licki et al. (2015). The measurements were performed in the laboratory plant with NO_x concentration up to 1700 ppm and SO_2 concentration up to 1000 ppm. Such high NO_x and SO_2 concentrations were observed in the exhaust gases from marine high-power diesel engines fueled with different heavy fuel oils. In the first part of study, the simulated exhaust gases were irradiated by the EB from accelerator. The simultaneous removal of SO_2 and NO_x were obtained and their removal efficiencies strongly depend on irradiation dose and inlet NO_x concentration. For NO_x concentrations above 800 ppm, low removal efficiencies were obtained even if applied high doses. In the second part of study, the irradiated gases were directed to the seawater scrubber for further purification. The scrubbing process enhances removal efficiencies of both pollutants. The SO_2 removal efficiencies above 98.5% were obtained with irradiation dose greater than 5.3 kGy. For inlet NO_x concentrations of 1700 ppm the NO_x removal efficiency about 51% was obtained with dose greater than 8.8 kGy.

4.2.2.2 PIXE and PIGE Applications to Coal Burning Power Stations

Because of coal burning in power plants large number of toxic elements are released into the atmosphere. The amount released depends on the control system, furnace design, mineral content of coal and existing emission control standards (Valković et al. 1992). The pathways of toxic elements from coal-burning power plants to man as well as dose-response effects could be considered by applications of PIXE and μ-PIXE as analytical techniques. The authors have demonstrated that, in addition to concentration levels, information on concentration profiles (surface enrichment) of elements on fly ash particles, having diameter of a few micrometers or less, is essential (Valković 1983).

X-ray and γ-ray spectroscopies cover a wide range of analytical methods, which can be successfully, used in the analysis of coal and coal ash samples. X-rays and γ-rays emitted from coal or coal ash samples are the result of spontaneous decays or processes induced by sample bombardment. In both cases measured X-ray and γ-ray spectra carry information on elemental composition of sample investigated (Bellagamba et al. 1993). Samples can be collected at any of points of interest within the power plant. The General configuration of the modern coal-fired power plant is shown in Fig. 4.6.

The use of X-ray emission spectroscopy for the study of properties of fly ash particles was done already in the early 1980s by Valković (1983). The elemental composition of bulk ash was done by exposing the samples to radiation from radioactive sources, X-ray tube and synchrotron radiation. Proton micro-beam was used to measure the concentration profiles of different elements in individual ash particles. It was shown that X-ray emission spectroscopy could be used for the study of the elemental composition of fly ash. The bulk composition of fly ash could be conveniently determined by sample excitation with X-rays and by radioactive sources. Up to 20 elements could be determined in a single measurement.

In the same work, it was shown that magnetic and non-magnetic fractions of fly ash are different in elemental composition as well as in crystalline composition. The magnetic fraction was characterized by dominance of Fe over Ca and Sr, and by the presence of elements Ti, V, Ni and some others. The crystalline components were mainly pyrite, marcasite and szmolnokite.

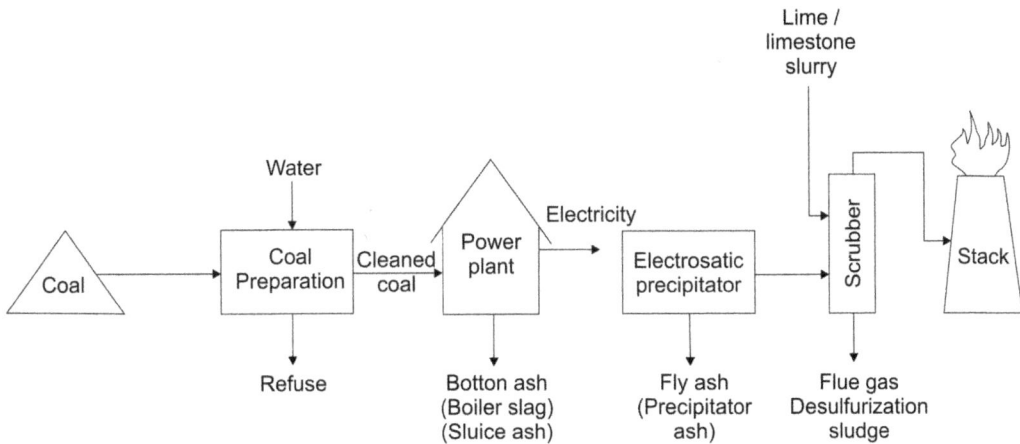

FIGURE 4.6
General configuration of modern coal-fired power plant.

Concentration profile measurements with proton microprobe showed that magnetic particles are characterized by K and Ca distributions peaking at the surface, while Fe distribution shows a maximum at the center of the particle. For a non-magnetic particle, Fe distribution follows the concentration profiles of Ca.

There are many reports on the bulk analysis of coal and coal ash samples for trace element content. In the work presented by Valković et al. (1992), the 3 MeV proton beam from the Van de Graaff accelerator at the Ruder Boskovic Institute in Zagreb, Croatia was used. Measured x-ray spectra were analyzed by using thin/thick target PIXE software. Quality control was performed by analysis of certified reference materials: NBS 1632 and BCR 40. Additional information on coal and coal ash composition could be obtained by XRF measurements, PIGE and RBS measurements as well a direct gamma spectrometry. One should keep in mind that the reference materials for coal and coal ash are homogeneous only down to 50–150 mg scale. Therefore, their use for thin PIXE targets is questionable (Figs. 4.7 and 4.8).

The spatial distribution of trace elements in individual coal fly ash particles is an important factor in assessing their environmental and human health impact. In addition, some of the surface properties of fly ash particles might have an influence on the ESP efficiency for their removal from flue gases. With the use of proton microprobe (proton beam focused to about 1 μm) one can obtain information on the distribution of elements in fly ash particles. Analytical information on single particles should be considered in the light of the absence of any relevant standard reference material. Such a reference material could be produced at least for some types of particles.

4.2.3 Food Irradiation

Food irradiation is a process in which food products are exposed to a controlled amount of radiant energy to kill harmful bacteria such as *Escherichia coli*, Listeria and Salmonella. The process can also control insects and parasites, reduce spoilage and inhibit ripening and sprouting. Since 2011, food irradiation has come into greater focus because many other pathogen intervention technologies have been unable to provide sustainable solutions to address pathogen contamination in foods, see for example (Machi 2011).

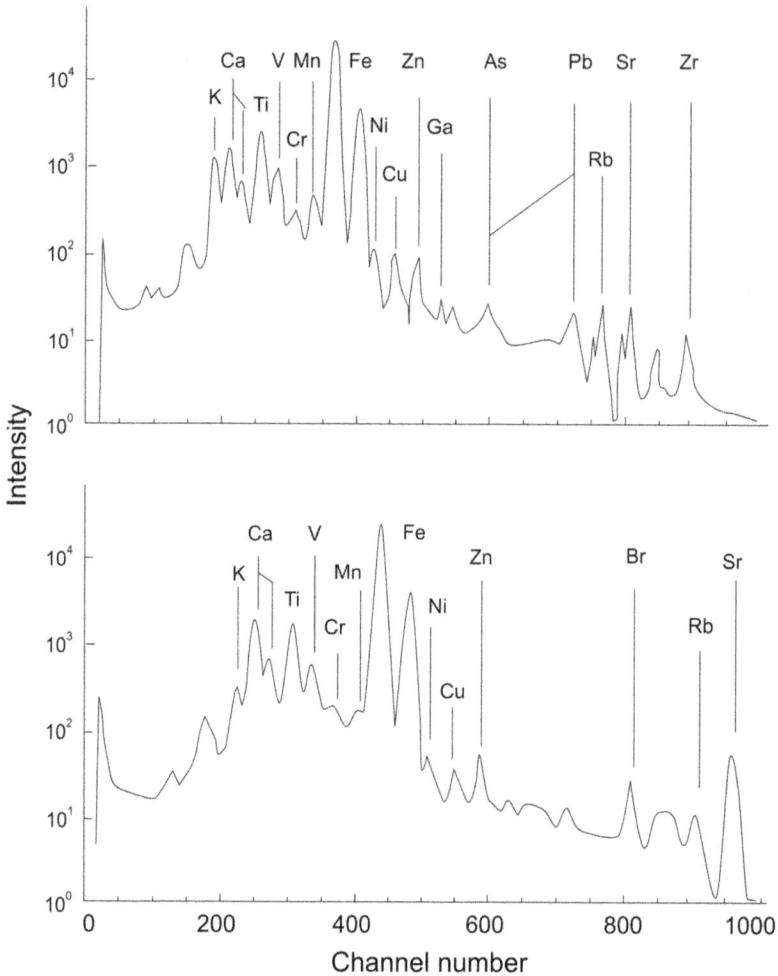

FIGURE 4.7
X-ray spectra from 3 MeV proton bombardment of top: Coal SRM BCR 40; and bottom: Coal SRM NIST 1632. Both spectra are measured by "thick" window Si(Li) detector.

There are 26 approved facilities in EU, but only 6 are equipped with e-beam accelerators. The most common is commercial irradiation of dried aromatic herbs and spices. CEN, the European Committee for Standardization, is an association that brings together the National Standardization Bodies of 34 European countries. CEN is one of three European Standardization Organizations (together with CENELEC and ETSI) that have been officially recognized by the European Union and by the European Free Trade Association (EFTA) as being responsible for developing and defining voluntary standards at the European level. For the detection of irradiated foods, the following standards apply:

- EN-1786 for detection of irradiated food containing bone by ESR spectroscopy.
- EN-1787 for detection of irradiated food containing cellulose by ESR spectroscopy.

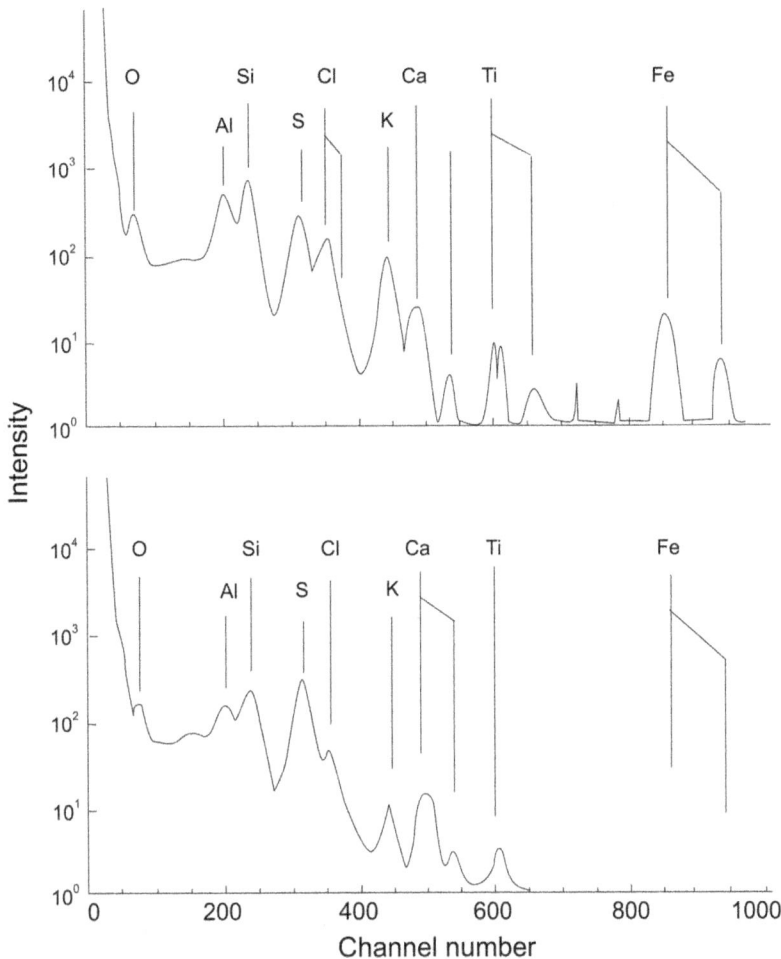

FIGURE 4.8
X-ray spectra from 2 MeV proton bombardment of top: Coal SRM BCR 40, and bottom: Coal SRM NIST 1632. Both spectra are measured by "thin" window Si(Li) detector.

- EN-13708 for detection of irradiated food containing crystalline sugar by ESR spectroscopy.
- EN-1788 for thermoluminescence detection of irradiated food from which silicate minerals can be isolated.
- EN-13751 for detection of irradiated food using photo-stimulated luminescence.

In the USA the Food and Drug Administration has approved the irradiation of meat and poultry and allows its use for a variety of other foods, including fresh fruits, vegetables and spices.

IBA's X-ray solution for food, eXelis®, is based on the industry-proven Rhodotron® E-beam accelerators and the eXelis® X-ray target system. One of the unique properties of X-rays is their narrow angular distribution concentrated in the direction of the product

TABLE 4.3

eXelis® X-ray generators

MODEL	eXelis® 50	eXelis® 100	eXelis® 200	eXelis® 300	eXelis® 1000
Max. Av. Current/mA	2	4	10	35	80
Energy options/MeV	5–7–10	5–7–10	5–7–10	5–7–10	5–7
Treatment mode	E or X	E or X	E or X	E and/or X	X

being treated. This provides excellent DUR and short treatment times, especially when compared to other radiation technologies.

The eXelis® treats food on pallets with an excellent DUR avoiding costly de-palletization and re-palletization process. By keeping your food in its initial secure packaging from entry to exit one can be sure that no additional re-contamination occurred after the treatment. The modular design of eXelis® enables to adapt the cost of a food treatment center to the capacity needs. In such a way one pays only for the power needed and one can increase the system's capacity by adding power modules when required with short downtime. See Table 4.3 for different eXelis® X-ray generators available.

4.2.4 Radiation Sterilization

The following items are usually sterilized: tissue transplants, health care products, pharmaceuticals, cosmetic products, packaging for cosmetics, etc. The main sterilization technologies are:

1. EtO
   Ethylene oxide (EtO) sterilization is a low-temperature gaseous process widely used to sterilize a variety of healthcare products, such as single-use medical devices. Using a vacuum-based process, EtO sterilization can efficiently penetrate surfaces of most medical devices and its lower temperature makes it an ideal process for a wide variety of materials.

   - Requires special permeable packaging for the gas to enter the package to be sterilized.
   - An aeration period is required to allow the gas to escape.
   - Residues left on the sterilized product are potentially carcinogenic and mutagenic.
   - EtO is explosive, toxic and harmful to the environment.

2. Gamma Radiation

   - Requires ^{60}Co radioactive source.
   - Products are usually processed in totes, carriers or pallets.
   - There are issues related with the supply, transport and disposal of radioisotope.
   - High product penetration

3. Electron Beam Irradiation

- The electrical power required.
- Cheapest sterilization technology.
- Usually, high-energy EB sterilizes products packaged in boxes.
- Low product penetration.

4. X-ray Irradiation

- The electrical power required.
- Offers much more penetration than EB and slightly better penetration than gamma irradiation.
- Allows products to be treated directly on pallets with excellent dose uniformity.

Historically, steam autoclaves or dry ovens were used to sterilize reusable medical items. However, they can easily damage many types of plastics used in single-use medical items, such as syringes, surgical gloves, artificial joints, etc. As a result, 40–50% of all disposable medical products manufactured in North America are currently radiation sterilized, primarily at ^{60}Co irradiation facility.

Recently there has been an initiative to replace ^{60}Co sources with high-power electron accelerators, mainly because of security concerns, since these sources can potentially be turned into a dirty bomb. EBs can be used directly or as X-ray sources, depending on the size-mass-density of the product. Electron linacs for medical sterilization were first implemented in the 1990s. Niowave Inc. (https://www.niowaveinc.com) has done some feasibility studies and showed that higher power (7–8 MeV and tens of KW) even greater, penetration machines have similar, or than ^{60}Co, and comparable cost. Niowave Inc. also irradiated many products and they were confirmed to be sterile.

A special case of sterilization is blood irradiation. Irradiating donated blood components before transfusion is the only accepted method to prevent transfusion-associated graft-versus-host-disease (TA-GVHD), an extremely dangerous blood complication with fatality rates reaching 90%. The difference in susceptibility to radiation between red cells and lymphocytes is the key for using X-ray and gamma ray treatments for blood products. A typical radiation dose required to deactivate harmful donor T-lymphocytes while leaving the other blood components is 15–50 Gy. High activity ^{137}Cs sources are the primary method for blood irradiation in approximately 80% of American hospitals that irradiate blood.

Cesium sources are typically used in the form of cesium chloride, which is extremely dangerous if used as a radiological weapon. It is easily dispersible, water soluble, highly reactive, and remains detrimental to humans and the environment for over 100 years. It is not surprising that ^{137}Cs sources are being phased out. Alternative methods of blood irradiation are urgently needed. Blood irradiation using x-ray tubes and up to 3 MeV electron linacs are currently being developed.

IBA's X-ray solution for sterilization centers, eXelis®, is based on the industry-proven Rhodotron® E-beam accelerators and an X-ray target. The eXelis® makes X-ray sterilization competitive against medium and large-size equivalent gamma-ray facilities. Today there are X-ray sterilization centers in Europe, Japan and North America and the first X-ray facility dedicated to sterilization is operational in Europe. One of the unique

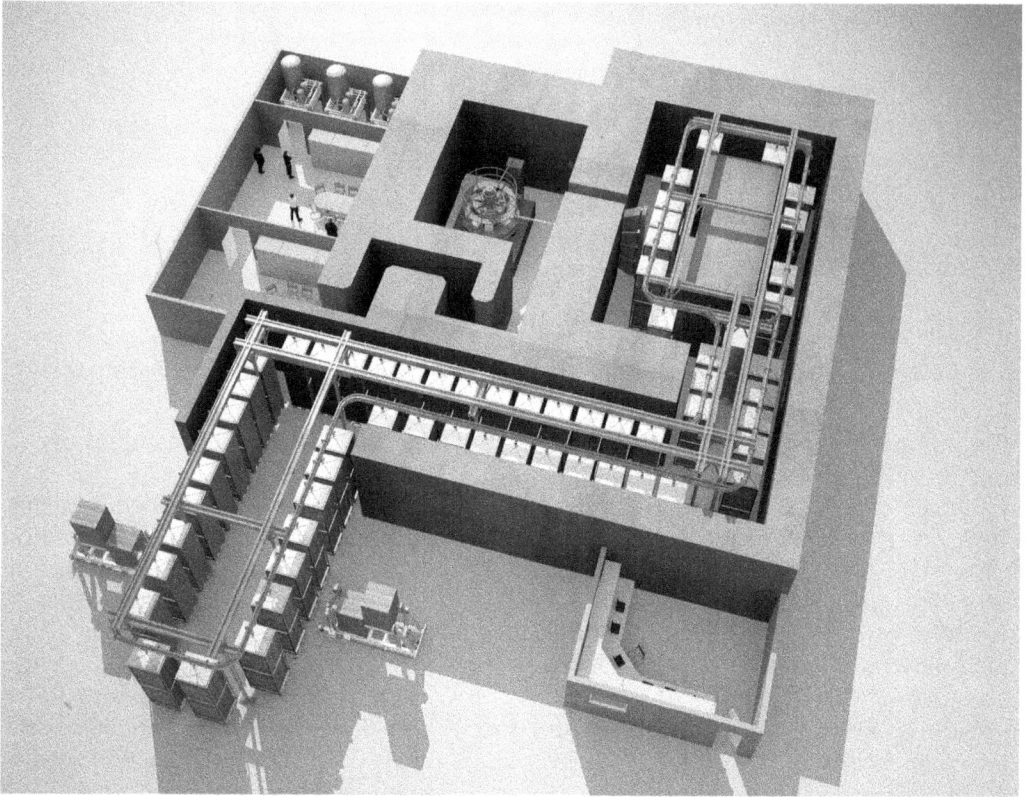

FIGURE 4.9
Full pellet sterilization in a typical eXelis® center. #1. Technical room, #2. Control room, #3. X-ray generator – eXelis®, #4. X-ray target, #5. Overhead conveyor, #6. Top/bottom exchange station, #7. Pallet loading–unloading area. (Courtesy IBA.)

properties of X-rays is their narrow angular distribution concentrated in the direction of the product being sterilized and thus allowing excellent dose uniformity ratios (DUR) and short treatment times compared with other radiation technologies.

X-ray sterilization systems irradiate full pallet loads by moving them continuously in front of the X-ray source. The loads are irradiated on opposite sides as they pass in front of the X-ray target at both high and low elevations. A typical eXelis® center for full pallet sterilization is shown in Fig. 4.9.

More than 50% of all single-use medical devices manufactured in the world today are sterilized with ionizing energy from photon and electron sources. A virtual animation of the X-ray center shown in Fig. 4.9 is available at www.iba-industrial.com/animis/sterilization/, curtesy of IBA.

Today a wide range of medical devices are sterilized by electron beam (E-beam) processing using IBA's Rhodotron® accelerators. Sterilization by E-beam is a mature technology compatible with many medical devices. The success it has encountered over the past years is mainly due to the very competitive sterilization cost per volume. So far, more than 35 Rhodotrons® have been installed worldwide.

E-beam processing is compatible with a majority of disposable medical products. This includes syringes, sutures, needles, catheters, gloves, dressings, dialysis units, infusion sets, prosthesis, cannulas, surgical blades, bags, containers, etc. The penetrating properties of

E-beam allow sterilizing medical devices through standard carton packaging. Thanks to the short treatment time, the potential impact of radiation on medical devices is very limited.

E-beam is the most cost-effective sterilization process irradiation. Usually, packaged medical devices have an average density of 0.15 g/cm^3. With an energy of 10 MeV, thicknesses of about 20–25 cm can be penetrated in a single pass and 60 cm in a double-side irradiation. In contrast to gamma, most of the power in the EB is directed towards the product. Accordingly, E-beam dose rates are much higher than gamma (curtesy of IBA).

4.3 Electron Beam Welding

Electron-beam welding is a fusion welding process in which a beam of high-velocity electrons is applied to two materials to be joined. The work pieces melt and flow together as the kinetic energy of the electrons is transformed into heat upon impact.

EB welding is a fusion welding process whereby electrons are generated by an electron gun and accelerated to high speeds using electrical fields. This high-speed stream of electrons is tightly focused using magnetic fields and applied to the materials to be joined. The beam of electrons creates kinetic heat as it influences with the work pieces, causing them to melt and bond together.

For precision welding requirements, the choice is usually between EB welding and laser beam welding. Sometimes other types of fusion welding, such as GMAW or GTAW, might be an option, but arc welding processes don't have the penetration, small heat-affected area, pinpoint precision, and weld purity of EB and laser welding. EBs and lasers can be focused and aimed with the exceptional accuracy required to weld the smallest of implantable medical devices, and yet also deliver the tremendous amounts of power required to weld large spacecraft parts. EB and laser welding are versatile, powerful, automatable processes. Both can create beautiful welds from a metallurgic and an esthetic perspective. Both can be cost-effective.

EB welding was developed in the late 1950s. It was quickly embraced by high-tech industries, such as aerospace, for the precision and strength of its resultant welds. An EB can be very accurately placed, and the weld can retain up to 97% of the original strength of the material. It is not an exaggeration to state that EB welding, in terms of the quality of the weld, is unbeatable: It is the top dog of welding processes.

EB welding is simple to explain. A tungsten filament is heated and power is applied to the point that the filament gives off electrons. These free electrons are accelerated and focused using electrical fields and magnetic "lenses". This stream of fast-moving electrons has tremendous kinetic energy. When these electrons strike a metal part, the kinetic energy is transferred to the molecular lattice of the material, heating it almost instantaneously. The power delivered by an EB can be massive up to 10,000 kW/mm^3. In fact, an EB welding system can throw enough power to simply vaporize metal (a process called EB machining). EB welding machines generally come in two power classifications, low voltage (60 kV) or high voltage (150 kV). A typical high-voltage machine rated to 7,500 watts can produce a weld in steel 2 in. deep with a width of approximately 10% of the penetration depth.

The logistics of operating an EB welding system are not simple, however. The process has to happen in a vacuum; otherwise, air/gas particles scatter and diffuse the electrons. A vacuum requires a vacuum chamber, so the size of a part to be welded is limited by the size of the chamber. Vacuum chambers can be small or large, but the larger the chamber, the longer it will take to establish the proper vacuum level, which is at a minimum 1.0×10^{-3} torr. The use of a vacuum, as well as the presence of X-radiation (a byproduct of the beam), precludes human handling, so the entire process has to be externally controlled, generally using CNC tables.

EB welding has been fully automated for decades. The collaboration of all these technologies (high voltage, vacuum and high-tech automation) means that EB welding requires well-trained operators and very competent maintenance and that the setup and running of an EB welding system can be expensive. EB welding is a fusion welding process and thus requires a precise fit between the parts being welded, as a filler material is generally not used or required. The parts must also be securely fixed to a motion-controlled table to precisely move the areas to be welded into contact with the EB. Most EB welding machines utilize a fixed beam with the part being manipulated under it via CNC.

Secure fixing also minimizes the effects of shrinkage and warping during welding. The EB has to be carefully calibrated and focused and timed with the CNC motion to deliver a consistent weld with uniform penetration and minimal porosity. Each welding cycle involves loading the welding chamber, pumping down the vacuum, welding the part, and then venting the vacuum.

The chokepoints are the pumping down of the vacuum chamber and the loading/unloading of the parts. Hence, it is imperative that the engineers and technicians involved maximize the number of parts to be welded each cycle and optimize the movement of the CNC table. When this is all done correctly, EB welding can achieve very high quality and high cost-effectiveness.

EB welding systems can weld all weldable metals and some metals that are not typically welded. EB welds are incredibly strong and pure. Impurities in the weld are vaporized, and welding in a vacuum means there are no gases or air to react and cause oxides. EB welding can also join dissimilar materials that would otherwise be unweldable due to differences in melting points, which result in intermetallic compounds that cause brittleness. The precise nature of the EB and tight heat-affected area allow EB welding to basically melt the lower-temperature material onto the unmelted, higher-temperature material, resulting in a compact, vacuum-tight weld. It can be a bit cumbersome to deal with, but the products of EB welding are first-class in all respects.

EB Welding is a highly efficient and precise welding method increasingly used within the manufacturing chain and of growing importance in different industrial environments such as the aeronautical and aerospace sectors. This is because, compared to other welding processes, EBW induces lower distortions and residual stresses due to the lower and more focused heat input along the welding line. The paper by Chiumenti et al. (2016) describes the formulation they adopted for the numerical simulation of the EBW process as well as the experimental work carried out to calibrate and validate it. The numerical simulation of EBW involves the interaction of thermal, mechanical and metallurgical phenomena. For this reason, the numerical framework couples the heat transfer process to the stress analysis to maximize accuracy. An in-house multi-physics FE software was used to deal with the numerical simulation. The definition of an ad hoc moving heat source was proposed to simulate the EB power surface distribution and the corresponding absorption within the work-piece thickness. Both heat conduction and heat radiation models were considered to dissipate the heat through the boundaries of the component. Titanium–alloy Ti_6Al_4V is the target material of this work. From the experimental side, the EB welding machine, the vacuum chamber characteristics and the corresponding operative setting are detailed. Finally, the available facilities to record the temperature evolution at different thermo-couple locations as well as to measure both distortions and residual stresses are described (Chiumenti et al. 2016).

4.4 Other Applications

The irradiation of pulp is of interest from different perspectives. Mainly it is required when a modification of cellulose is needed. Irradiation could bring many advantages, such as chemical savings and, therefore, cost savings and a reduction in environmental pollutants. In the work by Henniges et al. (2013), pulp and dissociated celluloses were analyzed before and after irradiation by EB. The focus of the analysis was the oxidation of hydroxyl groups to carbonyl and carboxyl groups in pulp and the degradation of cellulose causing a decrease in molar mass. For that purpose, the samples were labeled with a selective fluorescence marker and analyzed by gel permeation chromatography (GPC) coupled with multi-angle laser light scattering (MALLS), refractive index (RI), and fluorescence detectors. Degradation of the analyzed substrates was the predominant result of the irradiation; however, in the microcrystalline samples, oxidized cellulose functionalities were introduced along the cellulose chain, making this substrate suitable for further chemical modification (Henniges et al. 2013).

In the work by Jusri et al. (2018), the pretreatment process of lignocellulosic biomass (LCB) to produce biofuel has been conducted by using various methods including physical, chemical, physicochemical as well as biological. The conversion of the bioethanol process typically involves several steps which consist of pretreatment, hydrolysis, fermentation and separation. In their project, microcrystalline cellulose (MCC) was used in replacement of LCB since cellulose has the highest content of LCB for the purpose of investigating the effectiveness of new pretreatment method using radiation technology. Irradiation with different doses (100–1000 kGy) was conducted by using EB accelerator equipment at Agensi Nuklear Malaysia. Fourier Transform Infrared Spectroscopy (FTIR) and XRD analyses were studied to further understand the effect of the suggested pretreatment step to the content of MCC. Through this method namely IRR-LCB, an ideal and optimal condition for pretreatment prior to the production of biofuel by using LCB may be introduced (Jusri et al. 2018).

The International Irradiation Association, IIA, is a not-for-profit organization that aims to support the global irradiation industry and scientific community. A core objective of the Association is to support members in advancing the safe and beneficial use of irradiation. Communication, Education and Representation are key activities undertaken by the Association.

Membership of the Association is diverse and includes corporations, research institutes, universities and government-affiliated departments around the world. Increasingly the Association is forging relationships with regional, national or technology-specific organizations whose aims are aligned with those of IIA.

3D Printing has had an undeniable impact on the manufacturing world. Parts made faster, with less material waste, reduced machining time, and shorter time-to-market are just some of the benefits attainable with 3D printing. For example, Sciaky's EB Additive Manufacturing (EBAM) is a one-of-a-kind metal 3D printing technology that delivers on the key benefits mentioned above and excels at producing large-scale, high-value metal parts. It's no secret that large-scale forgings and castings can take many, many months to complete. EBAM, on the other hand, can produce high quality, large-scale metal structures, up to 19' in length, made of titanium, tantalum and nickel-based alloys in a matter of days, with very little material waste.

Starting with a 3D model from a CAD program, Sciaky's EB gun deposits metal (via wire feedstock), layer by layer, until the part reaches near-net shape and is ready for finish machining (see Sciaky 2021). Sciaky's IRISS® (Interlayer Real-time Imaging & Sensing System) is a patented Closed-Loop Control that provides consistent part geometry, mechanical properties, microstructure, metal chemistry and more from the first part to the last. Gross deposition rates range from 3.18 to 11.34 kg of metal per hour, depending upon the selected material and part features.

4.5 Ion Implantation

A widespread commercial application of particle accelerators is for ion implantation. Accelerator beams are used for ion implantation into metals, alloying a thin surface layer with foreign atoms to concentrations impossible to achieve by thermal processes, making for dramatic improvements in hardness and in resistance to wear and corrosion. Traditional hardening processes require high temperatures causing deformation; ion implantation on the other hand is a "cold process", treating the finished product. The ion-implanted layer is integrated in the substrate, avoiding the risk of cracking and delamination from normal coating processes.

Ion beam processing has for several years been well established in the semiconductor industry. In recent years ion implantation of tool steels ceramics and even plastics has gained increasing industrial awareness. The development of ion implantation to a commercially viable surface treatment of tools and spare parts working in production type environments is very dependent on technical merits, economic considerations, competing processes and highly individual barriers to acceptance for each particular application. Some examples of this will be discussed. The development of the process is very closely linked with the development of high current accelerators and their ability to efficiently manipulate the samples being treated, or to make sample manipulation superfluous by using special beam systems like the PSII. Furthermore, the ability to produce high beam currents (mA) of a wide variety of ions is crucial. Previously, it was broadly accepted that ion implantation of tools on a commercial basis generally had to be limited to nitrogen implantation. The development of implanters that can produce high beam currents of B, C, Ti, Cr and other elements ions is rapidly changing this situation, and today an increasing number of commercial implantation are performed with these ions although nitrogen is still successfully used in the majority of commercial implantations. Overall, the recent development of equipment makes it possible to a higher extent than before to tailor the implantation to a specific situation Furthermore, very interesting results have been obtained recently by implanting nitrogen at elevated temperatures, which yields a relatively deep penetration of the implanted ions. Direct nitrogen implantation and implantation combined with nitriding of aluminum have both shown interesting tribological potential (Straede 1992).

Ion implantation is used in the following areas of activity: semiconductors – CMOS fabrication, SIMOX, cleaving silicon, MEMS, metals, harden cutting tools, artificial human joints, ceramics and glasses, harden surfaces, modify optics, etc.

Ion implantation is performed with different ion energies and different doses depending on the task to be performed. Doses delivered are functions of beam intensities and exposure times; therefore accelerators are classified as follows:

- Low energy/high current: This group is composed of high current implanters, with ion energies of a few hundred eV to tens of keV. They are variable energy accelerators, single gap with currents to 50 mA.

- Medium energy/medium current: This is a group of original ion implanters, with variable energies of 50 to 300 keV range. Ion currents are in the 0.01–2 mA range, usually, multi-gap direct voltage machines using voltage-multiplier HV power supply.

- High energy/low current: This group has variable energy from 1 to 10 MeV and beam currents to hundreds of microampers. They can be either linacs or tandem charge-exchange columns, with both types using high-charge-states for upper energy range.

All of these accelerators are very dedicated machines. Vendors of this equipment are Varian Semiconductor bEquipment (USA), Axcelis Technology (USA) & SEN Corp., a joint venture in Japan with Sumitomo, Nissin Ion Equipment Company (Japan), Ulvac Technologies & IHI Corp (Japan), China Electronics Technology Group (China), Ibis technology (USA), Advanced Ion Beam Technology (USA) HVEE B.V. (The Netherlands), National Electrostatic Corporation (USA), Danfysik (Denmark).

Commercial ion implantation systems have been capable of delivering dopant ions at 200 eV since the mid-1990. However, the shift to 100 eV, approximately at the boundary between implantation and deposition (and overlapping on many etching processes), is a new regime for machine design and operation. The implementation of the precise control of dose and penetration depth, hallmarks of ion implantation for many years, to penetration depths of a few nm will encourage the development of new machine technologies and processes.

Ion implantation is the name given to the technology process, which uses ion accelerators to direct beams of ions into materials. These ions are used to modify the target material, to create radiation damage and to sputter away surface atoms. Ion implantation is the primary technology used in the semi-conductor industry to introduce impurities into semi-conductors to form devices and very large-scale interpreted (VLSI) circuits. All VLSI manufacturing includes ion implantation steps. In recent years, many new applications for MeV ion implantation have been developed. Some of these applications include the implantation of ions into artificial hip joints and other biomedical parts for an increased lifetime, implantation of metal gears, bearings, shafts, etc., to improve wear properties, superconductor research and integrated circuit (IC) fabrication.

Ion implantation is a material modification process by which ions are accelerated, typically in a single-ended accelerator, and inserted into a solid. This process is used to change the physical, chemical, or electrical properties of the solid. NEC has manufactured Open Air ion implantation systems for beam energies in the 200–400 keV range. Single-ended or tandem accelerators can be used for beam energies of over 400 keV. NEC offers both high temperature (800 °C) and low-temperature implant chambers (Fig. 4.10).

Possible applications include semiconductor doping, silicon substrates, surface finishing, damage study, Sputtering, gemstone modification, and many more (Figs. 4.11 and 4.12).

FIGURE 4.10
Xiamen 400 kV deck: This is a picture of a 400 kV Ion Implanter system that includes a high-temperature implant chamber. Installed in China in 2014. (Courtesy of National Electrostatics Corp.)

FIGURE 4.11
Air insulated, 500 kV Ion implantation. (Courtesy IBA.)

FIGURE 4.12
Tandetron, 2.5 MeV, used for ion implantation. (Courtesy IBA.)

In the early 1960s, Moore (1964, 1965) proposed that:

    i. The number of transistors per IC would increase steadily (at a rate of doubling every 1–2 years).

   ii. The cost per function (or "component) would decrease with improvements in production yields as a capability for producing increasing complex circuits improves.

Moore did not indicate how this would be accomplished; only that he expected that it would, in the face of certain obstacles such as the need for higher resolution patterning and concern for chip heating with more complex circuits. Almost a decade later, the mainstream transistor design based on bipolar junctions was beginning to be challenged by metal-oxide-semiconductor (MOS) transistors.

In 1974, Robert Dennard et al. at IBM published a model for how one could systematically shrink the dimensions of MOS transistors while also increasing the switching speed while reducing the switching power levels (Dennard et al. 1974). The key features of "Dennard scaling" were the proportional shrinking of both lateral and vertical transistor dimensions along with decreases in the drive voltages, maintaining approximately constant local electric fields in the MOS transistor leading eventually to what was referred to as "well-temper of Complementary Metal Oxide Semiconductor (CMOS)".

It was a paper by Dennard et al. (1974) that caught the attention of the industry with a resulting profound effect on microelectronics. They noted that as the horizontal dimensions of a transistor were scaled by a factor, speed improved by that same factor. At a time when IBM's MOS memories used a minimum dimension of 5 microns, they projected shrinking to fractions of a micron. (A human hair is 50–100 microns in diameter) This was the first attempt to relate a geometry shrink to the resulting power reduction and performance improvement.

Ahmad and Akram (2017) present a useful introduction in this broad field. They describe the concept of low-energy/high-energy ion implantation and its key applications in materials science. They divide the ion energies into three broad categories:

i. Low-energy ion implantation: range ~1–200 keV. Advanced applications of low-energy ion implanters include modification of the physical, chemical, or electrical or magnetic properties of thin films and nanostructure materials through doping of atoms as well as defect production. Low-energy ion implanter can be used to weld nanowires, nanotubes, or integrate nanowires to make nano-devices, which are a unique application of ion implanter in nanotechnology.

ii. Medium-energy ion implantation: range ~300 keV to 50 MeV. Medium-energy ions are usually generated and accelerated by accelerators such as Van de Graaff or pelletron accelerators. In these accelerators, one or two types of different ion sources are usually attached, which can generate almost all types of ions from hydrogen to uranium. Medium-energy ions are effectively utilized for deep ion implantation purpose. It is possible to implant required ion species into the required depth of samples precisely. High-energy ions have greater penetrating capabilities in materials while maintaining a straight path.

iii. High-energy ion irradiation: range ~50 MeV to hundreds of MeV. High-energy ions, protons and heavy ions, are usually generated and accelerated from high-energy ion beam accelerators such as a cyclotron or high potential terminal voltage tandem electrostatic accelerators.

Solid-state image sensor technologies have advanced drastically over the last 4 decades and have had success in the market. The sales amount of image sensors achieved 4.2×10^9 pieces in 2015 mainly because of the exponential growth of the mobile phone market. Image sensor applications are spreading everywhere and besides mobile phones. During the image sensor evolution, various device technologies and process technologies have been developed. Among them, ion implantation technology is one of the most important process technologies for image sensors. From the opposite viewpoint, image sensors are a very important application for ion implantation technology development. First, many ion implantation steps are applied to fabricate specific structures, such as PPD (pinned photodiode), special isolation structure, and to tune transistors at pixels. Secondly, to obtain deep PD (photodiode), high-energy implantations with a precise angle control are required, together with high aspect ratio resist patterns. In addition, a precise impurity profile formation is required to achieve a good signal electron transfer (Teranishi et al. 2018).

Ion implantation, in its many forms, continues to play an enabling role by providing a direct, controllable and cost-effective method for delivery of dopants in specific device locations and in appropriate concentrations. In addition, the use of non-dopant ions, such as C^+, N^+, F^+, H^+, and others, now plays a major role for "materials modification" for materials and structures in the transistor and mid-level interconnect layers (Tsukamoto et al. 2010).

Lian et al. (2005) have performed ion implantations in bulk samples of single-crystal lanthanide titanate pyrochlores $A_2Ti_2O_7$ (A = Sm, Eu, Gd, Dy and Er) using 1.0 MeV Kr^+ at room temperature at an ion fluence of 5×10^{14} ions/cm/. $Er_2Ti_2O_7$ was also implanted by 1 MeV Kr^+ at an ion fluence of 1.74×10^{14} ions/cm². Ion implantation-induced microstructural evolution has been examined using cross-sectional TEM. For $Er_2Ti_2O_7$ irradiated by 1 MeV Kr^+ at a dose of 1.4×10^{14} ions/cm^2, a highly damaged layer consisting of nano-sized disordered fluorite domains was observed. On other hand, a complete amorphous layer starting from surface was created by 1 MeV Kr^+ implantation at a fluence of 5×10^{14} ions/cm^2. The critical amorphization doses of different titanate pyrochlore single crystals implanted with 1 MeV Kr^+ were determined by comparing the experimental damage profiles with those simulated using SRIM-2000. The critical amorphization doses generally increase as the A-site cation changes from Sm^{3+} to Er^{3+}, suggesting that the radiation resistance of titanate pyrochlores increases with the decreasing ionic radius of A-site cations.

In the experiments performed by Jiang et al. (2005) cadmium niobate pyrochlore ($Cd_2Nb_2O_7$) single crystals have been irradiated at 150, 300, 450 and 600 K using 1.0 MeV Au^{2+} ions over flounces ranging from 0.01 to 3.5 ions/nm^2. The relative disorder on the Cd sublattice in the as-irradiated $Cd_2Nb_2O_7$ has been analyzed based on in situ 3.0 MeV He+ RBS along the (100)-axial channeling direction. The results show that the crystal can be readily amorphized under the Au^{2+} irradiation at or below 450 K; however, the relative Cd disorder tends to saturate at 600 K, and full amorphization does not occur at doses up to 5 dpa. Isochronal annealing (20 min) also has been performed at temperatures from 180 to 295 K for samples irradiated at 150 K. Thermal recovery of the disorder has been observed below room temperature.

Looking ahead to the near (5 years) future, one major challenge for the use of ion beam processing of nm-scale structures will center on the delivery of energetic species, dopants and other atoms, at energies of ~100 eV. Commercial ion implantation systems have been capable of delivering dopant ions at 200 eV since the mid-1990s. However, the shift to 100 eV, approximately at the boundary between implantation and deposition (and overlapping on many etching processes), is a new regime for machine design and operation. The implementation of the precise control of dose and penetration depth, hallmarks of ion implantation for many years, to penetration depths of a few nm will encourage the development of new machine technologies and processes.

Delivering 100 eV ions to surfaces with tightly controlled incidence angle, energy and dose is challenged by the difficulties in maintaining space charge balance and controlling beam divergence for these slowly moving ions (Current 2017). One approach is to replace low-energy ions with energetic neutrals, an approach already used in commercial etch tools (Kuo et al. 2018). The controlled flux of energetic neutral atoms can be provided by combining a pulsed plasma volume and a separately pulsed extraction plate, which is perforated for many high-aspect-ratio channels above a grounded wafer. The energetic ions, which are extracted by the grid potential from the plasma boundary, are neutralized by pick up of electrons during grazing angle collisions with walls of the grid channels. Similar to plasma-immersion ion implanters (PHI) used for low-energy, high-dose doping, by controlling the pulsed grid voltage, one can deliver energetic neutrals to a wafer surface for implantation (at energies somewhat above ~100 eV), deposition (at energies at and below ~100 eV) and etching/oxidation (with the use of reactive species). Additional reactivity could be supplied by optical excitation of the neutral atom flow with scanned laser beams at selected photon energies.

Early experiments with single-shot ion implanters demonstrated the improvements in control of device properties, such as threshold voltage, with the use of carefully placed arrays of dopant atoms over a random distribution of dopant locations from a typical ion beam (Shinada et al. 2005). Recently, 3D arrays of single dopant atoms, constructed by manually opening of bonding sites on H-passivated Si surfaces with an STM tip and controlled epitaxial Si growth, have demonstrated MOS transistors with single-atom channel doping, low-resistivity monolayer lines and many properties required for construction and operation of quantum-entangled computers (Fuechsle et al. 2012). Bringing these advances to bear on larger-scale quantum computers and other entangled devices requires a high throughput single-atom delivery system with atomic resolution on the atom placement and dose. Multiple approaches are under development for sustained targeting of single and controlled small numbers of atoms into near-atomic resolution targets, referred to as "deterministic doping" (Jamison et al. 2017). One approach is to count the passage of single ions "on the fly" and use fast ion deflection and precision target stage motion to direct single and countably small numbers of ions to specific target locations. High throughput, commercial single-ion delivery systems will certainly follow in the coming decade as quantum entangled devices enter wide use for computation, encryption, data storage and other applications.

Ion implantation, first used for commercial IC fabrication in the early 1970s has evolved into the dominant means of doping semiconductor devices for digital electronics, power switches, optical sensors, large-area displays and is just now entering the world of high-throughput doping of high-efficiency PV cells. In the IC device area, doping applications for CMOS transistors have now been joined by an even larger number of non-dopant, materials modification implant steps, accounting for a total of 50–80 passes through some form of implant tool for each completed advanced IC device. Requirements for ion and neutral beam processing of nm-scale materials and devices are driving the development of new accelerator technologies for new applications using quantum-determined properties.

Woo et al. (2005) have discussed thick Si-on-insulator wafers formation by ion-cut process. The silicon-on-insulator (SOI) wafer fabrication technique has been developed by using ion-cut process, based on proton implantation and wafer bonding techniques. Proton implantation has been done at energies ranging from 70 to 466 keV for thick SOI wafers (200–5000 nm) fabrication. A preliminary experiments showed that effective doses were in the 6–9×10^{16} H^+/cm^2 range, and the annealing at 450–550 °C for 30 min was found to be optimum for wafer splitting. Direct wafer bonding was performed by joining two wafers together after creating hydrophilic surfaces by a modified RCA cleaning, and an IR inspection was followed to ensure a void-free bonding. The wafer splitting was accomplished by annealing at the predetermined optimum condition, and high-temperature annealing was then performed at 1100 °C for 60 min to stabilize the bonding interface. TEM observation revealed no detectable defect at the SOI structure, and the interface trap charge density at the upper interface of the BOX was measured to be low enough to keep "thermal" quality.

Mitchell et al. (2005) have studied the formation of optically active, metal silicides using ion implantation and oxidation. While Si-based ICs dominate the microelectronics marketplace, they cannot be fabricated with optical functionality since Si is indirect. Alternative materials have been used in such applications but the ability to integrate' optically active material directly onto a silicon substrate to co-opt the advances in Si technology and processing capabilities is the better solution. Many of the transition metals form silicides that are direct band gap semiconductors and, as such, may be integrated with Si to achieve the desired optical properties. Ion implantation of the transition metal into Si was used to form the desired silicide phase by reaction of the metal with the Si substrate. Depending upon the fluence the resulting implanted layer can consist of a two-phase region in which the silicide phase forms as isolated precipitates randomly oriented within a heavily dislocated Si matrix. Rutherford backscattering/ion channeling spectrometry was used to monitor this process as a function of temperature and time. A unique method for orienting the silicide precipitates to align them crystallographically with the Si substrate and eliminate the ion-induced dislocations that form during the initial implant is discussed. This method involves oxidation of the implanted region to segregate the silicide phase at the oxide interface.

Xiao et al. (2005) presented their results obtained by MeV Si ion bombardments of thermoelectric (TE) Bi_xTe_3/Sb_2Te_3 multilayer thin films for reducing thermal conductivity. Bi_xTe_3/Sb_zTe_3 multilayer thin films were grown as TE materials using electron-beam evaporation. Solid antimony (Ill) telluride and bismuth (Ill) telluride were used for the growth of Bi, Te_3 and Sb, Te_3 layers. RBS analysis showed that the grown antimony telluride film has the stoichiometry of Sb_2Te_3 and the bismuth telluride film is $Bi_{1.1}Te_3$. The grown multilayer films have a periodic structure consisting of alternating $Bi_{1.1}Te_3$ and Sb_2Te_3 layers. The $Bi_{1.1}Te_3/Sb_2Te_3$ multilayer thin films with five or seven layers were analyzed by RBS. The grown $Bi_{1.1}Te_3$ and Sb_2Te_3 monolayer thin films and $Bi_{1.1}Te_3/Sb_2Te_3$ multilayer thin films were bombarded by 5 MeV Si ions. The thermal conductivity of the grown films was measured by a 3ω-method thermal conductivity measurement system. The multilayer structure and MeV Si ion bombardment made the films to have a lower thermal conductivity.

Seetala et al. (2005) have studied the effect of ion-beam irradiation on reducing the ordering temperature of FePt and FePtAu nanoparticles. FePt and FePt (Au 14%) 4 nm particles dispersed on a Si-substrate were irradiated by 300 keV Al-ions with a dose of 1×10^{16} ions/cm at 43 °C using a water-cooled flange in order to minimize the vacancy migration and voids formation within the collision cascades. Partial chemical ordering has been observed in as-irradiated particles with coercivity of 60–130 oersted (Oe). Post-irradiation annealing at 220 °C enhanced chemical ordering in FePt nanoparticles with a coercivity of 3500 Oe, the magnetic anisotropy of 1.5×10^7 erg/cc, and thermal stability factor of 130. A much higher 375 °C post-irradiation annealing was required in FePtAu, presumably because Au atoms were trapped at Fe/Pt lattice sites at lower temperatures. As the annealing temperature increased, anomalous features in the magnetization reversal curves were observed that disappeared at higher annealing temperatures.

Surface hardening by ion implantation gives significant advantages of ion implantation over other coating procedures. A significant advantage associated with the use of ion implantation is that the treated surface is an integral part of the work piece and does not suffer from any possible adhesion problems that may be attributed to the coatings. Furthermore, the moderate heating associated with the process virtually eliminates any risks of distortion or oxidation effects. Ion implantation produces no dimensional changes in the work piece. When treated surfaces wear, the atoms trapped interstitially in the metal structures become dislodged and may diffuse deeper into the surface. As this process continues, the atoms continue to close up micro cracks. This discourages crack propagation in the work piece and leads to better abrasion resistance. It can also prevent the ingress of oxygen and other potentially corrosive compounds.

Implanted biomedical prosthetic devices are intended to perform safely, reliably and effectively in the human body for prolonged periods of time. Stability under the imposition of repetitive loading in a hostile environment places unique demands on the materials, designs and manufacturing methods used to create the implant. Materials used for orthopedic devices should possess good biocompatibility, adequate mechanical properties, and sufficient wear and corrosion resistance, and they should be manufacturable at a reasonable cost. Titanium–aluminum–vanadium (Ti–A1–V) and cobalt-chromium-molybdenum (Co–Cr–Mo) alloys possess these unique requirements and have found successful applications in the field of orthopedics as prosthetic and fracture fixation devices. These alloys are used extensively in hip and knee implants as an articular surface sliding against ultra-high-molecular-weight

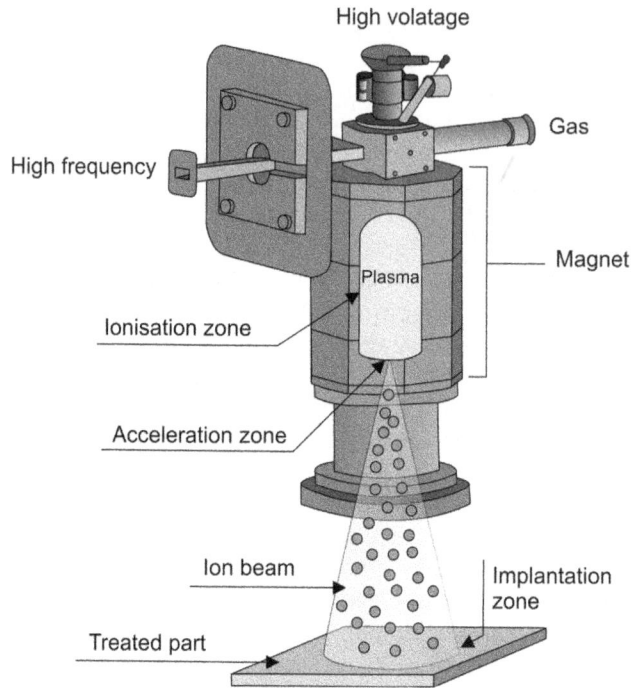

FIGURE 4.13
IonGUN, ECR (Electron Cyclotron Resonance) ion source.

polyethylene (UHMWPE). The presence of abrasive particles, such as bone, bone cement, or other foreign materials, can substantially increase wear and debris generation rates of the articulating surfaces *(1)*. By increasing the alloy's resistance to scratching, surface-hardened alloys can assist in reducing polymer wear (Shetty 2000).

The ionGUN ion implantation system enhances the surface properties of any materials. Surface hardening, increased corrosion resistance, low friction coefficient, are major effects observed on treated metals, glass, ceramics and polymers. Ion implantation technology is a cold physical cost-effective and environmentally friendly low-pressure process. Potentially relevant markets are numerous: mechanical parts, engines, cutting tools, biomedical implant, technical glasses, luxury and design, elastomers (Fig. 4.13).

A micro-accelerator of particles generates a highly energetic ion beam penetrating and modifying the surface of materials to enhance their properties without any coating. The penetration depth might reach up to 10 microns and the treatment effects are still measurable until 1mm. Depending on the nature of the implanted ions and the process parameters, you may obtain nano hammering effect, doping effect, surface amorphization, re-alloying, nano-structuring effect and even chemical modification if reactive gases are used. The part temperature never exceeds 80 °C: a cold metallurgy. The technology might be combined with other low-pressure technologies like PVD and PECVD processes to obtain even more breakthrough properties and performances: surface preparation for hard coatings, cobalt depletion, enhanced adhesion, etc.

The main process advantages are:

- Low-temperature surface treatment: any materials may be treated, even insulating and temperature sensitive ones with no bulk properties modification.

- No coatings: the material surface itself is modified and enhanced
- Environmentally friendly dry process

IONICS has a large equipment park available to lease for demonstration, feasibilities and production. Our engineering team can advise you. Co-development and projects with the Materia Nova R&D center to fine-tune your industrial solution are also possible. More characteristics that are technical are available in the product datasheet below. See also ion implantation information on the WALIBEAM website: www.walibeam.com.

4.6 Ion Beam Modification of Materials

Ion implantation may be used to introduce ion beam modification of materials (IBMM) which improves surface properties, such as wear resistance; surface hardness; fatigue life; corrosion resistance and electrical surface resistance of materials such as metals (including alloys), ceramics, glass and polymers and components made thereof. Wear and corrosion characteristics are often improved by a factor of 5–10 or more. The ion implantation process has the following advantages compared to other hardening and/or coating techniques; no dimensional changes occur, no heat distortion of the component, no coating, i.e., no adhesion problems, and it may be reapplied, e.g., after re-grinding. IBMM is used increasingly by industries in Japan, USA and Europe, and in high- and low-technology production areas. Numerous applications are reported, such as cutting and punching tools for metals; metal forming tools; cutting tools for paper, textiles, wood, leather, etc.; plastic molding tools; artificial prostheses (hip and knee joints); and ball bearings. These addressed problems are experienced in many different types of industry such as machine tooling industry; paper, textile and plastic production; automotive industry; aircraft industry, and surgical tools and components.

The development of physics theory and experimental equipment and methods in radiation material science during the second half of the 20th century, as well as actively seeking of new technology tools for solid-state electronics and semiconductor devices have led to the formation and application of radiation technologies in the semiconductor manufacturing process due to their high efficiency. Design and creation of high-performance, high-reliable, easy-to-use and maintain EB accelerators with energies up to 10 MeV, fundamental improvement of individual functional units, see work published by Pavlov and his team at Radiation Center of Institute-of-Physical Chemistry in Moscow (Pavlov et al. 1985, 2014, 2015, 2016; Pikaev et al. 1993) established leading position of radiation technologies based on electron accelerators in today's semiconductor industry in comparison with other methods of radiation treatment. The effectiveness of accelerated electrons able to replace the atoms from crystal lattice sites was shown by different research teams in relation to devices based on silicon and other materials, see Table 4.5.

The main features and advantages of EB technology for the modification of semiconductor devices are as follows:

i. Possibility of processing in air.
ii. Electron flux density adjustment by beam current and distance from the output window due to scattering in air.

   iii. Providing a large processing area due to operation in the scanning mode.

   iv. Very ample opportunities for the design and construction of under-beam equipment, including beam control systems and a variety of holders, automated sample delivery systems, freestanding vacuum/environment chamber with a beam input window for special investigations.

   v. Simultaneous pass-through processing of several semiconductor wafers, arranged one behind the other due to deep penetrating ability of accelerated electrons (range in Si about 1–2 cm).

   vi. Samples heating by beam and temperature control if necessary.

   vii. Small footprint and favorable cost of equipment, easy to use and maintain. Mentioned advantages make it possible to do a preliminary experiment quickly and realize large volume technology to provide needed combination of electrical parameters for specific device. Thus, the accelerator-based EB technologies could be effectively applied for manufacturing of competitive traditional and innovative solid-state devices, elaboration new approaches in culling methods and radiation hardness investigation.

The scientific and technical problems that could be studied with MeV ion beams include:

- MeV ion implantation to form buried deep-level insulator and semi-insulator layers.
- MeV ion beams processing for three-dimensional device fabrication and property modification of deeply buried interfaces.
- Synthesis of thin films, surface modification by MeV ion implantation.
- Concentration profiles of Hydrogen in solids (for example by using ^{15}N beam).
- Determination of oxygen in Si and other materials by charge particle activation analysis.
- Study of corrosion films on different surfaces (for example, oxygen concentration as a function of depth).
- Characterization of defects in semiconductors using nuclear analytical methods (PIXE, RBS, NRA, etc.).
- Microprobe mapping of elements (applications to interfaces, buried structures, etc.).
- Study of corrosion-erosion phenomena with thin layer activation (TLA) analysis (especially component of power plants, engines, etc.).
- Ion beam analysis of superconducting thin films, including areal density, average stoichiometric ratios, impurity concentrations, elemental depth profiles, etc.
- X-ray mirrors, multilayer analysis by backscattering.

In their study, Jin et al. (2019) used Ti + N ion implantation as a surface modification method for surface hardening and friction-reducing properties of Cronidur30 bearing steel. The structural modification and newly formed ceramic phases induced by the ion implantation processes were investigated by transmission electron microscopy (TEM), X-ray photoelectron spectroscopy (XPS), and grazing incidence XRD (GIXRD). The mechanical properties of the samples were tested by nano-indentation and friction experiments. The surface nano-hardness was also improved significantly, changing from ~10.5 GPa (pristine substrate) to ~14.2 GPa (Ti + N implanted sample). The friction coefficient of Ti + N ion-implanted samples was greatly reduced before failure, which is less than one-third of pristine samples. Furthermore, the TEM analyses confirmed a tri-lamellar structure at the near-surface region, in which amorphous/ceramic nano-crystalline phases were embedded into the implanted layers. The combined structural modification and hardening ceramic phases played a crucial role in improving surface properties, and the variations in these two factors determined the differences in the mechanical properties of the samples (Jin et al. 2019).

Tungsten is the main candidate material for plasma-facing armor components in future fusion reactors. In-service, fusion neutron irradiation creates lattice defects through collision cascades. Helium, injected from plasma, aggravates damage by increasing defect retention. Both can be mimicked using helium-ion-implantation. In a recent study (Das et al. 2019) on 3000 ppm helium-implanted tungsten (W-3000He), the authors assumed helium-induced irradiation hardening, followed by softening during deformation. This hypothesis was founded on observations of a large increase in hardness, substantial pile-up and slip-step formation around nano-indents and Laue diffraction measurements of localized deformation underlying indents. The authors test this hypothesis by implementing it in a crystal plasticity finite element (CPFE) formulation, simulating nano-indentation in W-3000He at 300 K. The model considers thermally activated dislocation glide through helium-defect obstacles, whose barrier strength is derived as a function of defect concentration and morphology. Only one fitting parameter was used for the simulated helium-implanted tungsten; defect removal rate. The simulation described the localized large pile-up remarkably well and predicted confined fields of lattice distortions and geometrically necessary dislocation underlying indents that agreed quantitatively with previous Laue measurements. Strain localization is further confirmed through high-resolution electron backscatter diffraction and TEM measurements on cross-section lift-outs from center of nano-indents in W-3000He.

4.7 Ion Beam Analysis of Materials

Ion beam analysis is a generic term used to describe a variety of accelerator techniques that mostly focus on interactions between MeV light elements (H, He) and materials. All IBA techniques are extremely sensitive and allow the detection of elements in the first nanometers (or deeper) of a material. Typical beam energies range from 2 MeV to 6 MeV and higher.

Common techniques include:

- RBS: used to measure heavy elements in a light matrix.
- ERD: used to measure light elements in a heavy matrix.
- PIXE: gives the trace and minor elemental composition.
- NRA: used to measure particular isotopes.

Channelling, used to measure inconsistencies with major and minor axes of crystals structures.

All these techniques are often done in micro beam geometry allowing mapping of elements on a micrometre scale.

NEC manufacturers IBA systems in different energies, depending on the isotopes to be studied:

1. MV systems are commonly used for basic RBS and ERD research.
2. MV systems are commonly used for the study of PIXE cross sections and NRA work.

Light element depth profiling by Elastic Recoil Detection Analysis (ERDA) is used in the following applications:

- Hydrogen detection in semiconductor materials.
- Hydrogen detection in high-strength metals.
- Detection of surface contaminants.
- Measurement of light element depth profiles following ion implantation or annealing.
- Depth profiling in corrosion films.
- Calibration of standards for other analysis techniques.

The use of ultra-low energy (ULE) ion implantation in the semiconductor industry has placed increasing pressure on SIMS depth profiling capabilities. For example, Novak et al. (2005) used NRA to calibrate implantation doses of boron ULE implants as an essential check of the SIMS accuracy. Comparison of NRA done in two different laboratories has revealed differences of up to 15%, making it essential to calibrate the measurements against an accurate standard. It is evident that the accepted SIMS depth profiling method for ULE B implants, low-energy oxygen bombardment with an oxygen leak, causes significant segregation of boron during the SIMS profile. High-resolution ERD measurements appear to be one of the few techniques other than SIMS that can resolve the near-surface details of the boron distribution. Careful comparison of the ERD and SIMS profiles allows refining of SIMS correction procedures to produce the most accurate measurement of the boron distribution. Similarly, ULE As implants show significant segregation of As toward the native oxide interface as a result of annealing.

Low dimension structures raise inevitably new technological challenges in materials science. The new structures must fulfill stringent requirements in composition, crystalline quality and interface sharpness among others. Fonseca et al. (2005) presented the results of Si/Ge quantum structures and FePt/C multilayer structures deposited at different temperatures by ion beam sputtering. Studying by high resolution back scattering the nanostructured magnetic and semiconducting materials the authors found the evidence for the presence of FePt nanoparticles embedded in the C matrix and Ge island, in Ge/Si multilayers structures. Size and stoichiometry of the nanoparticles and the multilayer periodicity were obtained using Rutherford backscattering at grazing angles of incidence. The strain state of the single crystalline layers was determined by tilt axis channeling.

Conventional focused ion beam systems employ a liquid-metal ion source (LMIS) to generate high-brightness beams such as Ga^+ beams. Recently there has been an increased need for focused ion beams in areas like biological studies, advanced magnetic-film manufacturing and secondary-ion mass spectroscopy (SIMS). Ji et al. (2005) reviewed the status of development on focused ion beam systems with ion species such as O_2^+, P^+

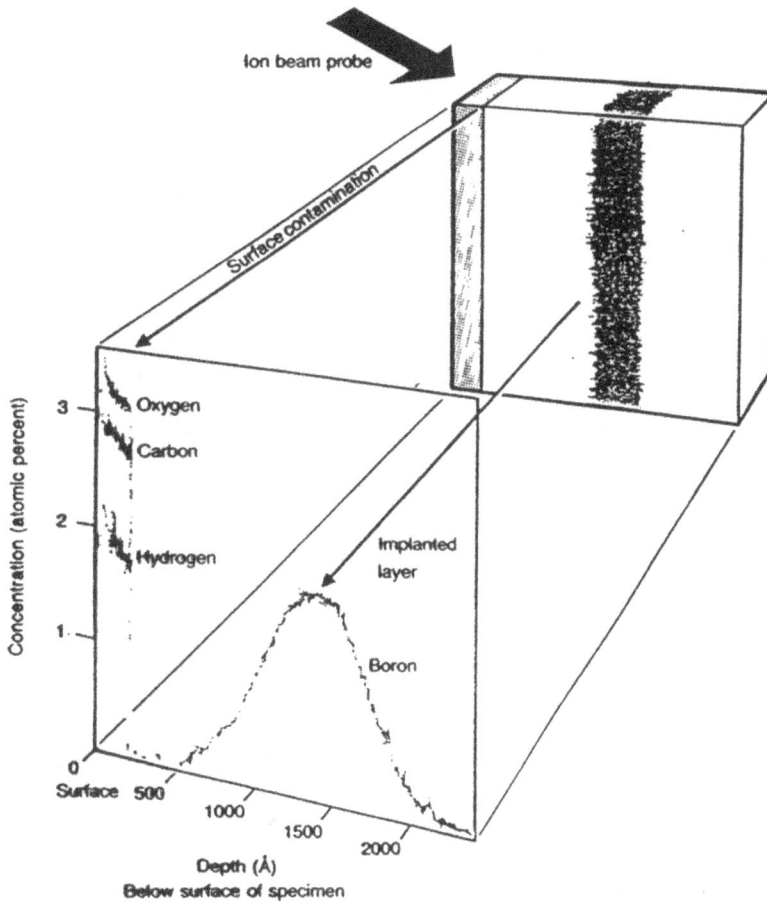

FIGURE 4.14
A concentration profile for a boron implanted silicon wafer as measured by ERDA.

and B$^+$. Compact columns for forming focused ion beams from low energy (~3 keV), to intermediate energy (~35 keV) were discussed. By using focused ion beams, a SOl MOSFET was fabricated entirely without any masks (Fig. 4.14).

The implantation was carried out at 40 keV to a dose of 1×10^{16} atoms cm^{-2}. Oxygen, carbon and hydrogen are present as surface contamination.

The setting and development of strength of Portland cement concrete depend upon the reaction of water with various phases in the Portland cement. Nuclear resonance reaction analysis involving the ^{1}H(^{15}N;α,γ)^{12}C reaction has been applied to measure the hydrogen depth profile in the few 100 nm thick surface layer that controls the early stage of the reaction. Specific topics that have been investigated include the reactivity of individual cementitious phases and the effects of accelerators and retarders (Livingston et al. 2010). The NRRA measurements are done at the 4 MeV Dynamitron tandem accelerator at the Ruhr-Universitat Bochum in Bochum, Germany. Typical particle currents at the target are about 15 nA. The 4.44 MeV γ-ray emitted in the ^{1}H(^{15}N;α,γ)^{12}C reaction is observed with a ϕ12 × 12 in. NaI(Tl) crystal with (49 ± 3)% photo peak efficiency connected to a 1024 channel multichannel analyzer.

Matteson et al. (2005) report the analysis of coupons of mineral flint and chert of archeological interest and various provenances in comparison to uranium-doped glass, mineral single crystal quartz and thermally grown silicon dioxide. The measurements are motivated by a desire to quantify the concentration of uranium in lithic sources of paleo Indian artifacts; uranyl oxide is suspected to be the origin of the observed UV fluorescence. RBS confirmed the concentration of uranium in these minerals to be on the order of tens of ppm as γ-ray spectroscopy analysis suggested previously. The level of uranium present does not correlate, however, with the observed fluorescence behaviours. Optical effects such as absorption and quenching due to other impurities are suspected to be the cause of the differences.

Determination of the strains in and around the quantum dots is important both to assess the results of the epitaxially grown thin film and to explain the optical performance of quantum dots samples. A series of InAs/GaAs quantum dots samples were fabricated by MBE in Stranski–Krastanov growth mode. Niu et al. (2005) report the ion-channeling studies of InAs/GaAs quantum dots presenting results of ion-channeling observations along (001) growth direction and (011) direction.

Researchers at the nuclear microprobe of the Ljubljana Tandetron laboratory have developed a proton beam writing system (Simčič et al. 2005). The system incorporates an existing scanning system with the external beam positioning and fast beam switching system. Microstructures are produced according to the digital plan in the B&W bitmap. When the beam passes over the unexposed regions of the digital plan, fast electrostatic deflectors sweep it off. Dose per pixel is controlled either by selecting the irradiation time or by the preset number of external counts. To achieve the production of three-dimensional microstructures, two different irradiation energies are applied at the same positions. By the variation of energy and the beam dose, various types of three-dimensional structures could be produced. Precise understanding of energy deposition in different layers and dose normalization are essential for regular exposure throughout the polymer. The contrast curve of SU-8 polymer for proton beam irradiation was measured by the authors as demonstrated by the selection of optical and SEM images of the resulting microstructures.

The first wall of an inertial fusion energy reactor may suffer from surface blistering and exfoliation due to helium ion fluxes and extreme temperatures. Tungsten is a candidate for the first wall material. Gilliam et al. (2005) conducted a study of helium retention and surface blistering with regard to helium dose, temperature and tungsten microstructure to learn how the damaging effects of helium may be diminished. Single crystal and polycrystalline tungsten samples were implanted with 1.3 MeV ^{3}He in doses ranging from $10^{19}/m^2$ to $10^{22}/m^2$. Implanted samples were analyzed by ^{3}He(d,p)^{4}He nuclear reaction analysis and neutron depth profiling techniques. Surface blistering occurred for doses greater than 10^{21} He/m^2 and was analyzed by scanning electron microscopy. Repeated cycles of implantation and flash annealing indicated that helium retention was reduced with decreasing implant dose per cycle. A carbon foil energy degrader could allow a continuous spectrum of helium implantation energy matching the theoretical models of He ion fluxes within the IFE reactor.

Hojo et al. (2005) reported on radiation effects on yttrium-stabilized zirconia (YSZ) irradiated with He or Xe ions at high temperatures. YSZ is a candidate for use as optical and insulating material in a nuclear reactor and/or fusion reactor. In the present study, in situ observations by an electron microscope were performed during helium and xenon ion irradiations of YSZ single crystals. Damage evolution in the YSZ crystals was observed during irradiation with 35 keV He$^+$ with a flux of 5×10^{13} ions/cm^2 s at 1073 K, 1273 K and 1473 K. The damage evolution was also observed with 60 keV Xe^{2+} with a flux of 5×10^{12} ions/cm^2 s at 1073 K and 1473 K. In these cases, bubble formation was clearly observed. Dislocation loops were formed by Xe ion irradiation at both temperatures but were not formed by He. This result shows clear dependence of loop formation on the ion species.

4.8 Micromechanical Manufacture

It is an accepted view that a new industrial revolution will be brought about by the establishment of advanced material processing and machining technologies that can create new materials down to the atomic and molecular levels. At this time, there are only two techniques for writing original patterns (as opposed to replicating them) at 0.1 μm and below: EBs and ion beams. EBs are at a mature state of development and have advantages in the absence of noise and in fast deflection capability. Ion beams, on the other hand, have demonstrated the absence of potentially faster-writing speed. Thus focused ion beam lithography is a serious candidate for future fine pattern writing.

Nowadays, technologies for the micro-miniaturization of mechanical structures are being developed within the fields that are commonly referred to as micro electro mechanical systems (MEMS) mainly in the USA, micro system technology (MST) in Europe, and micromachine technology in Japan. The substantial part of all these technologies is IC-based batch technologies from microelectronics. Microelectronics-based technologies enable the creation of micro-devices that incorporate simple mechanical components fabricated mainly from silicon (Kussul et al. 1996).

High energy ion implantation is often used in IC manufacturing, in forming new devices and advanced well structures such as the vertically modulated structures (VMS): VMSs are gaining widespread acceptance in the production manufacturing of DRAMs, SRAMs, logic devices, microprocessors and flash memory cards at the 0.5 μm level.

Triple well structures are used by several manufacturers to produce 16 Mb DRAMs and beyond, as reported by Fujii at al (1989). This well structure can be also applied to CMOS logic devices. The major application for mass production of this triple well technology is reported (Umezawa et al. 1992; Jinbo et al. 1992) to be for flash memories in order to achieve high packing density (30% reduction in chip size) and lower power supply voltage operation. The triple well structure was initially formed by deep well diffusions at high temperature using standard medium current ion implantation equipment. This same triple well structure can be achieved without the high-energy multiple chained implantations up to 3 MeV.

Microelectronics-based micromechanics is rather limited for the construction of 3D micro mechanisms with moving parts. Kussul et al. (1996) proposed the use of micro equipment for transfer the technologies of mechanical engineering to the micro domain. They showed that equipment precision increases linearly with decreasing size. To make micro equipment, a series of equipment generations with gradually decreasing dimensions was suggested. Miniaturization of equipment reduces power consumption and floor area occupied. Coupled with automation, it drastically reduces the cost of micro equipment. This in turn reduces the cost of micromechanical devices manufactured by micro equipment. Micro equipment-based manufacturing also increases throughput by the concurrent operation of large numbers of low-cost micro equipment pieces. The low cost and high productivity of micro equipment-based manufacturing widens the range of feasible micromechanical applications, both single-unit and mass. The designs for micro valve fluid filters, capillary heat exchangers, electromagnetic and hydraulic step motors that could be easily implemented by micromechanical engineering technologies are proposed. Hybrid technologies combining massively parallel micro equipment-based manufacturing and batch manufacturing may be a possibility.

Single-ion control was first introduced for the purpose of' micromachining with heavy ions. This is a very promising field, but for light ions, there are difficulties in detecting the arrival of individual ions at a target, secondary electron emission being relatively low. Two items deserve to be mentioned:

i. Perfect special depressions can be produced with radii of curvature between and some 100 pm by etching single heavy ion tracks. These spherical depressions can be used to focus ultrasonic waves in an ultrasonic microscope. Now even lines or arrays of these micro-mirrors can be produced with high precision using the single-particle techniques.

ii. The so-called single-pore membranes have been produced for many years. These membranes of about 50 mm diameter contain exactly one single etched track pore of 50 pm length and 5 pm diameter at their center, and their production represents one of the most economic uses of an ion beam.

Single-pore membranes are used to simulate a short section of the human circulatory system, where red blood cells of about 8 pm diameter have to squeeze themselves through capillaries of about 5 pm diameter. Healthy cells are flexible enough to do that, but there are diseases like diabetes that reduce the flexibility of the red blood cells.

When ions of sufficient electronic energy loss traverse a dielectric film or foil, they alter the chemical bonding along their nominally straight path within the material. A suitable etchant can quickly dissolve these so-;called latent tracks leaving holes of small diameter (~10 nm) but long length – several μm. By continuing the etching process gradually, increase of the diameter reproducibly and uniformly is achieved (Felter et al. 2005). The trackable medium can be applied as a uniform film onto large substrates. The small, monodisperse holes produced by this track etching can be used in conjunction with additional thin-film processing to create functional structures attached to the substrate. For example, Lawrence Livermore National Laboratory and Candescent Technologies Corporation (CTC) co-developed a process to make arrays of gated field emitters (~100 nm diameter electron guns) for CTC's Thin CRT™ displays, which have been fabricated to diagonal dimensions >13 in. (Felter et al. 2005).

The negatively charged nitrogen vacancy (NV⁻) defect center in diamond has gained much attention for applications in quantum optics due to its long coherence time at room temperature and its great potential for quantum sensing and quantum communications. Reliable formation of colour centers with long spin coherence times and the placement of colour centers into quantum registers are major challenges for applications (Lake et al. 2021). The authors reported depth-resolved photo-luminescence measurements of NV⁻ centers formed along the tracks of swift heavy ions in synthetic single-crystal diamonds that have been doped with 100 ppm nitrogen during the crystal growth period. Measurements show that NV⁻ centers are formed mostly within the regions of electronic stopping processes domination instead of the end of ion range. Thermal annealing after irradiation additionally increased NV⁻ yields. The formed centers could be isolated for exploration of colour center qubits in registers with an average qubit spacing of a few nanometers and of order of 100 colour centers per micrometer for 10–30 μm long chains (Lake et al. 2021).

4.9 Cluster Ion Beams and Applications

Research on gas cluster ion beam (GCIB) processes has started in the Ion Beam Engineering Experimental Laboratory, Kyoto University, Japan in 1988 after intense gas cluster formation had been confirmed by a room temperature supersonic nozzle (see Yamada and Toyoda 2009, for the review of GCIB facility and technology). GCIB process has become increasingly popular as add-on component for ultra-high vacuum techniques such as X-ray photoelectron spectroscopy (XPS), scanning probe microscopy (SPM) and secondary ion mass spectroscopy (SIMS) (Fig. 4.15).

Cluster ion-surface collisions have been found to produce low energy bombardment effects at very high density and GCIB processes exhibit unique non-linear effects that are useful for novel surface processing applications. The effects include low energy ion bombardment, lateral sputtering and, low-temperature thin film formation. GCIB processing has been successfully applied for shallow junction formation; for high rate etching; for surface smoothing of materials including metals, dielectrics, superconductors and diamond; and for high-k oxide and DLC thin film deposition (Yamada and Toyoda 2005).

Industrial applications of GCIB processes are conducted by several Japanese companies under the Nanotechnology program called "Advanced Nano-Fabrication Process Technology Using Quantum Beams" of NEDO/METI (New Energy and Industrial Technology Development Organization/the Ministry of economy and Technology Industry). In US, R&D especially for semiconductor applications are under way at Epion Corporation (https://epion.co/), International SEMATECH (Irwin and Klenow 1996) and their cooperated associations. Epion is the only company that is developing industrial GCIB equipment and has joined the NEDO/METI project. In nano-scale GCIB processes, the effect of cluster size (atoms/cluster) on surface processing, especially damage production and becomes important. In the project, GCIB equipment with a cluster size selection system has been developed. The difference in effects single ion beam and GCIB on the surface of the material is shown in Fig. 4.16.

GCIB is an ion beam consisting of several thousand gas atoms (molecules) such as argon, which enables ion beam etching with extremely low energy per atom and surface

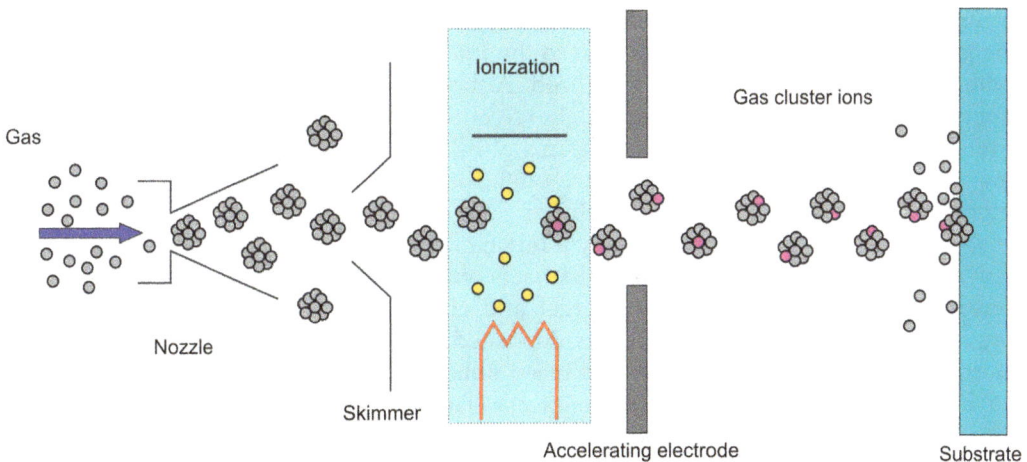

FIGURE 4.15
Overview of gas cluster ion beam instrument.

Single atomic ion beam   **Cluster ion beam**

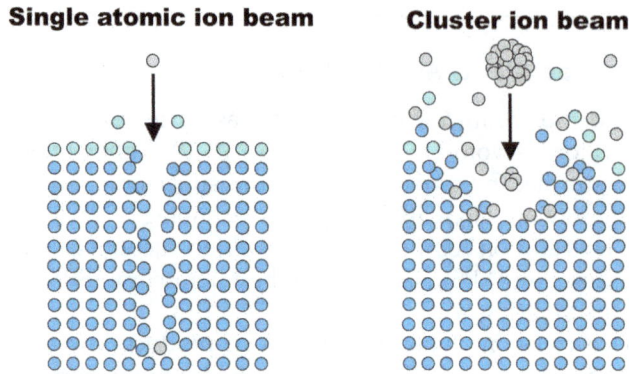

FIGURE 4.16
Comparison of the characteristics of ion beams of the past and cluster ion beams.

flattening after etching, not possible with other ion beam technologies. Next, we shall present some examples of GCIB applications.

Yamada et al. (2001) described the principles and an early experimental status of GCIB processing as a surface modification technique for practical industrial applications. Theoretical and experimental characteristics of GCIB processes and of related equipment development are described from the moment of neutral cluster formation, through ionization, acceleration and impact upon a surface. The impact of an accelerated cluster ion upon a target surface imparts very high energy densities into the impact area and produces non-linear effects that are not observed in the impacts of atomic ions. Unique characteristics of GCIB bombardment have been found to offer the potential for various industrial applications that cannot be achieved by conventional ion beam processing. Among prospective applications are included shallow ion implantation, high rate sputtering, surface cleaning and smoothing and low-temperature thin film formation. Sputtering effects produced by cluster ion impact are particularly interesting. High sputtering yields and lateral distribution of sputtered atoms cause surface smoothing effects that cannot be achieved with monomer ion beams. Surface smoothing to atomic levels is expected to become the first production use of GCIB.

A gas cluster is an aggregate of a few to several thousands of gaseous atoms or molecules, and it can be accelerated to the desired energy after ionization. Since the kinetic energy of an atom in a cluster is equal to the total energy divided by the cluster size, a quite-low-energy ion beam can be realized. Although it is difficult to obtain low-energy monomer ion beams due to the space charge effect, equivalently low-energy ion beams can be realized by using cluster ion beams at relatively high acceleration voltages. The low-energy feature and the dense energy deposition at a local area are important characteristics of the irradiation by gas cluster ions. The diameter of a gas cluster with a cluster size of several thousand is only a few nanometers, so that thousands of atoms or molecules penetrate the target in an area that is only a few nanometers in diameter, which causes multiple collisions between target and cluster atoms. Therefore, all of the impinging energy of a gas cluster ion is deposited at the surface region, and this dense energy deposition is the origin of enhanced sputtering yields, crater formation, shockwave generation, and other nonlinear effects (Toyoda and Yamada 2008). The authors described a GCIB system for 300-mm-diameter Si wafers. Surface smoothing, shallow doping, low-damage etching, trimming and thin-film formations are promising applications of GCIBs.

Swenson (2004) has described a new technique that allows the measurement of the average charge, energy and mass of large, multiply charged, gas clusters. For this study, a beam with averages of charge state +3.2, energy 64 keV and mass 10,400 atoms was produced in a high-intensity Ar gas cluster ion beam, and the stability of these clusters undergoing collisions with Ar gas atoms was measured using a gas cell target. At this energy, multiple collisions with Ar gas atoms are observed to progressively abrade the cluster particles. The mass loss is consistent with a simple theory that assumes thermalization of the collision energy followed by evaporation. The measurements are important for cluster-surface interactions because they allow cluster energy and dose to be measured for multi-charged clusters. The author has shown that higher energy clusters that cause larger craters are more affected by the cluster-gas collisions, and this effect can be beneficial when using a cluster beam to smooth a surface (Swenson 2005).

Bourelle et al. (2005) presented a sidewall polishing with a gas cluster ion beam for photonic device applications. A novel surface smoothing method is named gas cluster ion beam-chemical mechanical polishing (GCIB-CMP). This method employs GCIB with irradiation angles above 60° from the surface normal, and therefore can be applied to the smoothing of sidewalls of a patterned structure. Surface morphology is compared between various optical materials Si, SiO_2 and MgF_2 irradiated by SF_6 clusters. Si and SiO_2 surfaces are substantially smoothed at 0.3 nm and below in average roughness at incident angles around 80°. The authors estimated the performance improvement of Si and/or SiO_2 based photonic devices having smooth sidewall surfaces formed by the GCIB-CMP method.

Many studies have demonstrated that improving the surface smoothness and cleanliness of high voltage electrodes increases the voltage standoff capability, but none has specifically investigated the role of nano-scale and atomic level surface roughness. Using AFM imaging, Swenson et al. (2005) have studied the effect of GCIBs on oxygen-free Cu electrode material that is used in high gradient RF cavities. Using Ar clusters accelerated by 30 kV, with a dose of 6×10^{14} e cm^{-2}, they have effectively removed an asperity that was 3500 Å wide and 350 Å high. Subsequent processing with 5 kV acceleration reduced the surface roughness from an R_a (average roughness-average deviation from the mean) value of 13.2 Å to 4.8 Å. This demonstrates the effectiveness of GCIB for reducing submicron roughness to atom-level smoothness.

GCIBs provide new opportunities for bio imaging and molecular depth profiling with SIMS (Winograd 2018). These beams, consisting of clusters containing thousands of particles, initiate desorption of target molecules with high yield and minimal fragmentation. The review by Winograd (2018) emphasizes the unique opportunities for implementing these sources, especially for bio imaging applications. He has developed theoretical aspects of the cluster ion/solid interaction to maximize conditions for successful mass spectrometry. In addition, the author has reviewed the history of how GCIBs have become practical laboratory tools. Special emphasis was placed on the versatility of these sources, as size, kinetic energy, and chemical composition can be varied easily to maximize lateral resolution, hopefully to less than one μm, and to maximize ionization efficiency.

Tian et al. (2017) have applied GCIB-SIMS to an interesting medical problem. They showed the full potential of GCIB-SIMS for mapping intact lipids in biological systems with better than 10 μm lateral resolution. In their study, they investigated also the capability of GCIB-SIMS for imaging high-mass signals from intact cardiolipin (CL) and gangliosides in the normal brain and the effect of a controlled cortical impact model (CCI) of traumatic brain injury (TBI) on their distribution. A combination of enzymatic and chemical treatments was employed to suppress the signals from the most abundant phospholipids (phosphatidylcholine (PC) and phosphatidylethanolamine (PE)) and enhance the signals from the low abundance CLs and gangliosides to allow their GCIB-SIMS detection at 8 and 16 μm spatial resolution. Brain CLs have not been observed previously using other contemporary imaging mass spectrometry techniques at better than 50 μm spatial resolution. High-resolution images of naive and injured brain tissue facilitated the comparison of CL species across three multicellular layers in the CA1, CA3, and DG regions of the hippocampus. GCIB-SIMS also reliably mapped losses of oxidizable polyunsaturated CL species (but not the oxidation-resistant saturated and mono-unsaturated gangliosides) to regions including the CA1 and CA3 of the hippocampus after CCI. This work extends the detection range for SIMS measurements of intact lipids to above m/z 2000, bridging the mass range gap compared with MALDI. Further advances in high-resolution SIMS of CLs, with the potential for single cell or supra-cellular imaging, could be essential for the understanding of CL's functional and structural organization in normal and injured brain (Tian et al. 2017).

Several other investigators have also used a combination of GCIB-SIMS. For example, see Hill et al. (2011) who used the DC beam buncher-TOF SIMS instrument, J105 3D Chemical Imager (Ionoptika, the UK). They employed this instrument to image the naive and CCI brain sections with and without EDC/PLC treatment. The instrument is equipped with a 40 keV C_{60}^{+} source and a 70 keV gas cluster ion source (GCIB), described in detail by Hill et al. (2011). For this study, GCIB with CO_2 gas feeding is more beneficial for high mass species (e.g., CLs, gangliosides) and high spatial resolution.

Atomic layer etching (ALE) of metals using GCIBs and acetylacetone vapor was demonstrated by Toyoda and Uematsu (2019). Since GCIBs are an aggregate of thousands of gas atoms or molecules, the energy/atoms or energy/molecules can be easily reduced down to several eV, which is ideal for the energetic beam in the removal steps of ALE. In addition, GCIB creates the transient high-temperature and high-pressure condition on a target surface, surface reactions between the adsorbed molecules and target atoms are enhanced even at low substrate temperature. The authors have reported ALE with oxygen GCIB using acetic acid vapor and acetylacetone (acac) as an organic vapor. ALE of Cu films with 5 kV O_2-GCIB and acac were demonstrated successfully. From the etching experiment with continuous O_2-GCIB irradiation, etching for metals (Cu, Co, Ni, Ru, Ta) are also enhanced by the addition of acac during O_2-GCIB irradiation.

According to Wang et al. (2019), GCIB is a promising technique for preserving molecular structures during ion sputtering and successfully profiling biological and soft materials. However, although GCIB yields lower damage accumulation compared with C_{60}^{+} and monoatomic ion beams, the inevitable alteration of the chemical structure can introduce artifacts into the resulting depth profile. To enhance the ionization yield and further mask damage, a low-energy O_2^{+} (200–500 V) co-sputter can be applied. While the energy per atom (E/n) of GCIB is known to be an important factor influencing the sputter process, the manner through which E/n affects the GCIB-O_2^{+} co-sputter process remains unclear. In this study, polyethylene terephthalate (PET) was used as a model material to investigate the sputter process of 10–20 kV $Ar_{1000-4000}^{+}$ (E/n = 2.5–20 eV per atom) with and without O_2^{+} co-sputter at different energies and currents. Time-of-flight secondary ion mass spectrometry (ToF-SIMS) with Bi_3^{2+} as the primary ion was used to examine surfaces sputtered at different fluences. The sputter craters were also measured by alpha-step and atomic force microscopy in quantitative imaging mode. The SIMS results showed that the steady state cannot be obtained with E/n values of less than 5 eV per atom due to damage accumulation using single GCIB sputtering. With a moderate E/n value of 5–15 eV per atom, the steady state can be obtained, but the ~50% decay in intensity indicated that damage cannot be masked completely despite the higher sputter yield. Furthermore, the surface Young's modulus decreased with increasing E/n, suggesting that depolymerization occurred. At an E/n value of 20 eV per atom, a failed profile was obtained with rapidly decreased sputter rate and secondary ion intensity due to the ion-induced crosslink. With O_2^+ co-sputtering and a moderate E/n value, the oxidized species generated by O_2^+ enhanced the ionization yield, which led to a higher ion intensity at a steady state in general. Because higher kinetic energy or current density of O_2^+ led to a larger interaction volume and more structural damage that suppressed molecular ion intensity, the enhancement from O_2^+ was most apparent with low-energy–high-current (200 V, 80 µA cm^{-2}) or high-energy–low-current (500 V, 5 µA cm^{-2}) O_2^+ co-sputtering with 0.5 µA cm^{-2} GCIBs. In these cases, little or no intensity drop was observed at the steady state.

The physics of the surface interaction of the cluster beam strongly depends upon gas composition, cluster size, cluster size distribution and beam energy. Typical argon GCIBs are composed of clusters ranging from several hundred to several thousand atoms in size. It has been shown that Ar clusters can be used to smooth surfaces at a sub-nanometer level. Argon cluster beam smoothing typically occurs in the energy range between 15 and ~30 keV (Greer et al. 2020). As such, the average energy per atom is of the order of 10 eV/atom upon cluster impact with the surface and subsequent dissociation. Ion cluster beams formed with reactive gases such as oxygen and nitrogen can be formed, but at somewhat lower current densities than those obtainable with argon. Upon impact, reactive gas clusters undergo strong chemical reactions at the substrate surface. An extension of this chemical interaction is to utilize reactive clusters in an ion beam-assisted, thin-film physical vapor deposition process. This has been demonstrated with relatively low energy (E < ~10 keV) oxygen clusters in an electron-beam evaporator to form extremely low resistivity indium–tin oxide films on room-temperature substrates.

Carbon-60, or just C_{60}, is a type of fullerene molecule, consisting of sixty carbon atoms formed into a hollow sphere, not unlike the shape of a soccer ball. Ionoptika (2021) produced the first C_{60} ion beam, in collaboration with the University of Manchester, and since then more than 140 units have been sold worldwide. Compared to monatomic ion beams, C_{60} beams result in a much "gentler" sputtering action, greatly reducing the damage caused to sub-surface layers. Preferential sputtering – normally an issue for monatomic beams – is largely non-existent for C_{60}, while the etch rate of is also relatively consistent across different material types. This makes C_{60} a very powerful, very consistent sputter source. As an analytical beam, the gentle sputtering action of C_{60} also reduces the fragmentation of larger molecules, resulting in an enhanced molecular signal intensity. For techniques such as SIMS, this can be incredibly important, as it significantly increases the usable mass range of the technique.

4.10 Thin Layer Activation

TLA is a method for the measurement of wear, corrosion and erosion. A thin layer of atoms in the material surface is made radioactive by bombardment with a beam of charged particles. If the material is subsequently worn away from the surface, total activity is reduced. The total activity level can be accurately monitored and, since the activity to depth relationship is known, the amount of material worn away can be directly measured.

The depth of the layer can be varied, usually between 10 μm and 300 μm and most materials can be activated. The characteristics of TLA can be summarized as follows: on-line measurements: very high sensitivity (up to 10 nm/h); measurements in real conditions; exact location of wear; convenient for inaccessible parts; can be applied to all materials (except plastics). Because of these properties, TLA is being used in the car industry, aeronautics, chemical industry, lubricants, mechanical industry, new materials and tribology. Some typical applications are wear on engine components, shaft and bearings, paper mill components and machine cutting tools, corrosion on pipes and plant, turbine blades.

For examples of the TLA method applications in industry see IAEA (1997) which contains IAEA INIS (International Nuclear Information System) database on TLA. The INIS Database, covering all aspects of the peaceful applications of nuclear science and technology, is a bibliographical database containing over 2.15 million references with abstracts ranging from 1970 to the present.

IAEA carried out in 1991–1995 a coordinated research project (CRP) on the TLA method and its application in industry. This CRP contributed to a better understanding of the TLA method and the development of new measuring and application techniques. In 1997, the IAEA issued IAEA-TECDOC-924 (IAEA 1997), based on the CRP results and describing its main achievements. Since the range of potential applications has expanded, and techniques have become more accurate and technically and economically competitive IAEA published an update of IAEA-TECDOC-924, see IAEA (2020). This publication presents different aspects of the methods and techniques, as well as typical case studies; highlights important achievements of the technology in research and development; and demonstrates the present and potential value of the industrial applications of the technology.

When a beam of accelerated ions enters a material, the particles rapidly lose energy and penetrate to a well-defined depth (see Fig. 4.17). A small number of charged particles

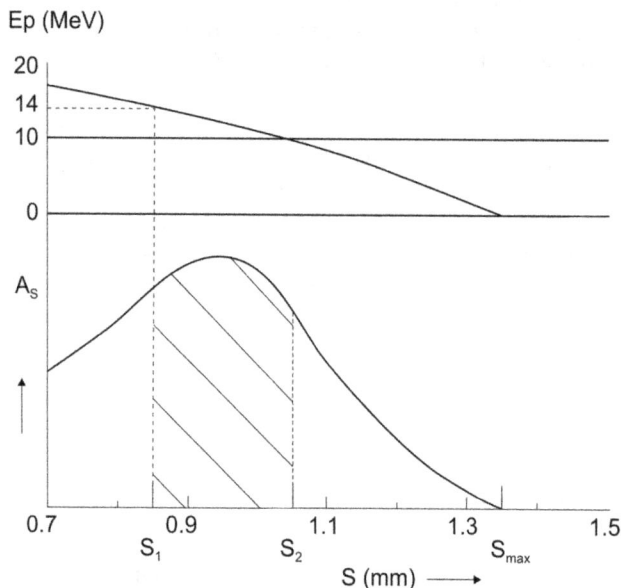

FIGURE 4.17

The scheme of TLA: Dependence of proton energy (upper part) and activity concentration (lower part) on depth for irradiation of iron by 30 MeV protons. Only the last part of the range is shown. The hatched area indicates the depth range of approximately constant activity concentration. (Modified from IAEA, 1997.)

interact with atomic nuclei of the material, induce a nuclear reaction and produce radioactive isotopes. The concentration of radioactive atoms produced within the under-surface layer is very low. The activation produces only a very low level of radioactivity, typically a few μCi (1 μCi = 3.7 × 10⁴ Bq). Once the radioactive label is created under the surface, it decays with the emission of gamma rays.

The parameters of a label must correspond to the problem under consideration in the best way. Its radionuclide composition must be utmost simple to have reliable measurement results in industrial conditions. The depth of the radioactive layer must be comparable with the expected loss. The activity depth distribution must be precisely known in depth as it is used to convert the decrease of counting intensity into linear or mass destruction parameters.

Ranges of charged particles in different elements are well known and available information (Williamson et al. 1966) permit to select the appropriate particle energy for a specific problem. Ranges in compounds and alloys may be calculated according to Bragg's formula:

$$1/R_A = \sum C_i/R_i$$

where R_A is the range in alloy, R_i, is the range in element i and C_i, is the concentration of element i. The depth of the radioactive layer depends on the particle energy and beam incidence angle:

$$d = [R(E_0) - R(E_{th})]\sin(\theta) = R_0 \sin(\theta)$$

where E_0 and E_{th}, are particle energy and reaction threshold, R is the particle range of energy E_0 in a certain material and θ is the angle between the beam axis and the surface. The TLA technique allows to adjust the thickness of the label and thereby the sensitivity of the measurement. Q-reaction values and thresholds are tabulated in many books (see e.g. Qtool 2011: Calculation of Reaction Q-values and Thresholds) but one must bear in mind that real effective thresholds include the values of the Coulomb barrier.

The activation of chemical elements and the main construction materials in the majority of cases is well studied, see for example Albert et al. (1987) and Konstantinov and Krasnov (1971). Some additional recommendations are given in (IAEA 2020). These recommendations may serve only as a rough guide since a large variety of problems, materials and irradiation means may lead to another choice of radionuclides and solutions.

For the purpose of TLA mainly charged particle accelerators (cyclotron, van de Graaff) are used in the medium energy range (5–30 MeV) and in special cases when volumetrical irradiation is necessary also nuclear reactors or neutron generators, but in this case the rules of radiation protection probably does not allow an in-situ measurement because of the high level of radioactivity.

This method helps to increase the lifetime and reliability of various machines, installations and technological processes. The most promising areas for TLA applications are as follows:

1. Automobile and engine industry piston rings (running surface, flanks) piston ring groves piston skirt all kinds of bearings: crankshafts, camshafts, connecting rods, etc. camshaft, camheads cylinder liner valve seat, valve shaft guidance fuel injection nozzle, injection pump, fuel tubes all kinds of gear wheels gear and gear-parts.

2. Pumps: all kinds of sealing surfaces of the housing, blades, roller, wheelers, etc.

3. Turbines: blades, distance pins, shaft-bearings.

4. Refrigeration systems: compressor parts, piston skirts, cylinder wall, shaft bearings, rod bearings, blades, roller wheels and valves.

5. Printing machines: needles, guidance, bearings.

6. Textile machinery: knitting machines: guidance, needles, connecting rods and bearings, mills-bearings, shafts, guidance.

7. Railway: wheel surfaces, brake discs, brake shoes, part of rails

8. Chemical industry: reactor vessel, tubes and pipes, valve system, nozzle

9. Oil industry: test of anti-wear and anti-corrosion properties of lubricants, transport pipe-lines.

10. Machinery: fabricating tools (turning tools, etc.), bearings and other machine parts.

11. Other: cooling systems with liquid metals.

Let us describe some of the applications reported in the literature. TLA is a convenient tool for activating thin surface layers in order to study real-time the surface loss by wear, corrosion or erosion processes of the activated parts, without disassembling or stopping running mechanical structures or equipment. The research problem is the determination of the irradiation parameters to produce point-like or large area optimal activity-depth distribution in the sample. Different activity-depth profiles can be produced depending on the type of the investigated material and the nuclear reaction used. To produce activity that is independent of the depth up to a certain depth is desirable when the material removed from the surface by wear, corrosion or erosion can be collected completely. By applying dual energy irradiation, the thickness of this quasi-constant activity layer can be increased or the deviation of the activity distribution from a constant value can be minimized. In the main, parts made of metals and alloys are suitable for direct activation, but by using secondary particle implantation the wear of other materials can also be studied in a surface range a few micrometers thick (Takács et al. 2007). In most practical cases, activation of a point-like spot (several mm) is enough to monitor the wear, corrosion or erosion, but for special problems relatively large surfaces areas of complicated spatial geometry need to be activated uniformly. Two ways are available for fulfilling this task, (i) production of large-area beam spot or scanning the beam over the surface in question from the accelerator side, or (ii) a programmed 3D movement of the sample from the target side. Taking into account the large variability of tasks occurring in practice, the latter method is preferred.

4.11 PET Applications in Industry

PET is also being developed for industrial applications and is expected to become an accepted technology in the future, especially in the chemical engineering industry. The most important current developments are in the earth sciences and especially in minerals exploration, extraction and process control.

As early as the late 1980s the application of PET for the non-invasive use to obtain quantitative, 2D/3D images of the flow and distribution of fluids inside process units, whose steel walls may be up to several centimeters thick was proposed Jonkers et al. (1990). With the aid of a NeuroECAT positron tomographer, the PET technique has been utilized to image important (model) processes in the petrochemical industry, using physical labeling of the phase to be imaged. First, the displacement of a brine/surfactant phase, labeled with ^{66}Ga-EDTA, in a piece of reservoir rock was imaged. Secondly, the dehydration of water-in-oil emulsions was monitored dynamically by labeling the water phase with ^{68}Ga-EDTA.

The second study in particular demonstrates that in the presence of noisy data the image reconstruction method utilized strongly influences the results obtained. With the advent of PET in nuclear medicine, the availability of short-lived positron-emitting nuclides such as ^{11}C ($t_{1/2}$ = 20 min), ^{13}N ($t_{1/2}$ = 10 min) and ^{15}O ($t_{1/2}$ = 2 min) has increased considerably, allowing the investigation of industrially important reactions by chemical labeling. Utilizing the NeuroECAT in a special mode, the catalytic oxidation of carbon monoxide could be imaged in a model tubular reactor by using ^{11}C-labeled CO, providing information about the kinetics of the individual reaction steps and interactions and about the degree of occupation of catalytically active sites (Jonkers et al. 1990).

PET has great potential as a non-invasive flow imaging technique in engineering, since 511 keV gamma rays can penetrate a considerable thickness of (e.g.) steel. The RAL/ Birmingham multiwire positron camera was constructed in 1984, with the initial goal of observing the lubricant distribution in operating aero-engines, automotive engines and gearboxes, and has since been used in a variety of industrial fields; see Parker et al. (1994). The major limitation of the camera for conventional tomographic PET studies is its re- stricted logging rate, which limits the frequency with which images can be acquired. Tracking a single small positron-emitting tracer particle provides a more powerful means of observing high-speed motion using such a camera. Following a brief review of the use of conventional PET in engineering, and the capabilities of the Birmingham camera, the paper by Parker et al. (1994) describes recent developments in the Positron Emission Particle Tracking (PEPT) technique and compares the results obtainable by PET and PEPT using, as an example, a study of axial diffusion of particles in a rolling cylinder.

PET has been adapted at the University of Birmingham for use in imaging industrial pro- cesses (Leadbeater 2009). PEPT technique was used to follow the motion of a single particle labeled with a positron-emitting radioactive isotope. The decay of the isotope and subsequent annihilation of the positron resulted in two back-to-back gamma rays, the detection of which defines a line along which the tracer particle is located. Triangulation of a small number of successive events allowed the particle to be accurately located in three dimensions on a short timescale. The dynamic behaviour of the material could be studied in detail using this method. Primarily, the work of the author was focused upon developing a modular positron camera that could be arranged in custom geometries around the system under study. This camera was transportable and has been used to investigate a number of applications in situ, opening a wide range of potential applications that had previously proved impossible to study using the PEPT technique. The modular positron camera resulting from this work offers a novel instrument with the potential to deliver new information on industrial processes. As well as proven suc- cessful operation on a number of applications, the performance and limitations of the camera have been investigated and described.

4.12 Accelerators in Automobile Industry

Many industrial applications of particle accelerators are particularly useful for the automotive industry. Indeed, components such as tires, foam, ball bearings, gears, camshafts and tie-rod ends are produced using either EB thermal processing or irradiation. Modern ion implantation systems make the advanced electronic systems in cars possible, see Table 4.4. Surface hard- ening can benefit any object that undergoes stress, extreme temperature or experiences wear. Most of the moving parts in a car could benefit from surface hardening and from new methods like ion implantation and material irradiation. Firestone was among first to devote a

TABLE 4.4

Electron beam irradiation machine manufacturers

No.	Company	Address	Product	web
1	Wasik Associates, Inc	Dracut, MA 01826-1500 USA	E-beam systems 300 keV to 1 MeV	wasik.com
2	WuXi El Pont Radiation Technology Co. Ltd	WuXi, Jiangsu 214151, China	E-beam accelerators 0.5–10 MeV	www.elpont.net/en
3	Mevex	Stittsville, ONK2S 1B4, Canada	1–40 MeV 1–120 kW	www.mevex.com
4	CGN Dasheng Electron Accelerator Technology Co., Ltd	China	Curtain: 0.2–0.35 MeV HF-HV: 0.5–5 MeV Resonant: 0.5–2.5 MeVLinac: 4–15 MeV	www.cgndea.com
5	IBAlon Beam Applications	Louvain-la-Neuve, Belgium	Turn-key irradiation solutions-different	www.iba-industrial.com
6	NJCWAY Company Ltd	Beijing 100084 China	1–10 MeV, 50 kW linacs + other	nutech.com
7	ITHP Subsidiary of the Alcen Group	Drele46500 Thegra France	Pulsed Electron Beam Sterilization	https://www.ithpp-alcen.com/en/
8	Sciaky, Inc.	4915 W 67th St, Chicago, IL, USA	welding systems and EBAM® industrial 3d printing machines.	https://www.sciaky.com/
9	PTR-Precision Technologies, Inc.	Am Erlenbruch 963505 Langenselbold · Germany	Electron Beam Welding equipment.	https://www.ptr-ebeam.com/
10	Cambridge Vacuum Engineering (CVE)	Waterbeach, Cambridge CB25 9QX UK	Manufacturers of Electron Beam Welders, 50–200 kV, to 100 kW.	https://www.camvaceng.com/
11	TECHMETA	74370 Épagny Metz-Tessy France	electron-beam welding machines	https://techmeta-engineering.com/
12	Pro beam	Zeppelinstr. 26 82205 Gilching Germany	Electron beamWelding, drilling, hardening	https://www.pro-beam.com/en/
13	Orion	9 Kosinskaya str., Moscow, 111538, Russia	electron-beam, vacuum, laser, molecular-beam.	http://www.orion-ir.ru/en/company/
14	Omegatron	Hitachinaka, Ibaraki, 312-0058, Japan	electron guns and ion guns, precision instruments and machinery	http://www.omegatron.co.jp
15	Energy Sciences Inc., ESI	42 Industr. Way Wilmington, MA 01887 USA	Electron beam + total solution approach	https://www.ebeam.com/
16	Electron Crosslinking AB	Halmstad Sweden	Electron beam accelerators, technology, dosimetry	https://www.crosslinking.com/
17	VIVIRAD S.A.	23, Rue Princip. 67117 Strasbourg France	Electron Beam Systems from 300 KeV to 10 MeV,	https://www.vivirad.com/

considerable effort to the better understanding of the possible applications of such radiation to the production of tires or tire components (Hunt and Alliger 1979). A number of chemical reactions can occur when elastomeric compounds are exposed to high-energy waves, but crosslinking and degradation are the most important. The degree to which the crosslinking reaction predominates depends upon the nature of the rubber, compounding ingredients and the dosage.

In general, the effects achieved by the radiation of a rubber are quite similar to those resulting from heat. However, radiation cure or precure of compounds offers the advantage that the degree of crosslinking can be better controlled. Uniform crosslinking is possible since the high-speed electrons penetrate uniformly throughout the sample. Curing with heat on the other hand may result in a greater degree of crosslinking on the surface of the sample than the center because of low heat conductivity. In general, radiation can be used to advantage to crosslink partially rubber tire components so that they retain better their shape and dimension during tire assembly and final cure or vulcanization. Added advantages of radiation precure include a reduction of material usage, substitution of synthetic for natural rubber without loss in strength and the fact that partially cross-linked components will not thin out or become displaced during construction and vulcanization of the tire (Tables 4.5 and 4.6).

TABLE 4.5

Accelerator-based electron beam technologies were realized recently for different bipolar devices (Adopted from Pavlov et al. 2016)

Device type	Task realized	Irradiation fluence (cm^{-2}) dose (kGy)	Limiting parameter
Compensated p^-/n^- structure of presice Zener diode (6.5V@ 7.5 mA).	U_F reduction from 0.64 to 0.54 V for Au-doped structures.	$3-10^{17}$ cm^{-2} (320 °C, 2 h anneal)	Irradiation time
Pulsed p^-nn^- diode (20 μm n-epi base) (1 A@ 200 V)	t_α reduction to 5 ns	2000 kGy (350 °C, 1 h anneal)	U_F increasing
High frequency bipolar transistor with comb-like 'npn' structure (5A@200V)	t_s reduction from 700 to 200 ns	200–300 kGy 390 °C, 1 h anneal)	Gain reduction to from 60 to 50
Darlington bipolar transistor	Gain, its spread and t_s reduction to optimum values	100–200 kGy (300–390 °C, 1–2 h anneal)	Gain reduction
Insulated gate transistor IGBT	Switching time reduction	150–250 kGy (330 °C, 1 h anneal)	Gain reduction
Fast recovery diode 50A@1700V	Reverse recovery time reduction up to 200 ns	200 kGy (330 °C, 1 h anneal)	Reverse current, softness
Fast recovery diode 1, 5, 15A@400, 600V	t_{rr} reduction to 30–120 ns (for different types)	400 kGy (330 °C, 1 h anneal)	U_F and reverse current increasing
Welding diode 7.1kA@400V, 600 V	t_{rr} reduction to 3 ns, frequency increasing up to 12 kHz	250 kGy (250 °C, 4 h anneal)	Middle softness; U_F and reverse current increasing
HC HV Thyristor (different types)	Precise regulation parameters (spread less than 5%)	In situ lifetime control on test structures (<10 kGy)	On-state voltage drop increasing
mc-Silicon solar cell 15% efficiency	Radiation degradation investigation efficiency decreasing to 1%	$1 \cdot 10^{12}$–$5 \cdot 10^{16}$ cm^{-2}.	–
Superluminescent heterostructure diode	Culling structures with hidden defects detection	<40 (+ time to failure)	–

TABLE 4.6

Summary

Industrial process	Procedures	Car part
Electron beam material irradiation (radiation processing)	Cross-linking of polymers	High-performance eleectrical wires, Insulation, Plastic PE foam, tyre components
	X-rays curing of composites	Fender, Body frame, Panels
Electron beam material processing(thermal processing)	Welding	Gears, Drive rings, Tie-rod ends, Turbocharger, Airbag Igniter Cartridges
	Heat treating (surface hardening)	Cam shaft
Ion implantation	Ion implantation into semiconductors	Electronics
	Ion implantation into metals (Surface hardening)	Cam shaft, Valves, Pistons, Crankshaft, Rocker Arms, Brake Pads, Brake Disks, Brake Calliper.

By now, it is established that replacing steel with X-ray cured carbon composites can reduce car energy consumption by 50%. Carbon–fiber-reinforced composites were cured in molds using X-rays derived from a high-energy, high-current EB. X-rays could penetrate the mold walls as well as the fiber reinforcements and polymerize a matrix system. Matrix materials made from modified epoxy-acrylates were tailored to suitably low viscosity so that fiber wetting and adhesion could be attained. Techniques similar to vacuum-assisted resin transfer molding (VARTM) and conventional vacuum bagging of wet lay-ups were used. Inexpensive reinforced polyester molds were used to fabricate vehicle fenders. Moderately low-dose X-ray exposure was sufficient to attain functional properties, such as resistance to heat distortion at temperatures as high as 180°C. The matrix system contained an impact additive that imparted toughness to the cured articles. "Class A" high gloss surfaces were achieved. Thermo-analytical techniques were used on small-sized samples of X-ray-cured matrix materials to facilitate the selection of a system for use in making prototypes of vehicle components. X-rays-penetrated metal pieces that were placed within layers of carbon-fiber twill, which were cured and bonded into a structure that could be mechanically attached without concern over fracturing the composite. X-ray curing is a low-temperature process that eliminates residual internal stresses which are imparted by conventional thermo-chemical curing processes (Herer et al. 2009).

References

Ahmad, I. and Akram, W. 2017. Introductory Chapter: Introduction to Ion Implantation, Ion Implantation – Research and Application, Ahmad, I., IntechOpen, 10.5772/intechopen.68785.

Albert, P., Blondiaux, G., Debrun, J. L., Giovagnoli, A., and Valladon, M. 1987. Thick target yields for the production of radioisotopes. In Handbook on Nuclear Activation Data, Technical Reports Series No. 273: 537–630. IAEA, Vienna, Austria.

Bellagamba, B., Caridi, A., Cereda, E., Braga Marcazzan, G. M., and Valković, V. 1993. PIXE Application of the study of trace element behaviour in coal combustion cycle. *Nucl. Instrum. Methods Nucl. Res. B* **75**: 222–229.

Bourelle, E., Suzuki, A., Sato, A., Seki, T., and Matsuo, J. 2005. Sidewall polishing with a gas cluster ion beam for photonic device applications. *Nucl. Instrum. Methods Phys. Res. B* 241: 622–625.

Bungau, C., Barlow, R., Cywinski, R., and Tygier, S. Accelerator driven systems for energy production and waste transmutation. TUPP147: 1854–1856. Proceedings of 11th biennial European Particle Accelerator Conference, EPAC08, Genoa, Italy.

Chao, A. W. and Chou, W. Eds. 2011. *Reviews of Accelerator Science and Technology, Volume 4: Accelerator Applications in Industry and the Environment.* World Scientific Publishing Co., Singapore.

Chao, A. W. and Chou, W. Eds. 2016. *Reviews of Accelerator Science and Technology, Volume 8: Accelerator Applications in Energy and Security.* World Scientific Publishing Co., Singapore.

Cherif, S. H. and Valković, V. 1996. An Overview of Accelerator Based Technologies, IAEA TECDOC, IAEA, Vienna, Austria.

Chiumenti, M., Cervera, M., Dialami, N., Wu, B., and Jinwei, L. 2016. Numerical modeling of the electron beam welding and its experimental validation. *Finite Elem. Anal. Des.* 121: 118–133.

Chmielewski, A.G., Licki, J., Pawelec, A., Tymiński, B., and Zimek, Z., 2004. Operational experience of the industrial plant for electron beam flue gas treatment. *Rad. Phys. Chem.* 71: 439–442.

Chmielewski, A. G. 2019. Recent progress in the radiation technology development achieved by INCT, Poland – R&D, service, collaboration, education and promotion. Presented at IAEA Consultancy Meeting held on 11–12 November 2019, Vienna, Austria.

Cleland, M., Galloway, R., Montoney, D., et al. 2009. Radiation Curing of Composites for Vehicle Components and Vehicle Manufacture. International Topical Meeting on Nuclear Research Applications and Utilization of Accelerators 4-8 May 2009, Vienna – Paper AP/IA-04.

Costa, J. P. C., Assis, M., Teodoro, V., et al. 2020. Electron beam irradiation for the formation of thick Ag film on Ag_3PO_4. *R. Soc. Chem. Adv.* 10: 21745–21753.

Current, M. I. 2017. Perspectives in low-energy ion (and neutral) implantation. Proceedings of International Workshop on Junction Technology-2017, Uji, Japan.

Current, M. I. 2019. The role of ion implantation in CMOS scaling: a tutorial review. *AIP Conf. Proc.* 2160: 020001; 10.1063/1.5127674

Das, S., Yu, H., Tarleton, E., et al. 2019. Hardening and Strain Localisation in Helium-Ion-Implanted Tungsten. *Sci. Rep.* 9: 18354. 10.1038/s41598-019-54753-3

De Forest, L. 1907. Space Telegraphy. US Patent 879,532 (January 29, 1907).

Dennard, R. H., Gaensslen, F. H., Yu, H.-N., Rideout, V. L., Bassous, E., and LeBlanc, A. R. 1974. Design of ion-implanted MOSFET's with very small physical dimensions. *IEEE J. Solid-State Circuits* 9: 256–268.

EPA. 2015. Emission Control Technologies. https://www.epa.gov/sites/default/files/2015-07/documents/chapter_5_emission_control_technologies.pdf

Felter, T. E., Musket, R. G., and Bernhardt, A. F. 2005. Field emitter arrays and displays produced by ion tracking lithography. *Nucl. Instrum. Methods Phys. Res. B* 241: 346–350.

Fleming, J. A. 1905. Instrument for Converting Alternating Electric Currents into Continuous Currents. US Patent 803684 (April 29, 1905).

Fonseca, A., Franco, N., Alves, E., Barradas, N. P., Leitao, J. P., Sobolev, N. A., Banhart, D. F., Presting, H., Ulyanov, V. V., and Nikiforov, A. I. 2005. High resolution back scattering studies of nanostructured magnetic and semiconducting materials. *Nucl. Instrum. Methods Phys. Res. B* 241: 454–458.

Fuechsle, M., Miwa, J. A., Mahapatra, S., Ray, H., Lee, S. O. Warschkow, O., Hollenberg, L. C. L., Klimeck, G., and Simmons, M. Y. 2012. A single atom transistor. Nature Nanotechnology (Feb. 19, 2012): 1–5.

Fujii, S., Ogihara, M., Shimizu, M., et al. 1989. A 45 ns 16-Mbit DRAM with triple-well structure. *IEEE J. Solid State Circuits* 24: 1170–1175.

Garnett, R. W. and Sheffield, R. L. 2015. Overview of accelerator applications in energy. *Rev. Accel. Sci. Technol.* 8: 1–25.

Gilliam, S. B., Gidcumb, S. M., Forsythe, D., Parikh, N. R., Hunn, J. D., Snead, L. L., and Lamaze, G. P. 2005. Helium retention and surface blistering characteristics of tungsten with regard to first wall conditions in an inertial fusion energy reactor. *Nucl. Instrum. Methods Phys. Res. B* 241: 491–495.

Greer, J., Fenner, D., Hautala, J., et al. 2020. Etching, smoothing, and deposition with gas-cluster ion beam technology. *Surf. Coat. Technol.* 133: 273–282.

Hamm, R. W. and Hamm, M. E. 2011. The beam business: accelerators in industry. *Phys. Today* 64(6): 46–51.

Hamm, R. W. and Hamm, M. E., Eds. 2012. *Introduction to the Beam Busines, in Industrial Accelerators and their Applications* (World Scientific, Singapore, 2012), ISBN-13 978-981-4307-04-8.

Hamm, R. W. 2012. Review of Industrial Accelerators and Their Applications. Paper AP/IA-12 International Topical Meeting on Nuclear Research Applications and Utilization of Accelerators, IAEA Vienna, May 4–8, 2009.

Henniges, U., Hasani, M., Potthast, A., et al. 2013. Electron beam irradiation of cellulosic *Mater. Opport. Limit. Mater.* 2013(6): 1584–1598.

Herer, A., Galloway R., Cleland, M. R., et al. 2009. X-ray-cured carbon–fiber composites for vehicle use. *Radiat. Phys. Chem.* 78(7): 531–534.

HERTIS. 2021. Hybrid Exhaust-gas-cleaning Retrofit Technology for International Shipping project (https://www.rtu.lv/en/hep/accelerator-technologies/international-projects/hertis).

Hill, R., Blenkinsopp, P., Thompson, S., Vickerman, J., and Fletcher, J. S. 2011. A new time-of-flight SIMS instrument for 3D imaging and analysis. *Surf. Interface Anal.* 43(1–2): 506–509.

Hojo, T., Yamamoto, H., Aihara, J., Furuno, S., Sawa, K., Sakuma, T., and Hojou, K. 2005. Radiation effects on yttria-stabilized zirconia irradiated with He or Xe ions at high temperature.*Nucl. Instrum. Methods Phys. Res. B* 241: 536–542.

Hunt, J. D. and Alliger, G. 1979. Rubber – application of radiation to tire manufacture. *Radiat. Phys. Chem. (1977)* 14(1–2): 39–53.

IAEA. 1987. Electron beam processing of combustion flue gases. IAEA-TECDOC-428. International Atomic Energy Agency. Vienna, Austria.

IAEA. 1997. The thin layer activation method and its application in industry. IAEA-TECDOC-924. International Atomic Energy Agency. Vienna, Austria.

IAEA. 2004. Emerging applications of radiation processing. Proceedings of a technical meeting held in Vienna, 28–30 April 2003. IAEA-TECDOC-1386. International Atomic Energy Agency. Vienna, Austria.

IAEA. 2015. Utilization of accelerator based real time methods in investigation of materials with high technological importance. IAEA Radiation Technology Reports No. 4. Proceedings of an IAEA technical meeting, International Atomic Energy Agency. Vienna, Austria.

IAEA. 2019. Improvement of the reliability and accuracy of heavy ion beam analysis. Technical Reports Series No. 485. International Atomic Energy Agency. Vienna, Austria.

IAEA. 2020. Radiotracers techniques for wear, erosion and corrosion measurement. IAEA-TECDOC-1897. International Atomic Energy Agency. Vienna, Austria.

Idris, S., Hairaldin, S. Z., Tajau, R., et al. 2014. An experience of electron beam (EB) irradiated gemstones in malaysian nuclear agency. *AIP Conf. Proc.* 1584: 170–174.

Ionoptika. 2021. www.ionoptila.com/ion-beams-their-applications/

Irwin, D. A. and Klenow, P. J. 1996. Sematech: purpose and performance. *Proc. Natl. Acad. Sci. USA* 93: 12739–12742.

Jamison, D. N., Lawrie, W. L. L., Robinson, S. G., Jakob, A. M., Johnson, B. C., and McCallum, J. C. 2017. Deterministic doping. *Mater. Sci. Semicond. Process.* 62: 23–30.

Jiang, W., Weber, W. J., and Boatner, L. A. 2005. Accumulation and recovery of disorder in Au^{2+}-irradiated $Cd_2Nb_2O_7$. *Nucl. Instrum. Methods Phys. Res. B* 241: 372–376.

Ji, Q., Leung, K.-N., and King, T.-J. 2005. Development of focused ion beam systems with various ion species. *Nucl. Instrum. Methods Phys. Res. B* 241: 335–340.

Jin, J., Wang, W., and Chen, X. 2019. Microstructure and mechanical properties of Ti + N ion implanted cronidur30 Steel. *Materials* 12(3): 427. Published online 2019 Jan 30. 10.3390/ma12030427

Jinbo, T., Nakata, H., Hashimoto, K., et al. 1992. A 5-V-only 16-Mb flash memory with sector erase mode. *IEEE J Solid State Circuits* 27: 1547–1533.

Jonkers, G., van den Bergen, E. A., and Vonkeman, K. A. 1990. Application of positron emission tomography in industrial research. 19. Japan conference on radiation and radioisotopes,

Tokyo (Japan); 14–16 Nov 1989. Nippon Aisotopu Hoshasen Sogo Kaigi Hobunshu. CODEN NAHHE 19: 157–173.

Jusri, N. A. A., Azizan, A., Ibrahim, N., Mohd Salich, R., and Abd Rahman, M. F. 2018. Pretreatment of cellulose by electron beam irradiation method. *IOP Conf. Ser. Mater. Sci. Eng.* 358: 012006.

Kawamura, K., Aoki, S., Kimura, H., Adachi, K., Kawamura, K., Taketomo, E., Katayama, T., and Kaido, C. 1980. On the removal of NOx and SOx in exhaust gas from the sintering machine by electron beam irradiation. *Radiat. Phys. Chem. (1977)* 16 (2): 133–138.

Kim, J. J., Ha, J. M., Lee, H. M., Raza, H. S., Park, J. W., and Cho, S. O. 2016. Effect of electron-beam irradiation on organic semiconductor and its application for transistor-based dosimeters. *ACS Appl. Mater. Interfaces* 8(30): 19192–19196.

Konstantinov, I. O. and Krasnov, N. N. 1971. Determination of the wear of machine parts by charged particle surface activation. *J. Radioanal. Chem.* 8(357): 1971.

Kussul, E. M., Rachkovskij, D. A., Baidyk, T. N., and Talayev, S. A. 1996. Micromechanical engineering: a basis for the low-cost manufacturing of mechanical microdevices using micro-equipment. *J. Micromech. Microeng.* 6: 410–425.

Kuo, T.-C., Shih, T.-L., Su, Y.-H., Lee, W.-H., Current, M. J., and Samukawa, I. S. 2018. Neutral beam and ICP etching of HKMG MOS capacitors: Observations and plasma-induced damage model. *J. Appl. Phys.* 123: 161517 (6 pp).

Lake, R. E., Persaud, A., Christian, C., Barnard, E. S., Chan, E. M., Bettiol, A. A., Tornut, M., Trautmann, C., and Schenkel, T. 2021. Direct formation of nitrogen-vacancy centers in nitrogen doped diamond along trajectories of swift heavy ions. *Appl. Phys. Lett.* 118: 084002 (5 pp).

Leadbeater, T. W. 2009. The Development of positron imaging systems for applications in industrial process tomography. PhD thesis, The University of Birmingham. https://core.ac.uk/download/pdf/40041877.pdf

Lee, K., Lee, G., Han, K., Han, J., and Chung, K. 2008. Cellulose modification study by e-beam irradiation & its applications. 2008 IEEE 35th International Conference on Plasma Science, Karlsruhe, 2008, pp. 1-1, 10.1109/PLASMA.2008.4590926

Lian, J. L. Wang, L. M., Ewing, R. C., Boatner, L. A. 2005. Ion beam implantation and cross-sectional TEM studies of lanthanide titanate pyrochlore single crystals. *Nucl. Instrum. Methods Phys. Res. B* 241: 365–371.

Licki, J., Pawelec, A., Zimek, Z., and Witman-Zając S. 2015. Electron beam treatment of simulated marine diesel exhaust gases. *Nukleonika* 60(3): 689–695.

Liu, B., Dong, Z. L., Hua, Y., Fu, C., Li, X., Tan, P. K. and Zhao, Y. 2018. Electron-beam radiation induced degradation of silicon nitride and its impact to semiconductor failure analysis by TEM. *AIP Adv.* 8(115327): 11. 10.1063/1.5051813

Livingston, R. A., Schweitzer, J. S., Rolfs, C., et al. 2010. Heavy ion beam measurement of the hydration of cementitious materials. *Appl. Radiat. Isot.* 68: 683–687.

Machi, S., Tokunaga, O., Nishimura, K., Hashimoto, S., Kawakami, W., Washino, M., Kawamura, K., Aoki, S., and Adachi, K. 1977. Radiation treatment of combustion gases. *Radiat. Phys. Chem.* 9 (1–3): 371–388.

Machi, S. 2011. Trends for electron beam accelerator applications in industry. *Rev. Accel. Sci. Technol.* 4: 1–10.

Maiorino, J. R., dos Santos, A., and Pereira, S. A. 2003. The utilization of accelerators in subcritical systems for energy generation and nuclear waste transmutation – the World Status and a Proposal of a National R&D Program. *Braz. J. Phys.* 33(2): 267–272.

Manutchehr-Danai, M. (Ed.). 2009. Heat Treatment and Irradiation of Topaz. In: *Dictionary of Gems and Gemology*. Springer, Berlin, Heidelberg. 10.1007/978-3-540-72816-0_10446

Matteson, S., Avara, M. J., Nguyen, C. V., and Kim, S. H. 2005. RBS characterization of uranium in flint and chert. *Nucl. Instrum. Methods Phys. Res. B* 241: 465–469.

Mitchell, L. J., Holland, O. W., Hossain, K., Smith, E. R., Golden, T. D., Duggan, J. L., and McDaniel, F. D. 2005. Formation of optically-active, metal silicides using ion implantation and/or oxidation. *Nucl. Instrum. Methods Phys. Res. B* 241: 548–552.

Moore, G. 1964. The Future of Integrated Electronics. Fairchild Semiconductor internal publication.

Moore, G. 1965. Cramming more components onto integrated circuits. *Electron. Mag.* 38(8): April 19, 1965.

Niu, H., Chen, C. H. Wang, H. Y., Wu, S. C., and Lee, C. P. 2005. Ion-channeling studies of InAs/GaAs quantum dots. *Nucl. Instrum. Methods Phys. Res. B* 241: 470–474.

Novak, S. W., Magee, C. W., and Buyuklimanli, T. 2005. Using accelerator techniques to verify details in SIMS profiles. *Nucl. Instrum. Methods Phys. Res. B* 241: 321–325.

NRC. 2021. Backgrounder on Irradiated Gemstones. https://www.nrc.gov/reading-rm/doc-collections/fact-sheets/irradiated-gemstones.html. Page Last Reviewed/Updated Wednesday, January 6, 2021.

Park, J.-H., Ahn, J.-W., Kim, K.-H., and Son, Y.-S. 2019. Historic and futuristic review of electron beam technology for the treatment of SO_2 and NOx in flue gas. *Chem. Eng. J.* 355: 351–366.

Parker, D. J., Hawkesworth, M. R., Broadbent, C. J., Fowles, P., Fryer, T. D., and McNeil, P. A. 1994. Industrial positron-based imaging: Principles and applications. *Nucl. Instrum. Methods Phys. Res. Sect. A Accel. Spectrom. Detect. Assoc. Equip.* 348(2–3): 583–592.

Pavlov, Yu. S. and Sharamentov, S. I. 1977. Transistorized Frequency Divider of Microwave Pulses. Instrum. Exp. Tech. 20: 149–152. II Conference on Plasma & Laser Research and Technologies IOP Publishing Journal of Physics: Conference Series 747 (2016) 012085 10.1088/1742-6596/747/1/012085.

Pavlov, Yu. S., Solovev, N. G., and Tomnikov, A. P. 1985. Syncronization circuit for shaping picosecond accelerated-electron pulses. *Instrum. Exp. Tech.* 28: 1335–1339.

Pavlov, Yu. S. and Lagov, P. B. 2015. Magnetic buncher accelerator for radiation hardness research and pulse detector characterization. 15th European Conference on Radiation and Its Effects on Components and Systems (RADECS) 336–338.

Pavlov, Yu. S., Dobrohotov, B. B., Pavlov, B. A., Nepomnyaschy, O. N., and Danilichev, V. A. 2014. Magnetic buncher accelerator UELV-10-10-T-1 for studing fluorescence and radiation physical researches Proceedings of XXIV Russian Particle Accelerators Conference (RuPAC'2014) Obninsk, Russia, 259–261.

Pavlov, Yu. S., Surma, A. M., Lagov, P. B., Fomenko, Y. L., and Geifman, E. M. 2016. Accelerator-based electron beam technologies for modification of bipolar semiconductor devices. II Conference on Plasma & Laser Research and Technologies, Journal of Physics: Conference Series 747: 012085. 10.1088/1742-6596/747/1/012085

Pikaev, A. K., Glazunov, P. Y., and Pavlov, Yu S. 1993. Radiation Center of Institute-of-Physical Chemistry in Moscow. *Radiat. Phys. Chem.* 42: 887–890.

Qtool. 2011. Calculation of Reaction Q-values and thresholds. https://t2.lanl.gov/nis/data/qtinfo.html

Research and Markets. 2019. Heat-Shrink Tubing Market by Voltage (Low, Medium, and High), Material (Polyolefin, Polytetrafluoroethylene, Fluorinated Ethylene Propylene), End-User (Utilities, Chemical, Automotive, Food & Beverage), Region – Global Forecast to 2024. https://www.researchandmarkets.com/r/ii3ru1.

Richardson, O. W. 1901. On the Negative Radiation from Hot Platinum. *Proc. Cambridge Philos. Soc.* 11: 286–295.

Rubbia, C., et al. 1995. Conceptual design of a fast neutron operated high-power Energy Amplifier. Report CERN/AT 95-44-ET.

Rubin, L. and Poate, J. 2003. Ion Implantation in Silicon Technology. The Industrial Physicist, June/July 2003, pp. 12–15.

Sciaky. 2021. Metal 3D Printing with Electron Beam Additive Manufacturing (EBAM®). https://www.sciaky.com/additive-manufacturing/electron-beam-additive-manufacturing-technology.

Seetala, N. V., Harrell, J. W., Lawson, J., Nikles, D. E., Williams, J. R., and Isaacs-Smith, T. 2005. Ion-irradiation induced chemical ordering of FePt and FePtAu nanoparticles. *Nucl. Instrum. Methods Phys. Res. B* 241: 583–588.

Shemwell, B. E., Ergut, A., and Levendis, Y. A. 2002. Economics of an integrated approach to control SO_2, NO_x, HCl, and particulate emissions from power plants. *J. Air & Waste Manag. Assoc.* 52(5): 521–534.

Shetty R. H. (2000) Surface Hardening of Orthopedic Implants. In: Wise D. L., Trantolo D. J., Lewandrowski K. U., Gresser J. D., Cattaneo M. V., Yaszemski M. J. (eds) *Biomaterials Engineering and Devices: Human Applications*. Humana Press, Totowa, NJ. 10.1007/978-1-59259-197-8_11

Shinada, T., Okamoto, S., Kobayashi, T., and Ohdomari, I. 2005. Enhancing semiconductor device performance using ordered dopant arrays. *Nature* 437: 1128–1131.

Simčić, J., Pelicon, P., Rupnik, Z., Mihelič, M., Razpet, A., Jenko, D., and Macek, M. 2005. 3D micromachining of SU-8 polymer with proton microbeam. *Nucl. Instrum. Methods Phys. Res. B* 241: 479–485.

Straede, Chr. A. 1992. Ion implantation as an efficient surface treatment. *Nucl. Instrum. Methods Phys. Res.* B68: 380–388.

Swenson, D. R. 2005. Analysis of charge, mass and energy of large gas cluster ions and applications for surface processing. *Nucl. Instrum. Methods Phys. Res. B* 241: 599–603.

Swenson, D. R. 2004. Measurement of averages of charge, energy and mass of large, multiply charged cluster ions colliding with atoms. *Nucl. Instrum. Methods B* 222: 61–67.

Swenson, D. R., Degenkolb, E., Insepov, Z., Laurent, I., and Scheitrum, G. 2005. Smoothing RF cavities with gas cluster ions to mitigate high voltage breakdown. *Nucl. Instrum. Methods Phys. Res. B* 241: 641–644.

Takács, S., Ditrói, F., and Tárkányi, F. 2007. Homogeneous near surface activity distribution by double energy activation for TLA. *Nucl. Instrum. Methods Phys. Res. B* 263: 140–143.

Teranishi, N., Fuse, G., and Sugitani, M. 2018. A review of ion implantation technology for image sensors. *Sensors* 18: 2358, 10.3390/s18072358.

Tian, H., Sparvero, L. J., Amoscato, A. A., Bloom, A., Bayır, H., Kagan, V. E. and Winograd, N. 2017. Gas cluster ion beam time-of-flight secondary ion mass spectrometry high-resolution imaging of cardiolipin speciation in the brain: identification of molecular losses after traumatic injury. *Anal. Chem.* 89: 4611–4619.

Toyoda, N. and Yamada, I. 2008. Gas cluster ion beam equipment and applications for surface processing. *IEEE Trans. Plasma Sci.* 36(4): 1471–1488.

Toyoda, N. and Uematsu, K. 2019. Atomic layer etching by gas cluster ion beams with acetylacetone. *Jpn. J. Appl. Phys.* 58, SEEA01 (pp 5).

Tsukamoto, K., Kuroi, T., and Kawasaki, Y. 2010. Evolution of ion implantation technology and its contribution to semiconductor industry. Ion Implantation Technology-2010, *AIP Proc* 1321: 9–16.

Umezawa, A., Atsumi, S., Kuriyama, M., et al. 1992. A 5-V-only operation 0.6 µm flash EEPROM with row decoder scheme in triple-well structure. *IEEE J. Solid State Circuits* 27: 1540–1545.

Valković, V. 1983. *Trace Elements in Coal, Vol. I and II*. CRC Press, Boca Raton, FL, USA.

Valković, V. 2006. Applications of nuclear techniques relevant for civil security. *J. Phys. Conf. Ser.* 41: 81–100.

Valković, V., Makjanić, J., Jakšić, M., Popović, S., Bos, A. J. J., Vis, R. D., Wiederspahn, K., and Verheul, H. 1984. Studies of fly ash particles by X-Ray emission spectroscopy. *Fuel* 63: 1357–1362.

Valković, V. and Zyszkowski, W. 1994. Accelerators in science and industry: focus on the Middle East & Europe. *IAEA Bullet.* 36: 24–29.

Valković, O., Jakšić, M., Caridi, A., Cereda, E., Haque, A. M. I., Cherubini, R., Moschini, G., Menapace, E., Markowicz, A., Makarewicz, M., and Valković; V. 1992. X-Ray and gamma spectroscopy of coal and coal ash samples. *Nucl. Instrum. Methods Nucl. Res. B* 66: 479–484.

Valković, V., Cereda, E., Menapace, E., Moschini, G., and Valković, O. 1992. Environmental and human health impact of coal burning power stations. *Int. J. PIXE* 2: 575–592.

Wang, S.-K., Chang, H.-Y., Chu, Y.-H., Kao, W.-L., Wu, C.-Y., Lee, Y.-W., You, Y.-W., Chu, K.-J., Hung, S.-H., and Shyue, J.-J. 2019. Effect of energy per atom (E/n) on the Ar gas cluster ion beam (Ar-GCIB) and O_2^+ cosputter process. *Analyst* 144(10): 3323–3333.

Weiser, M., and Kalbitzer, S. 1985. Observation of Quantum Mechanical Diffusion of ^{1}H in Ta at Low Temperatures. *Z. Phys. Chem.e Folge* 143: 183-106. R. Oldenbourg Verlag, München, Germany.

Williamson, C. F., Boujot, J. P., and Picard, J. 1966. Table of range and stopping power of chemical elements for charged particles of energy 0.05 to 500 MeV. Rapport CEA-R-3042, Saclay, France.

Winograd, N. 2018. Gas cluster ion beams for secondary ion mass spectrometry. *Ann. Rev. Anal. Chem.* 11: 29–48.

Woo, H. J., Choi, H. W., Kim, J. K., Kim, G. D., Hong, W., Choi, W. B., and Bae, Y. H. 2005. Thick Si-on-insulator wafers formation by ion-cut process. *Nucl. Instrum. Methods Phys. Res. B* 241: 531–535.

Wu, H. 2002. Electron Beam Application in Gas Treatment in China. FNCA Workshop, Takasaki, Dec. 16, 2002.

Xiao, Z., Zimmerman, R. L., Holland, L. R., Zheng, B., Muntele, C. I. D., and Ila, D. 2005. MeV Si ion bombardments of thermoelectric Bi_xTe_3/Sb_2Te_3 multilayer thin films for reducing thermal conductivity. *Nucl. Instrum. Methods Phys. Res. B* 241: 568–572.

Yamada, I., Matsuo, J., Toyoda, N., and Kirkpatrickc, A. 2001. Materials processing by gas cluster ion beams. *Mater. Sci. Eng.*: R34(6): 231–295.

Yamada, I. and Toyoda, N. 2005. Recent advances in R&D of gas cluster ion beam processes and equipment. *Nucl. Instrum. Methods Phys. Res. B* 241: 589–593.

Yamada, I. and Toyoda, N. 2009. Review of Cluster Ion Beam Facilities and Technology. Applications of Accelerators in Research and Industry 2008, (McDaniel, F. D. and Doyle, B. L., Eds.). Melville, New York, AIP Conference Proceedings, 1099: 13–16.

Zimek, Z. 2001. Electron accelerators for radiation processing: Criterions of selection and exploitation. International symposium on utilization of accelerators; Sao Paulo (Brazil); 26–30 Nov 2001. Report IAEA-SM-366/11, pp. 4–5. International Atomic Energy Agency, Vienna, Austria.

Zimek, Z. 1995. High power electron accelerators for flue gas treatment. *Radiat. Phys. Chem.* 45(6): 1013–1015.

Zwolińska, E., Sun, Y., Chmielewski, A. G., Pawelec, A., and Bułka, S. 2020. Removal of high concentrations of NOx and SO_2 from diesel off-gases using a hybrid electron beam technology. *Energy Rep.* 6: 952–964.

Additional reading

Bird, J. R. and Williams, S. J., Eds. 1989. *Ion Beams for Materials Analysis*. Academic Press, Sydney. ISBN: 978-0-12-099740-4.

Chatterjee, P. 2007. Device Scaling: The Treadmill that Fueled Semiconductor Industry Growth. IEEE SSCS Newsletter (Winter/January 2007).

Ditrói, F. 2002. Determination of the charged particle beam energy/intensity uncertainties at multitarget irradiations. *Nucl. Instrum. Methods Phys. Res.* B188: 115–119.

Eberle, D. C. C., Wall, C. M. M., and Treuhaft, M. B. B. 2005. Applications of radioactive tracer technology in the real-time measurement of wear and corrosion. *Wear* 259: 1462–1471.

Essig, G, and Fehsenfeld, P. 1980. Thin layer activation technique and wear measurements in mechanical engineering in Nuclar Physics Methods in Materials Research, Proceedings of the Seventh Divisional Conference Darmstadt, September 23–26, 1980. Bethge, K. (Ed) Springer Link. Pp 70–81.

Feldman, L. C., Mayer, J. W., and Picraux, S. T. 1982. *Materials Analysis by Ion Channeling*. Academic Press, New York.

Hamm, R. W. and Hamm, M. E., Eds. 2012. Introduction to the Beam Business. *Industrial Accelerators and their Applications*. World Scientific, Singapore. Pp 57–85.

Hoffmann, M., et al. 2001. ^{7}Be recoil implantation for ultra-thin-layer-activation of medical grade polyethylene: effect on wear resistance. *Nucl. Instrum. Methods Phys. Res.* B183: 419–424.

Kiraly, B., et al. 2009. Evaluated activation cross sections of longer-lived radionuclides produced by deuteron induced reactions on natural iron. *Nucl. Instrum. Methods Phys. Res.* B267: 15–22.

Krasnov, N. N. 1974. Thick target yield. *Int. J. Appl. Radiat. Isot.* 25: 223–227.

Kučera, J., Havranek, V., Smolik, J. Schwarz, J., Vesely, V., Kugler, J., Sykorova, I., and Santroch, J. 1999. INAA and PIXE of atmospheric and combustion aerosols. *Biol. Trace Elem. Res.* 71–72: 233–245.

Lacroix, O., Sauvage, T., and Blondiaux, G. 1997. Methodological aspects of recoil nuclei implantation technique applied in tribology or corrosion studies. Fourteenth Int. Conf. Appl. Accel. Res. Ind., vol. 392, AIP Publishing; pp. 969–972.

Lacroix, O., et al. 1997. Ultra-thin layer activation by recoil implantation of radioactive heavy ions: applicability in wear and corrosion studies. *Nucl Instrum. Methods Phys. Res.* B122: 262–268.

Lenauer, C., et al. 2015. Piston ring wear and cylinder liner tribofilm in tribotests with lubricants artificially altered with ethanol combustion products. *Tribol. Int.* 82: 415–422.

Mackova, A., Havranek, V., Vacık, J., Salavcova, L., and Spirkova, J. 2006. RBS, PIXE and NDP study of erbium incorporation into glass surface for photonics applications. *Nucl. Instrum. Methods Phys. Res. B*, 249: 856–858.

Murphy, B.T., Haggan, D.E., and Troutman, W.W. 2000. From circuit miniaturization to the scalable IC. *Proc. IEEE* 88(5): 691–703.

Naramoto, H., Yamamoto, S., and Narumi, K. 2000. RBS/NRA/channeling analysis of implanted immiscible species. *Nucl. Instrum. Methods Phys. Res. B* 161–163: 534–535.

Scharf, W., and Wawrzonek, L. 1985. Thin layer activity depth distribution for wear measurements of cast iron. *Nucl. Instrum. Methods Phys. Res.* B12: 490–495.

Stroosnijder, M. F., et al. 2005. Wear evaluation of a cross-linked medical grade polyethylene by ultra-thin layer activation compared to gravimetry. *Nucl Instrum. Methods Phys. Res.* B227: 597–602.

Svorcik, V., Kolarova, K., Slepicka, P., Mackova, A., Novotna, M., and Hnatowicz, V. 2006. Modification of surface properties of high and low density polyethylene by Ar plasma discharge, *Polym. Degrad. Stab.*, 91: 1219–1225.

Tesmer, J. R., Nastasi, M., Maggiore, C. J., Barbour, J. C., and Mayer, J. W., Eds. 1995. Handbook of Modern Ion-Beam Materials Analysis. Materials Research Society, Pittsburgh, PA, USA.

Tesmer, J. R., Nastasi, M., Maggiore, C. J., Barbour, J. C., and Mayer, J. W., Eds. 1995. Handbook of Modern Ion-Beam Materials Analysis. Materials Research Society, Pittsburgh, PA, USA.

Vincent, L., et al. 2002. Simplified methodology of the ultra-thin layer activation technique. *Nucl. Instrum. Methods Phys. Res.* B190: 831–834.

Ziegler, J. F., Ziegler, M. D., and Biersack, J. P. 2010. SRIM – the stopping and range of ions in matter. *Nucl. Instrum. Methods Phys. Res. B* 268: 1818–1823.

5

Miscellaneous Applications

5.1 Introduction

Terrorism is a major threat in the 21st century and an enduring challenge to human ingenuity. The vulnerability of societies to terrorist attacks results in part from the proliferation of chemical, biological and nuclear weapons of mass destruction, but is also a consequence of the highly efficient and interconnected systems that we rely on for key services such as transportation, information, energy and health care. In today's society, acts of terrorism must involve in some stages the illicit trafficking of either explosives, chemical agents, nuclear materials and/or humans. Therefore, the society must rely on the anti-trafficking infrastructure, which encompasses responsible authorities: their personnel and adequate instrumental base.

The border management system should keep pace with expanding trade while protecting from the threats of terrorist attacks, illegal immigration, illegal drugs and other contrabands. The border of the future must integrate actions abroad to screen goods and people prior to their arrival in the country, and inspections at the border and measures within the country to ensure compliance with entry and import permits. Agreements with neighbors, major trading partners and private industry with all extensive pre-screening of low-risk traffic, thereby allowing limited assets to focus attention on high-risk traffic. The use of advanced technology to track the movement of cargo and the entry and exit of individuals is essential to the task of managing the movement of hundreds of millions of individuals, conveyances and vehicles.

The list of materials, which are subject to inspection with the aim of reducing the acts of terrorism, includes explosives, narcotics, chemical weapons, hazardous chemicals and radioactive materials. To this, we should add also illicit trafficking with human beings.

Since the problem is fundamentally international, the solution by definition should be looked for by organized international effort. Although there are number of international organizations whose mandate is exactly that, to name only some: INTERPOL, Organization for the Prohibition of Chemical Weapons, OPCW, International Maritime Organization (IMO), IAEA, etc., we shall mention here an international effort that includes accelerators. This is an EC-funded H2020 project RADIATE, Research and Development with Ion Beams-Advancing Technology in Europe (Cordis 2021).

Altogether, 14 partners from public research and 4 SMEs are joining forces in the RADIATE project exchanging experience and best practice examples in order to structure the European Research Area of ion technology application. Besides further developing ion beam technology and strengthening the cooperation between European ion beam infrastructures, RADIATE is committed to providing easy, flexible and efficient access for researchers from academia and industry to the participating ion beam facilities. About

DOI: 10.1201/9781003033684-5

15.800 hours of transnational access in total is going to be offered free of charge to users who successfully underwent the RADIATE proposal procedure.

Joint research activities and workshops aim to strengthen Europe's leading role in ion beam science and technology. The collaboration with industrial partners will tackle specific challenges for major advances across multiple subfields of ion beam science and technology. RADIATE aims to attract new users from a variety of research fields, who are not yet acquainted with ion beam techniques in their research and introduce them to ion beam technology and its applicability to their field of research. New users will be given extensive support and training (see https://www.ionbeamcenters.eu/radiate/).

RADIATE project partners are: (1) Atomki, Hungary; (2) CAEN, Italy; (3) CIMAP CNRS, France; (4) ETH Zürich, Switzerland; (5) HZDR, Germany; (6) IMEC, Belgium; (7) INFN, Italy; (8) INSP (CNRS), France; (9) Ionoptika Ltd., the UK; (10) Ionplus AG, Switzerland; (11) IST, Portugal; (12) JSI, Slovenia; (13) JYU, Finland; (14) KU Leuven, Belgium; (15) Orsay Physics, France; (16) RBI, Croatia; (17) Surrey, the UK; (18) Uni BWM, Germany; and (19) University of Vienna, Austria.

ionbeamcenters.eu.Ionbeamcenters.eu offers unprecedented research capabilities for European and international scientists from a variety of research fields. During the project, ionbeamcenters.eu will become a platform to share information and software among researchers and will enable online access to data analysis software and ion beam-related databases and research data. The aim is to unify all national ion beam facilities, not just the ones involved in the RADIATE project, creating one single virtual European ion beam center at

5.2 Accelerator Applications in Security

Nuclear techniques have been applied in the detection of hidden explosives for a number of years. They work on the principle that nuclei of the chemical elements in the investigated material can be bombarded by penetrating nuclear radiation (mainly neutrons). As a result of the bombardment, nuclear reactions occur and a variety of nuclear particles, gamma and X-ray radiation are emitted, specific for each element in the bombarded material (see Valković 2006).

The problem of material (explosive, drugs, chemicals, etc.) identification can be reduced to the problem of measuring elemental concentrations. Nuclear reactions induced by neutrons can be used for the detection of chemical elements, their concentrations, and concentration ratios or multielemental maps, within the treated material (Tables 5.1 and 5.2).

Neutron scanning technology offers capabilities far beyond those of conventional inspection systems. The unique automatic, material-specific detection of terrorist threats can significantly increase the security at ports, border-crossing stations, airports, and even within the domestic transportation infrastructure of potential urban targets as well as protecting forces and infrastructure wherever they are located. Old control measures slow down the traffic flow, they are time-consuming and inefficient.

5.3 Cargo Inspection

Cargo inspection within the cargo container requires a list of different types of sensors including: contact switch, piezo-electric pressure sensor, geophone, microphone, magnetic sensor, induction loop (metal detector), infrared sensor (thermal radiation), break beam

TABLE 5.1

Nitro-compound explosives

Name	Molecular weight	C	H	N	O	Density (g/cm^3)
TNT	227.13	7	5	3	6	1.65
RDX	222.26	3	6	6	6	1.83
HMX	296.16	4	8	8	8	1.96
Tetryl	287.15	7	5	5	8	1.73
PETN	316.2	5	8	4	12	1.78
Nitroglycerin	227.09	3	5	3	9	1.59
EGDN	152.1	2	4	2	6	1.49
DNB	168.11	6	4	2	4	1.58
Picric acid	229.12	6	3	3	7	1.76
AN	80.05	–	4	2	3	1.59

TABLE 5.2

C/N and C/O concentration ratios

Name	C/O	H/N	C/N	O/N
NG	0.33	1.67	1	3
TNT	1.17	1.67	2.33	2
RDX	0.5	1	0.5	1
PETN	0.42	2	1.25	3
AN	0	2	0	1.5

device, image intensifier/night viewing device, photo camera, electronic still camera, TV or video camera (analog or digital), radar, X-ray devices, neutron sensor and chemical sensors. Modern cargo inspection systems are non-invasive imaging techniques using penetrating radiations (gamma and X-rays) in scanning geometry, with the detection of transmitted or radiation produced in the investigated sample. Fast scanning of standard containers (few minutes to less than 1 minute) should be performed to provide the customs officers with a high-resolution radiograph of the load. Unfortunately, this information is "non-specific" in that it gives no information on the nature of objects that do not match the travel documents. Moreover, there are regions of the container where x and gamma-ray systems are "blind". The so-called "Dual View Technology" offers the advantage of resulting in two projections (top view and side view) of the objects hidden inside the container. These systems are best when used in combination with other sensors. For example, SAIC has installed a combination of three portals: VACIS portal (Orphan et al. 2005), OCR portal and passive radiation portal in a port of Hong Kong (Orphan 2004). In such a way a 100% screening of containers is accomplished for more than 2 million containers per year. A throughput averaging 1960 containers per day is currently a common practice. Dose from gamma-ray exposure to the object inside the cargo container is <0.5 μSv (which is a 1/10th of a 10 hours long inter-continental flight) up to 100 μSv per inspection in the case of a dual-view system.

Cargo inspection by X-rays has become essential for seaports and airports. With the emphasis on homeland security issues, the identification of dangerous things, such as explosive items and nuclear materials, is the key feature of a cargo inspection system. In

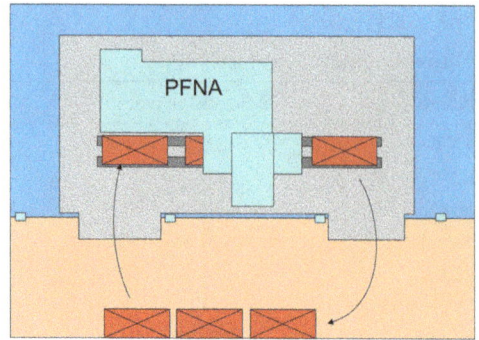

FIGURE 5.1
One possible arrangement for cargo container inspection at the port.

addition, new technologies based on dual-energy X-rays, neutrons and monoenergetic X-rays have been studied to achieve sufficiently good material identification. As most of the X-ray sources are based on RF electron linear accelerators (linacs), we have given a relatively detailed description of the principle and characteristics of linacs. Cargo inspection technologies based on neutron imaging, neutron analysis, nuclear resonance fluorescence, computer tomography are discussed in the article by Tang (2015). The author has summarized also the main equipment vendors and their products (Fig. 5.1).

Container terminals allow the transfer of intermodal cargo containers between ships, trucks and rail. Terminals allowing the transfer of containers to/from ships are called maritime container terminals, those that facilitate rail/truck transfers are known as inland container terminals. In a global economy, container terminals are under pressure to optimize their operations to drive efficiencies and the larger terminals are heavily automated with complex software systems (terminal operating systems) and cranes (usually gantry cranes) co-ordinating a ballet of container movements to minimize waiting time for cargo as well as reducing the number of man-hours and crane time required by any one container.

Maritime containers represent a significant part of the international trade and supply chains, which are the backbone to the world's economy. Containers transport involves numerous manufacturers, logistic nodes, operators, platforms and check-points (in particular containers ports). Improving their security requires an integrated research and development approach, including risk assessment, traceability, secure exchange between nations and across operators, and fast but effective screening.

The European Commission's security research is supporting the development of technologies for container security and supply chains in general. The goal is to develop new security solutions that fully meet the requirements of the end-users and to improve the competitiveness of European economic players.

It is also important to note that the US author ities (DHS: Department of Homeland Security) have adopted new legislation that – as from 2012 – impose a 100% scanning in foreign ports of containers bound to the USA. The European Commission continues to advocate the internationally recognized multi-layered risk-based approach including mutual recognition of trade partnership programs for enhancing and protecting the international supply chain. In the framework of this approach, the Commission is ready to work with the USA to find possible technological solutions to help address the security concerns linked to container security.

The ContainerProbe portal hardware when coupled with a to-be-developed "Data Fusion" system (software and global connectivity) will comprise an overall product solution called ContainerNet. It is designed to be a global system for 100% Risk Screening of inter-modal containers while they are in motion. It will have the following suspected risk screening capabilities: A ContainerProbe Poster has been submitted to the Cordis/FP7 security partnering website.

The presence of the following categories of materials is of interest to any of the applied inspection protocols:

i. Misdeclared Hazardous Materials: illegal waste exports or imports, hazardous materials causing many annual maritime insurance claims, accumulated pest poisons.

ii. Contraband Materials: smuggled goods to avoid import duties and restrictions, narcotic drugs, weapons for criminals, illegal immigrants.

iii. Terrorism Materials: explosives and precursors, weapons of mass destruction, fissile materials.

The demand for this type of detection capability with high throughput has been declared by the EU, USA and other nations as a consequence of the rising policy of Civil Security.

The statistics of container traffic are presented in the yearly UNCTAD (United Nations Conference on Trade and Development) reports. The Review of Maritime Transport has been published annually since 1968. The flagship report provides an analysis of structural and cyclical changes affecting seaborne trade, ports and shipping, as well as an extensive collection of statistics. The last Review available is UNCTAD (2020). According to its Executive Summary at the beginning of 2020, the total world fleet amounted to 98,140 commercial ships of 100 gross tons and above, equivalent to a capacity of 2.06×10^9 dwt. In 2019, the global commercial shipping fleet grew by 4.1%, representing the highest growth rate since 2014, but still below levels observed during the 2004–2012 period. Gas carriers experienced the fasted growth, followed by oil tankers, bulk carriers and container ships. The size of the largest container vessel in terms of capacity went up by 10.9%. The largest container ships are now as big as the largest oil tankers and bigger than the largest dry bulk and cruise ships, and probably the container ship sizes have reached a peak. The industry had a number of difficulties due to restrictions relating to the outbreak of COVID-19.

More stringent environmental requirements continue to shape the maritime transport sector. Carriers need to maintain service levels and reduce costs, and at the same time ensure sustainability in operations. Greenhouse gas emissions from international shipping continue to rank high on the international policy agenda. Progress was made at IMO towards the ambition set out in its initial strategy on reduction of greenhouse gas emissions from ships. These include ship energy efficiency, alternative fuels and the development of national action plans to address greenhouse gas emissions from international shipping.

The potential threats to port security come on many different levels from range of groups and individuals with very different aims and objectives. Theft, fraud, corruption, drugs trafficking and people trafficking are all major issues for port security, especially when their motivation can be traced back to organized crime or terrorist groups. The combination of sophisticated organized crime, the heightened risk of terrorist activity and the ongoing threat of low-level crime ensures that ports are under threat 24 hours per day, 365 days per year. Well-organized criminal gangs now have access to enough money, knowledge and skills to develop considerably more complex ways of operating (van Unnik, 2004). Faced with such innovative and complex criminal activity, port security needs to raise and improve.

Risk assessment screening of all containers must rapidly process a dynamic global database of registered information. Precise tracking of containers is a logistics function. Physical measurement is essential to ensure that the contents which start the journey remain the same. Only penetrating radiation can interrogate a container without intrusion; overcoming the limitations of passive radiation detection.

Depending on X-ray-like imaging transfers the risk assessment to human monitors are prone to erratic performance. Detailed images of all (100%) container contents at a major port have been demonstrated to be economically unacceptable and consequently abandoned for the time being.

A very short but intense burst or flash of pulsed neutrons through the container from the neutron generator, being as long as the container, will cause prompt gamma emissions from the bulk contents. The slow-moving containers on a train or automated vehicle at 5 m/s (~20 kph) appear effectively motionless. The full energy gamma spectrum collected during the following two seconds provides a profile of the elemental composition of the bulk contents.

The already established ISPS (The International Ship and Port Facility Security Code is an amendment to the Safety of Life at Sea (SOLAS) Convention (1974/1988) on minimum security arrangements for ships, ports and government agencies) database for customs and security plus logistical tracking of each and every container in transit will be used by ContainerNet to perform a fast match of contents' gamma profile with contents classification. Additional detection of neutrons caused by active neutron "pings" can be used to immediately indicate fissile materials. The transit time within the port provides a buffer for security database correlations. Transit from inter-modal transport node to node provides further off-line time for deeper background investigations in disparate databases via ContainerNet. Containers with anomalies to the matching process are diverted to the more detailed scanning systems which produce images and elemental information for analysis of suspected security threats.

Within the framework of three EU projects (EURITRACK, ERITR@C and C-BORD) and one NATO project (SfP-980526) a technology of container inspection with new neutron probe used as a secondary sensor has been developed for detailed investigation of the suspected container volume as determined by an X-ray system. Demonstration facility for first two projects has been constructed and mounted at the container terminal Brajdica, port of Rijeka, Croatia. Some 150 containers have been inspected in combination with Smith Heimann 300 unit.

The EURITRACK inspection concept is as follows: (1) X-ray radiography to detect and localize suspicious items inside the cargo container; (2) Tagged neutron inspection after focusing the beam on the area of interest; (2.1) Neutron flight-path spectrum to determine the depth of interest inside the cargo container; (2.2) Gamma-ray spectra corresponding to different "slices" along the tagged neutron beam; (3) Gamma spectra unfolding with a linear combination of elemental gamma spectra (C, O, N, Fe...); (4) C, N and O relative proportions in a triangle representation to sort explosives and narcotics from benign materials.

Some of the highlights are described in Perot et al. (2009), Valković (2009), Obhodas et al. (2010) and El Kanawati et al. (2011). Following the implementation of abovementioned projects, several disadvantages and imperfections of using tagged neutrons have been identified:

    i. EURITRACK-like system requires site occupying ~500 m² of the active container terminal storage area. Loss of income to the Container terminal and Port of

Rijeka during the implementation of EURITRACK and ERITR@C projects was estimated to ~50,000.00 €/annum.

ii. No functional sensor (Smiths Heimann-Tagged neutron system) communication has been established.

iii. The neutron beam intensity has been limited to ~10^7 n/s in 4π requiring irradiations of suspected area/spot/voxel in the order of 15 minutes since only a fraction of neutron generator output was used with detection of associated alpha particles.

iv. Only a small percentage of port leaving containers could be screened.

v. Questionable determination of nitrogen and subsequent construction of C–N–O diagrams.

A follow-up project entitled "effective Container inspection at BORDer control points" (C-BORD 2020), finalized in November 2018, aimed to develop and test a complex, affordable solution for inspecting containers and large cargo on EU borders, especially in sea ports (https://www.cbord-h2020.eu/). The project assumes C-BORD system to work in a large scope of non-intrusive container control (Non-Intrusive Inspection, NII), to be able to detect illegal drugs, tobacco, explosive materials, chemical warfare agents, trafficking in human beings and Special Nuclear Materials (SNM). The project included five control methods: RTG scanner, radiometric gates, neutron system, and photo fission system and fume analyzer. It was funded under the framework of European program for research and innovation Horizon 2020. Eighteen partners from seven EU countries were involved in the project including scientists, European companies and final users (customs service).

C-BORD project aimed to show, that five of the already available methods can be integrated in a system working with two inspection lines and operated from one decision center. Such integration would tighten the control system without making the inspections more cumbersome for shippers and without needlessly blocking the flow of goods. The tested system would be applicable primarily to inspect containers in sea ports and land crossings.

The first line of the inspection was an industrial RTG scanner, gates monitoring radiation and system for chemical analysis of fumes from inside the container. The first two of those systems are currently the most technologically refined and are available in commercial versions.

Industrial scanner screens the container with X-rays, making an image of its inside with the precision up to a fraction of a millimeter. This way, an experienced customs officer can analyze the contents of the container, especially identify suspicious cargo, which differ from typical characteristics of products mentioned in customs declaration, creating so called anomalies. Moreover, the radiometric gates system allows to detect traces of radiation characteristic to some materials containing radioactive materials. There are some potentially dangerous natural as well as synthetic materials, that emit specific gamma radiation, which – after energy analysis – can be identified even when the intensity is very low, or the radiation is attenuated by potential shielding. The shielding alone would be very well visible on the X-ray image taken on this line.

Another solution proposed for implementation in the first inspection line is a system for chemical analysis, which allows to analyze the composition of the air inside the container. This system is often called electronic dog, or sniffer. As a result of nearly real-time chemical analysis, it is possible to detect even trace amounts of volatile substances that are characteristic of typically smuggled materials, like intoxicating or explosive agents. Here, it is important to compare the results of the inspection with the information contained in cargo's customs declaration.

The team working on the C-BORD project assumed, that the measurement on the first inspection line would be analyzed "on the fly" by qualified customs officers. In an event of detecting suspicious traits, cargo would be directed to second, more technologically advanced line of non-intrusive control. This is a fundamental difference in comparison with current methods of inspection, wherein suspicious results of the control result in directing the cargo to time-consuming, costly for the shipper and, in many cases, troublesome direct inspection, that consists in opening the container and inspecting the cargo by a crew of customs officers (Sibczynski et al. 2017).

Second line of inspection were two systems of analyzing the cargo with the use of neutrons and high energy photon beam. The neutron activation system combines the phenomenon of non-elastic neutron scattering with related time-of-flight (TOF) spectroscopy. Technology allows for achieving effective and precise system of detection and identification of explosive materials or smuggled drugs, for example. Neutrons with an energy of 14 MeV are emitted isotopically from a tubular generator. The signal from atomic nuclei of the analyzed cargo, which was excited by the neutrons, is registered by a system of detectors and analyzed. The analysis allows to determine the relative amount of nitrogen, carbon and oxygen with good precision, and in turn, the most possible material (explosive, intoxicant and "harmless" material), in the activated part of the container, pointed at during the X-ray inspection beforehand (Fontana et al. 2017; Pino et al. 2018).

The additional system utilizes the phenomenon of nuclear photofission induced by photons of energy up to 9 MeV and is dedicated to detection of smuggled SNM. A beam of high-energy gamma photons is obtained using linear electron accelerator. Electrons, after slowing down in a heavy target, emit photons that can induce when colliding nuclei of certain isotopes, especially uranium and plutonium, which are of potential importance for terrorist activities. Under the framework of the project, there were conducted tests of four devices for detection of nuclear materials. One of them is a special system of detectors that searches for a characteristic signal coming from both neutrons and gamma radiation. On this basis, it is able to ascertain, in a short time (20 minutes) and in a safe and non-invasive way, whether suspicions related to the nature of cargo assumed on the first inspection line were justified.

C-BORD project demonstrated, that from the technical standpoint, it is possible to build and integrate a multi-component inspection system. The tests were conducted both in a road control point in Hungary and in container ports in Gdańsk and Rotterdam. They showed the effectiveness of the constructed system as well as good performance of user-friendly software developed for this project. Fontana et al. (2018) describe a distributed DAQ system developed for the C-BORD project.

In the frame of the C-BORD project, a Rapidly Relocatable Tagged Neutron Inspection System (RRTNIS) has been developed for a nonintrusive inspection of cargo containers, aiming at explosives and other illicit goods detection (Sardet et al. 2021). Twenty large-volume NaI detectors were used to determine the elements composing inspected materials from their specific gamma-ray spectra signatures induced by fast neutrons. The RRTNIS inspection is focused on a specific suspect area selected by X-ray radiography. An unfolding algorithm decomposed the energy spectrum of this suspect area on a database of pure element gamma signatures. A classification is then performed between inorganic materials, such as metals, ceramics, or chemicals, and organic materials such as wood, fabrics or plastic goods. Concerning organic materials, the obtained elemental proportions of carbon, nitrogen and oxygen allow discriminating explosives from illicit drugs and benign substances. The final laboratory tests were performed at CEA, Saclay, France. Simulants of explosives and illicit drugs have been hidden at different depths inside iron or wood cargo materials, which are representative of the different neutron and gamma attenuation properties expected to encounter in real cargo containers. Hundreds of experiments have been performed, showing that a few kilograms of explosives or narcotics stimulants can be detected by the RRTNIS in 10-minute inspections.

Further development of this initiative is ongoing in the framework of the project ENTRANCE, the system will be tested in the real seaport environment and at functioning border crossings to EU.

Many of the disadvantages of the real-world usage of discussed systems, especially in huge ship container terminals could be avoided by design, construction and building an autonomous and flexible inspection system: Autonomous Ship Container Inspector System. This could be accomplished by:

- The incorporation of inspection system on the container transportation devices (straddle carriers, yard gentry cranes automated guided vehicles, trailers – to be decided by container terminal operator). In such a way, no part of the port operational area will be used for inspection. The inspection scenario will include container transfer from ship to transportation device with inspection unit mounted on it, inspection during container movement to the container location (either to exit the truck and/or train or terminal yard position).

- Use of neutron generator without associated alpha particle detection. The area analyzed will be determined by mounting neutron generator inside shielding and collimator defining narrow fan beam of neutrons. This will allow the use of higher neutron intensity (5×10^9–10^{10} n/s in 4π).

- The inspected container will be stationary in the "inspection position" on the transportation device while the "inspection unit" will move along side with speed of 12 m/minute, resulting in ~1 minute inspection time, T_i, of the whole container (2 TEU). The efforts will be directed towards reduction of T_i to as low as possible value. A detached part of "inspection unit" will be mounted on the gentry' crane spreader detecting radionuclides and SNM using fast neutron bombardment and detection of induced gamma and neutron radiation.

- It is proposed to design, construct and build trailer inspection unit using the following analytical methods simultaneously: neutron radiography, X-ray radiography, neutron activation analysis, (n, γ) and (n, n'γ) reactions, neutron absorption, neutron backscattering, X-ray backscattering. Neutron-based techniques will take the advantage of using "smart collimators" for neutrons, both emitted and detected. The inspected voxel will be defined by intersections/union of neutron generator and detector solid angles.

In addition, the information on container weight will be obtained during the container transport and foreseen screening. One of the essential parameters to be measured is the density of material in the container. Based on this measurement one should be able to evaluate the "average container density" and calculate its weight.

- Inspection unit: It is assumed only a hybrid gamma ray - neutron system that based not only in a typical FNGR system (radiography system) will offer better detections capabilities. In the design of inspection unit considerations of incorporations of following analytical methods to be used simultaneously.
  - Neutron radiography.
  - Gamma-ray radiography.
  - Neutron activation analysis, (n, γ) and (n, n'γ) reactions.
  - Neutron absorption.

The last two techniques will take the advantage of using "BM smart collimators" for neutrons, both emitted and detected. The inspected voxel is defined by intersections/ union of neutron generator and detector solid angles.

There are several proposed radiography systems (Yousri et al. 2012; Reijonen et al. 2007), however, the new system must be based on the AC6015XN air cargo scanner which has developed by CSIRO and Nuctech (Cutmore et al. 2009) because this is the most promising device. Although the fact that the AC6015XN is a triple beam device and there are 3 R-values the Australian scientists use only the 2 R-value. However, the third R-value could offer extra and sometimes valuable information.

Attention should be paid to the choice of the X-ray sources. Which combination of the X-ray beams is the best among the typical energies 3, 4, 5, 6, 7, 8 or 9 MeV, or some other values should be investigated. Smaller detectors especially for the X-ray array will offer better inspection capabilities (5 × 5 mm). Also with smaller neutron detectors (e.g., 10 × 10 mm), one can "see" more details on the neutron radiography mode which is a significant enhancement on the 20 × 20 mm detectors used today.

In the case of neutrons source the only solution is DT neutron generator (because 14 MeV neutrons have relevant acceptable penetration capabilities). However, the typical generators have a short live tubes (typical 500–2000 hours), which makes the associated alpha particle, API, method inappropriate. Therefore, one should use the neutron generators using plasma generators with at least 25,000-hour operational life. In this scenario, the inspection unit will be "U" shaped construction with three arms: right, bottom and left.

Left arm will contain radiation units: neutron generator and X-ray source (6 MeV linac) inside shielding box, both having "smart collimators" in order to produce collimated beams.

The bottom arm will contain the battery of gamma detectors (5" × 5" NaI or BGO – number, size and type to be determined). From previous measurements (Sudac et al. 2008) it is known that for the detection of 100 kg of explosive in iron matrix of density 0.2 g/cm^3 54,485 s were needed with neutron beam intensity of 10^7 n/s and by using two 3" × 3" NaI (Tl) gamma-ray detectors inside lead shielding thickness 5 cm. With the neutron beam intensity increased for three orders of magnitude this measuring time would be reduced to 54 s. By using battery of thirteen 5" NaI detectors, this can be reduced to some 3 s.

Right arm will contain neutron detectors array for neutron radiography containing probably some 700 neutron plastic detectors dimensions 20 × 20 × 75 mm (to be determined); X-ray detector array a number of 10 × 10 × 50 mm CsI(Tl) detectors (exact size, number and type to be determined). Both arrays will be stationary with respect to radiation sources and mounted inside the appropriate shielding. This arm will also have a beam transmission detector equipped with a smart collimator and n-γ separation circuit, this detector be at fixed position with respect to neutron source at all times.

The size of inspected voxel is defined by "smart collimators" in front of detectors and neutron source. With two gamma rays, batteries in the bottom arm 24 positions of the inspection unit would be required to cover 40" container length. If in addition 3 vertical positions are included this will result in 3 × 24 = 72 scans, each lasting 3 s. It follows that the complete inspection can be done for some 3–5 minutes. It should be mentioned that this estimation is valid for the detection of 100 kg of explosives (and other threat materials) in iron matrix with a density of 0.2 g/cm³ or organic matrix of density 0.1 g/cm³. Results for containers inspected in the port of Rijeka, Croatia indicated that the average container density was around 0.1 g/cm³, with organic type cargo represented by 62%, metallic with 15%. It follows that for containers with higher density cargo the measurement time could be longer, which could be decided from the first evaluation of container average density (Figs. 5.2–5.7).

During this project implementation considerations should be made on the three variants of inspection units: "U", "L" and "I" shaped. One should first start with "U" shaped inspection unit because it can be realized with the applications of present-day knowledge. In this case, the inspection unit is U-shaped so that the container can be brought inside it. The inspected container will be stationary in the "gantry crane inspection position" while the "inspection unit" will move along side with speed of 1–2 m/min, resulting in ~10 min inspection time of the whole container (Figs. 5.8 and 5.9).

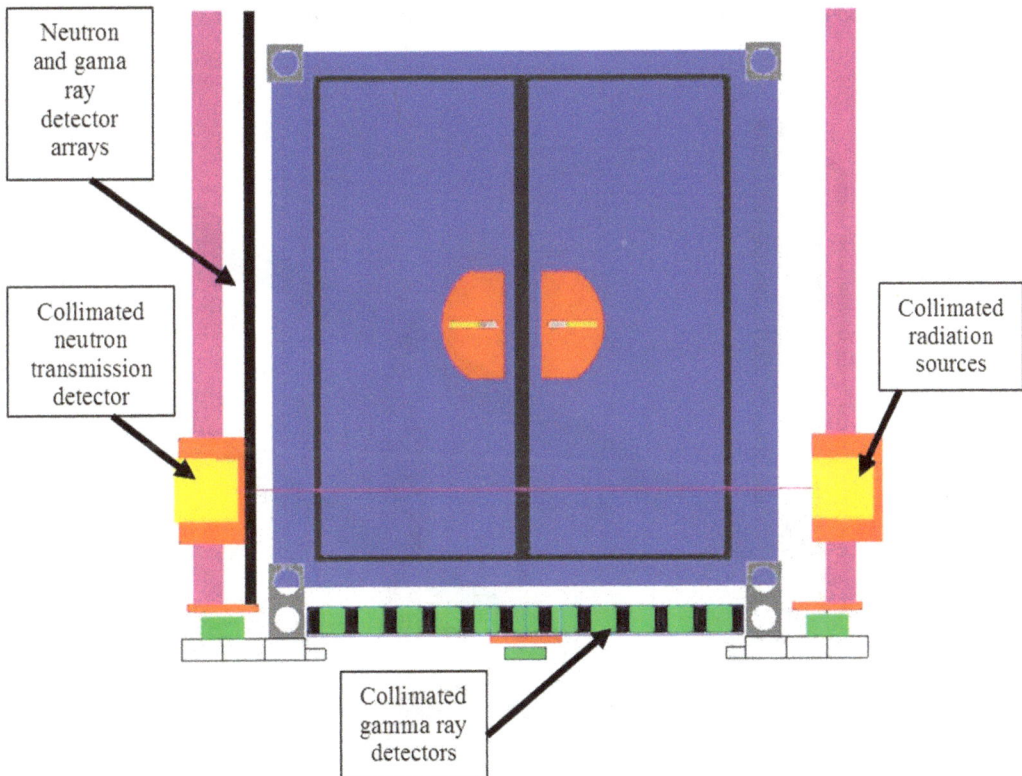

FIGURE 5.2
Schematic presentation of U-shaped inspection unit; density determination by using neutron transmission measurement.

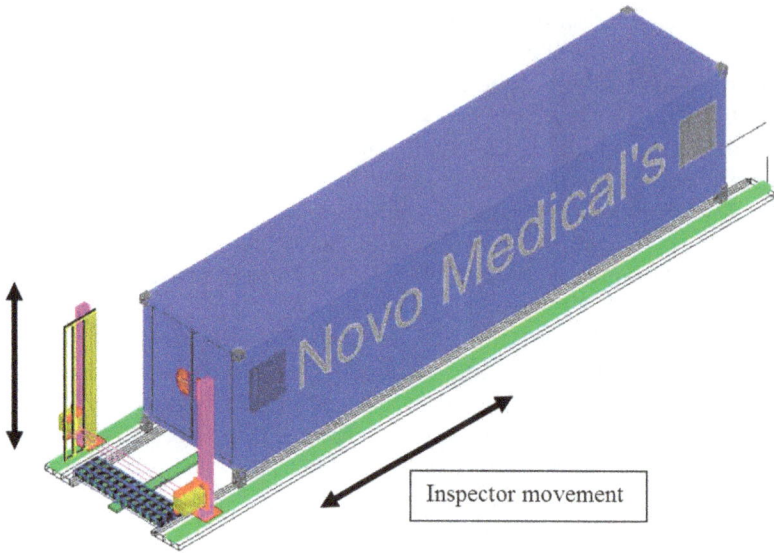

FIGURE 5.3
U-shaped inspection unit will slide on rails alongside the inspected container.

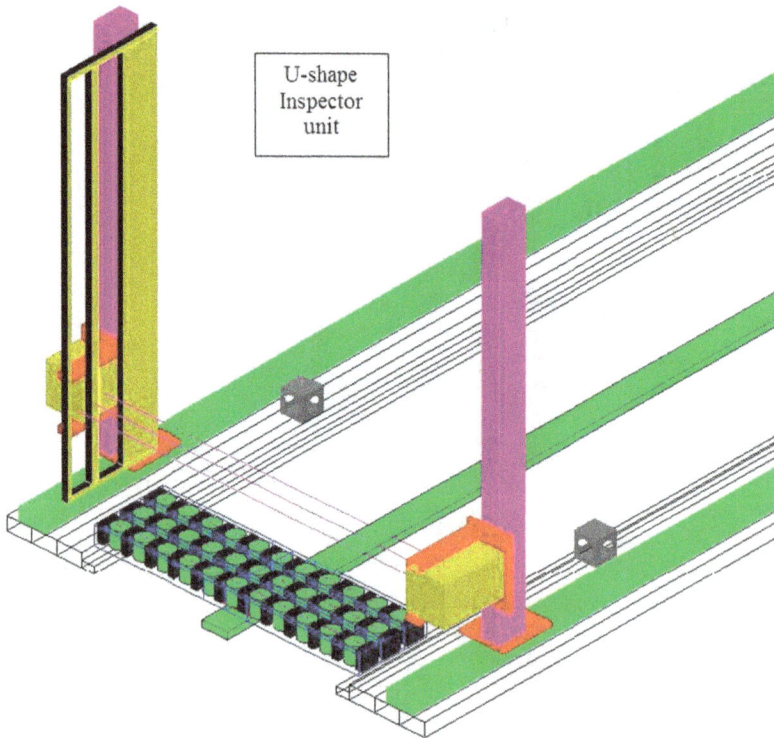

FIGURE 5.4
Components of "U" shaped inspector: collimated radiation sources (neutrons and X-rays), collimated gamma-ray detectors (bottom), collimated neutron transmission detector, neutron and X-ray detector's arrays.

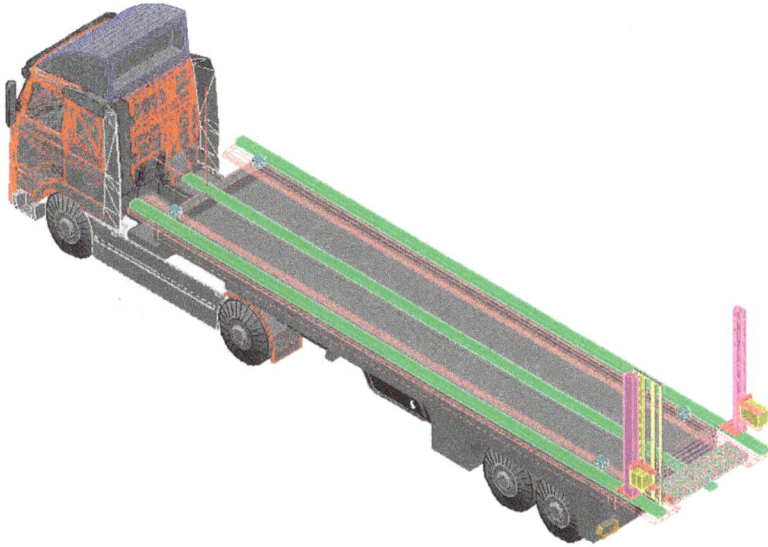

FIGURE 5.5
Inspector mounted on the transport track.

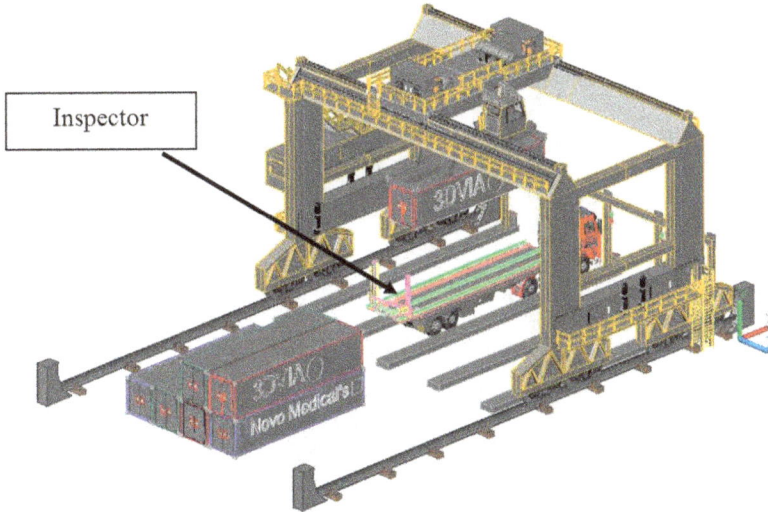

FIGURE 5.6
Loading the container onto truck-mounted inspection unit.

The chemical elements present in the investigated voxel should be determined by measuring resulting gamma spectra by the 3 arrays, each containing 12 bottoms placed gamma detectors inside shielding and equipped with "smart collimators" for C, O (+ other elements) relative concentration determination.

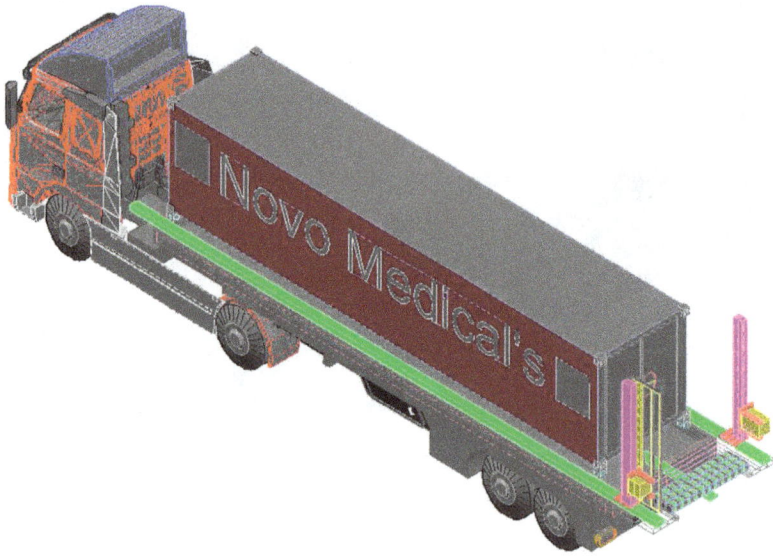

FIGURE 5.7
On the way to the yard or any other transfer point.

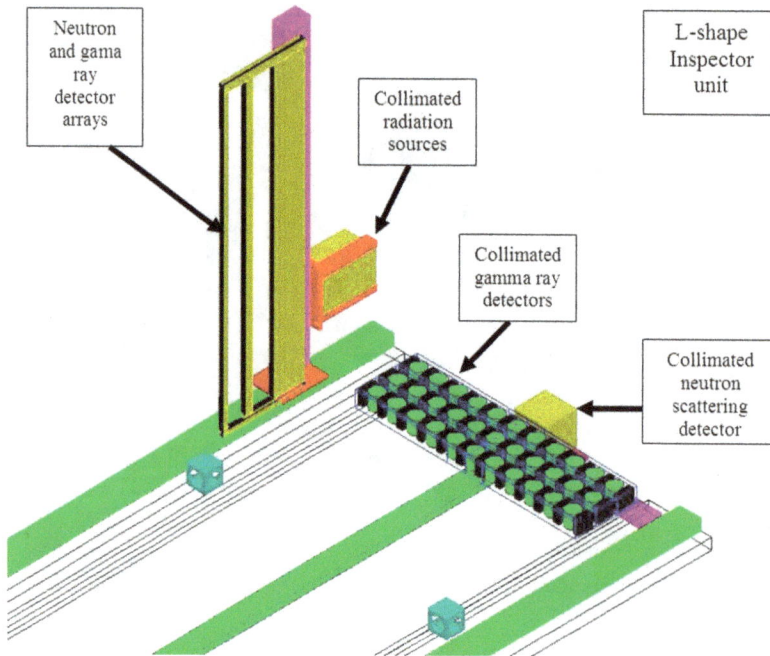

FIGURE 5.8
Components of "L" shaped inspector: collimated radiation sources (neutrons and X-rays), collimated gamma-ray detectors (bottom), collimated neutron transmission detector, neutron and X-ray detector's arrays.

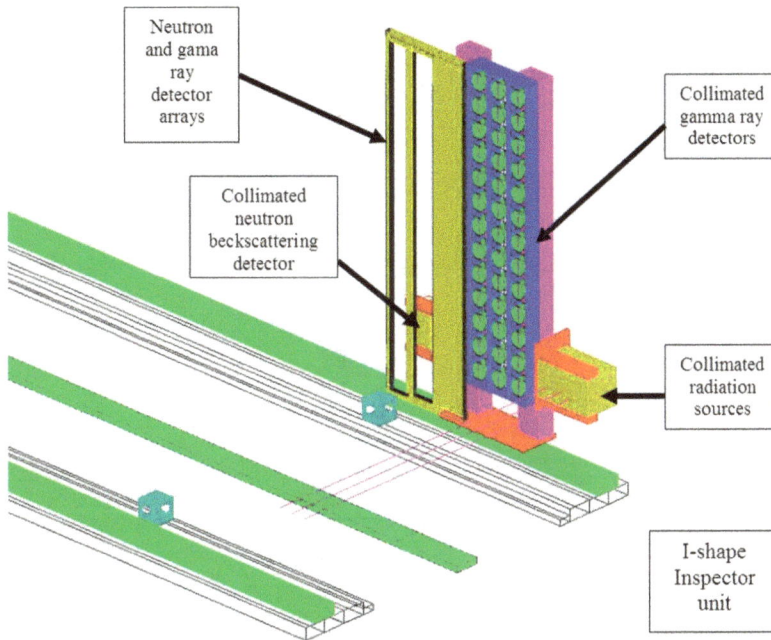

FIGURE 5.9
Components of "I" shaped inspector: collimated radiation sources (neutrons and X-rays), collimated gamma-ray detectors, collimated neutron transmission detector, neutron and X-ray detector's arrays.

Other parameters of components used are:

- Neutron beam intensity (5×10^9–10^{10} n/s in 4π) but collimated using "smart collimators".
- Neutron detector (transmission detector) also collimated by using "smart collimator".
- Gamma source collimated – bremstrahlung source: 6 MeV linac (to be decided).
- Two source positions: bottom and top of the container or continuous up-down movement.

Construction of C–O–ρ(density) triangle diagrams for material identification as shown in Fig. 5.10. The sum triangle graph should be compared with individual γ-detector triangle graphs in the data analysis process while cross-checked with radiographs.

Until now, cargoes have been shipped based on the container weights declared in the advance booking information provided by shippers. Vessel stowage plans and port operations are typically based on these pre-declared weights which can vary significantly from the actual mass of the cargo transported. The consequent risk to the health and safety of seafarers and port operatives is clearly apparent. Weighbridges located at the port gate are easy to implement, but present significant challenges in establishing the true tare weight of the vehicle and in differentiating the weights of multiple container loads. Within the port, reach stackers and fork-lift trucks provide a relatively low-cost path for indirect weight measurement to be integrated into vehicle systems, e.g., inferred from hydraulic pressure. However, this is likely to be of lower accuracy than direct measurement systems and it is questionable whether it would deliver full compliance with

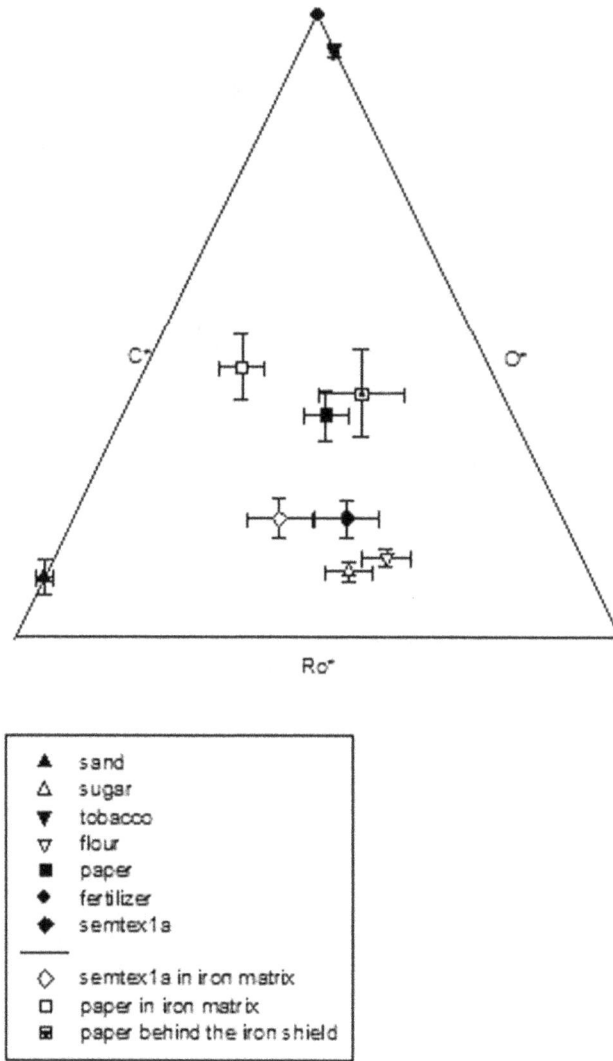

FIGURE 5.10
Triangle diagram for various materials inside the empty container together with the paper and explosive semtex1a in iron matrix and paper behind the iron shield. (After Valković et al. 2016.)

any genuinely robust regulation. Rubber-tired gantry cranes and straddle carriers provide a good opportunity for the implementation of direct weight measurement of individual containers within the port environment but before final vessel loading operations. Such systems may offer a highly flexible solution with minimal disruption to existing port operations and container logistics.

The proposed inspection unit will support the key recommendations for the implementation of container weight verification technology that container weight should be measured as close as possible to the point of lifting. In addition to enabling the verification of individual container weights, the proposed inspection unit will also enable automatic estimation of center of gravity; something that will be extremely useful in improving the safety of stacking operations.

5.3.1 Detection of Explosives

The list of explosive compounds has more than 100 items including some improvised primary explosives like acetone peroxide, diazodinitrophenol (ddnp/dinol), double salts, hmtd, leadazide, lead picrate, methyl ethyl ketone peroxide (MEKAP), mercury fulminate, "milk booster", nitromannite, sodium azide and some others. Instruction on how to prepare them can be found in the open literature and on the internet. However, the most often used explosives are: Trinitrotoluene (TNT), Pentaerythritoltetranitrate (PETN), Cyclotrimetilentrinitramin (RDX), Trinitrophenylmethylnitramine (Tetryl), Tetrytol and Hexatol. TNT, commonly known as TNT, is a constituent of many explosives, such as amatol, pentolite, tetrytol, hexatol, torpex, tritonal, picratol, ednatol and composition B. It has been used under such names as Triton, Trotyl, Trilite, Trinol and Tritolo. In a refined form, TNT is one of the most stable of high explosives and can be stored over long periods of time. It is relatively insensitive to blows or friction. TNT is used in pressed and cast form. Pressed TNT can be used as a booster or as a bursting charge for high-explosive shells and bombs. PETN is one of the strongest known high explosives. It has also been used under such names as Pentrit, Nitropenta, Niperyt and TEN. It is more sensitive to shock or friction than TNT or tetryl, thus it is never used alone as a booster. It is primarily used in booster and bursting charges of small-caliber ammunition, in upper charges of detonators in some land mines and shells, in shaped charges, and as the explosive core of primacord. It is also used as an explosive compound in plastic explosives (PEP).

At present, there are three methods for the detection of explosives: X-ray imaging, vapor detection and nuclear techniques. Presently, there is no single technology that can address all the questions concerning the detection of explosives and, indeed, other illicit contraband, the philosophy that emerges is that of an integral system combining methodologies. Such a system could contain a nuclear technology device, a vapor detector and an X-ray imaging device, all backed by an intelligence-gathering system. Nuclear techniques are based on the fact that explosives, drugs and various other materials, contain chemical elements such as hydrogen, carbon, nitrogen and oxygen, in distinctly different quantities.

Looking at the ratios of some of these elements, and in particular at the carbon/oxygen and carbon/nitrogen ratios, it is easy to see that a characteristic signature of all the known explosives can be obtained. This signature will be noticeably different from any that the material in which the explosives are hidden might give out. The problem of identifying explosives is thus reduced to the problem of elemental identification. The penetrating ability of neutrons and γ-rays offers an effective way of measuring the hydrogen, carbon, nitrogen, oxygen, phosphorus, sulfur and chlorine content of the suspect material. The elements of interest can be identified either from the energy of the scattered neutrons or from the emission of characteristic γ-rays.

Using a combination of fast and thermal neutrons or only fast neutrons, several explosive detection techniques were developed. These techniques could also be used, among other things, for detecting hidden drugs (the carbon/nitrogen ratio of all drugs is very different from that of explosives as well as from most innocuous materials); chemical weapons verification (the oxygen/nitrogen ratio in chemical weapons is quite different to that of conventional explosives. Furthermore, chemical weapons contain other elements such as phosphorus, that can also be identified); and the online measurement in coal-fired power plants of factors such as the coal's calorific value, the percentage of volatile matter, and sulfur content; all useful for the efficient operation of the plant and for controlling the emissions.

Long-life sealed tube neutron generators appear very attractive when used in industrial systems for a large number of reasons. For on-site real-time process control, high-energy neutrons, through their capability to penetrate deeply into materials, enable the dosing of a significant volume of materials in a short time, depending on the nature of the elements being sought. In aluminum production, an analyzer located at the output of a fluorated alumina silo enables short-time neutron activation analysis of the bulk material for production. Its goal is to give a feedback signal related to Fluorine and Sodium content with an accuracy of 0.1%. On-line elemental analyzers for other bulk materials such as coats or cements may be designed with, a pulsed neutron generator. An accurate determination of Ca, Si, Al, Fe, O, through a combination of gamma lines detected from activation and from prompt capture mechanisms was recently reported. In glass production, neutron analysis of Si, O, and other elements could be completed at two points in the production line, at raw material entry and in the mix. In steel production, a continuous analysis of Fe, Ca, Si and Al is also of great interest.

We have a long experience in neutron inspection of ship containers. Therefore, I shall present some of our works on using C–O–ρ triangle graphs in the identification of materials hidden inside containers (Sudac et al. 2007, 2008). The 14 MeV tagged neutron technique has been used to confirm the presence of the explosive Semtex#1a within the inspected container. Two 3″ × 3″ NaI(Tl) were put at the transmission position in the cone of the tagged neutrons beam. Each detector was shielded with the 5 cm thick lead (Sudac and Valković 2006). The container around the target was filled with the iron boxes of density 0.2 g/cm³.

We have also illustrated that various threat materials (including the explosives) could be detected inside the container by measurement of three variables, C and O concentration and density of investigated material (Sudac et al. 2008). Carbon and oxygen were detected, but not the nitrogen. It can be shown that 100 kg of Semtex#1a can be detected in the iron matrix of density 0.2 g/cm³ by using the triangle graph like the one shown in Fig. 5.10.

The presence of an iron matrix around the target does not change the position of the material inside the triangle plot. It is well known that the organic matrix has a big influence on the neutron beam because of the attenuation, as well as termalization, of the fast neutrons. In order to demonstrate the usefulness of the neutron system, it was necessary to investigate if and how organic matrix changes the position of the signal of the target material inside the triangle diagram.

The organic cargo was simulated with wooden plates (dry wood has formula $C_{22}H_{31}O_{12}$). The wooden plates of 0.53 g/cm³ density and a total thickness of about 27 cm were put in front of the target material in one measurement and close to the container wall in the other one. The total thickness of the wood plates was about 60 cm along the tagged neutron beam axis, which means that the effective density is about 0.25 g/cm³ (because of the empty space between the plates). But, the average density along the container was even less, around 0.1 g/cm³.

Some of the experimental results are shown in Fig. 5.11 showing the triangular diagram where the measured points are located in the three regions. The first region corresponds to the paper points. The second region corresponds to the semtex1a explosive and the third region belongs to the flour. The division is not rigid. Some overlap exists especially between semtex1a and flour. Generally, the effect of the wood plates is to lower down the position of the material in the diagram. It could be seen that the wooden plates which are far from the target (Case 2) have less influence on the target material position in the diagram than the wooden plates which are close to the target (Case 1 and Case 3). Even in the case when some overlap between the regions could be expected, the identification of the paper, semtex1a and flour is possible provided that the error on estimating the

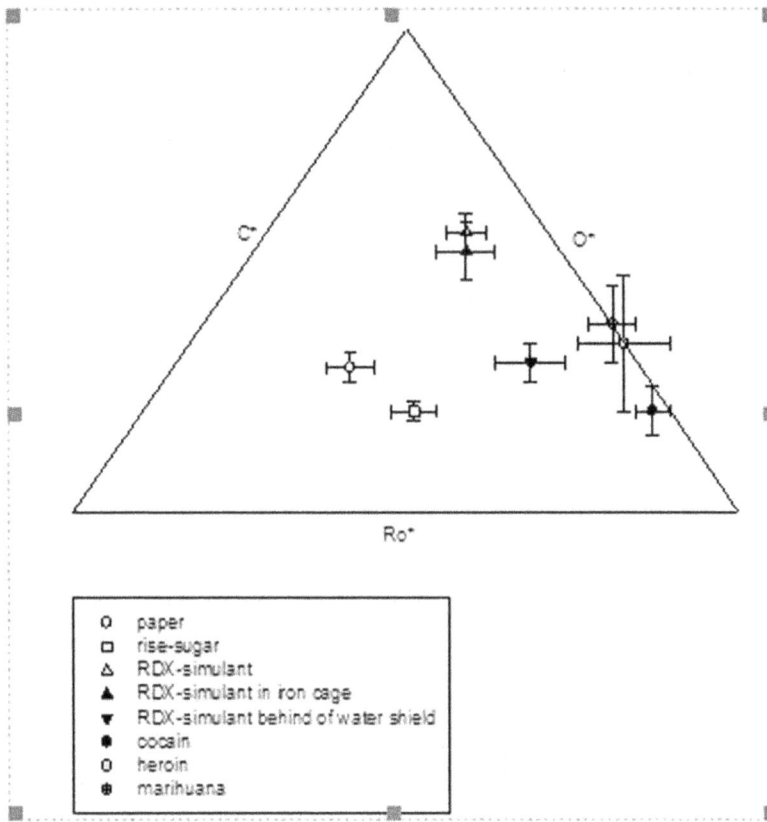

FIGURE 5.11
Triangle diagram for the transmission detectors. (Following Sudac et al. 2010.)

surrounding organic matrix density is not too large. By knowing the matrix density and applying the appropriate calibration it could be found how much the position of the target inside the diagram is lowered down in comparison with the target case alone.

Axes in the figure are defined as:

$$Ro^* = \frac{\left(\frac{N}{\bar{N}}\right)^2}{\left(\frac{N}{\bar{N}}\right)^2 + \left(\frac{C}{\bar{C}}\right)^2 + \left(\frac{O}{\bar{O}}\right)^2} \cdot C^* = \frac{\left(\frac{C}{\bar{C}}\right)^2}{\left(\frac{N}{\bar{N}}\right)^2 + \left(\frac{C}{\bar{C}}\right)^2 + \left(\frac{O}{\bar{O}}\right)^2} \cdot O^* = \frac{\left(\frac{O}{\bar{O}}\right)^2}{\left(\frac{N}{\bar{N}}\right)^2 + \left(\frac{C}{\bar{C}}\right)^2 + \left(\frac{O}{\bar{O}}\right)^2}$$

where N' is the number of counts in the tagged neutron peak, C is the number of counts in 4.44 MeV carbon peak, O is the number of counts in the 6.13 MeV oxygen peak and $\bar{N}', \bar{C}, \bar{O}$ are average values.

Similar measurements were done during the Eritr@ck project with the EURITRACK system which was installed in the container terminal Brajdica, seaport Rijeka. In that case, RDX simulant was successfully detected, but also some drugs like cocaine, heroin and marihuana by using the triangle diagram (see Fig. 5.11).

5.4 Nuclear Monitoring and Safeguards Activity

There are many long-lived radioactive nuclides generated by nuclear facilities that are very hard to detect by conventional alpha, beta or gamma-ray counting methods. As a result, their unintentional release occurs which, although of little significance biologically, provides information on the processes inside the nuclear facility. Many of these long-lived radioactive nuclides can now be detected by AMS methods and are being used for a variety of investigations.

Two isotopes that have been well developed are:

- ^{36}Cl ($T_{1/2} = 0.3 \times 10^6$ years). This nucleus is released by nuclear power plants and presumably reprocessing plants in detectable amounts by AMS.
- ^{129}I ($T_{1/2} = 16 \times 10^6$ years). This nucleus is released by nuclear power plants and by reprocessing plants as well. It can be detected easily by AMS at great distances in about 1 liter of sea water or on air filters.

Besides detecting long-lived radioactive nuclides such as ^{129}I and ^{36}Cl, it is possible to detect rare long-lived isotopes of uranium and thorium using AMS because of the almost complete absence of isobars.

As part of a nuclear safeguards protocol, the assay of ^{236}U, ^{233}U in uranium and ^{229}Th/^{230}Th in thorium is valuable for indicating the past use of these elements (Litherland 1995).

Several techniques have been used for the measurement of trace amounts of plutonium. Among them, the most commonly used method is alpha-spectrometry. With alpha-spectrometry, the isotope ratio of ^{239}Pu/^{240}Pu cannot be measured because the alpha-particles from both isotopes have nearly identical energies. The ^{239}Pu/^{240}Pu ratio often carries the most important information in a case study, as it reveals the original source (nuclear bombs or reactors of certain types) of plutonium. Among all the analytical methods available today, mass spectrometry seems to be the most promising one to fulfill this need. Both thermal ionization and inductively coupled plasma mass spectrometry (ICP-MS) have reasonable ionization efficiency for plutonium, but they cannot eliminate the hydride and other molecular interferences to yield reliable results. AMS, on the other hand, is capable of counting and identifying an individual atom without any molecular interference. However, the ionization and transmission efficiencies of plutonium in a Tandetron-based AMS system are expected to be low. But the unique feature of AMS, which is necessary for trace-amount plutonium detection, warrants an effect to determine these efficiencies experimentally so that the usefulness of AMS for measuring the plutonium ratio in environmental samples can be established (Litherland 1995).

The nuclear safeguards system which is used to monitor compliance with the Nuclear Non-proliferation Treaty relies to a significant degree on the analysis of environmental samples. Undeclared nuclear activities can be detected through the determination of the isotopic ratios of uranium and plutonium in such samples. It is necessary to be able to measure plutonium at the femtogram level in this application, and measure the full suite of uranium isotopes ($^{233-238}$U) where the total uranium content may be at the nanogram level.

Hotchkis et al. (2010) described the capability of AMS system at Australian Nuclear Science and Technology Organisation, ANSTO. The commissioning of a fast isotope cycling system for actinides has led to improved precision, with reproducibility of 4% for actinide isotope ratios. The background level for the key rare isotope ^{236}U is found to be

8.8 fg, for total uranium content in the ng range, and is limited by ^{236}U contamination rather than ion misidentification. For plutonium, the background is at the low fg level.

AMS with tritium is a method able to detect even very small nuclear releases. The application of this method for monitoring the tritium concentration at nuclear power plants (CANDU-type) and at a nuclear fuel reprocessing plants and as a diagnostic tool in fusion experiments is often applied. AMS measurements of the tritium, created in D–D fusion reactions can give information about the spatial distribution of triton fluxes to the vessel wall of a Tokamak, about the plasma confinement and about the interaction of the heating neutral deuterium beam with the confined and rotating plasma.

A chapter by Zendel et al. (2011) in *Handbook of Nuclear Chemistry* (Vértes et al. 2011) summarizes the nuclear safeguards verification measurement techniques. In Zendel et al.'s chapter, the "nuclear safeguards" verification system and procedures and measurement methods that allow the safeguards inspectorates/authorities to verify that nuclear materials or facilities are not used to further undeclared military activities are described. These procedures and methods provide the strong technical basis upon which the safeguards inspectorates/authorities issue their conclusions and receive the broadest international acceptance regarding the compliance of participating states with their obligations.

On 31 January 1961, the Board of Governors approved the Agency's safeguards system (IAEA 1961). Under Article III. A.5. of the Statute the Agency is authorized "to establish and administer safeguards designed to ensure that special fissionable and other materials, services, equipment, facilities and information made available by the Agency or at its request or under its supervision or control are not used in such a way as to further any military purpose; and to apply safeguards, at the request of the parties, to any bilateral or multilateral arrangement, or, at the request of a State, to any of that State's activities in the field of atomic energy". Since then the Agency's safeguards system has been improved many times as the political and technical situation changed. However, all the time an important part of the IAEA safeguards system was its "analytical laboratory", SAL.

Boulyga et al. (2015) mass spectrometric techniques available on-site in Seibersdorf, Austria SAL location. The lab is equipped with mass spectrometric techniques, such as thermal ionization mass spectrometry (TIMS), secondary ion mass spectrometry (SIMS), and ICP-MS belonging to the most powerful methods for the analysis of nuclear material and environmental samples collected during inspections. Each of the currently applied techniques provides definite merits (e.g., precision, accuracy, time-effectiveness, high sensitivity, spatial resolution, reduced molecular interference, etc.) for a specific safeguards related application.

IAEA has never acquired an accelerator on its Laboratory's premises therefore accelerator-based analytical techniques (including AMS) are not adequately represented in its laboratories. This is seen in the present edition of the Development and Implementation Support Programme for Nuclear Verification 2020–2021 (IAEA-Safeguards 2020). We quote,

- Environmental sample analysis techniques:
  - Development of new methodologies for the age determination of particles in environmental samples using large geometry (LG) – SIMS.
  - Development of new methodologies for using the SEM and ToF – SIMS capability to characterize individual particles collected on environmental swipe samples.

- Implementation of the Laser-Ablation ICP-MS technique for the measurement of particles containing low levels of Pu or mixed U/Pu.

- Analysis support and NWAL coordination:
  - Ensure efficient and effective operation of the Network of Analytical Laboratories (NWAL), including participation in inter-laboratory comparison exercises.
  - Expand the NWAL, with the main focus on overall capacity and quality assurance of particle analysis of environmental samples.
  - Increase capacity for the safe and secure shipment of nuclear material samples and disposal of analytical residues.

We could humbly suggest IAEA the acquisition of an modest accelerator whose main purpose would be analytical applications of all accelerator-based analytical techniques.

For the radioactive waste containers survey, the most sensitive detection method for alpha emitters dispersed in a large matrix is based on triggered fission reactions by short neutron pulses. it is now possible to detect 1 mg of ^{239}Pu in a 200 litters container, in less than 15 minutes.

The elemental characterization of materials constituting radioactive waste is of great importance for the management of storage and repository facilities. To complement the information brought by gamma or X-ray imaging, the performance of a fast neutron interrogation system based on the associated particle technique (APT) has been investigated using MCNP simulations and performing proof-of-principle experiments by Carasco et al. (2010). APT provides a 3D localization of the emission of fast neutron-induced gamma rays, whose spectroscopic analysis allows identifying the elements present in specific volumes of interest in the waste package. Monte Carlo calculations show that it is possible to identify materials enclosed behind the thick outer envelope of a $\approx 1 \text{ m}^3$ cemented waste drum, provided the excited nuclei emit gamma rays with a sufficient energy to limit photon attenuation. Neutron attenuation and scattering are also predominant effects that reduce the sensitivity and spatial selectivity of APT, but it is still possible to localize items in the waste by neutron TOF and gamma-ray spectroscopy. Experimental tests confirm that the elemental characterization is possible across thick mortar slabs (Carasco et al. 2010).

5.4.1 Neutron Radiography

The neutron radiography is a method of non-destructive testing that is complementary to X and γ radiography. X-rays and gamma rays interact strongly with dense materials, but pass easily through lighter materials. However, neutron radiation passes easily through many dense materials, while interacting strongly with light elements such as hydrogen. As a result, many components that would be difficult to inspect using X-ray imaging or gamma rays are more easily analyzed using neutrons. It gives essential results that none of the other methods can give. This relates to the material transparency when exposed to the neutron flux. Light atoms, such as hydrogen and carbon, are well observed.

Also known as neutron imaging or neutron tomography, neutron radiography was developed and codified as an imaging technique decades ago, but has yet to be fully adopted by the wider NDT community due to over-reliance on aging nuclear reactor facilities (see e.g., Chen 2011).

This technique is useful for: non-destructive testing of thin films of glue in honeycomb or fiber glass epoxy structures, for example; inspection of new materials such as composites, carbon fibers or ceramics; non-destructive inspection of certain metallic or concrete block assembly; detection of moisture or corrosion at their very early stage: hydroxydation detection in materials; study of oil films or lubricants and pyrotechnic device inspection.

Neutron radiography and tomography are proven techniques for the nondestructive testing of manufactured components in the aerospace, energy and defense sectors, not to mention numerous research applications. Neutron tomography is presently underutilized in part because of a lack of accessible, high-intensity neutron sources. Currently, these services are mainly supplied to the NDT community via a handful of neutron imaging-capable nuclear research facilities open for commercial use because only such facilities can supply enough neutrons for imaging. By developing a high-intensity neutron source that does not rely on a reactor, Phoenix (2021) is bringing newfound reliability and ease of access to this powerful testing method.

At the present time, neutron radiography is one of the main NDT techniques able to satisfy the quality-control requirements of explosive devices used in space programs. Most of the detonating devices of the Ariane space program have been systematically submitted to NR examination at the CEA facilities for more than 20 years. The detection of cracks of 0.1 mm thickness in the explosive charge is common and the efficiency of the technique enables easily distinguishes differences in the compression of the explosive even through different metallic containers such as lead, aluminum or steel. The ability to detect compounds containing hydrogen atoms is also used to inspect oil levels and insulating organic materials. Neutron radiography also facilitates the checking of adhesive layers in composite materials, surface layers (polymers, varnishes, etc.). All types of O-rings and joints containing hydrogen can be observed even through a few centimeters thickness of steel (CEA 2021).

Neutron imaging is a non-invasive method for material research on the macroscopic level. It is carried out at laboratories equipped with powerful neutron sources, suitable neutron beam lines and neutron detection systems. Decades ago neutron radiography began capturing images with film techniques. These techniques yielded excellent spatial resolution even over large fields of view. In recent years, improvements in the detection techniques and their digitization have been the main forces driving successes in neutron imaging. Several detector options have been developed, implemented and used in practical applications in order to achieve digital information from the neutron transmission process which is needed for a quantitative evaluation of image data by sophisticated methods like neutron tomography, phase-contrast imaging, neutron interferometry and time-dependent studies.

The most common approach in digital neutron imaging is a conversion of the neutron field information into visible light by a scintillation process, where a neutron converter is needed because neutrons do not excite directly due to their neutral charge. Low-level light signals can be observed either with sensitive camera systems or by using amorphous silicon-based semiconductor plate devices. However, these now-established detection techniques are still limited in respect to spatial and time resolution. The best possible spatial resolution which can be achieved today is available by a system built at PSI (Paul Scherrer Institute, Villingen, Switzerland) with about 10 μm pixel size. Recently, it was upgraded with a tilted option for an increased resolution by a factor of 4 in one direction. Scintillator-based techniques are limited by the dissipation of the secondary particles. This limitation has motivated the search for new detector options. One approach is a pixilated system where the readout per incoming neutron can be used to calculate precisely the position of its impact. Such devices are realized as the TIMEPIX system already.

The system was tested successfully at PSI and other neutron imaging facilities. Other options might be to use MEDIPIX devices with neutron-absorbing/converting materials. A similar pixel detector, EIGER, has been developed at PSI with satisfactory performance in the field of neutron imaging (Lehmann et al. 2011).

A Coordinated Research Project (CRP) on neutron imaging – a non-destructive tool for materials testing – has been run by IAEA in the period 2003–2006 (see IAEA 2008). List of participants included scientists from Malaysia, South Africa, Romania, Bangladesh, Germany, USA, Switzerland, Russian Federation, Brazil and India. The results of CRP have been presented at numerous conferences and in number of publications (Mikerov et al. 2004; Barmakov et al. 2004; Bogolubov et al. 2005; Pugliesi et al. 2005; Hassanein et al. 2006; Dinca and Pavelescu 2006).

Another unique property of neutron radiography is that contrast tagging can be done with neutrons, unlike other forms of electromagnetic radiation. Because of the way neutrons interact with materials, some elements act as "neutron absorbers", which eagerly react to neutron radiation and are thus extremely opaque to neutron imaging. One such element is gadolinium. In nondestructive testing methods such as dye penetrant testing, flaws and surface discontinuities such as hairline cracks, gaps, leaks and fatigue cracks are detected using capillary action. The dye acts as a contrast agent for these applications, making the flaws more easily visible to the naked eye. By creating a gadolinium-rich liquid solution and applying it to an object, the object will be "dyed" in a similar way, but only neutrons will be able to "see" where the dye settles.

Gadolinium tagging is critical to jet engine turbine blade quality assurance, which is one of neutron imaging's most significant NDT niches. These turbine blades are cast around ceramic molds extruded out of lightweight metal that actually has a lower melting point than the maximum temperature of its operational environment – so if they aren't properly cooled, they can actually melt while the engine is running! To counteract this, the turbine blades are cast with cooling channels running through them that air can flow through. In the casting process, bits of ceramic remnants can become stuck in the blades' cooling channels, which present a significant risk of catastrophic engine failure and danger to the passengers. Rigorous quality assurance has to root out these types of defects as comprehensively as humanly possible – however, for both X-ray and neutrons, the attenuation rates between the metals and ceramic remnants are too similar for the contrast to be easily apparent in an X-ray or neutron image. This is where gadolinium tagging comes in. By washing the turbine blades in a gadolinium solution, gadolinium seeps into the porous structures of the ceramic remnants but washes out of everywhere else, dyeing them so that they can be easily identified on a neutron image as indicative of a defect! (for details see Ascenzo 2021).

The textural and geometrical properties of the pore networks (i.e., such as pore size distribution, pore shape, connectivity and tortuosity) provide a primary control on the fluid storage and migration of geofluids within porous carbonate reservoirs. These properties are highly variable because of primary depositional conditions, diagenetic processes and deformation. This issue represents an important challenge for the characterization and exploitation plan in this type of reservoir.

In the study reported by Zambrano et al. (2019), the complementary properties of neutrons and X-ray experiments are carried out to better understand the effects of pore network properties on the hydraulic behaviour of porous carbonates. Neutrons have unique properties and are particularly suitable for this study due to the sensitivity of neutrons to hydrogen-based fluids. The used methodology combines dynamic neutron radiography (NR), integrated X-ray and neutron tomography (XCT, NCT) and computational fluid dynamics simulations (lattice-Boltzmann method) of porous carbonate reservoir analogs from central and southern Italy.

A very useful and comprehensive study of data helpful in the design of filters for thermal neutron radiography has been reported by Barton (2001). Many existing facilities might benefit from the optimization of temporary or permanent filters designed to transmit thermal neutrons, attenuate fast neutrons, attenuate gamma and soften the Maxwellian spectrum to maximize detail contrast. Each situation will require a different filter design, but in each case the design should consider directly measured filter data, rather than relying on fundamental cross-section libraries and computer calculations. A review of measured data shows that quasi-single crystal sapphire (Al_2O_3) at ambient temperature can perform better than liquid nitrogen cooled silicon (Si), quartz (SiO_2) and bismuth (Bi) for discrimination between thermal and fast neutrons or for softening the Maxwellian spectrum. If discrimination against gamma radiation is also needed measurements for bismuth showing thermal neutron transmission, which is poor relative to Al_2O_3 should be considered. The use of a thick bismuth filter cooled by liquid nitrogen is found to be attractive primarily for cold neutron beam optimization, not for thermal neutron beams. Single crystal lead fluoride (PbF_2) is an alternative to quasi-single crystal bismuth in those situations that demand gamma reduction (Barton 2001).

5.5 Cultural Heritage

Cultural heritage is the legacy of physical artifacts and intangible attributes of a group or society that are inherited from past generations, maintained in the present and restored for the benefit of future generations. Physical or "tangible cultural heritage" includes buildings and historical places, monuments, artifacts, etc. that are considered worthy of preservation for the future. Scientific studies of art and archeology present a necessary complement for cultural heritage conservation, preservation and investigation. As cultural heritage objects are frequently unique and non-replaceable, non-destructive techniques are mandatory and, hence, nuclear techniques have a high potential to be applied to study these valuable objects. Nuclear techniques, such as neutron activation analysis, X-ray fluorescence (XRF) analysis or ion beam analysis, have a potential for non-destructive and reliable investigation of precious materials, such as ceramics, stone, metal or pigments from paintings. Such information can help to repair damaged objects adequately, distinguish fraudulent artifacts from real artifacts and assist archeologists in the appropriate categorization of historical artifacts.

The IAEA as a leading supporter of the peaceful use of nuclear technology assists laboratories in its Member States to apply and develop nuclear methods for cultural heritage research for the benefit of socioeconomic development in emerging economies (IAEA 2011). The list of scientific and technical papers on the applications of nuclear techniques to cultural heritage studies is huge, therefore we shall list only some of the references in order to guide the reader to the more detailed texts, see for example Dran et al. (2004), Spoto (2000), Adriaens and Dowsett (2006), Duval et al. (2004), Budnar et al. (2004), Zucchiatti et al. (2002, 2003, 2004), Neelmeijer and Mäder (2002, 2004), Denker and Opitz-Coutureau (2004), Van Ham et al. (2002, 2004, 2005), Pappalardo et al. (2004), Jembrih-Simbürger et al. (2004), De Ryck et al. (2003) and more, see also Aditional reedings.

Particle beams are used for non-destructive analysis of works of art and ancient relics. The conservation and preservation of cultural heritage are one of the main concerns within developed parts of the world, including North America and Europe, today. For example, Adriaens (2005) gives an overview of research in or associated with the pan-European network COST Action G8, which aims at achieving a better preservation and conservation of cultural heritage by increasing the knowledge of art and archeological objects through advanced chemical and physical analyses. The paper is focused on the use of various analytical techniques for the examination of cultural heritage materials and includes research examples on painted works of art, ceramics, glasses, glazes and metals. In addition, attention is drawn to advances in analytical instrumentation, for example, the development of portable techniques to perform analyses on site, and to the need for collaboration between people directly involved in the field of cultural heritage and analytical scientists.

It is clear that precious cultural heritage objects should remain unaltered after they are exposed to analytical investigation. Therefore non-destructive methods are of crucial importance for investigations. This simply means that the (unavoidable) side-effects of an irradiation must not be noticeable on the object of interest, now or in the future. This can be primarily achieved by reducing the intensity of the irradiation to very low levels. In order to obtain meaningful analytical information, the low primary irradiation has to be balanced by a correspondingly high detection efficiency of the secondary signal one wants to analyze. Great strides in this direction have been undertaken in recent times, opening up many possibilities to analyze valuable pieces of art.

Ion beam analysis is today the most versatile technique for investigating objects of cultural significance. This is due to the flexibility of ion beams, where the beam species (protons, alphas, heavy ions), the energy, the intensity and the size of the beam (sub-millimeter to sub-micron) can be varied in a suitable way. In addition, the efficiency and resolution of detector systems for X-rays, γ-rays, and charged particles have greatly improved over the years. An important aspect for ion beam analyses of art objects is the use of an external beam because often these objects cannot be brought inside the accelerator vacuum system. A multitude of different ion beam techniques is now available: NRA, PIXE, PIGE, RBS, ERDA.

The most common method for cultural heritage investigation is XRF. Due to its different depth sensitivity, it is complementary to PIXE, and is sometimes combined with it. Portable instruments make XRF a very valuable method for studying objects which cannot be moved to an accelerator (Table 5.3).

If the absolute age of an object containing organic carbon is of interest, ^{14}C dating is often used. Since this requires taking a small piece of material from the object, it is not a truly non-destructive method. However, counting ^{14}C atoms directly by AMS, rather than counting the infrequent β-decays (the original method), has increased the detection efficiency of ^{14}C by a factor of about 10^6. This then allows ^{14}C measurements to be performed on very small samples, sometimes as low as a few micrograms of carbon, with negligible effects on the sampled object. The age range extends back to some ten half-lives of ^{14}C, i.e., to about 50,000 years.

There are several books covering this interesting area of accelerator (nuclear physics in a broader sense) applications, including Janssens and Van Grieken (2004), IAEA (2011), Macková et al. (2016) and others.

Irradiation conservation of cultural heritage has been discussed by Ponta (2008). The use of non-invasive and non-destructive methods is highly relevant for cultural heritage objects in particular, due to their uniqueness and the often high cost of material as well as immaterial value. It is, however, of great importance to gain a simple overview of their material distribution, the manufacturing techniques, the provenance and the current

TABLE 5.3

Ion Beam Analysis Facilities in Europe (as reported by Macková et al. 2016)

Country	Institution	web
Austria	NSIL, IAEA, Seibersdorf	https://www.seibersdorf-laboratories.at/en/home
Belgium	ALTAÏS, Namur	http://www.unamur.be/sciences/physique/larn/
Croatia	Ruđer Bošković Institute, Zagreb	http://cems.irb.hr/en/
Czech Republic	CANAM, Nuclear Physics Institute, Řež	http://canam.ujf.cas.cz/
Finland	University of Helsinki, Ion beam analysis laboratory, Helsinki	https://tuhat.halvi.helsinki.fi/portal/en/equipment/ion-beam-analysis
France	AGLAE, Accélérateur Grand Louvre d'Analyses Elémentaires, Paris	http://c2rmf.fr/analyser/un-laboratoire-de-haute-technologie-pour-les-collections-des-musees/aglae
	CENBG Centre d'Études Nucléaires de Bordeaux Gradignan	http://www.cenbg.in2p3.fr/
	SAFIR Système d'Analyse par Faisceaux d'Ions Rapides, Institut des NanoSciences de Paris	http://www.insp.jussieu.fr/Systeme-d-Analyse-par-Faisceaux-d.html
Germany	HZDR Helmholtz Zentrum Dresden-Rossendorf, Ion Beam Analysis Group, Dresden	http://www.hzdr.de/
Greece	NCSR National Centre for Scientific Research DEMOKRITOS, Athens	http://www.demokritos.gr/
Italy	INFN LABEC, Florence	http://labec.fi.infn.it/
	Laboratori Nazionali di Legnaro, AN2000 accelerator, INFN Legnaro	http://www.lnl.infn.it/index.php/en/accelerators-3/an-2000
Hungary	MTA Atomki, Laboratory of Ion Beam Applications, Institute for Nuclear Research, Debrecen	http://w3.atomki.hu/index_en.html
	MTA Wigner FK, Department for Materials Science by Nuclear Methods, Wigner Research Centre for Physics, Budapest	http://wigner.mta.hu/en
Romania	Horia Hulubei National Institute for Nuclear Physics and Engineering, Bucharest-Magurele	http://www.nipne.ro/
Serbia	FAMA Facility for Modification and Analysis of Materials with Ion Beams, Vinča Institute of NuclearSciences, Belgrade	http://www.vin.bg.ac.rs/010/index.html
Slovak republic	Advanced Technologies Research Institute, Trnava	http://www.mtf.stuba.sk/english/institutes/advanced-technologies-research-institute/
Slovenia	MIC Microanalytical Centre, Jožef Stefan Institute (JSI), Ljubljana	http://www.rcp.ijs.si/mic/
Spain	Centro Nacional de Aceleradores, CNA, is a joint centre of Universidad de Sevilla, Junta de Andalucía and CSIC, Seville	http://acdc.sav.us.es/cna/
UK	Ion Beam Centre, University of Surrey, Guildford	http://www.surrey.ac.uk/ati/ibc/
Ukraine	ISSPMT Institute of Solid State Physics, Materials Science and Technologies, National Academy of Sciences of Ukraine, Kharkov	http://www.kipt.kharkov.ua/kipt_sites/isspmst/main_site/ENG34Main.html
	Institute of Applied Physics, National Academy of Sciences of Ukraine, SUMY	http://iap.sumy.org/viewexpbase/en/?id=15

condition (e.g., the determination of possible damage) by transmission imaging techniques. While X-ray imaging is often sufficient for such investigations, there are numerous cases where the method reaches its limits. Here, the complementarities of neutrons applied in a similar manner can provide new insights into the object studied. The better transmission for metals and the higher contrast for organic materials can already be exploited in the simple neutron radiography mode. More advanced are the neutron tomography methods, which are available at the neutron imaging facilities of the PSI, Switzerland, on a wide dimensional scale range. Virtual slices through the objects at arbitrary positions enable deeper perceptions and also dimensional determinations in full 3D. Further methodical improvements have been developed. Narrowing the neutron energy band allows information on the microcrystalline structure to be obtained directly, and using phase-contrast and dark-field techniques by means of a grating interferometer device enables a deeper understanding of the material compositions and structures. The examples in this paper are chosen to demonstrate the application range of neutron imaging and the performance of the different set-ups (Mannes et al. 2014).

This tutorial review by Aucouturier and Darque-Ceretti (2007) intended to show, through demonstrative examples chosen from the recent literature, how surface characterization conducted by modern investigation methods leads to very rich information on a cultural heritage artifact's history, fabrication techniques and conservation state. The application of surface science to conservation science is of great help for the definition of a conservation and restoration policy for museums and archeological objects. A brief description of the investigation methods is given, together with relevant references for more detailed information.

Next, we shall describe, with some details, the accelerator used in the Louvre (Roberts and Tuttle 2015). The museum's particle accelerator is placed in a basement, 15 m below the towering glass pyramid of the Louvre Museum in Paris. The "Accélérateur Grand Louvre d'analyse élémentaire, (AGLAE)" has been a part of the museum since 1988. Researchers use AGLAE's beams of protons and alpha particles to find out what artifacts are made of and to verify their authenticity. The amounts and combinations of elements an object contains can serve as a fingerprint hinting at where minerals were mined and when an item was made.

For example, AGLAE was used to check whether a saber scabbard gifted to Napoleon Bonaparte by the French government was actually cast in solid gold (as it was!), to identify the minerals in the hauntingly lifelike eyes of a 4500-year-old Egyptian sculpture known as The Seated Scribe (black rock crystal and white magnesium carbonate veined with thin red lines of iron oxide).

AGLAE facility is unique in that the activities are 100% dedicated to cultural heritage. It is the only particle accelerator that has been used solely for this field of research. The techniques used at AGLAE include particle-induced X-ray and gamma-ray emission spectrometry, which can identify the slightest traces of elements ranging from lithium to uranium. Because its beams work outside the vacuum, allows researchers to study objects of any size and shape.

There is enormous literature on the analysis of different gemstones embedded in archeological artifacts, especially jewelry. Two stones received most of the attention: emerald and lapis lazuli. Jewelry is often a valuable archeological finding, as such it has been analyzed by many techniques including nuclear analytical methods discussed in this text. Combined external-beam and μ/Raman characterization garnet stones are discussed by Calligaro et al. (2002), while emeralds are discussed in papers by Calligaro et al. (2000) and Kržič et al. (2013).

Lapis lazuli is the mixture of minerals with the most important component being lazurite (up to 40%), having formula $Na_7Ca(Al_6Si_6O_{24})(SO_4)(S_3)\cdot H_2O$; possible colors ultramarine, midnight blue, bluish-green (Mindat 2021). The nuclear analytical techniques can identify the provenance of the mineral and its impurities by the determined multi-element composition. For the discussion of this problem and related topics see Wyart et al. (1981), Lo Giudice et al. (2009), Re et al. (2011, 2013, 2015), Angelici et al. (2015), Bacci et al. (2009), Schmidt et al. (2009), Favaro et al. (2012), Calligaro et al. 2014), Grassi et al. (2004), and others.

Analysis of glass been done at Institute Jožef Stefan in Ljubljana, Slovenia for a long period of time, see Šmit et al. (2002, 2004, 2005, 2009, 2012, 2014).

Archeological applications cannot be generalized, and any comprehensive review would necessarily be lengthy. By now, there is a large number of AMS radiocarbon dating in archeology. Like most applications of AMS ^{14}C, the benefits to archeology almost entirely come from the reduction of sample size of a factor of at least 1,000. This has permitted the dating of important or unique objects, but of arguably more value, it has allowed far greater selection to be made on exactly what is chosen for dating.

Three categories of AMS archeological achievements should be mentioned (IAEA 1995): (a) Improvements in dating (where "conventional" ^{14}C dating already done), (b) Dating of "difficult" sites through judicious selection of material, (c) Resolving major issues.

There is a strong interest in the possibility of reliably dating mortar from buildings that are medieval or older because it is often difficult to find suitable organic material (timber, charcoal) for dating. Although the method of ^{14}C dating by AMS that requires less than 1 milligram of carbon has made it easier to collect a sufficient amount of organic sample material for ^{14}C dating, this material rarely dates the initial stages of a building. This is because repeated rebuilding and changing of timbers as a consequence of rot, wars and devastating fires often lead to younger ages than that of the original stone building. In other cases, reuse of older timber or charcoal included in mortar from the burning of lime will give too high ages due the age of the wood at the time of construction.

Mortar contains carbon from carbon dioxide absorbed from the atmosphere at the time of construction, which in principle makes it ideal for ^{14}C dating. To produce lime, limestone (fossil calcium carbonate) is heated to above 1000 °C to liberate carbon dioxide and produce quicklime (calcium oxide). The quicklime is then slaked with water to form calcium hydroxide which is mixed with water and sand to form mortar. Calcium hydroxide in mortar reacts with carbon dioxide from the air forming calcium carbonate, hardening the mortar. The ^{14}C content of a mortar sample can thus give a measure of the time elapsed since the time of hardening. As the carbon in the mortar originates from atmospheric carbon dioxide the radiocarbon age arrived at can be converted to calendar years using normal calibration procedures.

There are, however, well-known risks associated with this method. The mortar may contain old limestone, either from incompletely burnt limestone or from carbonaceous sand used as filler, leading to ages that are too old. Conversely, delayed hardening in thick walls or later recrystallization of the carbonate involving younger carbon dioxide can lead to dates that are too young. In the literature, examples are found of both agreement and disagreement between mortar dating's and known ages (Delibrias and Labeyrie 1964; Delibrias et al. 1964; Baxter and Walton 1970a, 1970b; Folk and Valastro 1979; Malone et al. 1980; Van Strydonck et al. 1983, 1986).

At present, a mortar dating collaboration is in progress between the conventional ^{14}C Dating Laboratory in Helsinki and the AMS ^{14}C Dating Laboratory in Aarhus. The Helsinki laboratory has been working on mortar dating for several years and obtained convincing results by conventional ^{14}C dating, see for example (Sonninen et al. 1989). The risk of contamination from unburnt limestone is reduced by mechanical separation. The mortar samples are gently crushed and wet sieved using a mesh width of 65 microns. The grains of mortar are small, 1–10 microns, and they easily pass through the sieve, whereas the fractions of calcite crystals of the unburnt limestone are so much larger and harder that they can be well separated.

The study of old documents is done by the analysis of material on which they are written or by analysis of writing material itself (Giuntini et al. 1995; Del Carmine et al. 1996). The same is done with paintings, where the more attention is paid to the analysis of paint (Mando et al. 2005; Ballirano and Maras 2006; Miliani et al. 2008).

Ancient metallurgy is often the subject of investigations (Constantinescu et al. 2012b) and nuclear analytical methods can make a significant contribution to the subject. Of special interest is the production and the use of brass (Bayley 1990; Istenič and Šmit 2007), plated metallic objects (Šmit et al. 2008; Jezeršek et al. 2010) and gold items (Constantinescu et al. 2009, 2012a).

5.6 Environment Protection

Accelerator-based analytical techniques are used in monitoring environmental pollution and for identifying the pollution sources. Because of their multi-elemental capabilities, and the possibility of measuring concentration profiles, they have been extensively used for air pollution studies. One of the major sources of air pollution is coal burning for electrical power generation and heating, despite improvements in combustion operations and the use of gas cleaning devices such as electrostatic precipitators and electron accelerators. Moreover, electrical precipitators may show minimal collection efficiency for particles in the 0.1–1.0 micrometer size range. Such particles have longer atmospheric residence times and greater effects on health and air quality than would an equal mass of larger particles. They contain unusually large surface concentrations of potentially toxic trace elements which increase with decreasing particle size, owing to mechanisms of fly ash formation.

PIXE and XRF are very effective techniques in atmospheric aerosol investigation; therefore, many authors extensively use them. In the work by Calzolai et al. (2008) an inter-laboratory comparison of the results obtained analyzing several samples (collected on different substrata) with both techniques is presented: the samples were analyzed by PIXE (in Florence, at the 3MV Tandetron accelerator of LABEC laboratory) and by XRF (in Genoa and Milan, where two Oxford XRF instruments are operational). The results of the three sets of measurements are in good agreement for all the analyzed samples. The aim of this work was also to compare PIXE and XRF performance in atmospheric aerosol analysis with the routine set-up currently in use at the three laboratories, to determine the best technique to be applied depending on the substratum used for aerosol sampling and the main elements of interest for each specific research project. Results of the comparison between the minimum detection limits of both techniques are shown for all the measured elements, for different substrata (Teflon, polycarbonate and cellulose mixed esters), Calzolai et al. (2008).

The Mezquital valley is a vast area near Mexico city that has been irrigated with wastewater from Mexico city for more than 50 years. At present, this water source continues to be used while new irrigation areas are being incorporated according to rural demand. The research study by Solis et al. (2005) was conducted to evaluate the relationship between the accumulation of metals in soils and plants and the physico-chemical properties of soils irrigated in this manner for 50 and 100 years, respectively. Soil properties such as pH and total organic carbon (TOC) were determined by conventional methods. Plant and soil total trace metals Fe, Co, Ni, Cu, Zn and Pb were determined using PIXE. Lower pH and TOC contents were obtained for soils irrigated during 100 years, indicating a higher metal bioavailability. This is not reflected in plant content for most of the reported elements, but Zn and Pb show a higher absorption in 100 years old plots (26–79%) than in 50-year-olds plots, indicating a pH dependence (Solis et al. 2005).

Chmielewski (2011) discussed in details the emerging applications of electron beam for environment protection, in particular the treatment of marine diesel fuel gas, residual ballast water treatment, sewage sludge treatment. Chmielewski et al. (2010) tested the electron beam treatment technology for purification of exhaust gases from the burning of heavy fuel oil (HFO) with sulfur content approximately 3 wt% at the Institute of Nuclear Chemistry and Technology laboratory plant. The study was conducted to determine the sulfur dioxide (SO_2), oxides of nitrogen (NOx) and polycyclic aromatic hydrocarbon (PAH) removal efficiency as a function of temperature and humidity of irradiated gases, absorbed irradiation dose, and ammonia stoichiometry process parameters. In the test performed under optimal conditions with an irradiation dose of 12.4 kGy, simultaneous removal efficiencies of approximately 98% for SO_2 and 80% for NO_x were recorded. The simultaneous decrease of PAH and one-ringed aromatic hydrocarbon (benzene, toluene and xylenes [BTX]) concentrations was observed in the irradiated flue gas. Overall removal efficiencies of approximately 42% for PAHs and 86% for BTXs were achieved with an irradiation dose 5.3 kGy. The decomposition ratio of these compounds increased with an increase of absorbed dose. Thus, the electron beam flue gas treatment (EBFGT) technology ensures simultaneous removal of acid (SO_2 and NO_x) and organic (PAH and BTX) pollutants from flue gas emitted from the burning of HFO.

On 1 January 2020, a new limit on the sulfur content in the fuel oil used on board ships came into force, marking a significant milestone to improve air quality, preserve the environment and protect human health (International Maritime Organization; IMO 2020). The rule limits the sulfur in the fuel oil used on board ships operating outside designated emission control areas to 0.50% m/m (mass by mass) – a significant reduction from the previous limit of 3.5%. Within specifically designated emission control areas the limits were already stricter (0.10%). This new limit was made compulsory following an amendment to Annex VI of the International Convention for the Prevention of Pollution from Ships (MARPOL).

Cargo ships are big emitters because they were using HFO derived as a residue from crude oil distillation. HFO had a much higher sulfur content which, following combustion in the engine, ended up in ships' emissions. For example ship with two stroke Diesel engines, 6–14 pistons (up to 81 MW) has a consumption of approximately 250 tons per day had an emission of NO_x (1500 ppm), SO_x (600 ppm), CO (60 ppm), VOC (180 ppm).

Now, the vast majority of ships are using very low sulfur fuel oil (VLSFO) to comply with the new limit, and no safety issues have to date been reported to IMO.

The approved alternative to IMO (2020) is off-gases purification (open sea and ECA). For 1.5% S in fuel for SOx ECAs, ship has to be equipped in FGT SO_x ≤ 6 g/kWh (recalculated as SO_2). As an example, FGD system (upper) + SCR (bottom) was installed on the board proposed by Chmielewski (2017).

Project ARIES (Accelerator Research and Innovation for European Science and Society) held on 1 December 2017. meeting in CERN, Switzerland on the topic "ARIES Meets Industry – accelerator application to the ship exhaust gases treatment". The above mentioned presentation (Chmielewski 2017) was delivered together with:

- (Torims 2017) – How to bring particle accelerator on board of ship?
- (Zimek 2017) – Ideas of compact accelerator development for the exhaust gases treatment and what could be bought today in the particle accelerator market for the ship needs?
- (Mosconi 2017) – Italian Coast Guard experience in ensuring compliance with the existing requirements for the ship emissions.
- (Kuczkowski 2017) – Remontowa Ship Yard experience and challenges installing the scrubbers on board of the ships.
- (Bocchetti 2017) – Experience and policy of Grimaldi for ship exhaust gas emission level moderation. Company perspective and needs.

5.6.1 Radiation Treatment of Wastewater, Ballast Tank Waters and Sediments

Maritime traffic is the single largest contributor to air pollution – a single cruise ship emits as much pollution as one million cars. Several technologies are being explored to reduce the pollutants in the exhausts of ships' diesel engines. Accelerator scientists have proposed a solution that involves breaking down particulate matter as well as molecules of sulfur and nitrogen oxides with an electron-beam accelerator of a few hundred kilovolts, before safely extracting them using water. The ARIES (Accelerator Research and Innovation for European Science and Society) Horizon 2020 project, coordinated by CERN, is working on a real-scale test of this technology. A first test was performed recently on an old and rusty Soviet-era Latvian tugboat named Orkāns ("storm" in Latvian), moored at the Riga shipyard on the Baltic Sea. The small vessel, procured by the Riga Technical University in Latvia, has an old but powerful engine that could easily be made available for the duration of the tests.

A long pipe, equipped with several detectors, connected the tugboat to an accelerator-on-a-truck that was provided by the Fraunhofer FEP of Dresden in Germany. On the truck, the exhausts were treated in a specially built chamber, with the electrons from the accelerator inducing molecular excitation, ionization and dissociation to break down the pollutant molecules. Before finally being released into the air, the pollutants from the exhausts were washed out using water in a small "wet scrubber", designed and built by the INCT of Warsaw in Poland, who originally proposed this novel accelerator-based approach.

The compact system developed in the EU-funded ARIES project offers a unique solution to tackle hazardous emissions from maritime transport, where the use of low-grade diesel fuel means just one large container ship can produce the same amount of sulfur pollution as 50 million cars.

This hybrid particle accelerator-based system is unique in that it acts on both sulfur and nitrogen as well as particulate matter with a considerable reduction in cost and size compared to other pollution-reduction solutions, according to Vretenar (2019). Field trials conducted with an old tugboat in Latvia last year confirmed earlier laboratory tests, suggesting an optimized version of the system could remove as much as 98% of sulfur oxide (SO_x) and nitrogen oxide (NO_x) emissions from ship exhaust.

The ARIES solution uses a compact electron-beam accelerator operating at a high voltage to break down sulfur and nitrogen oxide molecules as exhaust gases pass through a plasma chamber. The dissolved pollutants then react with water vapor in the flue gases, forming droplets that are washed out together with particulate matter using jets of sea-water in a small conventional wet scrubber. The polluted water is then stored on board for safe disposal in port.

A major advantage of this solution is that it can be easily retrofitted to the world's large fleet of cargo ships, cruise liners and passenger ferries as well as installed in new-build vessels as the maritime industry prepares to meet increasingly stringent environmental and emissions standards. The results of this part of the ARIES project are a major step toward sustainable transportation and the establishment of a new green economy. The next challenges are to demonstrate the economic and technical viability of the solution under the real thermal, environmental, mechanical and safety conditions of an operating ship, and to have the technology certified by maritime classification societies (Vretenar 2019).

Project details are following: Acronym: ARIES, Project No.: 730871. Participants: Switzerland (Coordinator), Sweden, Austria, Belgium, France, Germany, Hungary, Italy, Latvia, Malta, The Netherlands, Poland, Portugal, Romania, Slovakia, Slovenia, Spain, and the UK. Total costs: € 10,176,755, EU contribution: € 10,000,000. Duration: May 2017 to April 2021.

A planned follow-up initiative called HERTIS will address challenges mentioned in (Vretenar 2019). The project partners will further refine the technology and install a full-scale accelerator and scrubber pilot system aboard a cargo ship operated by a major European shipping company, laying the foundations for the future commercialization of an optimized version of the pollution-reducing solution.

In another corner of the planet, in Taegu, Korea, a pilot plant to clean waste water from dyeing complex has been in successful operation by EB-TECH Co. since 1998. The irradiation plant has the capacity of 1,000 m^3/day using an accelerator of 1 MeV × 40 mA. Waste water is irradiated under the beams as a thin layer curtain. Dose given to water is from 1 to 2 kGy. Irradiated waste water is then treated biologically to decrease biochemical oxygen demand (BOD) and chemical oxygen demand (COD). By combining radiation and biological treatment dye molecules are more efficiently decomposed because radiolytic products of water such as OH radicals convert the molecules into biodegradable structure. The EB-TECH Co. has proposed an industrial scale plant based on the pilot experiments of which capacity is 20,000m^3/day using 2 units of 400 kW (1 MeV) accelerator. Total construction cost is estimated as US$ 4–4.5 million. Treatment cost including construction and operation is calculated lower than that of UV and ozone.

According to the International Convention for the Control and Management of Ship's Ballast Water and Sediments (BWM Convention), see for example (IMO 2019) all ships in international traffic are required to manage their ballast water and sediments to a certain standard, according to a ship-specific ballast water management plan. All ships have to carry a ballast water record book and an international ballast water management certificate. The ballast water management standards are being phased in over a period of time. New ships must meet the ballast water treatment standard. Existing ships should exchange ballast water mid-ocean but they will need to meet the ballast water treatment standard by the date of a specified renewal survey (Fig. 5.12).

The International Convention for the Control and Management of Ships' Ballast Water and Sediments, 2004 (BWM Convention), entered into force globally on 8 September 2017. Eventually, most ships will need to install an on-board ballast water treatment system.

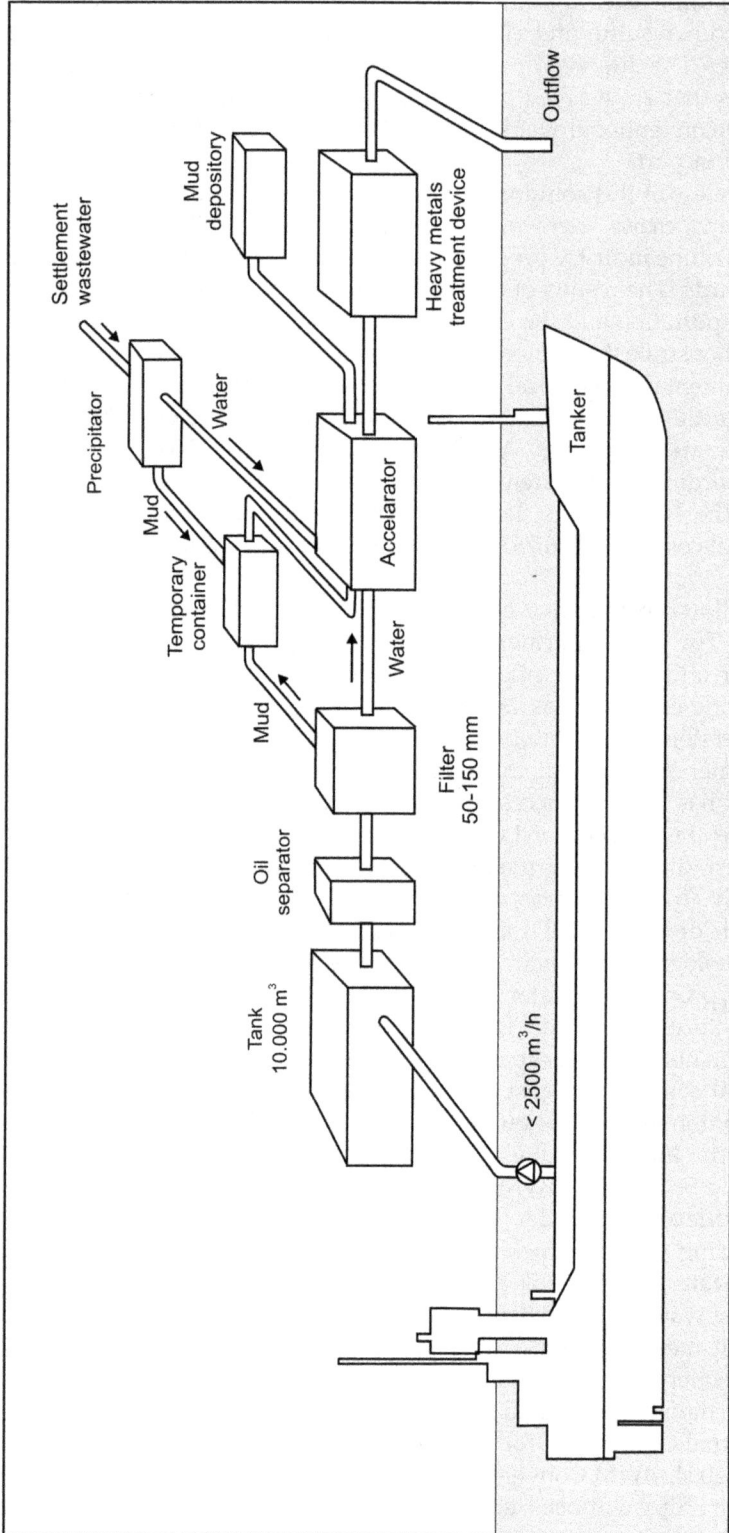

FIGURE 5.12
Proposed solution for the treatment of ship ballast water.

It is established that electron beam technology can be effectively used for dangerous pathogens removal (including *Vibrio cholerae*, *Escherichia coli* and *Enterococci*) and removal of invasive species which may displace native sealife this harm local ecosystem. The question that emerges is: how to bring particle accelerator on board of ship? This problem was discussed in some details by Torims (2017). According to the author following steps need to be done:

- IMO: acceptance;
- European Commission: marine equipment certification;
- Flag States: set the rules;
- Class Societies: RO's;
- Engine manufacturers: particular requirements for each engine type;
- Accelerators manufacturers: to build:
- Shipyards and ship repair companies: somebody has to install all this!

One has to be aware that in the process there are many rules to comply with and there many rule makers – so, to succeed, one has to involve rule makers in the prototype development process as well. In the same time, one needs to comply with physics rules which are not so easy to be changed.

5.6.2 Radiation Processing of Sewage Sludge

Disinfection of municipal sewage sludge in installation equipped with electron accelerator is advocated by Chmielewski and Zimek (2019). The schematic of the components of such an installation is shown in Fig. 5.13. The presence for health threatening pathogen causes a need to hygienize sewage sludge before it can be used as fertilizer. Electron beam treatment not only kills pathogenic microorganisms and parasites and their eggs but also decomposes toxic organic substances in the sludge designated for agriculture utilization.

Irradiation is a quick and credible method of disinfection of municipal sewage sludge. Sludge disinfection with electron accelerator can be used as soil fertilizer immediately after treatment and no large land areas are needed to their disposal for long time. The heavy metal content in treated sludge must be under the limits acceptable for agricultural applications. In spite of that the investment cost of such facility is relatively high, the operating cost are moderate and may be reduced by biogas manufacturing and heat and electricity generation. Sewage sludge can be used as a biomass in the methane fermentation process (renewable source of energy) and obtained digestate can be used as a fertilizer after hygienization by electron beam.

IAEA is implementing a CRP (Coordinated Research Program, #F23033) on Radiation Inactivation of Bio-hazards Using High Powered Electron Beam Accelerators; start date. 13 April 2018 – expected end date: 12 April 2022. The objectives are to formulate guidelines to enhance and strengthen use of electron beam accelerators for treatment of biohazards of concern under changing conditions such as at high dose rates, different ambient conditions and varying substrates in applications such as radiation sterilization, hygienization of bio-solids, sanitizing infectious hospital waste or toxic effluents and eliminating deliberate biohazards. Participating countries are Argentina, Brazil, Canada, Egypt, Germany, Hungary, Kuwait, Malaysia, Poland, Portugal, Republic of Korea, Tunisia, Turkey and the USA.

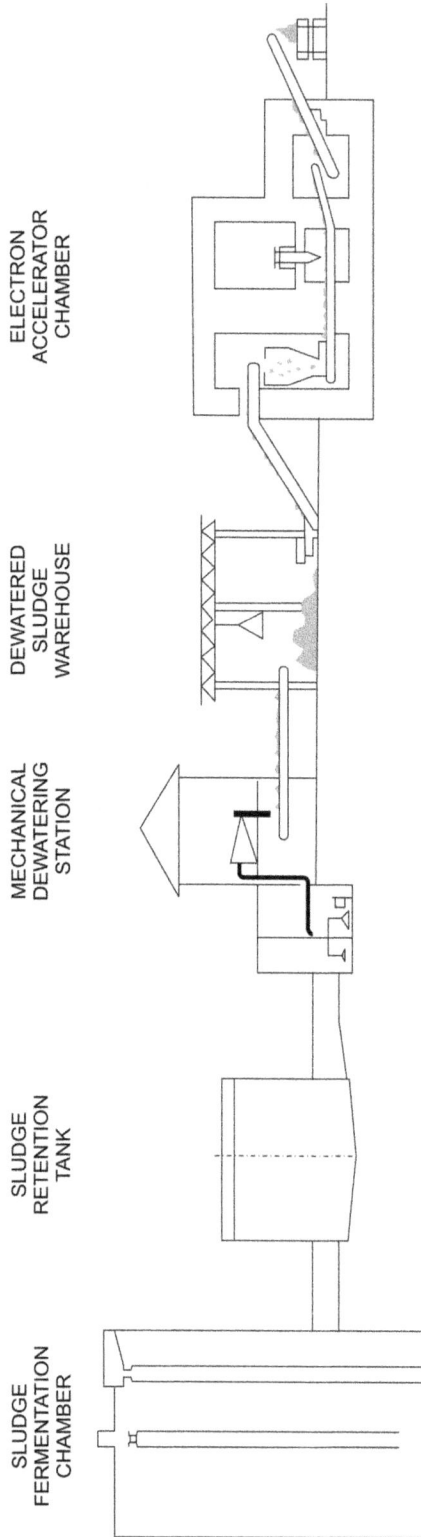

SLUDGE
FERMENTATION
CHAMBER

SLUDGE
RETENTION
TANK

MECHANICAL
DEWATERING
STATION

DEWATERED
SLUDGE
WAREHOUSE

ELECTRON
ACCELERATOR
CHAMBER

FIGURE 5.13
Schematic presentation of installation for disinfection of municipal sewage equipped with electron accelerator.

5.7 Neutral Particle Beams

Neutral particle beams (NPBs) are called in science fiction novels and movies "death rays". NPBs involve accelerating streams of atomic or subatomic particles to nearly light speed and shooting them downrange at a target. Once the particles collide with the target, they began to unravel it at the atomic and molecular level, breaking the bonds that create its shape. At a visual level, it would appear that the target is melting. The kinetic energy transferred from the particles to the target also heats it up, contributing to the melting of the target material.

On 13 July 1989, the BEAM experiment Aboard Rocket (BEAR) linear accelerator was successfully launched and operated in space (O'Shea et al. 1990). The flight demonstrated that a neutral hydrogen beam could be successfully propagated in an exo-atmospheric environment. The accelerator, which was the result of an extensive collaboration between Los Alamos National Laboratory and industrial partners, was designed to produce a 10 mA (equivalent), 1 MeV neutral hydrogen beam in 50 μs pulses at 5 Hz. The major components were a 30 keV H-injector a 1 MeV radio frequency quadrupole, two 425 MHz RF amplifiers, a gas cell neutralizer, beam optics, vacuum system and controls. The design was strongly constrained by the need for a lightweight rugged system that would survive the rigors of launch and operate autonomously. Following the flight, the accelerator was recovered and operated again in the laboratory.

NPBs have technological and engineering challenges similar to those of other directed energy weapons (DEW). NPBs require lots of energy, a mechanism to accelerate and focus the particles or subatomic particles, and another mechanism for aiming them. Apparently active in this field are the Department of Defenses in the USA, Russia and China.

Some insight into early Russian efforts is presented in the report authored by Wells (1982). Soviet research at that time was focused on the injection of NPBs into magnetically confined plasmas supposed to be the most effective method of heating these plasmas to obtain the conditions for thermonuclear fusion. Theoretically, such beams may also be used as an exo-atmospheric particle-beam weapon (PBW).

Physicists at Nuclear Physics Institute (NPI) in Novosibirsk were the first to develop the two key components for the production of intense neutral beams: (i) an intense, high-brightness, low-emittance negative ion source and (ii) a high-efficiency plasma beam neutralizer. The negative ion beam current output of the surface-plasma negative ion source, originally developed in 1971 with a planotron and a Penning-type discharge chamber, was increased dramatically by introducing cesium vapor into the arc discharge of the source. The importance of the high-brightness, low-emittance surface-plasma source lies in its use for high-current accelerators and its possible use for PBWs.

The plasma beam neutralizer developed at NPI was able to convert the high-intensity negative ion beams into neutral beams with an efficiency as high as 85%. The formation of an 80 cm long stable plasma target with plasma thickness of 2.2×10^{15} atoms/cm^2, which is the optimum for the neutralization of 1 MeV ^{2}H$^-$ ions.

The report by Wells (1982) was based on a review of the Soviet open-source literature on the production and transport of negative-ion and neutral beams. It did not find direct evidence of a program to develop exo-atmospheric particle-beams in the Soviet Union at that time. However, there is always a possibility that such a program existed in conjunction with the nuclear fusion research described in the literature. One has to keep in mind that there is a high degree of commonality to both nuclear fusion and PBW systems of many components and basic physical problems in the generation, acceleration and transport of intense particle beams which could give a country the basis for considering the development of an exo-atmospheric PBW.

Efforts in China described by Zhou and Qian (2014), A possibility of having a PBW, a DEW, and using it for attacking targets by high energy particle-beam attracted some attention and encouraged research in theory and experiment. PBW can be classified into charged and neutral PBW according to the electrical property of particles. In the article, authors start from the basic principles, including the acceleration, propagation, and interaction, which provides a theoretical support for design and construction of the PBW system. Besides, the target tracking system is described and discussed in detail, in which the estimation of trajectory is based on the model-filter process to adjust shoot direction in real-time.

Actually, much of the efforts made by Russia and China in this field is hidden under a strict umbrella of military secrets, while US Department of Defense is open, see for example (a) the solicitation by Missile Defense Agency (MDA) for program year 2018 and (b) the official link for this solicitation: (a) https://www.sbir.gov/sbirsearch/detail/1482629), (b) https://www.acq.osd.mil/osbp/sbir/solicitations/index.shtml.

The objective is to develop the enabling technologies required to field an operational exo-atmospheric NPB system. This topic seeks lightweight, compact, energy efficient, radiation hardened, components significantly more advanced than technologies demonstrated in the 1989 experiment (O'Shea et al. 1990) to enable development of NPB system(s) capable of operating in sub-orbital (pop-up) or orbital (space based) mode. A conceptual NPB system would generate, accelerate, focus, and direct a stream of highly energetic electrically neutral atomic particles, traveling at near the speed of light, unperturbed by the earth's magnetic field, at exo-atmospheric targets. Particle interactions with target matter can cause damage and generate measurable emissions allowing target characterization. Desired NPB enabling technologies include:

- Lightweight, compact and energy-efficient particle accelerators.
- Compact power sources.
- Particle neutralizers that exhibit minimal scatter, have extended operational life and have minimal impact on the operating environment.
- Anion sources, extractors and injectors.
- Beam transport, collimator, focusing, steering, sensing and tracking components.
- Sensors capable of detecting emissions from targeted objects.

Proposed efforts may seek to develop any of the above components within the context of a proposed system concept, with emphasis on achieving low size, weight and power.

Although NPBs have been under development, mainly for military reasons, industrial participation in the NPB development program in Los Alamos has resulted in the commercial production of accelerators, such as the small 1 MeV RFQ linac flown abroad a test rocket a few years ago. This industrial participation is continuing in several more recent NPB programs, with complete linac systems now being built as demonstration units by commercial aerospace companies.

According to Toole and Willis (1990), the goal of the NPB technology program is to develop a DEW system that can function as an effective component in a Strategic Defense System. The NPB should have the capability of being used as both a weapon and discriminator platform. It can kill missiles and reentry vehicles in the boost, post-boost and mid-course portion of an ICBM trajectory as well as discriminated objects during the

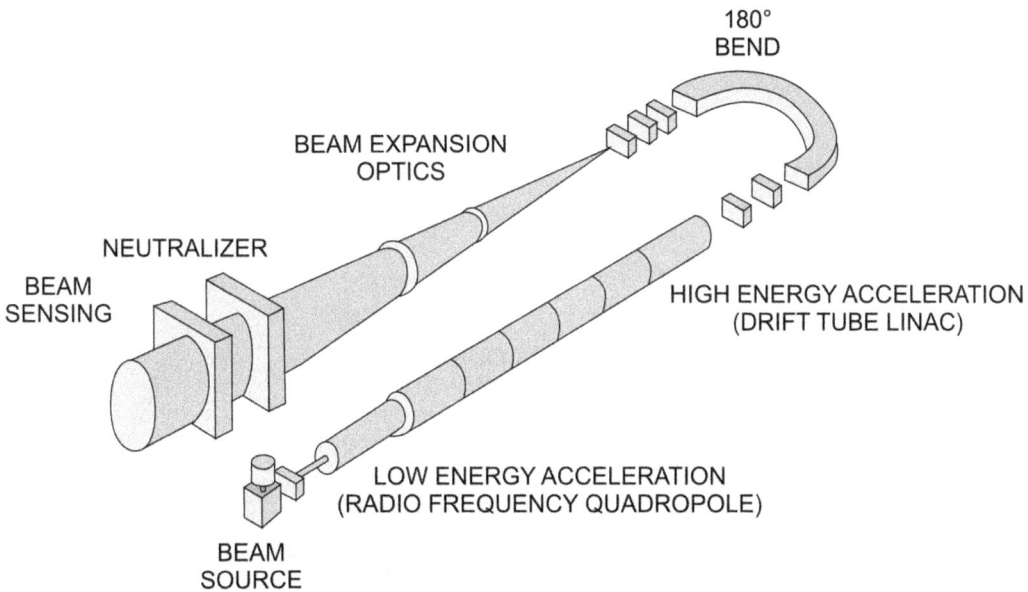

FIGURE 5.14
NPB beamline components. (After Toole and Willis 1990.)

midcourse phase. Objects from those boosters and "buses" not engaged in the boost and post boost phase would be engaged once the reentry vehicles and decoys have been deployed, i.e., during the midcourse phase of the trajectory. The NPB can be used to provide a passive, active and interactive discrimination capability against these targets. Passive discrimination is accomplished by viewing visible, ultraviolet (UV) and/or infrared (IR) emissions from targets and decoys using on-board acquisition sensors. Active discrimination is accomplished by illuminating targets and decoys with a laser tracker on-board the NPB. Interactive discrimination is accomplished by illuminating the target with the NPB which results in the emission of X-rays and neutrons which are proportional to the mass of the target. These emitted particles are measured by a free-flying detector to determine the mass of the objects. During the discrimination process target state vectors (position and velocity) can be determined which can be handed over to space-based or ground-based interceptors. All the sensors are on the same platform. This reduces data processing since sensor-to-sensor correlation is not required (Figs. 5.14 and 5.15).

The Pentagon wants a space-based DEW that would destroy enemy missiles shortly after takeoff. The weapon, called a NPB, would be tested from orbit in 2023. The MDA wants a total of $380 million through the 2023 fiscal year to develop the DEW.

Particle beams can be more effective than lasers, which only burn the surface of their targets. Particle beams of sufficient power can penetrate beyond the surface of an enemy missile, igniting its fuel supply, melting its mechanical components and frying its electronics. Particle beams are also capable of bypassing laser-deflection measures like brightly polished, mirror-like surfaces. Particle beams could be used to destroy ballistic missiles in the so-called "boost phase"-shooting them down seconds after launch, while they are still accelerating and before they release their warheads. During that stage, which only lasts about five minutes, the missiles are moving relatively slowly and are producing

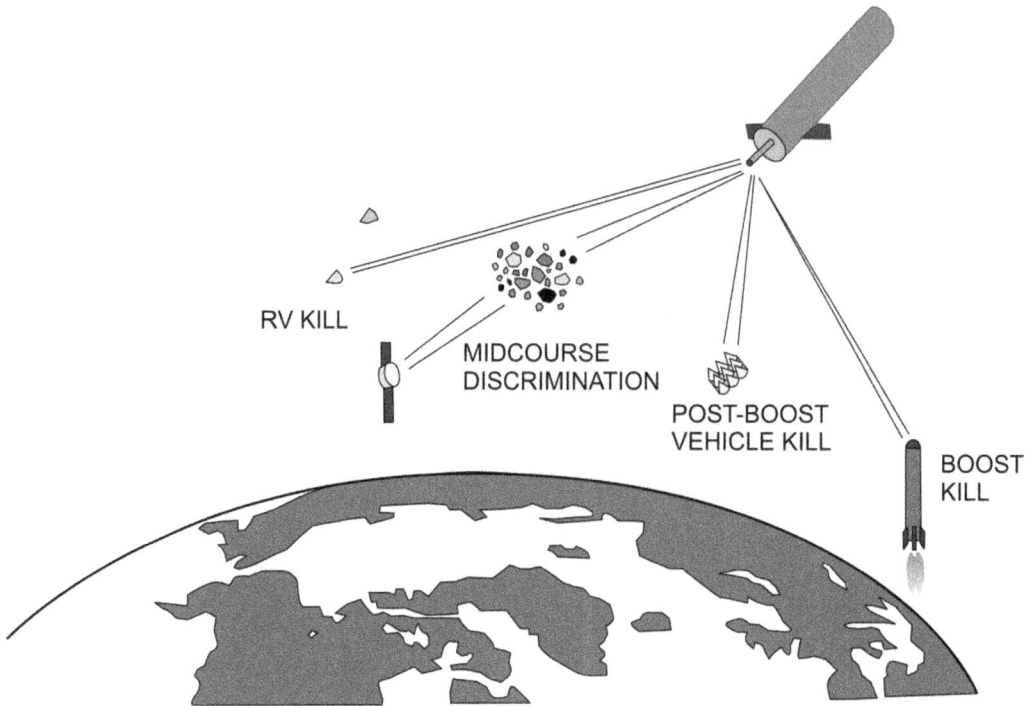

FIGURE 5.15
Space based NPB concept. (After Toole and Willis 1990.)

a massive heat signature that makes them easier to spot and track. For the NPB to work, it will need to maintain a coherent beam over the 1,000 kilometers or so from low-earth orbit to the ground. Its power source needs to be able to hold that beam, but be light and small enough to be launched into orbit. And it will need to process real-time launch detection and tracking data from a constellation of moving satellites.

In a practical invention, Lee (2004) provided a lithography technology using collimated NPBs. The method forms a patterned image on a resist formed on a wafer through a NPB lithography comprising the steps of (a) introducing collimated radical NPBs into a lens with penetrating holes to focus the radical NPBs, (b) passing the radical NPBs that have passed through the lens through a mask located underneath the lens, and (c) introducing the radical NPBs onto the resist formed on the wafer to simultaneously perform both exposure process and development process.

NPBs that heat fusion plasmas in the DIII-D National Fusion Facility have been improved by an engineering upgrade that allows them to adjust their energy in real time. DIII-D is a tokamak, a device that uses magnetic fields to contain high-temperature plasmas for research and development of fusion energy. Changing the particle energy of the beams as the plasma evolves reduces interactions with electromagnetic plasma waves. The reduced interactions let the injected particles remain within the plasma for a longer time and provide greater heating compared to a fixed (and higher) energy.

This technique provides a new tool to control high temperature, magnetically confined, fusion plasmas. Input heating to the plasma can now respond to the presence of plasma waves to ensure that the injected particles remain within the plasma for the longest possible time. The new tool keeps the plasma hotter and better suited to generating fusion.

In recent experiments, the DIII-D team has shown that reducing the injection energy of NPBs late in a plasma discharge can actually increase plasma heating and current drive. Fusion experiments previously used NPBs operating at the highest possible energy, and that energy was fixed for the duration of the experiment. The beam voltage sets the energy of the injected particles, and greater energy leads to hotter plasmas. In some conditions, however, these high-energy deuterium particles excite electromagnetic waves in the plasma and those waves can drive the beam particles out of the plasma prematurely, reducing their ability to heat the plasma to fusion conditions. The engineering change allows the beam to produce the same total power level at different values of particle energy. Injecting high power, but low energy, beams into a plasma can greatly reduce the number of electromagnetic waves produced in the plasma (//www.energy.gov/science/fes/articles/neutral-particle-beams-work-better-working-smarter). See publications discussing related problems Pace et al. 2017; Pace et al. 2018; Pawley et al. 2017; Rauch et al. 2017.

References

Aucouturier, M. and Darque-Ceretti, E. 2007. The surface of cultural heritage artefacts: physico-chemical investigations for their knowledge and their conservation. *Chem. Soc. Rev.* 36(10): 1605–1621.

Adriaens, A. 2005. Non-destructive analysis and testing of museum objects: an overview of 5 years of research. *Spectrochim. Acta Part B* 60: 1503–1516.

Adriaens, A. and Dowsett, M. G. 2006. Applications of SIMS to cultural heritage studies. *Appl. Surf. Sci.* 252(19): 7096–7101.

Angelici, D., Borghi, A., Chiarelli, F., et al. 2015. μ-XRF analysis of trace elements in Lapis Lazuli-Forming minerals for a provenance study. *Microsc. Microanal.* 21(2): 526–533.

Ascenzo, W. 2021. Neutron radiography and turbine blades. *Quality Magazine June* 8. 2021. https://www.qualitymag.com/articles/96564-neutron-radiography-and-turbine-blades)

Bacci, M., Cucci, C., Del Federico, E., et al. 2009. An integrated spectroscopic approach for the identification of what distinguishes Afghan lapis lazuli from others. *Vibrat. Spect.* 49: 80–83.

Ballirano, P. and Maras, A. 2006. Mineralogical characterization of the blue pigment of Michelangelo's fresco "The Last Judgment". *Am. Mineral.* 91: 997–1005.

Barmakov, Yu. N., Bogolubov, E. P., Koshelev, A. P., et al. 2004. Detection quantum efficiency and spatial resolution of neutron CCD-detectors. *Nucl. Instrum. Methods* B213: 241–245.

Barton, J. P. 2001. Filters for thermal neutron radiography. *Nondestruct. Test. Eval.* 16(2–6): 95–110.

Baxter, M. S. and Walton, A. 1970a. Glasgow University Radiocarbon Measurements III. *Radiocarbon* 12: 496–502.

Baxter, M. S. and Walton, A. 1970b. Radiocarbon dating of mortars. *Nature* 225: 937–938.

Bayley, J. 1990. The production of brass in Antiquity with particular reference to Roman Britain, in 2000 Years of Zinc and Brass (Ed. Craddock, P. T.). British Museum Occasional paper 50: 7–27. ISBN: 978-0861590506.

Bocchetti, D. 2017. Experience and policy of Grimaldi for ship exhaust gas emission level moderation. Company perspective and needs. Presentation at ARIES Meets Industry – accelerator application to the ship exhaust gases treatment. 1 Dec 2017, CERN, Switzerland.

Bogolubov, E., Bugaenko, O., Kuzin, S., et al. 2005. CCD detectors for fast neutron radiography and tomography with a cone beam. *Nucl. Instrum. Methods.* A542: 187–191.

Boulyga, S., Konegger-Kappel, S., Richterb, S., and Sangélya, L. 2015. Mass spectrometric analysis for nuclear safeguards. *J. Anal. At. Spectrom.* 30: 1469–1489.

Bridarolli, A., Odlyha, M., Burca, G., Duncan, J. C., Akeroyd, F. A., Church, A., and Bozec, L. 2021. Controlled environment neutron radiography of moisture sorption/desorption in nanocellulose-treated cotton painting canvases. *ACS Appl. Polym. Mater.* 3(2): 777–788.

Budnar, M., Simcic, J., Rupnik, Z., Ursic, M., Pelicon, P., Kolar, J., and Strlic, M. 2004. In-air PIXE set-up for automatic analysis of historical document inks. *Nucl. Instrum. Methods B* 219: 41–47.

Calligaro, T., Dran, J.-C., Poirot, J.-P., Querré, G., Salomon, J., and Zwaan, J. C. 2000. PIXE/PIGE characterisation of emeralds using an external micro-beam. *Nucl. Instrum. Methods Phys. Res. B* 161-163: 769–774.

Calligaro, T., Colinart, S., Poirot, J.-P., and Sudres, C. 2002. Combined external-beam and μ-Raman characterization of garnets used in Merovingian jewelry. *Nucl. Instrum. Methods Phys. Res. B* 189: 320–327.

Calligaro, T., Coquinot, Y., Pichon, L., et al. 2014. Characterization of the lapis lazuli from the Egyptian treasure of Tôd and its alteration using external μ-PIXE and μ-IBIL. *Nucl. Instrum. Methods Phys. Res. B* 318: 139–144.

Calzolai, G., Chiari, M., Lucarelli, F., et al. 2008. PIXE and XRF analysis of particulate matter samples: an inter-laboratory comparison. *Nucl. Instrum. Methods Phys. Res. Sect. B Beam Interact. Mater. Atoms.* 266(10): 2401–2404.

Carasco, C., Perot, B., Mariani, A., El Kanawati, W., Valković, V., Sudac, D, and Obhodas, J. 2010. Material characterization in cemented radioactive waste with associated particle Technique, *Nucl. Instr. Methods Phys. Res. A* 619: 419–426.

C-BORD. 2020. C-BORD Project—Effective Container Inspection at BORDer Control Points—Effective Container Inspection at BORDer Control Points, Nov. 2020, [online] Available: https://www.cbordh2020.eu/

CEA. 2021. Neutron Radiography: Non Destructive Testing. https://www-llb.cea.fr/neutrono/nr1.html

Chen, W. 2011. Introduction to Neutron Radiography. Submitted as coursework for Physics 241, Stanford University, Winter 2011.

Chmielewski, A. G., Ostapczuk, A., and Licki, J. 2010. Electron beam technology for multipollutant emissions control from heavy fuel oil-fired boiler. *J. Air Waste Manag. Assoc.* 60: 932–938.

Chmielewski, A. G. 2011. Electron accelerators for environmental protection. *Rev. Accel. Sci. Technol.* 4(1): 147–159.

Chmielewski, A. G. 2017. Electron beam treatment of marine diesel exhaust gases. Presentation at ARIES Meets Industry – accelerator application to the ship exhaust gases treatment. 1 Dec 2017, CERN, Switzerland.

Chmielewski, A. G. and Zimek, Z (Eds.). 2019. *Electron Accelerators for Research, Industry and Environment – the INCT Perspective.* Editorial Series on Accelerator Science, Institute of Electronic Systems, Warsaw University of Technology, Warsaw, Poland.

Constantinescu, B., Bugoi, R., Cojocaru, V., et al. 2009. Elemental analysis through X-ray techniques applied in archaeological gold authentication – the case of Transylvanian gold and of the Dacian bracelets. *Spectrochim. Acta B* 64: 1198–1203.

Constantinescu, B., Vasilescu, A., Stan, D., et al. 2012a. Studies on archaeological gold items found in Romanian territory using X-Ray-based analytical spectrometry. *J. Anal. At. Spectrom.* 27: 2076–2081.

Constantinescu, B., Vasilescu, A., Radtke, M., and Reinholz, U. 2012b. SR-XRF and micro-PIXE studies on ancient metallurgy of thirteen Dacian gold bracelets. *Appl. Phys. A* 109: 395–402.

Cordis. 2021. RADIATE, Research and Development with Ion Beams – Advancing Technology in Europe. https://cordis.europa.eu/project/id/824096. Download 09.06.2021.

Cutmore, N. G., Liu, Y., Sowerby, B. D., et al. 2009. Recent developments in fast neutron radiography for the interrogation of air cargo containers", Proc. IAEA Conf., Vienna: May 2009, SM/EN-01, p. 136.

Del Carmine, P., Giuntini, L., Hooper, W., Lucarelli, F., and Mandò, P. A. 1996. Further results from PIXE analysis of inks in Galileo's notes on motion. *Nucl. Instrum. Methods Phys. Res. B* 113: 354–358.

Delibrias, G. and Labeyrie, J. 1964. Dating of old mortars by the carbon-14 method. *Nature* 201: 742–743.

Delibrias, G., Guillier, M. T., and Labeyrie, J. 1964. Saclay natural radiocarbon measurements I. *Radiocarbon* 6: 233–250.

De Ryck, I., Adriaens, A., Pantos, E., and Adams, F. 2003. A comparison of microbeam techniques for the analysis of corroded ancient bronze objects, *Analyst* 128: 1104–1109.

Denker, A. and Opitz-Coutureau, J. 2004. Paintings – high-energy protons detect pigments and paint-layers, *Nucl. Instrum. Methods B* 213: 677–682.

Dinca, M., and Pavelescu, M. 2006. Calculation for a neutron imaging system based on a CCD camera. *Roman. J. Phys.* 51(3–4): 363–370.

Donzella, A., Boghen, G. Bonomi, G., Fontana, A., Formisano, P., Pesente, S., Sudac, D., Valković, V., and Zenoni, A. 2006. Simulation of a tagged neutron inspection system prototype. *J. Phys. Conf. Ser.* 41: 233–240.

Dowsett, M. G., and Adriaens, A. 2004. The role of SIMS in cultural heritage studies, *Nucl. Instrum. Methods B* 226: 38– 52.

Dran, J. C., Salomon, J., Calligaro, T., Walter, Ph. 2004. Ion beam analysis of art works: 14 years of use in the Louvre. *Nucl. Instrum. Methods B* 219–220: 7–15.

Duval, A., Guicharnaud, H., and Dran, J. C. 2004. Particle induced X-ray emission: a valuable tool for the analysis of metal point drawings, *Nucl. Instrum. Methods B* 226: 60–74.

El Kanawati, W., Perot, B., Carasco, C., Eleon, C., Valković, V., Sudac, D. J., and Obhodas, J. 2011. Conversion factors from counts to chemical ratios for the EURITRACK tagged neutron inspection system. *Nucl. Instrum. Methods Phys. Res. A* 654: 621–629.

Favaro, M., Guastoni, A., Marini, F., Bianchin, S., and Gambirasi, A. 2012. Characterization of lapis lazuli and corresponding purified pigments for a provenance study of ultramarine pigments used in works of art. *Anal. Bioanal. Chem.* 402: 2195–2208.

Folk, R. L. and Valastro, S. Jr. 1979. Dating of lime mortar by 14C: radiocarbon dating. *Proceedings of the Ninth International Conference*, eds. Berger, R. and Suess, H. E., pp. 721–732. California Press.

Fontana, C. L., Carnera, A., Lunardon, M., et al. 2017. Detection System of the First Rapidly Relocatable Tagged Neutron Inspection System (RRTNIS), Developed in the Framework of the European H2020 C-BORD Project. Conference on the Application of Accelerators in Research and Industry, CAARI 2016, 30 October – 4 November 2016, Ft. Worth, TX, USA. Physics Procedia 90: 279–284.

Fontana, C. L., Carnera, A., Lunardon, M., et al. 2018. A distributed data acquisition system for nuclear detectors. Int. J. Mod. Phys. Conf. Ser. Vol. 48: 1860118 (7 pages).

Giuntini, L., Lucarelli, F., Mandò, P. A., Hooper, W., and Barker, P. H. 1995. Galileo's writings: chronology by PIXE. *Nucl. Instrum. Methods Phys. Res. B* 95: 389–392.

Grassi, N. 2009. Differential and scanning-mode external PIXE for the analysis of the painting "Ritratto Trivulzio" by Antonello da Messina. *Nucl. Instrum. Methods Phys. Res. B* 267: 825–831.

Grassi, N., Migliori, A., Mando, P. A., and Calvo del Castillo, H. 2004. Identification of lapis-lazuli pigments in paint layers by PIGE measurements. *Nucl. Instrum. Methods Phys. Res. B* 219–220: 48–52.

Hassanein, R., De Beer, F.C., Kardjilov, N., and Lehmann, E. 2006. Scattering Correction Algorithm for Neutron Radiography and Tomography Tested at Facilities with Different Beam Characteristics. *Physica B* 385–386: 1194–1196.

Hotchkis, M. A. C., Child, D. P., and Zorko. B. 2010. Actinides AMS for nuclear safeguards and related applications. *Nucl. Instrum. Methods Phys. Res. Sect. B Beam Interact. Mater. Atoms.* 268(7–8): 1257–1260.

IAEA. 1961. The agency's safeguards system. IAEA Report INFCIRC/26. IAEA, Vienna, Austria.

IAEA. 1995. IAEA meeting on accelerator mass spectrometry. Zagreb, Croatia, April 19–21. https://inis.iaea.org/collection/NCLCollectionStore/_Public/35/066/35066061.pdf

IAEA. 2008. Neutron Imaging: A Non-Destructive Tool for Materials Testing Report of a co-ordinated research project 2003–2006. IAEA-TECDOC-1604. International Atomic Energy Agency, Vienna, Austria.

IAEA. 2011. Nuclear Techniques for Cultural Heritage Research. IAEA Radiation Technology Series No. 2. International Atomic Energy Agency, Vienna, Austria.

IAEA-Safeguards. 2020. Development and Implementation Support Programme for Nuclear Verification 2020-2021. Report STR-393. IAEA, Vienna, Austria.

Istenič, J. and Šmit, Ž. 2007. The beginning of the use of brass in Europe with particular reference to the southeastern Alpine region. In Metals and mines: studies in archaeometallurgy (Eds. La Niece, S., Hook, D. R., Craddock, P. T.). Pp 140-147. ISBN: 978-1904982197.

IMO. 2019. *International Maritime Organization*. Implementing the Ballast Water Management Convention. https://www.imo.org/en/MediaCentre/HotTopics/Pages/Implementing-the-BWM-Convention.aspx

IMO. 2020. International Maritime Organization, IMO 2020 – cutting sulphur oxide emissions. https://www.imo.org/en/MediaCentre/HotTopics/Pages/Sulphur-2020.aspx

Janssens, K. and Van Grieken, R. (Eds.). 2004. *Non-destructive Micro Analysis of Cultural Heritage Materials*. Elsevier, Amsterdam.

Jembrih-Simbürger, D., Neelmeijer, C., Mäder, M., and Schreiner, M. 2004. X-ray fluorescence and ion beam analysis of iridescent Art Nouveau glass – authenticity and technology, *Nucl. Instrum. Methods B* 226: 119–125.

Jezeršek, D., Šmit, Ž., and Pelicon, P. 2010. External beamline setup for plated target investigation. *Nucl. Instrum. Methods Phys. Res. B* 268: 2006–2009.

Kržič, A., Šmit, Ž., Fajfar, H., et al. 2013. The origin of emeralds embedded in archaeological artefacts in Slovenia. *Geologija* 56: 29–46.

Kuczkowski, P.: 2017. Remontowa Ship Yard experience and challenges installing the scrubbers on board of the ships. Presentation at ARIES Meets Industry – accelerator application to the ship exhaust gases treatment. 1 Dec 2017, CERN, Switzerland.

Kudrya, V. P. and Maishev, Yu. P. 2018. Applications of the technology of fast neutral particle beams in micro- and nanoelectronics. *Russ. Microelectron.* 47: 332–343.

Lee, H.-J. 2004. Neutral particle beam lithography. US patent WO/2004/075269.

Lehmann, E., Tremsin, A. S., Grünzweig, C., et al. 2011. Neutron imaging – detector options in progress. *J. Instrum.* 6(1): C01050 (15 pp).

Litherland, A. E. 1995. AMS Monitoring of Nuclear Facilities and Safeguard Related Activities. In IAEA meeting on Accelerator Mass Spectrometry. P. 45–46. https://inis.iaea.org/collection/NCLCollectionStore/_Public/35/066/35066061.pdf

Lo Giudice, A., Re, A., Calusi, S., et al. 2009. Multitechnique characterization of lapis lazuli for provenance study. *Anal. Bioanal. Chem.* 395: 2211–2217.

Macková, A., MacGregor, D., Azaiez, F., Nyberg, J., and Piasetzky, E. (Eds.). 2016. Nuclear Physics for Cultural Heritage. Appendix A: European facilities using nuclear techniques to study cultural heritage. Published by Nuclear Physics Division of the European Physical Society.

Malone, G., Valastro, S. Jr. , and Vazela, A. G. 1980. Carbon-14 chronology of mortar from excavation in the Medieval Church of Saint-B'enigne, Dijon, France. *J. Field Archaeology* 7: 329–343.

Mando, P. A., Fedi, M. E., Grassi, N., and Migliori, A. 2005. Differential PIXE for investigating the layer structure of paintings. *Nucl. Instrum. Methods Phys. Res. B* 239: 71–76,

Mannes, D., Lehmann, E., Masalles, A., et al. 2014. The study of cultural heritage relevant objects by means of neutron imaging techniques. *Non-Destruct. Test. Condition Monit.* 56(3): 137–144.

Mikerov, V. K., Zhitnik, I. A., Barmakov, Ju. N., et al. 2004. Prospects for efficient detectors for fast neutron imaging. *Appl. Radiat. Isot.* 61(4): 529–535.

Miliani, C., Daveri, A., Brunetti, B. G., and Sgamellotti, A. 2008. CO_2 entrapment in natural ultramarine blue. *Chem. Phys. Lett.* 466: 148–151.

Mindat. 2021. https://www.mindat.org/min-2357.html. Accessed 27 May 2021.

Mosconi, M. 2017. Italian Coast Guard experience in ensuring compliance with the existing requirements for the ship emissions. Presentation at ARIES Meets Industry – accelerator application to the ship exhaust gases treatment. 1 Dec 2017, CERN, Switzerland.

Nebbia, G., Pesente, S., Lunardon, M., Moretto, S., Viesti, G., Cinausero, M. Barbui, M., Fioretto, E., Filippini, V., Sudac, D., Nad, K., Blagus, S., and Valković, V. 2005. Detection of hidden explosives in different scenarios with the use of nuclear probes. *Nucl. Phys. A* 752 (2005): 649C–658C.

Neelmeijer, C. and Mäder, M. 2002. The merits of particle induced X-ray emission in revealing painting techniques. *Nucl. Instrum. Methods B* 189: 293–302.

Neelmeijer, C. and Mäder, M. 2004. Reverse painting on glass as seen by the proton beam. *Nucl. Instrum. Methods B* 226: 126–135.

Obhođaš, J., Sudac, D., Blagus, S., and Valković, V. 2007. Analysis of an object assumed to contain "Red Mercury". *Nucl. Instrum. Methods Phys. Res. B* 261: 922–924.

Obhodas, J., Sudac, D., Valković, V., Baricevic, M., Franulovic, A., Perot, B., Carasco, C., Alain, M., Raoux, A.-C., and El Kanawati, W. 2010. Analysis of contenerized cargo in the ship container terminal, *Nucl. Instrum. Methods Phys. Res. A* 619: 460–466.

Orphan, V. J. 2004. Cargo and Vehicle Inspection Solutions. Contribution to MTS R&T Coordination Conference, November 16–17. 2004. Washington, D.C., USA.

Orphan, V., Muenchau, E., Gormley, J., and Richardson, R. 2005. Advanced cargo container scanning technology development. *Appl. Radiat. Isot.* 63(5–6): 723–732.

O'Shea, P. G., Butler, T. A., Lynch, M. T., McKenna, K. F., Pongratz, M. B., and Zaugg, T. J. 1990. A linear accelerator in space· the beam experiment aboard rocket. Proceedings of the Linear Accelerator Conference 1990, Albuquerque, New Mexico, USA. https://accelconf.web.cern.ch/l90/papers/th454.pdf

Pace, D. C., Collins, C. S., Crowley, B. et al. 2017. Control of power, torque, and instability drive using in-shot variable neutral beam energy in tokamaks. *Nucl. Fusion* 57(1): 014001. 10.1088/0029-5515/57/1/014001

Pace, D. C., Austin, M. E., Bardoczi, L., et al., 2018. Dynamic neutral beam current and voltage control to improve beam efficacy in tokamaks. *Phys. Plasmas* 25: 056109 (13 p). 0.1063/1.5016160

Pappalardo, G., Costa, E., Marchetta, C., et al. 2004. Non-destructive characterization of della Robbia sculptures at the Bargello museum in Florence by the combined use of PIXE and XRF portable systems, *J. Cult. Herit.* 5: 183–188.

Pawley, C. J., Crowley, B., Pace, D. C., et al. 2017. Advanced control of neutral beam injected power in DIII-D. *Fusion Eng. Des.* 123(3): 453–457.

Perot, B., Carasco, C., Valković, V., Sudac, D., and A. Franulovic, A. 2009. Detection of illicit drugs with the EURITRACK System, application of accelerators in research and industry: 12th International Conference. *AIP Conf Proc.* 1099: 565–569.

Pérot, B., Carasco, C., Eléon, C., et al. 2021. Sea Container Inspection with Tagged Neutrons.

Pesente, S., Lunardon, M., Nebbia, G., Viesti, G., Sudac, D., and Valković, V. 2007. Monte Carlo analysis of tagged neutron beams for cargo container inspection. *Appl. Radiat. Isot.* 65: 1303–1396.

Phoenix. 2021. https://phoenixwi.com/wp-content/uploads/2019/06/PhoenixNeutronRadiography.pdf

Pino, F., Fontana, C. L., Lunardon, M., et al. 2018. Advances on the development of the detection system of C-BORD's rapidly relocatable tagged neutron inspection. *Int. J. Modern Phys. Conf. Ser.* 48: 1860125 (9 pages).

Ponta, C. C. 2008. Irradiation conservation of cultural heritage. *Nucl. Phys. News* 18: 22–24.

Pugliesi, R., Andrade, M. L. G., Pereira, M. A. S., and Pugliesi, F. 2005. Neutron Induced electron Radiography. 5th International Topical Meeting on Neutron Radiography. Max Planck Institute, 07/2004. *Nucl. Instrum. Methods A* 542(1–3): 81–86.

Rauch, J., Pace, D. C., Crowley, B., et al. 2017. Upgrade to DIII-D National Fusion Facility PCS and neutral beam systems: in-shot variation of neutral beam particle energy external link. *Fus. Sci. Technol.* 72(3): 500–504.

Re, A., Lo Giudice, A., Angelici, D. 2011. Lapis lazuli provenance study by means of micro-PIXE. *Nucl. Instrum. Methods Phys. Res. B* 269(20): 2373–2377.

Re, A., Angelici, D., Lo Giudice, A., Maupas, E., et al. 2013. New markers to identify the provenance of lapis lazuli: trace elements in pyrite by means of micro-PIXE. *Appl. Phys. A* 111(1): 69–74.

Re, A., Angelici, D., Lo Giudice, A., et al. 2015. Ion beam analysis for the provenance attribution of lapis lazuli used in glyptic art: The case of the "Collezione Medicea". *Nucl. Instrum. Meth. Phys. Res. B*, 348: 278–284.

Reijonen, J., Andresen, N., Gicquel, F., et al., Eds. 2007. Development of advanced neutron/gamma generators for imaging and active interrogation applications, optics and photonics in global homeland security III. *Proc. SPIE* 6540: 65401P.

Roberts Jr., G. and Tuttle, K. 2015. The accelerator in the Louvre. Symmetry Magazine 2015. US Department of Energy, Office of Science. https://www.symmetrymagazine.org/article/may-2015/the-accelerator-in-the-louvre05/14/15

Sardet, A., Perot, B., Carasco, C., et al. 2021. Performances of C-BORD's tagged neutron inspection system for explosives and illicit drugs detection in cargo containers. *IEEE Transac. Nucl. Sci.* 68(3): 346–353.

Schmidt, C. M., Walton, M. S., and Trentelman, K. 2009. Characterization of Lapis Lazuli pigments using a multitechnique analytical approach: implications for identification and geological provenancing. *Anal. Chem.* 81: 8513–8518.

Sibczynski, P., Dziedzic, A., Grodzicki, K., et al. 2017. C-BORD – an overview of efficient toolbox for high-volume freight inspection. 2017 IEEE Nuclear Science Symposium and Medical Imaging Conference (NSS/MIC), pp. 1–3.

Solis, C., Andrade, E., Mireles, A., Reyes-Solis, I. E., Garcia-Calderon, N. M.C. Lagunas-Solar, M. C. C. U. Piña, C. U., and Flocchini, R. G. 2005. Distribution of heavy metals in plants cultivated with wastewater irrigated soils during different periods of time. *Nucl. Instrum. Methods Phys. Res. B* 241: 351–355.

Sonninen, E., Erametsa, P., and Jungner, H. 1989. Dating of mortar and bricks: an example from Finland: Archaeometry. *Proceedings of the 25th International Symposium*, Ed. Maniatis, Y., pp. 99–107. Amsterdam-Oxford-New York-Tokyo.

Spoto, G. 2000. Secondary ion mass spectrometry in art and archaeology. *Thermochim. Acta* 365: 157–166.

Stan-Sion, C., Roth, J., Krieger, K., et al. 2007. AMS – Sensitive tool used as nuclear safeguard and to diagnose fusion experiments. *Nucl. Instrum. Methods Phys. Res. Sect. B Beam Interact. Mater. Atoms.* 259(1): 694–701.

Sudac, D., Blagus, S., and Valković, V. 2004. Chemical composition identification using fast neutrons. *Appl. Radiat. Isot.* 61/1 (2004) 73–79.

Sudac, D., Blagus, S., and Valković, V. 2005. Inspections for contraband in a shipping container using fast neutrons and the associated alpha particle techniques: proof of principle. *Nucl. Instrum. Methods Phys. Res. B* 241: 798–803.

Sudac, D. and Valković, V. 2006. Vehicle and container control for the presence of a "Dirty Bomb". *Proc. SPIE* 6204: 620402–620410.

Sudac, D., Blagus, S., and Valković, V. 2007. The limitations of associated alpha particle technique for contraband container inspections. *Nucl. Instrum. Methods* 263(1): 123–126.

Sudac, D., Pesente, S., Nebbia, G. Viesti, G., and Valković, V. 2007. Identification of materials hidden inside a container by using the 14 MeV tagged neutron beam. *Nucl. Instrum. Methods Phys. Res. B* 261: 321–325.

Sudac, D., Matika, D., and Valković, V. 2008. Identification of materials hidden inside a sea-going cargo container filled with an organic cargo by using the tagged neutron inspection system. *Nucl. Instrum. Methods Phys. Res. A* 589(1): 47–56.

Sudac, D., Baričević, M., Obhodaš, J., Franulović, A., and Valković, V. 2010. The use of triangle diagram in the detection of explosive and illicit drugs. *Proc. SPIE* 7666: 76662V.

Šmit, Ž., Istenič, J., and Knific, T. 2008. Plating of archaeological metallic objects – studies by differential PIXE. *Nucl. Instrum. Methods Phys. Res. B* 266: 2329–2333.

Šmit, Ž., Milavec, T., Fajfar, H., Rehren, Th., Lankton, J. W., and Gratuze, B. 2014. Analysis of glass from the post-Roman settlement Tonovcov grad (Slovenia) by PIXE-PIGE and LA-ICP-MS. *Nucl. Instrum. Methods Phys. Res. B* 311: 53–59.

Šmit, Ž., Knific, T., Jezeršek, D., and Istenič, J. 2012. Analysis of early medieval glass beads – Glass in the transition period. *Nucl. Instrum. Methods Phys. Res. B* 278: 8–14.

Šmit, Ž., Pelicon, P., Holc, M., and Kos, M. 2002. PIXE-PIGE characterization of medieval glass. *Nucl. Instrum. Methods Phys. Res. B* 189: 344–349.

Šmit, Ž., Janssens, K., Schalm, O., and Kos, M. 2004. Spread of façon-de-Venise glassmaking through central and western Europe. *Nucl. Instrum. Methods Phys. Res. B* 213: 717–722.

Šmit, Ž., Janssens, K., Bulska, E., Wagner, B., Kos, M., and Lazar, I. 2005. Trace element fingerprinting of façon-de-Venise glass. *Nucl. Instrum. Methods Phys. Res. B* 239: 94–99.

Šmit, Ž., Stamati, F., Civici, N., Vevecka-Priftaj, A., Kos, M., and Jezeršek, D. 2009. Analysis of Venetian type glass fragments from the ancient city of Lezha (Albania). *Nucl. Instrum. Methods Phys. Res. B* 267: 2538–2544.

Tang, C. 2015. Low energy accelerators for cargo inspection. *Rev. Accel. Sci. Technol.y* 8: 143–163. 10.1142/S179362681530008X

Toole, M. T. and Willis, J. C. 1990. Paper Session I-A - Neutral Particle Beam Overview. The Space Congress® Proceedings. 16. https://commons.erau.edu/space-congress-proceedings/proceedings-1990-27th/april-24-1990/16

Torims, T. 2017. How to bring particle accelerator on board of ship? Presentation at ARIES Meets Industry – accelerator application to the ship exhaust gases treatment. 1 Dec 2017, CERN, Switzerland.

UNCTAD. 2020. *Review of Maritime Transport 2020*. United Nations Publications, New York, NY, USA.

Valković, V. 2006. Applications of nuclear techniques relevant for civil security. *J. Phys. Conf. Ser.* 41: 81–100.

Valković, V. 2009. New/future approaches to explosive/chemicals detection. Application of Accelerators in research and industry. 12th International Conference. AIP Conf Proc. 1099: 574–577.

Valković, V., Sudac, D., Blagus, S., Nađ, K., Obhođaš, J., and Vekić, B. 2007. Fast neutron inspection of sea containers for the presence of 'dirty bomb'. *Nucl. Instrum. Methods Phys. Res. B Beam Interact. Mater. Atoms.* 263(1): 119–122.

Valković, V., Sudac, D., and Matika, D. 2010. Fast neutron sensor for detection of explosives and chemical warfare agents. *Appl. Radiat. Isot.* 68: 888–892.

Valković, V., D. Sudac, K. Nad, and J. Obhodas. 2016. Container inspection in the port container terminal by using 14 MeV neutrons. *IEEE Trans. Nuc. Sci.* 63, No.3 (2016): 1536–1543.

Van Ham, R., Van Vaeck, L., Adriaens, A., Adams, F., Hodges, B., and Groenewold, G. 2002. Inorganic speciation in static SIMS: a comparative study between monatomic and polyatomic primary ions, *J. Anal. At. Spectrom.* 17: 753–758.

Van Ham, R., Van Vaeck, L., Adams, F., and Adriaens, A. 2004. Systematization of mass spectra for the speciation of inorganic salts with static secondary ion mass spectrometry (S-SIMS), *Anal. Chem.* 76 (9): 2609–2617.

Van Ham, R., Van Vaeck, L., Adams, F., and Adriaens, A. 2005. Feasibility of pigment analyses in Medieval paintings and manuscripts with TOFSIMS, *Anal. Bioanal. Chem.* 383: 991–997.

Van Strydonck, M., Dupas, M., and Dauchot-Dehon, M. 1983. Radiocarbon dating of old mortars. *PACT J.* 8: 337–343.

Van Strydonck, M., Dupas, M., Dauchot-Dehon, M., Pachiaaudi, C., and Maréchal, J. 1986. The influence of contaminating (Fossil) carbonate and the variations of kl3C in mortar dating. *Radiocarbon* 28(2A): 702–710.

Van Unnik, H. 2004. Making port security work. *Secur. Surveill. Detect.* 61: 90–92.

Vértes, A., Nagy, S., Klencsár, Z., Lovas, R. G., and Rösch, F. (Eds). 2011. *Handbook of Nuclear Chemistry*. Springer, Boston, MA. USA.

Vretenar, M. 2019. Bringing particle accelerators on ship. *Accelerating News CERN Accelerating Science.* https://acceleratingnews.web.cern.ch/article/bringing-particle-accelerators-ships

Wells, N. 1982. Production of Neutral Beams from Negative Ion Beams Systems in the USSR. Published by The Rand Corporation. Report R-2909/1-ARPA.

Wyart, J., Bariand, P., J. Filippi, J. 1981. Lapis lazuli from Sar-e-Sang, Badakhshan, Afghanistan. *Gems Gemmol.* 17: 184–190.

Yan, X., Yang, L., Zhang, X and Zhan, W. 2017. Concept of an accelerator-driven advanced nuclear energy system. *Energies* 10(7): 944 (pp 13).

Yousri, A. M., Osman, A. M., Kansouh, W. A., Reda, A. M., Bashter, I. I., and Megahid, R. M. 2012. Scanning of cargo containers by gamma-ray and fast neutron radiography. *Armenian J. Phys.* 5(1): 1–7.

Zambrano, M., Hameed, F., Anders, K., Mancini, L., and Tondi, E. 2019. Implementation of dynamic neutron radiography and integrated X-Ray and Neutron tomography in porous carbonate reservoir rocks. *Front. Earth Sci.* 7: article 329 (pp. 15).

Zendel, M., Donohue, D. L., Kuhn, E., Deron, S., and Bíró, T. 2011. Nuclear Safeguards Verification Measurement Techniques, pp 2893–3015. In: Vértes, A., Nagy, S., Klencsár, Z., Lovas, R. G., Rösch, F. (Eds) *Handbook of Nuclear Chemistry*. Springer, Boston, MA.

Zhou, C. and Qian, W. 2014. Particle-beam weapons system. Proceedings 2014 IEEE International Conference on Control Science and Systems Engineering, pp. 81–84. Yantai, China, 29–30. 12. 2014.

Zimek, Z. 2017. Ideas of compact accelerator development for the exhaust gases treatment and what could be bought today in the particle accelerator market for the ship needs? Presentation at ARIES Meets Industry – accelerator application to the ship exhaust gases treatment. 1 Dec 2017, CERN, Switzerland.

Zucchiatti, A., Bouquillon, A., Lanterna, G., Lucarelli, F., Mando, P. A., Prati, P. J. Salomon, J., and Vaccari, M. G. 2002. PIXE and A-PIXE analyzes of glazes from terracotta sculptures of the della Robbia workshop. *Nucl. Instrum. Methods B* 189: 358–363.

Zucchiatti, A., Bouquillon, A., Castaign, J., and Gaborit, R. 2003. Elemental analyses of a group of glazed terracotta angels from the Italian Renaissance as a tool for the reconstruction of a complex conservation history, *Archaeometry* 45 (3): 391–404.

Zucchiatti, A., Prati, P., Bouquillon, A., Giuntini, L., Massi, M., Miglior, A. Cagnana, A., and Roascio, S. 2004. Characterisation of early medieval frescoes by A-PIXE, SEM and Raman spectroscopy, *Nucl. Instrum. Methods B* 219. 20–25.

Additional reading

Bugoi, R. V. Cojocaru, B. Constantinescu, F. Constantinescu, D. and Grambole, F. 2003. Herrmann, Micro-PIXE study of gold archaeological objects, *J. Radioanal. Nucl. Chem.* 257 (2): 375– 383.

Caforio, L., Fedi, M. E., Mando, P. A., et al. 2014. Discovering forgeries of modern art by the 14C bomb peak. *Eur. Phys. J. Plus*, 129: 1–5.

Cagno, S., Favaretto, L., Mendera, M., et al. 2012. Evidence of early medieval soda ash glass in the archaeological site of San Genesio (Tuscany), *J. Arch. Sci.* 39: 1540–1552.

Coeuré, P., and Tran, Q. K. 1999. Procédés radiochimiques pour la conservation des biens culturels. *Actualite Chimique* 11: 141–143.

Constantinescu, B., Oberländer-Târnoveanu, E., Bugoi, R., Cojocaru, V., and Radtke, M. 2010. The Sarmizegetusa Bracelets. *Antiquity J.* 84: 1028–1042.

Constantinescu, B. R. Bugoi, V. Cojocaru, D. Voiculescu, D. Grambole, F. , and Herrmann, D. Ceccato, Romanian ancient gold objects provenance studies using micro-beam methods: the case of "Pietroasa" hoard, *Nucl. Instrum. Methods B* 231 (2005) 541–545.

Cortella, L., Tran, Q. K., Głuszewski, W. J., Moise, I. V., and Ponta, C. C. 2011. Nuclear Techniques for Preservation of Cultural Heritage Artefacts (brochure). Using Nuclear Techniques for the Characterization and Preservation of Cultural Heritage Artefacts in the European Region. IAEA Technical Cooperation Project – RER 8015.

Dowsett, M. G. A. Adriaens, M. Soares, H. Wouters, V. V. N. Palitsin, R. and Gibbons, R. J. H. Morris, The use of ultra-low-energy dynamic SIMS in the study of the tarnishing on silver, *Nucl. Instrum. Methods B* 239 (2005) 51–64.

Denker, A. J. Opitz-Coutureau, M. Griesser, R. and Denk, H. Winter, nondestructive analysis of coins using high-energy PIXE. *Nucl. Instrum. Methods B* 226 (1–2) (2004): 163–171.

Enguita, O., A. Climent-Font, G. Garcia, I. Montero, M.E. Fedi, M. Chiari, and F. Lucarelli, Characterization of metal threads using differential PIXE analysis, *Nucl. Instrum. Methods B* 189 (2002) 328–333.

H. Fajfar, Ž. Šmit, and M. Kos. PIXE-PIGE analysis of colored historic glass. Glass Technol. Eur. J. Glass Sci. Technol. A 54 (2013): 218–225.

Ganio, M., Boyen, S., D. Brems, D., et al. 2012. Trade routes across the Mediterranean: a Sr/Nd isotopic investigation on Roman colourless glass. *Glass Technol. Eur. J. Glass Sci. Technol.* 53: 217–224.

Głuszewski, W., Zagórski, Z. P., Tran, Q. K., Khoï, Q., and Cortella, L. 2011. Maria Skłodowska Curie-the precursor of radiation sterilization methods. *Anal. Bioanal. Chem.* 400: 1577–1582.

Gonzalez, M. E., Calvo, A. M., and Kairiyama, E. 2002. Gamma radiation for preservation of biologically damaged paper. *Rad. Phys. Chem.* 63: 263–265.

Gozani, T. Detection of Explosives and Other Threats Using Accelerator-Based Neutron Techniques. https://s3.cern.ch/inspire-prod-files-e/e36c9417f994d794ce91c737e4cb2977, pp. 445–460.

Lesigyarski, D., Šmit, Ž., Zlateva-Rangelova, B., Koseva, K., and Kuleff, I. 2013. Characterization of the chemical composition of archaeological glass finds from South-Eastern Bulgaria using PIXE, PIGE and ICP-AES. *J. Radioanal. Nucl. Chem.* 295: 1605–1619.

Ljubomirova, V., Šmit, Ž., Fajfar, H., and Kuleff, I. 2014. Chemical composition of glass beads from the necropolis of Apollonia Pontica (5th-3rd c. BC). *Archaeol. Bulgarica* 18 (2014) 1–17.

Magaudda, G. 2004. The recovery of biodeteriorated books and archive documents through gamma radiation: Some considerations on the results achieved. *J. Cult. Heritage* 5: 113–118.

Moise, I. V., Virgolici, M., Negut, C. D., et al. 2012. Establishing the irradiation dose for paper decontamination. *Rad. Phys. Chem.* 81: 1045–1050.

Nunes, I., Mesquita, N., Verde, S. C., Trigo, M. J., Ferreira, A., Carolino, M. M., Portugal, A., and Botelho, M. L. 2012. Gamma radiation effects on physical properties of parchment documents: assessment of Dmax. *Rad. Phys. Chem.* 81: 1943–1946.

Perea, A. I. Montero, G. Demortier, A. Climent-Font, Analysis of the Guerrazar treasure by PIXE, in: G. Demortier, A. Adriaens (Eds.), Ion Beam Study of Art and Archaeological Objects, Office for the Official Publications of the European Union, Luxembourg, 2000, pp. 99–101.

Ramière, R. 2002. *La désinfection des biens culturels par irradiation gamma, in Les contaminants biologiques des biens culturels*. Roquebert, M. F. (Ed.). Elsevier, Paris.

Ramiere, R. and Tran, Q. K. 1989. Nucleart: Nuclear techniques applied to art. *Nucl. Europe* 7: 50.

Rasmussen, K. L., Kučera, J., L. Skytte, L., et al. 2013. Was he murdered or was he not? – Part I: Analyses of mercury in the remains of Tycho Brahe. *Archaeometry* 55: 1187–1195.

Santra, S. D., Mitra, M., Sarkar, D., Bhattacharya, A., Denker, J. and Opitz-Coutureau, J. 2005. Rauschenberg, Analysis of some coins by energy dispersive X-ray fluorescence (EDXRF) and high energy particle induced X-ray emission (PIXE) techniques. *Nucl. Instrum. Methods* B229 (3–4): 465–470.

Stojanović, M., Šmit, Ž., Glumac, M., and Mutić, J. 2015. PIXE–PIGE investigation of Roman Imperial vessels and window glass from Mt. Kosmaj, Serbia (Moesia Superior). *J. Arch. Sci. Rep.* 1: 53–63.

Šmit, Ž., Uršič, M., Pelicon, P., et al. 2008. Concentration profiles in paint layers studied by differential PIXE. *Nucl. Instrum. Methods Phys. Res. B* 266: 2047–2059.

Šmit, Ž., Tartari, F., Stamati, F., Vevecka-Priftaj, A., and Istenič, J. 2013. Analysis of Roman glass from Albania by PIXEPIGE method. *Nucl. Instrum. Methods Phys. Res. B* 296: 7–13.

Tran, Q. K., Ramiere, R., Giniergillet, A. 1990. Impregnation with radiation-curing monomers and resins. *Adv. Chem. Ser.* 225: 217–233.

6

New Trends

6.1 Introduction

New trends in accelerator applications should be carefully monitored because they can develop into a major driving force for industrial and other development. In particular, the following subjects could see more new ideas in the near future: micro-machining using ion beams; production of new materials; safeguards, mine, hidden explosive detection and similar applications; applications in the nuclear fuel cycle (accelerator-assisted fission plants, heavy iron inertial fusion, incineration of nuclear waste); and applications in the coal fuel cycle (prospecting, characterization of the burning process phases, improvement of fly ash, SO_x and NO_x from flue gases).

Accelerator technologies may help developments and electricity production by fusion reactors, treat nuclear waste and allow for safer operation of fission reactors. As accelerator science and technology progressed over the past several decades, the accelerators themselves have undergone major improvements in multiple performance factors: beam energy, beam power and beam brightness. Consequently, accelerators have found applications in a wide range of fields in our life and in our society.

Let us mention Volume 8 of *Reviews of Accelerator Science and Technology* edited by Chao and Chou (2016) dedicated to applications in energy and security, two of the most important and urgent topics in today's world. This volume makes an effort to provide a review as complete and up to date as possible of these broad and challenging subjects. It also has a review article on accelerator science and technology in Canada with a focus on the TRIUMF laboratory.

Garnett and Sheffield (2015) present an overview of accelerator applications and accelerator technology in energy. Applications span a broad range of cost, size and complexity and include large-scale systems requiring high-power or high-energy accelerators to drive subcritical reactors for energy production or waste transmutation, as well as small-scale industrial systems used to improve oil and gas exploration and production.

6.2 Accelerator-Driven System

Commercial thermal reactors (LWR; CANDU) currently operate in once-through fuel cycles and are producing a large stockpile of radioactive waste, mainly long lived transuranic elements (Pu, minor actinides (MA) such as Am, Np, Cm, etc.) and long-lived fission fragments, such as ^{99}Tc and ^{129}I, which will require final underground geological

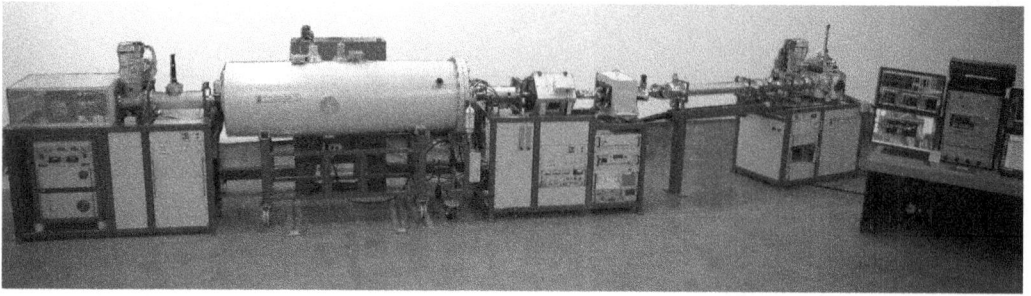

FIGURE 6.1
A picture of 3SDH 1 MV Pelletron system with RF source (left) and analysis endstation (right). (Courtesy of National Electrostatics Corp.)

repositories for thousands of years. Accelerators can be used to make neutrons to break down the waste. This could reduce the amount of time the waste needs to be stored by a factor of 1000 and the volume by a factor of 100.

NEC 3SDH 1 MV Pelletron system with RF source and analysis endstation was designed with the intended purpose of aiding in fusion research. It is capable of Ion Beam Analysis (IBA) techniques such as RBS, ERD, PIXE and NRA. Further detectors could be added to the endstation to allow for other techniques. Installed in Japan in 2014, see Fig. 6.1.

Maiorino et al. (2003) have presented a summary of the world status (at that time) on R&D related to the utilization of Accelerator-Driven System (ADS) for energy generation and transmutation of long lived nuclear waste. A proposal to start a Technical Working Group (TWG) in Brazil to prepare a Road Map having as a final goal an experimental facility to utilize an accelerator in basic and applied research; products and services and R&D in energy generation and transmutation was presented.

The buildup of radioactive stock piles, besides the concern of waste disposal, also brings the issue of proliferation, mainly due to Pu isotopes. To overcome these issues, the next generations of nuclear reactors are considering concepts that, coupled with a closed fuel cycle, aim at reducing the burden on geological storage and non-proliferation.

Due to their inherent safety features and waste transmutation potential, accelerator driven subcritical reactors (ADSRs) are the subject of research and development in almost all countries around the world. The neutrons needed to sustain fission are generated by the spallation process resulting from high-energy protons affecting a target element installed at the center of the core. In the paper by Bungau et al. (2008) the possible benefits of Fixed Field Alternating Gradient, FFAGs, as accelerator drivers for ADSR systems are analyzed. FFAGs afford fast acceleration, as there is no need of synchronization between RF and magnets, high average current with large repetition rate and large acceptance and are believed to be the best candidate for an ADSR unit. The mentioned study also focuses on the Monte Carlo studies of the reactor core design. The impact of the subcriticallity, target material and proton beam energy on the ADSR performance was also examined. Entirely novel ADSR configurations involving multiple accelerator drivers and associated spallation targets within the reactor core have also been considered. Calculations carried out using the GEANT4 simulation code present an attempt to improve the initial design proposed by Rubbia et al. (1995).

An ADS, which combines a particle accelerator with a subcritical core, is commonly regarded as a promising device for the transmutation of nuclear waste, as well as a potential scheme for thorium-based energy production. So far, the predominant choice of the accelerator for ADS is a superconducting linear accelerator (linac). The article by Pan and Dai (2015) gives a brief overview of ADS based on linacs, including the motivation, principle, challenges and research activities around the world. The status and future plan of the Chinese ADS (C-ADS) project is highlighted and discussed in depth as an example.

The article by Heidet et al. (2015) is a review of several accelerator–reactor interface issues and nuclear fuel cycle applications of accelerator-driven subcritical systems. The systems considered have the primary goal of energy production, but that goal is accomplished via a specific application in various proposed nuclear fuel cycles, such as breed-and-burn of fertile material or burning of transuranic material. Several basic principles are reviewed, starting from the proton beam window including the target, blanket, reactor core and up to the fuel cycle. They focus on the impact of the energy required to run the accelerator and associated systems on the potential electricity delivered to the grid. ADS feature many of the constraints and issues associated with critical reactors, with the added challenges of subcritical operation and coupling to an accelerator. Reliable accelerator operation and avoidance of beam trips are critically important. One interesting challenge is measurement of blanket subcriticality level during operation. The authors also review the potential benefits of ADS in various nuclear fuel cycle applications. Ultimately, accelerator-driven subcritical systems with the goal of transmutation of transuranic material have lower 100,000-year radioactivity than a critical fast reactor with recycling of uranium and plutonium.

The review paper by Calabretta and Méot (2015) summarizes projects and studies on circular accelerators proposed for driving subcritical reactors. The early isochronous cyclotron cascades, proposed about 20 years ago, and the evolution of these layouts up to the most recent solutions or designs based on cyclotrons and fixed field alternating gradient accelerators, are reported. In addition, the newest ideas and their prospects for development are discussed.

A summary of the world status on R&D related to the utilization of ADS for energy generation and mainly for transmutation of long-lived nuclear waste is presented by Maiorino et al. (2003). A proposal to start a Technical Working Group (TWG) in Brazil to prepare a Road Map having as a final goal an experimental facility to utilize an accelerator in basic and applied research, as well as products and services and R&D in energy generation and transmutation were presented.

The concept of ADS, where safety and transmutation were considered, was developed in the late eighties in the Brookhaven National Laboratory. However, the first conceptual molten-salt thermal energy amplifier and transmutation system was proposed in Los Alamos (Bowman et al. 1992) where the terminology by which ADS are known in the USA was introduced: Accelerator Transmutation of Waste (ATW).

In 1995 a group from CERN (Rubbia et al. 1995) presented the basic concept of the Fast Energy Amplifier. This was a sub critical nuclear system based on the U-Th fuel cycle, imbibed in melted lead (coolant), with an external source of neutrons coming from spallation in Pb induced by protons from a 3-stage cyclotron (1 GeV, 12.5 A), which actually raised the international interest on hybrid systems. Energy Amplifier proposed by Carlo Rubbia team (Carminati et al. 1993; Rubbia et al. 1995, see also Andriamonje et al. 1995) is based on the use of an accelerator acting as a source of "low-flux" neutrons, to breed the fission of thorium.

In Los Alamos, Bowman and his co-workers have been working for some time on a new approach for commercial nuclear energy production without a long-term high-level waste stream and for transmutation of both fission product and higher actinide commercial nuclear waste using a thermal flux of accelerator-produced neutrons in the 10^{16} n/cm^2-s range. Continuous neutron fluxes at this intensity, which is approximately 100 times larger than is typically available in a large-scale thermal reactor, appear practical owing to recent advances in proton linear accelerator technology and to the spallation target-moderator design. This large flux of thermal neutron makes possible a waste inventory in the transmutation system that is smaller by about a factor of 100 than competing concepts. The accelerator allows the system to operate well below criticality so that the possibility for a criticality accident is eliminated. No control rods are required. The successful waste transmutation would eliminate the need for nuclear waste storage on a geological time scale. The production of nuclear energy from ^{232}Th or ^{238}U is used to illustrate the general principles of commercial nuclear energy production without long-term high-level waste. There is sufficient thorium to meet the world's energy needs for many millennia (Bowman et al. 1992).

They described the new concept of energy amplification to extract nuclear energy with the help of accelerator induced nuclear cascades, extending to practical energy production the well-known "calorimeter" technique, widely used in High Energy Physics. The energy is produced from a nuclear fuel material disposed in a moderator medium through a process of breeding of a fissile element from a fertile element of the fuel material. After an initial phase, the rate between the concentrations of fissile and fertile elements reaches a substantial stability, resulting in a stable long-term energy production. The device must operate at relatively low neutron flux, in the 10^{14} cm^{-2} s^{-1} range, to ensure the correct performance of the breeding cycle and' to prevent the risk of criticality. Thorium as breeding fuel has considerable advantages when compared with uranium, in contrast with the full breeding based classic reactors for which the use of thorium presents serious difficulties. Thorium is more abundant than uranium, it generates much less transuranic actinides among the radioactive waste and the risk of nuclear proliferation is negligible. In the paper the basic concepts of the "Energy Amplifier, EA" are presented as well as alternative illustrative designs of the device. Some economic considerations are also discussed (Carminati et al. 1993; Rubbia et al. 1995) (Fig. 6.2).

The primary fuel is natural thorium, which is completely burnt after a number of fuel cycles through the EA. Actinides present in the fuel discharge bat the end of a fuel cycle

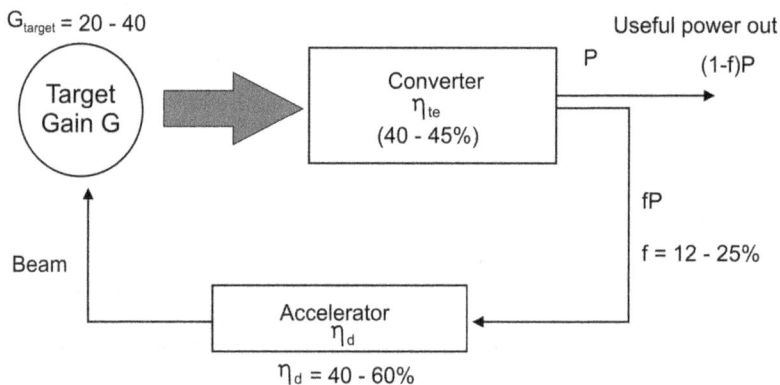

FIGURE 6.2

General principle of power generation with the help of an energy amplifier. The target has an energetic gain G which is converted into electricity with efficiency η_{te}. The power P produced in this way is partially recycled (fraction f) through the accelerator of efficiency η_d and re-injected as beam in the target, while the remaining fraction (1-f) P is delivered for further use. With the figures indicated, the range of such "useful" power is between 75% and 82% of the one produced by the converter. (After Carminati et al. 1993.)

are re-injected in the EA and become the seeds for subsequent cycle. This ensures a very efficient use of the primary fuel element. This objective is identical to the one eventually met by fast breeders. Compared to the consumption of natural fuel material, the EA is about 250 times more efficient than the present PWRs base on an open fuel cycle.

Later on, several experiments were conducted at CERN and several reports were published with some variants from Rubbia's idea. Fig. 6.3 illustrates the basic concept of the proposed Fast Energy Amplifier. The electric power generated is also used to run the accelerator (re-circulated power ≤ 5%). At each discharge of the fuel (every 5 years) the fuel is "regenerated". Actinides, mostly thorium and uranium are re-injected as new fuel in the EA, topped with fresh thorium. Fission fragments and the like are packaged and sent to the secular repository, where after ~1000 years the radioactivity decays to a negligible level.

Besides this process, often called the Rubbia's concept, several R&D initiatives related to ADS, partition and transmutation, etc., were developed in the world. Many of the initiatives and programs are discussed in the IAEA Technical Working Group on Fast Reactors meetings (IAEA 20022006).

ADSs are promising tools for the efficient transmutation of nuclear waste products in dedicated industrial installations, called transmuters. The MYRRHA (Multipurpose Hybrid Research Reactor for High-tech Applications) project at Mol, Belgium, placed itself on the path towards these applications with a multipurpose and versatile system based on a liquid Pb-Bi cooled fast reactor (57 MWt) which may be operated in both critical and subcritical modes. In the latter case the core is fed by spallation neutrons obtained from a 600 MeV proton beam hitting the LBE coolant/target. The accelerator

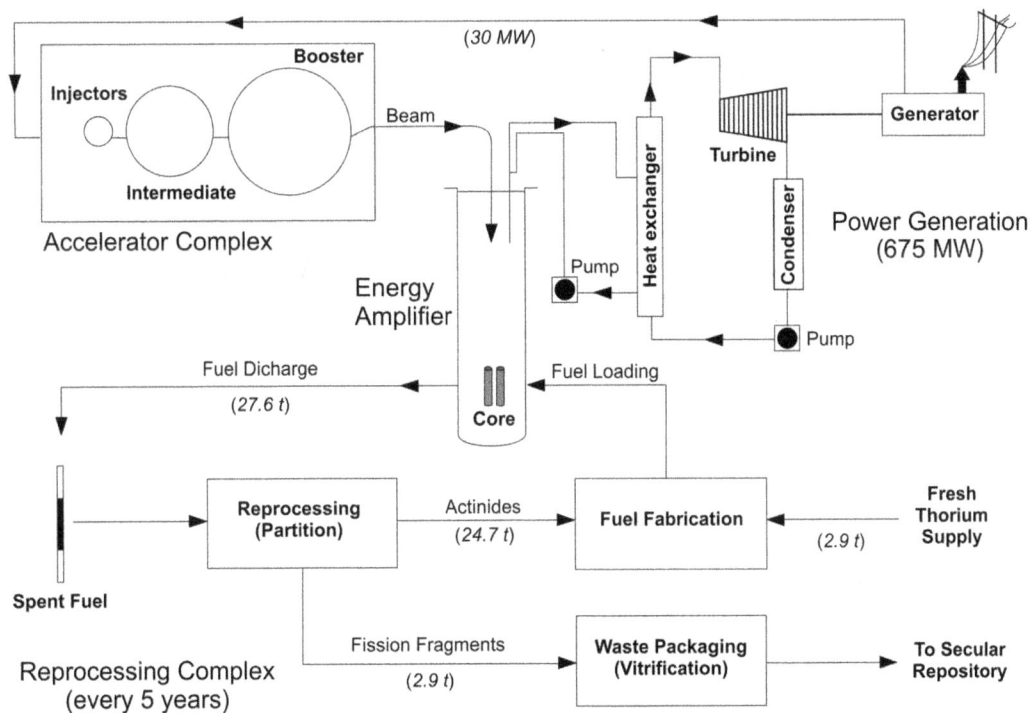

FIGURE 6.3
General lay-out of the Energy Amplifier complex. (After Rubbia et al. 1995.)

providing this beam is a high intensity CW superconducting linac, which is laid out for the highest achievable reliability (Vandeplassche et al. 2011). The combination of a parallel redundant and of a fault tolerant scheme should allow obtaining an MTBF value in excess of 250 hours that is required for optimal integrity and successful operation of the ADS. MYRRHA is expected to be operational in 2023. The forthcoming 2-year period is fully dedicated to R&D activities, and in the field of the accelerator they are strongly focused on the reliability aspects and on the proper shaping of the beam trip spectrum (D'Hondt et al. 2004, Abderrahim 2020).

The Myrrha project is part of the European Strategy Forum on Research Infrastructures, and is one of three new research reactors in the European Research Area of Experimental Reactors. The others are the Jules Horowitz Reactor at Cadarache in France and the Pallas reactor at Petten in the Netherlands. Construction of the Myrrha reactor is expected to begin in 2026, with full-operation from 2034.

MYRRHA will be the unique first large facility in the world that will allow the demonstration of the Accelerator Driven System (ADS) concept, thus initiating a large program of research in the field of spent fuel partitioning and transmutation.

The conceptual design of an ADS accelerator is under development for more than 10 years. These activities have been coordinated by the Accelerator Research Department of IPN Orsay at the European level through the successive EURATOM Framework Programs projects. They advocate the use of a solution based on a superconducting linear accelerator, to obtain an efficient and highly modular machine, with great potential in terms of reliability.

Since 2011 and the launch of the MAX project, supported by EURATOM FP7, these studies mainly focused on the case of the accelerator for MYRRHA in order to try to provide a consolidated design with a sufficient level of detail for the start of a possible detailed design phase before construction. The main MYRRHA accelerator beam characteristics are shown in Table 6.1.

China ADS project is described by Zhan (2017) as ADANES (Accelerator Driven Advanced Nuclear Energy System) performed under the leadership of Chinese Academy of Science (CAS). ADANES consist of the recycled fuel burner and the used fuel recycle. ADANES burner is optimized as nuclear waste transmutation, fissile material breeding and energy production in situ from traditional ADS. The fast core/blanket is designed to operate in the subcritical or critical state with or without accelerator driven respectively and burn "raw" recycled fuel which contains >50% fission products (FP). The fuel recycle is designed to remove >50% volatile FP by the extend "AIROX" and lanthanide FP by rare earth extraction from used fuel of LWR, then, form other residual as recycle fuel.

TABLE 6.1

MYRRHA accelerator beam characteristics

Particle	Protons
Energy	600 MeV
Current	4 mA
Time structure	CW, with 200 µs holes, ≥1 Hz repetition, For sub-criticality monitoring
Beam delivery To the reactor	Vertically from above through an achromatic beamline and window
Beam stability	Energy ± 1%, current ± 2%, position and size ± 10%
Beam shape on target	Circular r = 40 mm, flat power density, by AC scanning magnets
MTBF	>250 h

Consequently, ADANES is the ideal to close fuel cycle as utilizing fissile fuel ≥95% which means the fission energy could be sustainable for ~10000 yr. and minimizing radiotoxicity <4% with live-time <500 yr. In the approaches, the accelerator plays an important role of the burner starter which is 10%–15% duration of the burner operation due to the long refueling (~30 yr.), the much-simplified procedure of no water "raw" recycle fuel and the easier for storing small quantity waste (<4%) with lower decay heat by dry storage. Therefore, ADANES's nuclear power plant could use an accelerator to drive ~10 set of the fast reactor and transmute the MA about ~50% of those transmuted by traditional ADS in the same beam power.

Up to now, the key technical R&D made a significant breakthroughs. Injector I is 325 MHz option with ECR + RFQ + 2CM (7 spoke cavity) and Injector II is 162.5 MHz option with ECR + RFQ + 2CM (6 HWR cavity). The lower frequency is the lower RF power density, which is benefit to meet RAMI requirements in low energy part of linac. Both optional SCL injectors have extracted ~10 MeV and 1.1–2.7 mA CW proton beam. 25 MeV SCL [ECR + RFQ + 3CM (6HWR) + 1CM (7 spoke cavity)] is being assembled with the plan to extract the beam soon. The new concept of the granular target had been introduced, in which millimeter size solid grain, which is made by the target material, driven by gravity, the beam bombarding the grains from top to down to produce an intensive neutron and the deposited heat fluid with the grain out of target chamber, then treated off line. Therefore, the granular target power should jump to 10–100 MW, and withstand the impact by high power CW beam trip within 10 seconds as grain is discrete medium and selectable the different material of target. The prototype of 10–100 kW granular target has been test and results showed the agreement with design. There are 4 phases in Chinese ADS/ADANES burner development roadmap and new research sites are starting to be constructed. The 1st and 2nd phases will be finished around 2022, the burner is 10 MW scale which consists of the high average beam power ~4 MW, 400–600 MeV and 5–10 mA proton beam linac (Yan et al. 2017).

In India, ADSs have evoked considerable interest in the nuclear community of the world because of their capability to incinerate the MA and long-lived fission products radiotoxic waste and utilization of thorium as an alternative nuclear fuel. In Indian program, as described by Singh (2017), due to the vast thorium resources, ADS is particularly important as one of the potential routes for accelerated thorium utilization and the closure of the fuel cycle. The Department of Atomic Energy, India has envisaged the development of ADS in connection with its thorium utilization program. ADS consists of a high current proton accelerator, a spallation target and a sub-critical reactor. R&D is in progress for all the three systems. Efforts are on in India to develop such a system, one of the main components of which is a 1 GeV, high-intensity CW proton accelerator. The development is being done in phased manner and experimental facilities are set up to make ADS related studies. Initially, they have developed a 400 keV DC accelerator-based neutron source (10^8–10^9 n/second) for carrying out experiments on physics of ADS and for testing the simulations.

A facility, BRAHMMA (Beryllium oxide Reflected And High Density Polythene Moderated Multiplying Assembly), has been set up to measure flux distribution, flux spectra, total fission power, source multiplication and degree of sub-criticality. It uses 14 MeV neutrons produced through d + t reaction. It is a sub-critical assembly (k_{eff} = 0.87) of natural uranium. In order to increase the neutron yield to about 10^{11} n/second, a 400 keV, 1 mA RFQ for D+ ions has been built for replacing DC accelerator presently used for neutron generation. To perform these studies at higher energies, a Low Energy High Intensity (20 MeV, 30 mA) Proton Accelerator (LEHIPA) is also under development at BARC. This will also enable to understand accelerator physics issues (space charge and beam halo) associated with high-intensity beams. LEHIPA will be an intense neutron source.

An important component of ADS is a 1 GeV proton accelerator. It is planned to build it using superconducting technology from 3 MeV onwards and it is being done in collaboration with Fermilab. Resonators/cavities designed and built in India and processed at Fermilab, have met the specifications. R&D related to spallation targets and materials is also pursued at BARC. LBE will be used for spallation target. Computational codes have been developed for thermal hydraulics for LBE target simulations. Experimental loops for validation of thermal-hydraulics codes and corrosion studies on window materials have been built. BARC has also studied details of a one-way coupled reactor system that allows considerable reduction in the beam current. In this system, two sub-critical cores are used. The inner core is subcritical fast reactor with thermal neutron absorber liner surrounded by a gap and outer core is subcritical thermal reactor. The neutrons leaking from inner core can go to thermal core and are multiplied. However, neutron from thermal reactor cannot go to inner core due to absorber liner, that is why it is called one way coupled system, see also (Kapoor 2002).

Sugawara (2017) describes work on Japan ADS project. Partitioning and transmutation (P&T) technology of long-lived radioactive nuclides such as MA is a promising technology for reduction of the burden of the geological disposal of the high-level radioactive waste (HLW). The Japan Atomic Energy Agency (JAEA) has been continuously performing research and development on the P&T technology. The R&D on the P&T technology in JAEA is based on two concepts: one is the homogeneous recycling of MA in fast breeder reactors (FBRs) and the other is the dedicated MA transmutation cycle, "double-strata" strategy, using an ADS, see Lim et al. (2012).

Argonne National Laboratory, USA and Kharkiv Institute of Physics & Technology, Ukraine are collaborating on the design, the construction and the operation of an ADS supported by the Russian Research Reactor Fuel Return (RRRFR) program of the United States Department of Energy (Gohar 2017). RRRFR is a trilateral initiative among the United States, the Russian Federation and the International Atomic Energy Agency (IAEA) to repatriate high-enriched uranium fuels (fresh and irradiated) to Russia. The facility is planned to produce medical isotopes, train young nuclear professionals, support the Ukraine nuclear industry, and provide the capability for performing reactor physics, material research, and basic science experiments. This ADS facility uses a qualified proliferation-resistant low-enriched uranium (LEU) fuel and it is driven with an electron accelerator. The target design utilizes tungsten or natural uranium for neutron production through photonuclear reactions from the 100 KW electron beam using 100 MeV electrons. The neutron source intensity, spectrum and spatial distribution were studied as a function of the electron beam parameters to maximize the neutron yield and satisfy different engineering requirements. Physics, thermal-hydraulics and thermal-stresses analyses performed and iterated to minimize the maximum temperature and the thermal stresses in the target materials. The subcritical assembly design has the highest possible neutron flux intensity for such configuration with an effective neutron multiplication factor of <0.98. Different fuel designs and reflector materials considered for the subcritical assembly design. Shielding analyses defined the biological dose map around the facility during operation as a function of the heavy concrete shield thickness. Safety, reliability and environmental considerations are included in the design. The facility configuration can accommodate future design upgrades and new missions. In addition, it has unique features relative to the other international ADS facilities and it is suitable for studying ADS. Several horizontal neutron channels for performing basic research including a cold neutron source are built-in.

USA status on SRF Linacs Driving Subcritical GEM∗STAR Reactors is described by Johnson (2017). Scaling from the ORNL SNS 6% duty factor with 1.4 MW of beam on target to CW operation by appropriate upgrades of components, the traditionally considered ADS goal of 10 MW at 1 GeV is easily surpassed. However, studies have pointed out that even a few hundred trips of an accelerator lasting a few seconds could lead to unacceptable thermal stresses as each trip causes fission to be turned off in solid fuel structures found in conventional reactors. The newest designs are based on the GEM∗STAR concept (Bowman et al. 2010), however, take such trips in stride by using molten-salt fuel, where fuel pin fatigue is not an issue. Other aspects of the GEM∗STAR concept, which address all historical reactor failures, include an internal spallation neutron target and high-temperature molten salt fuel with continuous purging of volatile radioactive fission products such that the reactor contains less than a critical mass and almost a million times fewer volatile radioactive fission products than conventional reactors like those at Fukushima. GEM∗STAR is a reactor that without redesign will burn spent nuclear fuel, natural uranium, thorium, or surplus weapons material. It will operate without the need for a critical core, fuel enrichment, or reprocessing making it an excellent candidate for export. While conventional nuclear reactors are becoming more and more difficult to license and expensive to build, SRF technology development is on a steep learning curve and the simplicity implied by subcritical operation will lead to reductions in costs from regulatory hurdles and construction complexity.

6.3 Single Event Effects

The development of high-resolution charged particle microprobe has opened a door for the studies of effects produced by single ions. Although these studies require a sophisticated experimental setup, they have been initiated in a number of laboratories. Several areas of research have attracted attention:

- Single event upsets (SEU) in memory chips and semiconductor elements and systems.
- Effects of single ions on living cell.
- Micromachining.
- Scanning transmission ion microscopy.

6.3.1 Single Event Upsets in Logic and Memory Devices

Soft error or SEU is the phenomenon in which bit information of large-scale integrated circuit (LSI) memories is lost when charges induced by an energetic single-particle strike through a sensitive area of the memory exceeds a characteristic value known as the critical charge of the memory. SEU has become the most serious problem when high-density and high-speed memories are employed in a spacecraft's electronics system. Since radiation protection of these memories is almost impossible, due to the fact that space radiations that cause SEU have very high energies and due to the weight limitation of satellites, improvement in their SEU tolerance is a key to the application of highly integrated memories in space environment.

SEUs presented challenges for circuit performance and design for decades. Wallmark and Marcus (1962) predicted for the first time the ability for Single Event Transients (SETs) to induce soft errors in Integrated Circuits (ICs). This was confirmed to be a serious issue some thirty years later. Since then SEU have resulted in a variety of soft and hard errors in electronics including storage cells, logic circuits and microprocessors. SEU error propagation between interconnected very large scale integrated (VLSI) logic devices representative of a space-borne system has been studied in (Newberry 1992). The results show that up to 50% of the time a single transistor upset internal to a logic device can result in system failure. Catastrophic system failures were observed with occurrence frequencies in the range 0.2–1.2 failures/10^6 ions/cm^2 verifying upset propagation to system output with implications of error multiplication either within individual devices or as an error propagates through the system.

In the 90s microelectronic technologies reached the "deep submicron" era, allowing high density ICs working at frequencies faster than hundreds of MHz. This new paradigm changed the status of SETs to become a major source of reliability losses. Huge efforts have thus been made to characterize SETs in microelectronics, using experiments or by simulation, in order to reveal key factors leading to SET occurrence, propagation and capture in modern ICs. In this context, modeling and simulation are of primary importance to get accurate SET predictions.

Artola et al. (2015) focus on modeling SETs in innovative electronic devices which involves modeling steps at different scales, from ionizing particle to circuit response. Capabilities and intrinsic limitations of SET modeling, the incoming challenges in advanced devices and ICs, and the methodologies to improve SET simulation and prediction for future technologies are discussed. When an ion strikes on a semiconductor circuits it can cause a single event effect (SEE) affecting both sequential and combinational logic cells as well as multiple cells. To induce a SEU in the logic circuit, the SET has to propagate through the data path and be latched into a storage device. Therefore, only the final SET pulse at the input of the storage device matters (Rezgui et al. 2007, 2012).

Turowski et al. (2012) presented a three-dimensional NanoTCAD simulation of a 45 nm silicon-on-insulator (SOI) – static random-access memory (SRAM) cell indicating that the contributions of charge generation and transport in the surrounding oxides have a potentially significant effect on the SEU response in nanotechnology.

Power metal–oxide–semiconductor field-effect transistor (MOSFET) are widely used for the power-handling portion of electronic equipment. The phenomenon of single-event gate rapture (SEGR) in power MOSPET was first observed in 1987 (Fischer 1987). However, the catastrophic failure mode caused by SEGR in space radiation environments remains a critical issue for these devices (Kuboyama et al. 2012, and references therein).

Gerardin et al. (2012) analyzed the corruption of floating gate bits due to high-energy protons in 41 nm single-level NAND Flash memories. They found that proton-induced upsets at low doses are not negligible in deeply scaled single-level cell Flesh memories. Due to the combination of direct and indirect ionization effects, which may lead to threshold voltage shifts larger than 2 V. Upset cross-sections are around 10^{-19} cm^2, and increase with proton energy.

LaBel et al. (1991) in their study of heavy ion and proton-induced transient upset concluded that this process can influence the error rate of a fiber optic communication link in space. The worst-case would be during the dynamic operation of the system during some practical application. A receiver of the design used in these tests will dominate the error rate If the duration of transient disturbances is long (~5 µs) compared to the operating frequency, they will produce a dead time during which a significant number of bits can be lost for each SEU event.

With the continuous downscaling of CMOS technologies, the reliability has become a major bottleneck in the evolution of the next-generation systems. Technology trends such as transistor downsizing, use of new materials and system on chip architectures continue to increase the sensitivity of systems to soft errors. These errors are random and not related to permanent hardware faults. Their causes may be internal (e.g., interconnect coupling) or external (e.g., cosmic radiation). Wang and Agrawal (2008) presented a tutorial study of the radiation-induced SEU phenomenon caused by external radiation, which is a major source of soft errors. They summarized basic radiation mechanisms and the resulting soft errors in silicon.

Ciani et al. (2009) discuss diagnostic and error correction systems for avionic devices. The authors proposed a mitigation technique for SRAM-based FPGA avionics device. A fault-tolerant technique for a system present on a military aircraft has been analyzed and devised. The technique that has been developed is general purpose and can be introduced into any generic electronic device on an aircraft.

In the space environment, spacecraft designers have to be concerned with two main causes of SEEs: cosmic rays and high-energy protons. For cosmic rays, SEEs are typically caused by its heavy ion component. These heavy ions cause a direct ionization SEE, i.e., if an ion particle traversing a device deposits sufficient charge an event such as a memory bit flip or transient may occur. Cosmic rays may be galactic or solar in origin.

Protons, usually trapped in the earth's radiation belts or from solar flares, may cause direct ionization SEEs in very sensitive devices. However, a proton may more typically cause a nuclear reaction near a sensitive device area, and thus, create an indirect ionization effect potentially causing an SEE.

As spacecraft become driven more and more to reduce parameters such as power, weight, volume and cost, while requiring increased functionality, emerging commercial technologies – often vulnerable to SEEs – have come to the forefront. These technologies include high-speed and low-power CMOS and fiber optics. Types of ICs that utilize these technologies range from complex microprocessors to dense SRAMs. All mission-critical devices must be tested for SEE – one Latchup or Gate Rupture in a critical system can destroy a mission! (See NASA 2015).

NASA specifies the following SEEs as listed on the web page https://radhome.gsfc.nasa.gov/radhome/papers/seespec.htm:

- Single Hard Error (SHE): an SEU that causes a permanent change to the operation of a device. An example is a stuck bit in a memory device.

- Single Event Latchup (SEL): a condition that causes loss of device functionality due to a single event induced high current state. An SEL may or may not cause permanent device damage, but requires power strobing of the device to resume normal device operations.

- Single Event Burnout (SEB): a condition that can cause device destruction due to a high current state in a power transistor.

- Single Event Gate Rupture (SEGR): a single ion-induced condition in power MOSFETs that may result in the formation of a conducting path in the gate oxide.

- Single Event Effect (SEE): any measurable effect to a circuit due to an ion strike. This includes (but is not limited to) SEUs, SHEs, SELs, SEBs, SEGRs and Single Event Dielectric Rupture (SEDR).

- Multiple Bit Upset (MBU): an event induced by a single energetic particle such as a cosmic ray or proton that causes multiple upsets or transients during its path through a device or system.

NASA SEU guidelines require, wherever practical, to procure SEE immune devices. SEE immune is defined as a device having an $LET_{th} > 100$ MeV cm^2/mg, where LET_{th} is threshold LET, linear energy transfer. If device test data does not exist, ground testing is required. NASA provides a list of devices tested together with the test report (see NASA 2000).

6.3.2 Effects of Single Ions on Living Cells

Another application of great significance is the use of single ions to deliver precisely controlled ionization energy to selected regions of a living cell, thus enabling radiation effects to be followed with precision, overcoming many of the limitations of previous statistical studies. These single-ion applications offer major advances in previous work with collimated beams. Recently, a dramatic change took place in assessments of the hazard associated with public exposures to ionizing radiations. For many years, about half of the ionizing radiation dose received by the population at large was considered to originate from natural sources (cosmic, terrestrial and internal) and about half from manufactured sources (primarily medical diagnostic).

Now, however, about half of the population exposure is considered to originate from radon. This gaseous radioactive isotope is now recognized as being much more pervasive than previously acknowledged and a problem that was considered to be confined to special groups (such as uranium miners) has now confronted the community at large. Recognition that radon is more widespread than previously thought, along with a longstanding recognition of the biological effectiveness of alpha-particle radiation has led to estimates of significant numbers of human cancers being attributable to the alpha-particle irradiations from the breakdown of radon daughters. Such estimates are, however, fraught with considerable uncertainty and meaningful assessments of risk can only come about after a broad understanding of the biological effects of individual alpha particles over the LET range encompassed by the breakdown products of the radon daughters.

Single ions can be used to deliver precisely controlled ionization energy to selected regions of living cells, enabling radiation effects to be followed with precision, overcoming many of the limitations of statistical studies. These single-ion applications offer major advances on previous work with collimated beams. Here we presented some of the centers active in this field.

The ion microprobe SNAKE (superconducting nanoprobe for applied nuclear (Kern) physics experiments) at the University of Munich, Germany (Dollinger et al. 2003) enables to focus light and heavy ions delivered by the 14 MV Munich tandem accelerator. The irradiation setup at the ion microprobe SNAKE is used by Hauptner et al. (2006) to irradiate living cells with single energetic ions. The irradiation accuracy of 0.55 μm and respectively 0.40 μm allows irradiating substructures of the cell nucleus. By the choice of ion atomic number and energy, the irradiation can be performed with a damage density adjustable over more than three orders of magnitude. Immunofluorescence detection techniques show the distribution of proteins involved in the repair of DNA double-strand breaks. In one of the first experiments, the kinetics of the appearance of irradiation-induced foci in living HeLa cells was examined. In another experiments, a new effect was detected which concerned the interaction between irradiation events performed at different time points within the same cell nucleus.

The focusing heavy-ion microprobe at the Gesellschaft für Schwerionenforschung in Darmstadt (Germany) has been modified to enable the targeted irradiation of single, selected cells with a defined number of ions, see report (Heiss et al. 2006). With this setup, ions in the range from helium to uranium with LETs up to ~15,000 keV/μm can be positioned with a precision of a few micrometers in the nuclei of single cells that are growing in culture on a thin polypropylene film. To achieve this accuracy, the micro beam traverses a thin vacuum window with minimal scattering. Electron emission from that window is used for particle detection. The cells are kept in a specially designed dish that is mounted directly behind the vacuum window in a setup allowing the precise movement and the imaging of the sample with microscopic methods. An integrated software program that also controls the rapid deflection and switching of the beam locates the cells. In addition to detailed setup description, Heiss et al. (2006) describe the ability of the microprobe to reliably hit randomly positioned etched nuclear tracks in CR-39 with single ions as well as the ability to visualize the ion hits using immunofluorescence staining for 53BP1 as a marker of DNA damage in the targeted cell nuclei.

At the end let us describe a single ion hit facility that is being developed at the CEA/Pierre Süe Laboratory (LPS)/IPGP, Saclay, France (Daudin et al. 2007). This set-up is dedicated to the study of ionizing radiation effects on living cells, which will complete current research conducted on uranium chemical toxicity on renal and osteoblastic cells. The study of the response to an exposure to alpha particles will allow us to distinguish radiological and chemical toxicities of uranium, with a special emphasis on the bystander effect at low doses. Designed and installed on the LPS Nuclear microprobe, up to now dedicated to ion beam microanalysis, this set-up will enable us to deliver an exact number of light ions accelerated by a 3.75 MV electrostatic accelerator. An "in air" vertical beam permits the irradiation of cells in conditions compatible with cell culture techniques. Furthermore, cellular monolayers are kept in controlled conditions of temperature and atmosphere in order to diminish stress. The beam is collimated with a fused silica capillary tubing to target pre-selected cells. Motorization of the collimator with piezo-electric actuators enables fast irradiation without moving the sample, thus avoiding mechanical stress. An automated epifluorescence microscope, mounted on an antivibration table, allows pre- and post-irradiation cell observation. An ultra-thin silicon surface barrier detector has been developed and tested to be able to shoot a cell with a single alpha particle. For more on the Pierre Süe Laboratory microprobe see also article by Khodja et al. (2001). Similar facility is also operational in Lanzhow, China, see Sheng et al. (2013).

6.4 Micromachining

Two particular items deserve to be mentioned in the discussion of this exciting field. The first one relates to the known fact that perfect special depressions can be produced with radii of curvature 1–100 μm by etching heavy ion tracks. These spherical depressions can be used to focus ultrasonic waves in an ultrasonic microscope. Nowadays even lines or arrays of these micro-mirrors can be produced with high precision using the single-particle techniques.

The second one relates to the production of single-pore membranes that have been produced since early 1980s. These membranes of about 50 mm diameter contain exactly one single etched track pore of about 50 μm length and 5 μm diameter at their center. Their production represents one of the economic uses of an ion beam.

Irradiation of materials with heavy ions produces irreversible damages (tracks) that are a few nanometers in diameter along their trajectories in the membrane. These tracks are more prone to etching than the rest of the material and can turn into nanopores with the appropriate etching methods. Irradiation at various ion densities depending on the application, even down to one ion/membrane can be applied. Kaya and Keçeciz (2020) discuss fundamentals of track-etching technique and several sensor applications of single nanopore track-etched in polymer membranes. They included the background for the tracking process and the chemical etching technique for a better understanding of the background of this sensor paradigm. They especially focused on DNA, protein and nanoparticle sensors that work on the resistive-pulse principle. In the several cited studies in this paper, the detection and discrimination of the analytes were successfully executed by obtaining resistive-pulse signals under varying conditions of potential, concentration, etc. and commenting on the pulse amplitudes and pulse durations. The authors also gave examples of a different type of sensing that is based on the current-voltage characteristics of asymmetric nano-pores.

Current techniques for producing silicon membranes with well-defined pores allow the preparation of samples with extreme porosities, either highly porous or a single pore, but nothing in between. Highly porous ultrathin silicon membranes can be obtained by thermally annealing thin silicon films, which results in spontaneous pore formation (Striemer et al. 2007, Kim et al. 2008). The pore diameter can be controlled with high precision by choosing appropriate annealing conditions. This method does not allow control over the number of pores. Pores can also be fabricated one by one using ion beam sculpting (Li et al. 2001), focused ion beam drilling (Tong et al. 2004), e-beam lithography (Deshmukh et al. 1999), or e-beam drilling in a transmission electron microscope (Storm et al. 2003). Although these serial approaches can be applied to various silicon and even polymer materials, they are suitable only for the preparation of membranes with low numbers of pores. None of the above techniques give the possibility of tuning the geometry of the pores.

Vlassiouk et al. (2009) reported a versatile fabrication method for ultrathin, freestanding silicon nitride (SiN) membranes containing conical and double-conical pores prepared by the ion track-etching (ITE) technique, together with a study of their transport properties. The ITE technique is based on irradiating a film with accelerated heavy ions, and subsequent chemical etching of the irradiated sample. The number of heavy ions corresponds to the number of fabricated pores. The technique has been used for membrane production for more than four decades (Apel 2001); however, most of track-etched membranes are made in polymer films with thicknesses of at least several micrometers. In contrast, the ITE SiN membranes reported by Vlassiouk et al. 2009 have thicknesses of ~100 nm, and the pore diameter can be made as small as 1 nm. The membranes can also withstand very harsh chemical conditions not tolerated by polymer membranes, such as piranha treatment (a heated mixture of H_2SO_4 and H_2O_2) and high temperatures. The ultrathin nanoporous SiN membranes presented provide excellent separation capability based on electric charge and size of the permeate species.

The well-known SiN surface chemistry provides a convenient starting point for the chemical functionalization of the pore walls to tailor these separation capabilities. In addition, such membranes constitute an attractive platform for biosensing (Marconi et al. 2009). First, controlling the number of swift heavy ions that are used in the irradiation directly controls the number of pores. Single-ion irradiation available at several laboratories allows the preparation of single pore membranes. Second, the conical and double conical shapes of the pores with tunable opening angles, together with the ultrathin character of these membranes (free standing SiN membranes with 20-nm thickness are routinely fabricated by common microfabrication techniques), offer a versatile platform for biomolecule translocation and separation experiments, such as single-molecule DNA analysis.

Novoselov et al. (2004) first discovered so-called two-dimensional (2D) material. Since then a large number of 2D atomic crystal structures with extraordinary properties have been synthesized and prepared. In addition to graphene, transition-metal dichalcogenide (TND) monolayers are used as atomically thin semiconductors. In addition to using them individually 2D materials are being used by combining them into van der Waals heterostructures (Geim and Grigorieva 2013). Piling up atomically flat layers makes it possible to develop devices benefiting from properties of two or more individual 2D crystals. For example, the high conductivity of graphene in combination with the semiconducting MoS_2 (molybdenum disulfide) is used for the fabrication of flexible and transparent memory devices (Bertolazzi et al. 2013).

Focused ion beams, heavy ions in particular, have been used since the 1970s for milling patterns on surfaces. Unfortunately, these methods often damage the underlying substrate. Slow ions in high charge states avoid this problem because they carry large amounts of potential energy, being determined by the sum of ionization energies. The majority of this energy is released within a shallow area in the surface when the charged particle undergoes deexcitation. The realized energy is transferred mainly to the target's electronic system leading to a high degree of electronic excitation and ionization. In contrast to fast heavy ions, this leads to nanostructure formation on the surface without creating additional ionization and electronic excitation in the substrate along the ion's path (Aumayr et al. 2011).

Schwestka et al. (2020) in their study irradiated MoS_2/graphene heterostructures with highly charged xenon ions and observed nanometer-sized pores exclusively in the MoS_2 layer if it faces the ion beam. The remaining potential energy after transmission through the first layer is not sufficient to perforate a single layer of graphene. In addition, when the graphene faces the ion beam first, no damage is observed in the MoS_2 underneath. Their analysis of sample areas containing few-layer MoS_2 on graphene shows that under the conditions used by the authors, the highly charged ion perforation is limited to two layers. In summary, their result shows the extreme surface sensitivity of highly charged ions, establishing them as the useful tool for material property tailoring of van der Waals heterostructures down to monolayer precision (Schwestka et al. 2020).

Microelectronics-based micromechanics is rather limited for the construction of 3D micro mechanisms with moving parts. Kussul et al. (1996) proposed the use of micro equipment for transfer the technologies of mechanical engineering to the micro domain. They showed that equipment precision increases linearly with decreasing size. To make micro equipment, a series of equipment generations with gradually decreasing dimensions is suggested. Miniaturization of equipment will reduce power consumption and the floor area occupied. Coupled with automation, it will drastically reduce the cost of micro equipment. This in its turn will reduce the cost of micromechanical devices manufactured by micro equipment. Micro equipment-based manufacturing will also increase throughput by the concurrent operation of large numbers of low-cost micro equipment pieces. The low cost and high productivity of micro equipment-based manufacturing will widen the range of feasible micromechanical applications, both single-unit and mass. The authors proposed designs for micro valve fluid filters, capillary heat exchangers, electromagnetic and hydraulic step motors that could be easily implemented by micromechanical engineering technologies. Hybrid technologies combining massively parallel micro equipment-based manufacturing and batch manufacturing may be promising (Kussul et al. 1996).

6.5 New Developments in Medicine

Applications of accelerators in medicine are a very dynamic field. Some examples of new developments in medicine are presented next. For example, new accelerator-based technology being developed by the Department of Energy's SLAC National Accelerator Laboratory and Stanford University aims to reduce the side effects of cancer radiation therapy by shrinking its duration from minutes to under a second. Built into future compact medical devices, technology developed for high-energy physics could also help make radiation therapy more accessible around the world (Gnida 2018).

The SLAC/Stanford team has received crucial funding to proceed with two projects to develop possible treatments for tumors – one using X-rays, the other using protons. The idea behind both is to blast cancer cells so quickly that organs and other tissues do not have time to move during the exposure – much like taking a single freeze frame from a video. This reduces the chance that radiation will hit and damage healthy tissue around tumors, making radiation therapy more precise.

Proposed design of a device for X-ray therapy, which researchers hope will be able to deliver radiation powerful enough to blast cancer cells in under a second instead of minutes, see Fig. 6.4. The project, called PHASER, relies on a novel high-power

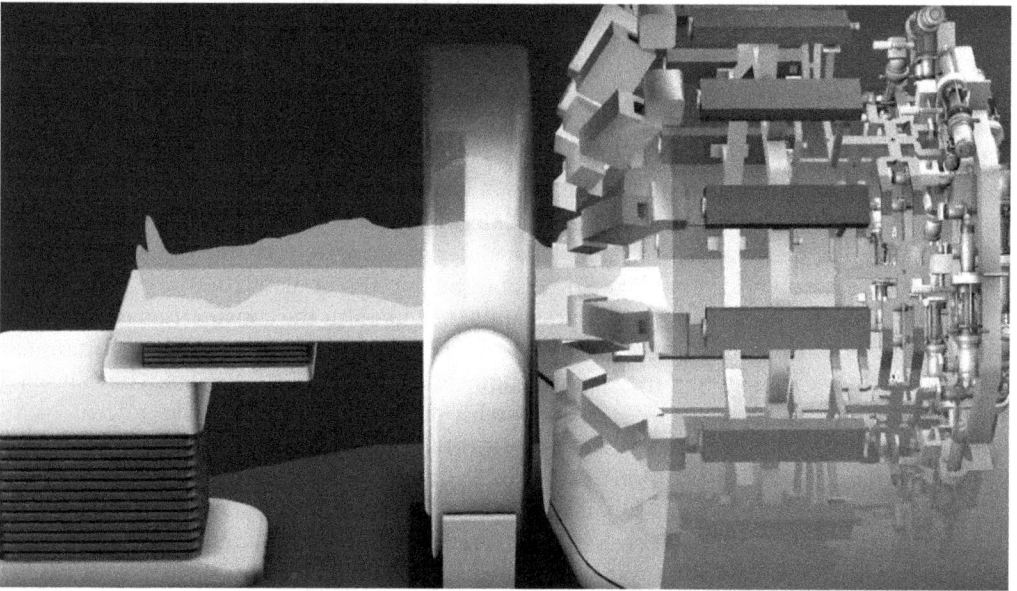

FIGURE 6.4
Proposed design of a future device for X-ray therapy, which will be able to deliver radiation powerful enough to blast cancer cells in under a second instead of minutes. (Credit: Greg Stewart/SLAC National Accelerator Laboratory.)

accelerator under development at SLAC (Maxim et al. 2012) will result in a flash delivery system for X-rays. Over the past few years, the PHASER team has developed and tested accelerator prototypes with special shapes and new ways of feeding radio-frequency fields into the tube. These components are already performing as predicted by simulations and pave the way for accelerator designs that support more power in a compact size.

Protons are less harmful to healthy tissue than X-rays because they deposit their tumor-killing energy in a more confined volume inside the body. However, large facilities are required to accelerate protons and adjust their energy. It also uses magnets weighing hundreds of tons that slowly move around a patient's body to guide the beam into the target. The goal is to design, fabricate and test an accelerator structure similar to the one in the PHASER project that will be capable of steering the proton beam, tuning its energy and delivering high radiation doses practically instantaneously; in other words to manipulate the proton beam that will make future devices simpler, more compact and much faster (Fig. 6.5).

In addition to making cancer therapy more precise, flash delivery of radiation also appears to have other benefits. It has also been established that healthy cells suffer less damage when the radiation dose is applied very quickly, and yet the tumor-killing effect is equal to or even a little bit better than that of a conventional longer exposure (Brooks et al. 2016). This has been established for test animals, if the effect holds for humans, it would be a completely new paradigm for the field of radiation therapy.

The present situation leaves millions of patients around the world only with palliative care because they do not have access to cancer therapy. This is the reason why the project is focusing on designing systems that are compact, power-efficient, economical, efficient to use in the clinical setting and compatible with existing infrastructure around the world. The proposed PHASER design is compact enough to fit into standard cargo containers. Its transportability could help make radiation therapy for cancer more accessible around the world (Gnida 2018).

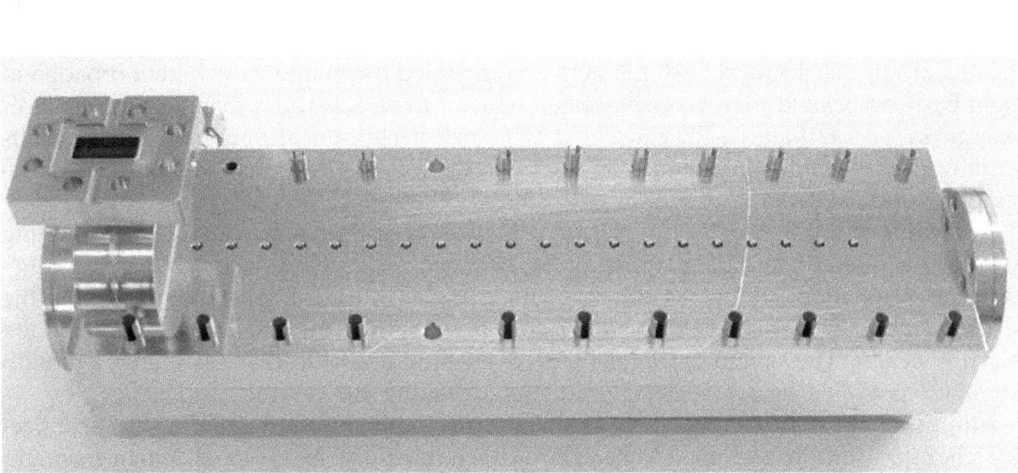

FIGURE 6.5
Prototype accelerator component for the PHASER project, which will use a novel accelerator design hundreds of times more powerful than accelerators in current devices for X-ray therapy. (Credit: SLAC National Accelerator Laboratory.)

6.6 Tabletop Particle Accelerators: Laser Plasma Accelerators

Rapid progress in the development of high-intensity laser systems has extended our ability to study light–matter interactions far into the relativistic domain, in which electrons are driven to velocities close to the speed of light. As well as being of fundamental interest in their own right, these interactions enable the generation of high-energy particle beams that are short, bright and have good spatial quality. Along with steady improvements in the size, cost and repetition rate of high-intensity lasers, the unique characteristics of laser-driven particle beams are expected to be useful for proton therapy treatment of cancers, materials characterization, radiation-driven chemistry, border security through the detection of explosives, narcotics and other dangerous substances and of course high-energy particle physics (Malka et al. 2008).

Laser wakefield accelerator (LWFA) are more powerful and smaller particle accelerators described by Palastro et al. (2020). They combine spherical aberration with a novel cylindrically symmetric echelon optic to spatial-temporally structure an ultrashort, high-intensity laser pulse that can overcome dephasing by propagating at any velocity over any distance. The ponderomotive force of the spatial-temporally shaped pulse can drive a wakefield with a phase velocity equal to the speed of light in a vacuum, preventing trapped electrons from outrunning the wake. Simulations in the linear regime and scaling laws in the bubble regime illustrate that this dephasingless LWFA can accelerate electrons to high energies in much shorter distances than a traditional LWFA – a single 4.5 m stage can accelerate electrons to TeV energies without the need for guiding structures. This is 10,000 times smaller than a proposed setup recording similar energy. This electron accelerator relies on a technique for sculpting the shape of laser pulses so that their peaks can travel faster than the speed of light (Palastro et al. 2020).

The physicists at the University of Maryland (Goers et al. 2015) have reported another effort. They demonstrated laser-plasma acceleration of high charge electron beams to the ~10 MeV scale using ultrashort laser pulses with as little energy as 10 mJ. This result is made possible by an extremely dense and thin hydrogen gas jet. Total charge up to ~0.5 nC is measured for energies >1 MeV. Acceleration is correlated to the presence of a relativistically self-focused laser filament accompanied by an intense coherent broadband light flash, associated with wave breaking, which can radiate more than ~3% of the laser energy in a ~1 fs bandwidth consistent with half-cycle optical emission. Their results enable truly portable applications of laser-driven acceleration, such as low-dose radiography, ultrafast probing of matter and isotope production.

According to Kurz et al. (2021), plasma wakefield accelerators are capable of sustaining GV cm^{-1} accelerating fields, surpassing the electric breakdown threshold in state-of-the-art accelerator modules by 3–4 orders of magnitude. In their paper, they report on the demonstration of a millimeter-scale plasma accelerator powered by laser-accelerated electron beams. They claim the acceleration of electron beams to 128 MeV, consistent with simulations exhibiting accelerating gradients exceeding 100 GV m^{-1}.

All-optical approaches to particle acceleration are currently attracting a significant research effort internationally. Although characterized by exceptional transverse and longitudinal emittance, laser-driven ion beams currently have limitations in terms of peak ion energy, bandwidth of the energy spectrum and beam divergence. In their paper, Kar et al. (2016) presented a laser-driven miniature device for simultaneous energy selection, collimation and post-acceleration of ions, which is an uniquely attractive combination compared with the methods explored so far for improving the quality of laser-driven ion beams (see for example Toncian et al. 2006; Kar et al. 2008; Schollmeier et al. 2008;Bartal et al. 2012;Busold et al. 2013). Their scheme exploits ultra-short (tens of ps), high-amplitude, unipolar electromagnetic pulses generated in the interaction of a high-power laser with a solid target. The technique is demonstrated in a proof-of-principle experiment using a 200 TW laser, achieving post-acceleration of protons to enhance their energy by ~5 MeV over less than a centimeter of propagation, that is, an accelerating gradient ~0.5 GeV m^{-1}. This is already beyond what can be sustained by conventional accelerator technologies and can be enhanced significantly at higher laser intensities.

Highly efficient optical generation of narrowband terahertz radiation enables un-explored technologies and sciences from compact electron acceleration to charge ma-nipulation in solids. State-of-the-art conversion efficiencies are currently achieved using difference-frequency generation driven by temporal beating of chirped pulses but remain far lower than desired or predicted. Jolly et al. (2019) showed that high-order spectral phase fundamentally limits the efficiency of narrowband difference-frequency generation using chirped-pulse beating and resolved this limitation by introducing a novel technique based on tuning the relative spectral phase of the pulses. For optical THz generation, they demonstrated a 13-fold enhancement in conversion efficiency for 1%-bandwidth, 0.361 THz pulses, yielding a record energy of 0.6 mJ and exceeding previous optically-generated energies by over an order of magnitude. Their results prove the feasibility of mJ-scale applications like terahertz-based electron accelerators and light sources and solve the long-standing problem of temporal irregularities in the pulse trains generated by interfering chirped pulses.

A new facility called CBETA (Cornell-Brookhaven ERL Test Accelerator) that combines some of the best traits of linear and circular accelerators has recently been constructed at Cornell University, Ithaca, NY, USA. ERL stands for energy-recovery linac having both linac-quality bunches and the large currents more typical of circular accelerators. This is world's first multi-turn superconducting radio-frequency (SRF) ERL, with a footprint of about 25 × 15 m, CBETA is designed to accelerate an electron beam to an energy of 150 MeV. CBETA recirculates multiple beams of different energies around the accelerator at one time. The electrons make four accelerating passes around the accelerator, while building up energy as they pass through a cryo-module with superconducting RF ac-celerating structures. In four more passes, they return to the superconducting cavities that accelerated them and return their energy back to these cavities – hence it is an Energy Recovery Linac. While this method conserves energy, it also creates beams that are tightly bound and are a factor of 1,000 times brighter than other sources (Hoffstaetter et al. 2018).

As a general rule, the longer the particle accelerator, the more powerful it is. Not any more, a team led by scientists at the Department of Energy's SLAC National Accelerator Laboratory has invented a new type of accelerator structure that delivers a 10 times larger energy gain over a given distance than conventional ones. This could make accelerators used for a given application 10 times shorter. The basis is to use terahertz radiation to boost particle energies (Othman et al. 2020).

In presently used accelerators, particles draw energy from a radio-frequency field fed into specifically shaped accelerator structures, or cavities. Each cavity can deliver only a limited energy boost over a given distance, so very long strings of cavities are needed to produce required energy beams. Since terahertz waves are 10 times shorter than radio waves, cavities in a terahertz accelerator can also be much smaller. In fact, the one de-scribed by Othman et al. (2020) is only 0.5 cm long.

The major difficulty in building these tiny cavity structures is machining them very precisely. Over the past few years, SLAC teams developed a way to do just that. Instead of using the traditional process of stacking many layers of copper on top of each other, they built the minute structure by machining two halves and bonding them together. The new structure produces a thousand times shorter particle pulses than those coming out of conventional copper structures. As a result, such an accelerator would produce beams that pulse at a higher rate producing more power over a given time period.

Othman et al. (2020) demonstrated coupling of an unprecedented RF power, up to 575 kW, into the mm-wave accelerator structure using quasi-optical setup. The standing wave accelerator structure consisted of a single cell copper cavity and a Gaussian to TM_{01} mode converter. The accelerator structure was powered by 110 GHz, 10 ns long RF pulses. These pulses were chopped from 3 ms pulses from a gyroton oscillator using a laser-driven silicon switch. They obtained an extremely high gradient of 230 MV/m corresponding to a peak surface electric field of 520 MV/m. These results were achieved after conditioning the cavity with more than 10^5 pulses.

An interesting reading could be found in a review article on laser-driven ion acceleration: A primer for laser-driven source development and potential applications (Schreiber et al. 2016).

Beside producing and accelerating electron and charged particle beams of paramount importance is also charged particle guiding and beam splitting. It is well established (Leibfried et al. 2005; Amini et al. 2010; Seidelin et al. 2006; Brown et al. 2011; Harlander et al. 2011) that for quantum information processing, planar chip-based ion traps have become a promising technology. Lesson learned from state-of-the-art manufacturing techniques used in semiconductor and printed circuit board technology used to create miniaturized traps with many micro-structured electrodes. Similarly, electromagnetic potentials customized in this way have been used in quantum manipulation experiments with neutral atoms in magnetic chip traps (Reichel and Vuletic 2011). The guiding and splitting of an electron beam by the effective time-independent (ponderomotive) force created by the microwave field have been demonstrated to work for very low electron energies, <5 eV (Hoffrogge et al. 2011; Hoffrogge and Hommelhoff 2011).

Zimmermann et al. (2021) reported the guiding and splitting of charged particle beams using ponderomotive forces created by the motion of charged particles through electrostatic optics printed on planar substrates. The lithographically produced electrodes' layout and the applied voltages, enabling the control of charged particle beams within precisely engineered effective potentials, can locally tailor shape and strength of the potential. They demonstrated guiding of electrons and ions for a large range of energies (from 20 to 5000 eV) and masses (from 5×10^{-4} to 131 amu) as well as electron beam splitting for energies up to the keV regime, as a proof-of-concept for more complex beam manipulation.

6.7 Autonomous Particle Accelerators

Particle accelerators are versatile tools used in production processes in industry, in medicine (tumor therapy in hospitals), in different branches of research and various applications. Consequently, there is growing demands on the stability and beam properties making a manual operation of accelerators increasingly challenging requiring the highest possible level of automation to support operations.

A new project of Deutsches Elektronen-Synchrotron (DESY) and Karlsruhe Institute of Technology (KIT) is in progress to the goal of having a fully autonomously operated accelerator (Burkart 2020). The cooperation "Autonomous Accelerator", which is supported by the Helmholtz Association and the two participating Helmholtz research centers within the framework of the Helmholtz Artificial Intelligence Cooperation Unit, brings "reinforcement learning" to the operation of two linear accelerators at DESY and KIT. Reinforcement learning involves measuring state values and adjusting control variables to determine their influence on each other, thus learning a control strategy that also takes into account its effects in the future. At the end, this will completely replace manual intervention.

The important characteristic of reinforcement learning is that the control system not only reacts but also plans into the future how to achieve a goal (Eichler 2020). "For this purpose, the control system can decide on the basis of the information collected so far, but it must also have enough range to literally 'conquer' new control regions in previously unknown terrain. The long-term goal of the research team is to operate an accelerator fully autonomously. First, the team is concentrating on controlling the density at which the electrons are distributed along the accelerated bunches. In addition to the length of these electron bunches - some of which pass the measuring device in less than a femtosecond – it is particularly the build-up effects in the particle bunches that make controlling this size very challenging; autonomous control is therefore essential for efficient and rapid optimization" (Eichler 2020).

For their experiments, the research team uses the test accelerators ARES (Accelerator Research Experiment at SINBAD) at DESY and FLUTE (Far-Infrared Linac and Test Experiment) at KIT. Both facilities are available for accelerator research within the framework of the "Matter and Technologies" program and offer sufficient test times for developing such algorithms. By using two similar, compact but not identical accelerators for the development of our artificial intelligence, we gain valuable experience on the transferability of our algorithms to other and larger accelerators.

Of the new types of accelerators, two really stand out and are already starting to be utilized by industry:

  i. Superconducting Linacs and Cyclotrons: The use of superconductivity leads to increases in efficiency and size reduction of systems for cancer therapy, radioisotope and neutron production (see Smirnov and Vorozhtsov 2018).
  ii. Fixed Field Alternating Gradient (FFAG) Cyclotron: FFAG Cyclotrons were developed as a neutron source for BNCT, a hadron source for radiotherapy, and for inspecting the contents of cargo containers. FFAGs could even be used as a new type of nuclear reactor, called an energy amplifier. One of the first non-scaling FFAG accelerator, which has many advantages because large and heavy magnets are avoided is described by Trbojevic et al. 2005.

Batsch et al. (2021) used a relativistic ionization front (RIF) to provide various initial transverse wakefield amplitudes for the self-modulation of a long proton bunch in plasma. They showed that with sufficient initial amplitude the phase of the modulation along the bunch was reproducible from event to event. The phase was not reproducible for lower initial amplitudes. They presented the results of experimental studies of the self/modulation (SM) phase for different timings of the RIF with respect to the proton bunch, measured after 10 m long plasma. These results demonstrate that the SM process can be seeded, in other words, the phase of modulation can be defined by the RIF and reproducible from event to event. The results obtained are of great importance for AWAKE (Advanced WAKefield Experiment) Run 2 to be performed at CERN (Muggli 2020).

6.8 New Electron Sources

6.8.1 Diamond Field-Emitter Array (DFEA)

Diamond Field-Emitter Arrays (DFEAs) are arrays of diamond pyramids with exquisitely sharp tips. The pyramid arrays are fabricated using standard silicon wafer processing techniques; therefore, they can be fabricated in arbitrary shapes to produce an inherently shaped beam. DFEAs produce electron beams by field emission with typical currents as high as 30 μA per pyramid at gradients of around 10 MV/m in the direct current (DC) regime. DFEAs emit under dc electric fields, as well as in time-varying electromagnetic fields in RF guns. These cathodes can produce an inherently shaped beam with little to no current loss, no production of x-rays due to the use of a mask, no shot-to-shot jitter and no drive laser required.

A field-emission cathode is of particular interest for use in an RF gun because it greatly simplifies the infrastructure needed to produce the beam, operates at lower fields and is less expensive to implement and operate than photocathodes. In addition, in an RF gun, field emission cathodes will produce inherently shorter bunches with lower energy spread than thermionic cathodes.

Nichols et al. (2020) reported the results of the experiment that demonstrated shaped beam production from a DFEA cathode in the 1.3 GHz RF gun at the Argonne Cathode Test-stand (ACT) at the Argonne National Laboratory, USA. The DFEA reported in their experiments was a square array of 25 pyramid emitters arranged in a 5 × 5 pattern with a 25 μm base and a 400 μm pitch. This cathode sample was chosen specifically to allow for imaging of individual emitted beamlets to most easily observe an emission pattern and determine the spacing between beamlets. Electron beams were produced from this cathode at gradients varying between 12 and 22 MV/m. Out of the 25 emitters, approximately 8 emitters were active at any given time. The bunch charge was measured using a Faraday cup, and beam images were observed on several yttrium-aluminum-garnet (YAG) screens along the beamline. The beamlets were transported along the beamline over a distance of more than 2.54 m (see also Andrews et al. 2020).

6.8.2 Halide Perovskites

Halide perovskite thin films $CsPbBr_3$ or $CsPbI_3$ are made by mixing precursor powders, $PbBr_2$ with CsBr, or PbI_2 with CsI, in the desired molar ratio in dimethyl sulfoxide or N, N-dimethylformamide and spin-coating on doped Si wafers (Liu et al. 2021). The authors reported efficient, regenerative and low-cost electron sources based on solution-processed halide perovskites thin films when they are excited with light with energy equal to or above their bandgap. Measurement of a quantum efficiency up to 2.2% and a lifetime of more than 25 h was reported. After degradation, the electron emission could be completely regenerated to its maximum efficiency by deposition of a monolayer of Cs. The electron emission from halide perovskites could be tuned over the visible and ultraviolet

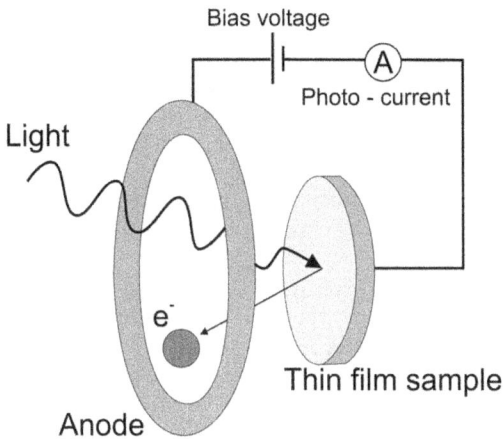

FIGURE 6.6

Schematic of the photoelectric effect measurement setup by Liu et al. (2021). Thin-film samples of $CsPbBr_3$ and $CsPbI_3$ were prepared on 2.54 cm^2 silicon substrates.

spectrum, while operating at vacuum levels with pressures at least two-orders higher than in state-of-the-art semiconductor electron sources. According to the authors, this efficient electron generation by the photoelectric effect in halide perovskite thin films could pave the path for the development of a disruptive technology for realizing low-cost, high efficiency, and spectrally tunable electron sources based on halide perovskite semiconductors, which can be easily integrated and operated at much higher pressures of vacuum than their classic semiconductor-based counterparts without appreciable surface pre-treatments (Fig. 6.6).

References

Abderrahim, H. A. 2020. MYRRHA project: an accelerator driven system (ADS) prototype (6 pp). https://www.osti.gov/etdeweb/servlets/purl/20202637.

Amini, J. M., Uys, H., Wesenberg, J. H., et al. 2010. Toward scalable ion traps for quantum information processing. *N. J. Phys.* 12: 033031 (15 pp).

Andrews, H., Nichols, K., Kim, D., et al. 2020. Shaped beams from diamond field-emitter array cathodes. *IEEE Trans. Plasma Sci.* 48(7): 2671–2675.

Andriamonje, S. Angelopoulosm, A., Apostolakism, A., et al. 1995. Experimental determination of the energy generated in nuclear cascades by a high energy beam. *Phys. Lett. B* 348(3–4): 697–709.

Apel, P. 2001. Track etching technique in membrane technology. *Radiat. Meas.* 34: 559–566.

Artola, L., Gaillardin, M., Hubert, G., Raine, M., and Paillet, P. 2015. Modeling single event transients in advanced devices and ICs. *IEEE Trans. Nucl. Sci.* 62(4): 1528–1539.

Aumayr, F., Facsko, S., El-Said, A. S., Trautmann, C., and Schleeberger, M. 2011. Single ion induced surface nanostructures: a comparaison between slow highly charged and swift heavy ions. *J. Phys. Condens. Matter* 23: 393001 (23 pp).

Bartal, T., et al. 2012. Focusing of short-pulse high-intensity laser-accelerated proton beams. *Nat. Phys.* 8: 139–142.

Batsch, F., Muggli, P., Agnello, R., et al. 2021. Transition between instability and seeded self-modulation of a relativistic particle bunch in plasma. *Phys. Rev. Lett.* 126: 164802-(6).

Brooks, A. L., Hoel, D. G. and Preston, R. J. 2016. The role of dose rate in radiation cancer risk: evaluating the effect of dose rate at the molecular, cellular and tissue levels using key events in critical pathways following exposure to low LET radiation. *Int. J. Radiat. Biol.* 92(8): 405–426.

Brown, K. R., Ospelkaus, C., Colombe, Y., et al. 2011. Coupled quantized mechanical oscillators. *Nature* 471: 196–199.

Burkart, F. 2020. DESY News: accelerate smarter with artificial intelligence. https://www.desy.de/news/news_search/index_eng.html?openDirectAnchor=1961

Bertolazzi, S., Krasnozhon, D., and Kis, A. 2013. Nonvolatile memory cells based on MoS_2. *Nano* 7: 3246–3252.

Bowman, C. D., Arthur, E. D., Lisowski, P. W., et al. 1992. Nuclear energy generation and waste transmutation using an accelerator-driven intense thermal neutron source. *Nucl. Instrum. Methods A* 320: 336–367.

Bowman, C. D., Vogelaar, R. B., Bilpuch, E. G., Howell, C. R., Tonchev, A. P., Tornow, W., and Walter, R. L. 2010. GEM∗STAR: the alternative reactor technology comprising graphite, molten salt, and accelerators. *Handbook of Nuclear Engineering*, 10.1007/978-0-387-98149-9_24, Springer Science and Business Media LLC.

Bungau, C., Barlow, R., Cywinski, R., and Tygier, S. 2008. Accelerator driven systems for energy production and waste transmutation. Proceedings of EPAC08, Applications of Accelerators, Technology Transfer and Relations with Industry, U03 Transmutation and Power Generation. Genoa, Italy. Published by the European Physical Society Accelerator Group (EPS-AG). TUPP147: 1854–1856.

Busold, S., et al. 2013. Focusing and transport of high-intensity multi-MeV proton bunches from a compact laser-driven source. *Phys. Rev. ST Accel. Beams* 16: 101302.

Calabretta, L. and Méot, F. 2015. Cyclotrons and FFAG Accelerators as Drivers for ADS. *Reviews of Accelerator Science and Technology* 8: 77–97.

Carminati, F., Klapisch, R., Revol, J. P., Roche, Ch., Rubio, J. A., and Rubbia, C. 1993. An Energy Amplifier for Cleaner and Inexhaustible Nuclear Energy Production Driven by a Particle Beam Accelerator. European Organisation for Nuclear Research, Report CERN/AT/93-47(ET).

Chao, A. W. and Chou, W. Eds. 2016. *Reviews of Accelerator Science and Technology, Volume 8: Accelerator Applications in Energy and Security*. World Scientific Publishing Co., Singapore.

Ciani, L., Catelani, M., and Veltroni, L. 2009. Single event upset (SEU): diagnostic and error correction system for avionics device. Proceedings of the XIX IMEKO World Congress Fundamental and Applied Metrology. CD-ROM. September 6–11, 2009, Lisbon, Portugal (p. 1380–1384). ISBN: 978-963-88410-0-1.

Daudin, L., Carrière, M., Gouget, B., Hoarau, J., and Khodja, H. 2006. Development of a single ion hit facility at the Pierre Süe Laboratory: a collimated microbeam to study radiological effects on targeted living cells. *Radiat. Protect. Dosimet.* 122(1-4): 310–312.

Deshmukh, M. M., Ralph, D. C., Thomas, M., and Silcox, J. 1999. Nanofabrication using stencil mask. *Appl. Phys. Lett.* 75: 1631–1633.

D'Hondt, P., Aït Abderrahim, H., Kupschus, P., et al. 2004. Myrrha, a Multipurpose Accelerator Driven System for R&d – Present Status. 10.1142/9789812702814_0021. https://ui.adsabs.harvard.edu/abs/2004sefi.conf.209D/abstract

Dollinger, G., Datzmann, G., Hauptner, A., Hertenberger, R., Koerner, H.-J., Reichart, P., and Volckaerts, B. 2003. The Munich ion microprobe: characteristics and prospect. *Nucl. Instrum. Methods Phys. Res. Sect. B* 210(1): 6–13.

Doyle, B. L., McDaniel, F. D., and Hamm, R. W. 2018. The Future of Industrial Accelerators and Applications. Report SAND2018-5903B.

Eichler, A. 2020. Accelerate smarter with artificial intelligence. News. www.research-in-germany.org › news › 2020/11 › 202…

Fischer, T. A. 1987. Heavy-ion induced gate rupture in power MOSFETs. *IEEE Trans. Nucl. Sci.* 34(6): 1786–1791.

Garnett, R. W. and Sheffield, R. L. 2015. Overview of accelerator applications in energy. *Rev. Accelerator Sci. Technol.* 8: 1–25.

Gerardin, S., et al. 2012. Proton-induced Upsets in 41-nm NAND floating gate cells. *IEEE Trans. Nucl. Sci.* 59(4): 838–844.

Geim, A. K. and Grigorieva, I. V. 2013. Van der Waals heterostructures. *Nature* 499: 419–425.

Gnida, M. 2018. *The Future of Fighting Cancer: Zapping Tumors in Less than a Second.* https://www6. slac.stanford.edu/news/2018-11-28-future-fighting-cancer-zapping-tumors-less-second.aspx.

Goers, A., et al. 2015. Multi-Mev electron acceleration by sub-terawatt laser pulses. *Phys. Rev. Lett.* 115, 194802. 10.1103/PhysRevLett.115.194802.

Gohar, Y. 2017. Ukraine NSC KIPT ADS project. Presentation at EuCARD2, Status of Accelerator Driven Systems Research and Technology Development. CERN, 7–9 February 2017.

Harlander, M., Lechner, R., Brownnutt, M., Blatt, R., and Hänsel, W. 2011. Trapped ion antennae for the transmission of quantum information. *Nature* 471: 200–203.

Hauptner, A., Cremer, T., Deutsch, M., et al. 2006. Irradiation of living cells with single ions at the ion microprobe SNAKE. *Acta Phys. Polon. Ser. A* 109: 273–278.

Heidet, F., Brown, N. R., and Tahar, M. H. 2015. Accelerator–reactor coupling for energy production in advanced nuclear fuel cycles. *Rev. Accel. Sci. Technol.* 8: 99–114. https://doi.org/10.1142/S1 793626815300066.

Heiss, M., Fischer, B. E., Jakob, B., Fournier, C., Becker, G., and Taucher-Scholz, G. 2006. Targeted irradiation of mammalian cells using a heavy-ion microprobe. *Radiat. Res.* 165: 231–239.

Hoffrogge, J., Fröhlich, R., Kasevich, M. A., and Hommelhoff, P. 2011. Microwave guiding of electrons on a chip. *Phys. Rev. Lett.* 106(19): 193001 (4 pp).

Hoffrogge, J. and Hommelhoff, P. 2011. Planar microwave structures for electron guiding. *N. J. Phys.* 13: 095012 (14 pp).

Hoffstaetter, G., et al. 2017. CBETA Design Report Cornell-BNL ERL Test Accelerator. arXiv:1706.04245 [physics.acc.ph].

Hoffstaetter, G. and Ryan, R. 2018. Small accelerator promises big returns. *CERN Courier* 58(2): 19–22.

Hoffstaetter, G. H., et al. 2018. CBETA, the 4-Turn ERL with SRF and Single Return Loop. 9th International Particle Accelerator Conference. Vancouver, British Columbia, Canada. 10.1842 9/JACoW-IPAC2018-TUYGBE2

IAEA. 2002. Working Material: Technical Meeting to "Review of National Programmes on Fast Reactors and Accelerator Driven Systems (ADS)". Karlsruhe, Germany, 22-26 April 2002. Report IAEA-Tm-25032. Reproduced by IAEA, Vienna, Austria.

IAEA. 2006. Working Material: Meeting of the Technical Working Group on Fast Reactors (TWG-FR). Beijing, China, 15–19 May 2006. Report IAEA-TM-28911Reproduced by IAEA, Vienna, Austria.

Jolly, S.W., et al. 2019. Spectral phase control of interfering chirped pulses for high-energy narrowband terahertz generation. *Nat. Commun.* 10: 2591. 10.1038/s41467-019-10657-4

Johnson, R. 2017. USA Progress on SRF Linacs Driving Subcritical GEM*STAR Reactors. Presentation at EuCARD2, Status of Accelerator Driven Systems Research and Technology Development. CERN, 7–9 February 2017.

Kapoor, S. S. 2002. Accelerator-driven sub-critical reactor system (ADS) for nuclear energy generation. *Pramana-J. Phys.* 59(6): 941–950.

Kar, S., et al. 2008. Dynamic control of laser-produced proton beams. *Phys. Rev. Lett.* 100: 105004.

Kar, S., et al. 2016. Guided post-acceleration of laser-driven ions by a miniature modular structure. *Nat Commun* 7: 10792. 10.1038/ncomms10792

Kaya, D. and Keçeciz, K. 2020. Track-etched nanoporous polymer membranes as sensors: a review. *J. Electrochem. Soc.* 167: 037543.

Khodja, H., Berthoumieux, E., Daudin, L., and Gallien, J. -P. 2001. The Pierre Süe Laboratory nuclear microprobe as a multi-disciplinary analysis tool. *Nucl. Instrum. Methods Phys. Res. Sect. B* 181(1–4): 83–86.

Kim E, Xiong, H., Striemer, C. C., Fang, D. Z., Fauchet, P. M., McGrath, J. L., and Amemiya, S. 2008. A Structure-permeability relationship of ultrathin nanoporous silicon membrane: a comparison with the nuclear envelope. *J. Am. Chem. Soc.* 130: 4230–4231.

Kuboyama, S., et al. 2012. Rediscovery of single-event gate rupture mechanism in power MOSFETs. *IEEE Trans. Nucl. Sci.* 59(4):749–754.

Kurz, T., Heinemann, T., Gilljohann, M. F., et al. 2021. Demonstration of a compact plasma accelerator powered by laser-accelerated electon beams. *Nat. Commun.*12: 2895 (9 pp).

Kussul, E. M., Rachkovskij, D. A., Baidyk, T. N., and Semion A Talayev, S. A. 1996. Micromechanical engineering: a basis for the low-cost manufacturing of mechanical microdevices using micro-equipment. *J. Micromech. Microeng.* 6: 410–425.

LaBel, K. A., Stassinopoulos, E. G., and Brucker, G. J. 1991. Transient SEUs in a Fiber optic system for space applications. *IEEE Trans. Nucl. Sci.* 38: 1546–1550.

Leibfried, D., Knill, E., Seidelin, E., et al. 2005. Creation of a six-atom 'Schrödinger cat' state. *Nature* 438: 639–642.

Li, J., Stein, D., McMullan, C., Branton, D., Aziz, M. J., and Golovchenko, J. A. 2001. Ion-beam sculpting at nanometre length scales. *Nature* 412:166–169.

Lim, J. Y., Pyeon, C. H., Yagi, T., and Misawa, T. 2012. Subcritical multiplication parameters of the accelerator-driven system with 100 MeV protons at the Kyoto University Critical Assembly. *Sci. Technol. Nucl. Install.* 2012: 395878 | 10.1155/2012/395878.

Liu, F., Sidhik, S., and Hoffbauer, M. A. 2021. Highly efficient photoelectric effect in halide perovskites for regenerative electron sources. *Nat. Commun.* 10.1038/s41467-021-20954-6. (8 pp).

Maiorino, J. R., dos Santos, A., and Pereira, S. A. 2003. The utilization of accelerators in subcritical systems for energy generation and nuclear waste transmutation – the world status and a proposal of a National R&D Program. *Braz. J. Phys.* 33(2): 267–272.

Malka, V., Faure, J., Gauduel, Y., et al. 2008. Principles and applications of compact laser–plasma accelerators. *Nat. Phys.* 4: 447–453. 10.1038/nphys966.

Marconi, A., Vlassiouk, I., and Smirnov, S. 2009. In Biosensing Using Nanomaterials, Biosensing with nanopores. Ed. Merçoçi, A. Wiley, Hoboken, N.J. Pp 459–490. Pluridirectional High-energy Agile Scanning Electron Radiotherapy (PHASER): a novel design for radiation treatment of cancer. https://biox.stanford.edu/research/seed-grants/pluridirectional-high-energy-agile-scanning-electron-radiotherapy-phaser-novel.

Maxim, P, Loo, B., and Tantawi, S. 2012. Pluridirectional High-energy Agile Scanning Electron Radiotherapy (PHASER): a novel design for radiation treatment of cancer. Interdisciplinary Initiatives Program Round 6 – 2012.

Muggli, P., for the AWAKE Collaboration. 2020. Physics to plan AWAKE Run 2. arXiv:1911.07534v2 [physics.acc-ph] 8 Apr 2020.

NASA. 2000. Radiation Effects & Analysis. https://radhome.gsfc.nasa.gov/radhome/RadDataBase/RadDataBase.html. Last Revised 21 September 2000.

NASA. 2015. NASA/GSFC Radiation Effects & Analysis Home Page. https://radhome.gsfc.nasa.gov/radhome/see.htm. Last Revised 5 November 2015.

Newberry, D. M. 1992. Single event upset error propagation between interconnecting VLSI devices. *IEEE Trans. Nucl. Sci.* NS-39: 446–449.

News. 2020. A milestone for MYRRHA. *Nuclear Engineering International*, 24 July 2020. https://www.neimagazine.com/news/newsa-milestone-for-myrrha-8042478.

Nichols, K. E., Andrews, H. L., Kim, D., et al. 2019. Demonstration of transport of patterned electron beam produced by diamond pyramid cathode in an rf gun. *Appl. Phys. Lett.* 116: 023502 (5 pp).

Novoselov, K. S., et al. 2004. Electric field effect in atomically thin carbon films. *Science* 306: 666–669.

Othman, M. A. K., Picard, J., Schaub, S., et al. 2020. Experimental demonstration of externally driven millimeter-wave particle accelerator structure. *Appl. Phys. Lett.* 117: 073502 p. 6.

Palastro, J. P., et al. 2020. Dephasingless laser wakefield acceleration. *Phys. Rev. Lett.* 124, 134802. 10.1103/PhysRevLett.124.134802.

Pan, W. and Dai, J. 2015. ADS based on linear accelerators. *Rev. Accel. Sci. Technol.* 8: 55–76.

Reichel, J. and Vuletic, V. (Eds). 2011. *Atom Chips*. Wiley, Weinheim, Germany.

Rezgui, S., Won, R., and Tien, J. 2012. SET characterizationand mitigation in 65 nm CMOS test structure. *IEEE Trans. Nucl. Sci.* 59(4):851–859.

Rezgui, S., Wang, J.J., Tung, E. C., McCollum, J., and Croncuist, B. 2007. New methodologies for SET characterization and mitigation inflash-based FPGAs. *IEEE Trans. Nucl. Sci.* 54(6): 2512–2524.

Rubbia, C., Rubio, J. A., Buono, S., et al. 1995. Conceptual Design of a Fast Neutron Operated High Power Energy Amplifier. European Organization for Nuclear Research, CERN, Geneva, Switzerland. Report CERN/AT/95-44(ET)

Schollmeier, M., et al. 2008. Controlled transport and focusing of laser-accelerated protons with miniature magnetic devices. *Phys. Rev. Lett.* 101: 055004.

Schreiber, J., Bolton, P. R., and Parodi, K. 2016. Invited Review Article: "Hands-on" laser-driven ion acceleration: a primer for laser-driven source development and potential applications. *Rev. Sci. Instrum.* 87: 071101 (10 pp).

Schwestka, J., et al. 2020. Atomic-Scale Carving of nanopores into a van der Waals heterostructure with slow highly charged ions. *ACS Nano* 14: 10536–10543.

Seidelin, S., Chiaverini, J., Reichle, R., et al. 2006. Microfabricated surface-electrode ion trap for scalable quantum information processing. *Phys. Rev. Lett.* 96(25): 253003 (4 pp).

Sheng, L., Du, G., Guo, J., Wu, R., Song, M., Yuan, Y. and Xiao, G. 2013. Focusing giga-electronvolt heavy ions to micrometers at the Institute of Modern Physics. *Rev. Sci. Instrume.* 84: 055113 (4 pp).

Singh, P. 2017. India ADS program. Presentation at EuCARD2, Status of Accelerator Driven Systems Research and Technology Development. CERN, 7–9 February 2017.

Smirnov, V. and Vorozhtsov. S. 2018. Feasibility study of a cyclotron complex for hadron therapy. *Nucl. Instrum. Methods Phys. Res. Sect. A Accel. Spectromet. Detect. Assoc. Equipment* 887: 114–121.

Storm, A. J., Chen, J. H., Ling, X. S., Zandbergen, H. W., and Dekker, C. 2003. Fabrication of solid-state nanopores with single-nanometre precision. *Nat Mater* 2: 537–540.

Striemer, C. C., Gaborskim, T. R., McGrath, J. L., and Fauchet, P. M. 2007. Charge- and size-based separation of macromolecules using ultrathin silicon membranes. *Nature* 445: 749–753.

Sugawara, T. 2017. Japan ADS project. Presentation at EuCARD2, Status of Accelerator Driven Systems Research and Technology Development. CERN, 7-9 February 2017.

Toncian, T., et al. 2006. Ultrafast laser-driven microlens to focus and energy-select mega-electron volt protons. *Science* 312: 410–413.

Tong, H. D., Jansen, H. V., Gadgil, V. J., Bostan, C. G., Berenschot, E., van Rijn, C. J. M., and Elwenspoek, M. 2004. Silicon nitride nanosieve membrane. *Nano Lett* 4: 283–287.

Trbojevic, D., Keil, E., and Sessler, A. 2005. Non-Scaling Fixed Field Gradient Accelerator (FFAG) Design for the Proton and Carbon Therapy. Utilization of Accelerators, Proceedings of an International Conference, Dubrovnik, 5–9 June 2005. IAEA Proceedings Series IAEA-CN-115, (8 pp). http://www-pub.iaea.org/MTCD/Publications/PDF/P1251-cd/papers/65.pdf.

Turowski, M., et al. 2012. Effect of carrier transport in oxides surrounding active devices on SEU in 45 nm SOI SRAM. *IEEE Trans. Nucl. Sci.* 59(4): 728–734.

Vandeplassche, D., Biarrotte J.-L., Klein H., and Podlech H. 2011. The MYRRHA linear accelerator. WEPS090 Proceedings of IPAC2011, San Sebastián, Spain, p. 2718–2720.

Vlassiouk, I., Apel, P. Y., Dmitriev, S. N., Healy, K., and Siwy, Z. S. 2009. Versatile ultrathin nanoporous silicon nitride membranes. *PNAS* 106 (50): 21039–21044.

Wallmark, J. T. and Marcus, S. M. 1962. Minimum size and maximum packing density of non-redunant semiconductor devices. *Proc. IRE.* 50(3): 286–298. Note: This Periodical ceased production in 1962. The current retitled publication is Proceedings of the IEEE.

Wang, F. and Agrawal, V. D. 2008. Single event upset: an embedded tutorial. VLSID'08: Proceedings of the 21st International Conference on VLSI Design. IEEE Computer Society, Washington, DC, United States. (p. 429–434).

Yan, X., Yang, L., Zhang, X., and Zhan, W. 2017. Concept of an accelerator-driven advanced nuclear energy system. *Energies* 10(7): 944 (p. 13).

Zhan, W. 2017. China ADS project. Presentation at EuCARD2, Status of Accelerator Driven Systems Research and Technology Development. CERN, 7–9 February 2017.

Zimmermann, R., Seidling, M., and Hommelhoff, P. 2021. Charged particle guiding and beam splitting with auto-ponderomotive potentials on a chip. *Nat. Commun.* 12: 390. Published online 2021 Jan 15. 10.1038/s41467-020-20592-4.

Index

For Product Safety Concerns and Information please contact our EU
representative GPSR@taylorandfrancis.com
Taylor & Francis Verlag GmbH, Kaufingerstraße 24, 80331 München, Germany

www.ingramcontent.com/pod-product-compliance
Lightning Source LLC
Chambersburg PA
CBHW080707220326
41598CB00033B/5333